Applied Numerical Methods

with MATLAB® for Engineers and Scientists

Fourth Edition

Steven C. Chapra

Berger Chair in Computing and Engineering
Tufts University

APPLIED NUMERICAL METHODS WITH MATLAB® FOR ENGINEERS AND SCIENTISTS,
FOURTH EDITION

Published by McGraw-Hill Education, 2 Penn Plaza, New York, NY 10121. Copyright © 2018 by McGraw-Hill
Education. All rights reserved. Printed in the United States of America. Previous editions © 2012, 2008, and
2005. No part of this publication may be reproduced or distributed in any form or by any means, or stored in a
database or retrieval system, without the prior written consent of McGraw-Hill Education, including, but not
limited to, in any network or other electronic storage or transmission, or broadcast for distance learning.

Some ancillaries, including electronic and print components, may not be available to customers outside the
United States.

This book is printed on acid-free paper.

2 3 4 5 6 7 8 9 LCR 21 20 19 18 17

ISBN 978-0-07-339796-2
MHID 0-07-339796-2

Chief Product Officer, SVP Products & Markets: *G. Scott Virkler*
Vice President, General Manager, Products & Markets: *Marty Lange*
Vice President, Content Design & Delivery: *Betsy Whalen*
Managing Director: *Thomas Timp*
Brand Manager: *Thomas M. Scaife, Ph.D.*
Director, Product Development: *Rose Koos*
Product Developer: *Jolynn Kilburg*
Marketing Manager: *Nick McFadden*
Market Development Manager: *Shannon O'Donnell*
Digital Product Developer: *Joan Weber*
Director, Content Design & Delivery: *Linda Avenarius*
Program Manager: *Lora Neyens*
Content Project Managers: *Jeni McAtee; Sandy Schnee*
Buyer: *Susan K. Culbertson*
Content Licensing Specialists: *Lorraine Buczek*
Cover Image: *Vitalii Nesterchuk/123RF©*
Compositor: *MPS Limited*
Printer: *LSC Communications*

All credits appearing on page or at the end of the book are considered to be an extension of the copyright page.

Library of Congress Cataloging-in-Publication Data

Chapra, Steven C., author.
   Applied numerical methods with MATLAB for engineers and scientists /
   Steven C. Chapra, Berger Chair in Computing and Engineering, Tufts University.
   Fourth edition. | New York, NY : McGraw-Hill Education, [2018] |
   Includes bibliographical references and index.
   LCCN 2016038044 | ISBN 9780073397962 (alk. paper) | ISBN 0073397962 (alk. paper)
   LCSH: Numerical analysis—Data processing—Textbooks. | Engineering
   mathematics—Textbooks. | MATLAB—Textbooks.
   LCC QA297 .C4185 2018 | DDC 518—dc23 LC record available at https://lccn.loc.gov/2016038044

The Internet addresses listed in the text were accurate at the time of publication. The inclusion of a website
does not indicate an endorsement by the authors or McGraw-Hill Education, and McGraw-Hill Education does
not guarantee the accuracy of the information presented at these sites.

mheducation.com/highered

To

My brothers,
John and Bob Chapra

and

Fred Berger (1947–2015)
who I miss as a good friend, a good man.
and a comrade in bringing the light of engineering
to some of world's darker corners.

ABOUT THE AUTHOR

Steve Chapra teaches in the Civil and Environmental Engineering Department at Tufts University, where he holds the Louis Berger Chair in Computing and Engineering. His other books include *Numerical Methods for Engineers* and *Surface Water-Quality Modeling*.

Steve received engineering degrees from Manhattan College and the University of Michigan. Before joining the faculty at Tufts, he worked for the Environmental Protection Agency and the National Oceanic and Atmospheric Administration, and taught at Texas A&M University and the University of Colorado. His general research interests focus on surface water-quality modeling and advanced computer applications in environmental engineering.

He has received a number of awards for his scholarly contributions, including the Rudolph Hering Medal, the Meriam/Wiley Distinguished Author Award, and the Chandler-Misener Award. He has also been recognized as the outstanding teacher at Texas A&M University (1986 Tenneco Award), the University of Colorado (1992 Hutchinson Award), and Tufts University (2011 Professor of the Year Award).

Steve was originally drawn to environmental engineering and science because of his love of the outdoors. He is an avid fly fisherman and hiker. An unapologetic nerd, his love affair with computing began when he was first introduced to Fortran programming as an undergraduate in 1966. Today, he feels truly blessed to be able to meld his love of mathematics, science, and computing with his passion for the natural environment. In addition, he gets the bonus of sharing it with others through his teaching and writing!

Beyond his professional interests, he enjoys art, music (especially classical music, jazz, and bluegrass), and reading history. Despite unfounded rumors to the contrary, he never has, and never will, voluntarily bungee jump or sky dive.

If you would like to contact Steve, or learn more about him, visit his home page at http://engineering.tufts.edu/cee/people/chapra/ or e-mail him at steven.chapra@tufts.edu.

CONTENTS

CHAPTER 15

CHAPTER 16

CHAPTER 17

CHAPTER 18

McGraw-Hill Connect®
Learn Without Limits

Connect is a teaching and learning platform that is proven to deliver better results for students and instructors.

Connect empowers students by continually adapting to deliver precisely what they need, when they need it, and how they need it, so your class time is more engaging and effective.

> 73% of instructors who use **Connect** require it; instructor satisfaction **increases** by 28% when **Connect** is required.

Analytics
Connect Insight®

Connect Insight is Connect's new one-of-a-kind visual analytics dashboard—now available for both instructors and students—that provides at-a-glance information regarding student performance, which is immediately actionable. By presenting assignment, assessment, and topical performance results together with a time metric that is easily visible for aggregate or individual results, Connect Insight gives the user the ability to take a just-in-time approach to teaching and learning, which was never before available. Connect Insight presents data that empowers students and helps instructors improve class performance in a way that is efficient and effective.

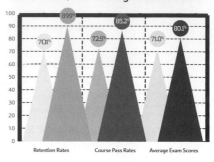

Connect's Impact on Retention Rates, Pass Rates, and Average Exam Scores

Retention Rates: 70.1% without Connect, 89.9% with Connect
Course Pass Rates: 72.5% without Connect, 85.2% with Connect
Average Exam Scores: 71.0% without Connect, 80.1% with Connect

without Connect / with Connect

> Using **Connect** improves retention rates by **19.8%**, passing rates by **12.7%**, and exam scores by **9.1%**.

Impact on Final Course Grade Distribution

	without Connect	with Connect
A	22.9%	31.0%
B	27.4%	34.3%
C	22.9%	18.7%
D	11.5%	6.1%
F	15.4%	9.9%

> Students can view their results for any **Connect** course.

Mobile

Connect's new, intuitive mobile interface gives students and instructors flexible and convenient, anytime–anywhere access to all components of the Connect platform.

Adaptive

THE **ADAPTIVE** **READING EXPERIENCE** DESIGNED TO TRANSFORM THE WAY STUDENTS READ

> More students earn **A's** and **B's** when they use McGraw-Hill Education **Adaptive** products.

SmartBook®

Proven to help students improve grades and study more efficiently, SmartBook contains the same content within the print book, but actively tailors that content to the needs of the individual. SmartBook's adaptive technology provides precise, personalized instruction on what the student should do next, guiding the student to master and remember key concepts, targeting gaps in knowledge and offering customized feedback, and driving the student toward comprehension and retention of the subject matter. Available on tablets, SmartBook puts learning at the student's fingertips—anywhere, anytime.

> Over **8 billion questions** have been answered, making McGraw-Hill Education products more intelligent, reliable, and precise.

www.mheducation.com

PREFACE

This book is designed to support a one-semester course in numerical methods. It has been written for students who want to learn and apply numerical methods in order to solve problems in engineering and science. As such, the methods are motivated by problems rather than by mathematics. That said, sufficient theory is provided so that students come away with insight into the techniques and their shortcomings.

MATLAB® provides a great environment for such a course. Although other environments (e.g., Excel/VBA, Mathcad) or languages (e.g., Fortran 90, C++) could have been chosen, MATLAB presently offers a nice combination of handy programming features with powerful built-in numerical capabilities. On the one hand, its M-file programming environment allows students to implement moderately complicated algorithms in a structured and coherent fashion. On the other hand, its built-in, numerical capabilities empower students to solve more difficult problems without trying to "reinvent the wheel."

The basic content, organization, and pedagogy of the third edition are essentially preserved in the fourth edition. In particular, the conversational writing style is intentionally maintained in order to make the book easier to read. This book tries to speak directly to the reader and is designed in part to be a tool for self-teaching.

That said, this edition differs from the past edition in three major ways: (1) new material, (2) new and revised homework problems, and (3) an appendix introducing Simulink.

1. New Content. I have included new and enhanced sections on a number of topics. The primary additions include material on some MATLAB functions not included in previous editions (e.g., `fsolve`, `integrate`, `bvp4c`), some new applications of Monte Carlo for problems such as integration and optimization, and MATLAB's new way to pass parameters to function functions.
2. New Homework Problems. Most of the end-of-chapter problems have been modified, and a variety of new problems have been added. In particular, an effort has been made to include several new problems for each chapter that are more challenging and difficult than the problems in the previous edition.
3. I have developed a short primer on Simulink which I have my students read prior to covering that topic. Although I recognize that some professors may not choose to cover Simulink, I included it as a teaching aid for those that do.

Aside from the new material and problems, the fourth edition is very similar to the third. In particular, I have endeavored to maintain most of the features contributing to its pedagogical effectiveness including extensive use of worked examples and engineering and scientific applications. As with the previous edition, I have made a concerted effort to make this book as "student-friendly" as possible. Thus, I've tried to keep my explanations straightforward and practical.

Although my primary intent is to empower students by providing them with a sound introduction to numerical problem solving, I have the ancillary objective of making this introduction exciting and pleasurable. I believe that motivated students who enjoy engineering and science, problem solving, mathematics—and yes—programming, will ultimately make better professionals. If my book fosters enthusiasm and appreciation for these subjects, I will consider the effort a success.

Acknowledgments. Several members of the McGraw-Hill team have contributed to this project. Special thanks are due to Jolynn Kilburg, Thomas Scaife, Ph.D., Chelsea Haupt, Ph.D., and Jeni McAtee for their encouragement, support, and direction.

During the course of this project, the folks at The MathWorks, Inc., have truly demonstrated their overall excellence as well as their strong commitment to engineering and science education. In particular, Naomi Fernandes of The MathWorks, Inc., Book Program has been especially helpful and Jared Wasserman of the MathWorks Technical Support Department was of great help with technical questions.

The generosity of the Berger family has provided me with the opportunity to work on creative projects such as this book dealing with computing and engineering. In addition, my colleagues in the School of Engineering at Tufts, notably Masoud Sanayei, Babak Moaveni, Luis Dorfmann, Rob White, Linda Abriola, and Laurie Baise, have been very supportive and helpful.

Significant suggestions were also given by a number of colleagues. In particular, Dave Clough (University of Colorado–Boulder), and Mike Gustafson (Duke University) provided valuable ideas and suggestions. In addition, a number of reviewers provided useful feedback and advice including Karen Dow Ambtman (University of Alberta), Jalal Behzadi (Shahid Chamran University), Eric Cochran (Iowa State University), Frederic Gibou (University of California at Santa Barbara), Jane Grande-Allen (Rice University), Raphael Haftka (University of Florida), Scott Hendricks (Virginia Tech University), Ming Huang (University of San Diego), Oleg Igoshin (Rice University), David Jack (Baylor University), Se Won Lee (Sungkyunkwan University), Clare McCabe (Vanderbilt University), Eckart Meiburg (University of California at Santa Barbara), Luis Ricardez (University of Waterloo), James Rottman (University of California, San Diego), Bingjing Su (University of Cincinnati), Chin-An Tan (Wayne State University), Joseph Tipton (The University of Evansville), Marion W. Vance (Arizona State University), Jonathan Vande Geest (University of Arizona), Leah J. Walker (Arkansas State University), Qiang Hu (University of Alabama, Huntsville), Yukinobu Tanimoto (Tufts University), Henning T. Søgaard (Aarhus University), and Jimmy Feng (University of British Columbia).

It should be stressed that although I received useful advice from the aforementioned individuals, I am responsible for any inaccuracies or mistakes you may find in this book. Please contact me via e-mail if you should detect any errors.

Finally, I want to thank my family, and in particular my wife, Cynthia, for the love, patience, and support they have provided through the time I've spent on this project.

Steven C. Chapra
Tufts University

Medford, Massachusetts
steven.chapra@tufts.edu

PEDAGOGICAL TOOLS

Theory Presented as It Informs Key Concepts.   The text is intended for Numerical Methods users, not developers. Therefore, theory is not included for "theory's sake," for example no proofs. Theory is included as it informs key concepts such as the Taylor series, convergence, condition, etc. Hence, the student is shown how the theory connects with practical issues in problem solving.

Introductory MATLAB Material.   The text includes two introductory chapters on how to use MATLAB. Chapter 2 shows students how to perform computations and create graphs in MATLAB's standard command mode. Chapter 3 provides a primer on developing numerical programs via MATLAB M-file functions. Thus, the text provides students with the means to develop their own numerical algorithms as well as to tap into MATLAB's powerful built-in routines.

Algorithms Presented Using MATLAB M-files.   Instead of using pseudocode, this book presents algorithms as well-structured MATLAB M-files. Aside from being useful computer programs, these provide students with models for their own M-files that they will develop as homework exercises.

Worked Examples and Case Studies.   Extensive worked examples are laid out in detail so that students can clearly follow the steps in each numerical computation. The case studies consist of engineering and science applications which are more complex and richer than the worked examples. They are placed at the ends of selected chapters with the intention of (1) illustrating the nuances of the methods and (2) showing more realistically how the methods along with MATLAB are applied for problem solving.

Problem Sets.   The text includes a wide variety of problems. Many are drawn from engineering and scientific disciplines. Others are used to illustrate numerical techniques and theoretical concepts. Problems include those that can be solved with a pocket calculator as well as others that require computer solution with MATLAB.

Useful Appendices and Indexes.   Appendix A contains MATLAB commands, Appendix B contains M-file functions, and new Appendix C contains a brief Simulink primer.

Instructor Resources.   Solutions Manual, Lecture PowerPoints, Text images in PowerPoint, M-files and additional MATLAB resources are available through Connect®.

PART ONE

Modeling, Computers, and Error Analysis

1.1 MOTIVATION

What are numerical methods and why should you study them?

Numerical methods are techniques by which mathematical problems are formulated so that they can be solved with arithmetic and logical operations. Because digital computers excel at performing such operations, numerical methods are sometimes referred to as *computer mathematics.*

In the pre–computer era, the time and drudgery of implementing such calculations seriously limited their practical use. However, with the advent of fast, inexpensive digital computers, the role of numerical methods in engineering and scientific problem solving has exploded. Because they figure so prominently in much of our work, I believe that numerical methods should be a part of every engineer's and scientist's basic education. Just as we all must have solid foundations in the other areas of mathematics and science, we should also have a fundamental understanding of numerical methods. In particular, we should have a solid appreciation of both their capabilities and their limitations.

Beyond contributing to your overall education, there are several additional reasons why you should study numerical methods:

1. Numerical methods greatly expand the types of problems you can address. They are capable of handling large systems of equations, nonlinearities, and complicated geometries that are not uncommon in engineering and science and that are often impossible to solve analytically with standard calculus. As such, they greatly enhance your problem-solving skills.

2. Numerical methods allow you to use "canned" software with insight. During

1

your career, you will invariably have occasion to use commercially available prepack-aged computer programs that involve numerical methods. The intelligent use of these programs is greatly enhanced by an understanding of the basic theory underlying the methods. In the absence of such understanding, you will be left to treat such packages as "black boxes" with little critical insight into their inner workings or the validity of the results they produce.

3. Many problems cannot be approached using canned programs. If you are conversant with numerical methods, and are adept at computer programming, you can design your own programs to solve problems without having to buy or commission expensive software.

4. Numerical methods are an efficient vehicle for learning to use computers. Because nu-merical methods are expressly designed for computer implementation, they are ideal for illustrating the computer's powers and limitations. When you successfully implement numerical methods on a computer, and then apply them to solve otherwise intractable problems, you will be provided with a dramatic demonstration of how computers can serve your professional development. At the same time, you will also learn to acknowl-edge and control the errors of approximation that are part and parcel of large-scale numerical calculations.

5. Numerical methods provide a vehicle for you to reinforce your understanding of math-ematics. Because one function of numerical methods is to reduce higher mathematics to basic arithmetic operations, they get at the "nuts and bolts" of some otherwise obscure topics. Enhanced understanding and insight can result from this alternative perspective.

With these reasons as motivation, we can now set out to understand how numerical methods and digital computers work in tandem to generate reliable solutions to mathemati-cal problems. The remainder of this book is devoted to this task.

1.2 PART ORGANIZATION

This book is divided into six parts. The latter five parts focus on the major areas of nu-merical methods. Although it might be tempting to jump right into this material, *Part One* consists of four chapters dealing with essential background material.

Chapter 1 provides a concrete example of how a numerical method can be employed to solve a real problem. To do this, we develop a *mathematical model* of a free-falling bungee jumper. The model, which is based on Newton's second law, results in an ordinary differential equation. After first using calculus to develop a closed-form solution, we then show how a comparable solution can be generated with a simple numerical method. We end the chapter with an overview of the major areas of numerical methods that we cover in Parts Two through Six.

Chapters 2 and 3 provide an introduction to the MATLAB® software environment. *Chapter 2* deals with the standard way of operating MATLAB by entering commands one at a time in the so-called *calculator,* or *command, mode.* This interactive mode provides a straightforward means to orient you to the environment and illustrates how it is used for common operations such as performing calculations and creating plots.

Chapter 3 shows how MATLAB's *programming mode* provides a vehicle for assembling individual commands into algorithms. Thus, our intent is to illustrate how MATLAB serves as a convenient programming environment to develop your own software.

Chapter 4 deals with the important topic of error analysis, which must be understood for the effective use of numerical methods. The first part of the chapter focuses on the *roundoff errors* that result because digital computers cannot represent some quantities exactly. The latter part addresses *truncation errors* that arise from using an approximation in place of an exact mathematical procedure.

1

Mathematical Modeling, Numerical Methods, and Problem Solving

> ## CHAPTER OBJECTIVES
>
> The primary objective of this chapter is to provide you with a concrete idea of what numerical methods are and how they relate to engineering and scientific problem solving. Specific objectives and topics covered are
>
> - Learning how mathematical models can be formulated on the basis of scientific principles to simulate the behavior of a simple physical system.
> - Understanding how numerical methods afford a means to generate solutions in a manner that can be implemented on a digital computer.
> - Understanding the different types of conservation laws that lie beneath the models used in the various engineering disciplines and appreciating the difference between steady-state and dynamic solutions of these models.
> - Learning about the different types of numerical methods we will cover in this book.

YOU'VE GOT A PROBLEM

Suppose that a bungee-jumping company hires you. You're given the task of predicting the velocity of a jumper (Fig. 1.1) as a function of time during the free-fall part of the jump. This information will be used as part of a larger analysis to determine the length and required strength of the bungee cord for jumpers of different mass.

You know from your studies of physics that the acceleration should be equal to the ratio of the force to the mass (Newton's second law). Based on this insight and your knowledge

Upward force
due to air
resistance

Downward
force due
to gravity

FIGURE 1.1
Forces acting
on a free-falling
bungee jumper.

of physics and fluid mechanics, you develop the following mathematical model for the rate of change of velocity with respect to time,

$$\frac{dv}{dt} = g - \frac{c_d}{m}v^2$$

where v = downward vertical velocity (m/s), t = time (s), g = the acceleration due to gravity ($\cong 9.81$ m/s^2), c_d = a lumped drag coefficient (kg/m), and m = the jumper's mass (kg). The drag coefficient is called "lumped" because its magnitude depends on factors such as the jumper's area and the fluid density (see Sec. 1.4).

Because this is a differential equation, you know that calculus might be used to obtain an analytical or exact solution for v as a function of t. However, in the following pages, we will illustrate an alternative solution approach. This will involve developing a computer-oriented numerical or approximate solution.

Aside from showing you how the computer can be used to solve this particular problem, our more general objective will be to illustrate (*a*) what numerical methods are and (*b*) how they figure in engineering and scientific problem solving. In so doing, we will also show how mathematical models figure prominently in the way engineers and scientists use numerical methods in their work.

1.1   A SIMPLE MATHEMATICAL MODEL

A *mathematical model* can be broadly defined as a formulation or equation that expresses the essential features of a physical system or process in mathematical terms. In a very general sense, it can be represented as a functional relationship of the form

$$\begin{matrix} \text{Dependent} \\ \text{variable} \end{matrix} = f\begin{pmatrix} \text{independent} \\ \text{variables} \end{pmatrix}, \text{parameters}, \begin{matrix} \text{forcing} \\ \text{functions} \end{matrix} \tag{1.1}$$

where the *dependent variable* is a characteristic that typically reflects the behavior or state of the system; the *independent variables* are usually dimensions, such as time and space, along which the system's behavior is being determined; the *parameters* are reflective of the system's properties or composition; and the *forcing functions* are external influences acting upon it.

The actual mathematical expression of Eq. (1.1) can range from a simple algebraic relationship to large complicated sets of differential equations. For example, on the basis of his observations, Newton formulated his second law of motion, which states that the time rate of change of momentum of a body is equal to the resultant force acting on it. The mathematical expression, or model, of the second law is the well-known equation

$$F = ma \tag{1.2}$$

where F is the net force acting on the body (N, or kg m/s^2), m is the mass of the object (kg), and a is its acceleration (m/s^2).

The second law can be recast in the format of Eq. (1.1) by merely dividing both sides by m to give

$$a = \frac{F}{m} \qquad (1.3)$$

where a is the dependent variable reflecting the system's behavior, F is the forcing function, and m is a parameter. Note that for this simple case there is no independent variable because we are not yet predicting how acceleration varies in time or space.

Equation (1.3) has a number of characteristics that are typical of mathematical models of the physical world.

- It describes a natural process or system in mathematical terms.
- It represents an idealization and simplification of reality. That is, the model ignores negligible details of the natural process and focuses on its essential manifestations. Thus, the second law does not include the effects of relativity that are of minimal importance when applied to objects and forces that interact on or about the earth's surface at velocities and on scales visible to humans.
- Finally, it yields reproducible results and, consequently, can be used for predictive purposes. For example, if the force on an object and its mass are known, Eq. (1.3) can be used to compute acceleration.

Because of its simple algebraic form, the solution of Eq. (1.2) was obtained easily. However, other mathematical models of physical phenomena may be much more complex, and either cannot be solved exactly or require more sophisticated mathematical techniques than simple algebra for their solution. To illustrate a more complex model of this kind, Newton's second law can be used to determine the terminal velocity of a free-falling body near the earth's surface. Our falling body will be a bungee jumper (Fig. 1.1). For this case, a model can be derived by expressing the acceleration as the time rate of change of the velocity (dv/dt) and substituting it into Eq. (1.3) to yield

$$\frac{dv}{dt} = \frac{F}{m} \qquad (1.4)$$

where v is velocity (in meters per second). Thus, the rate of change of the velocity is equal to the net force acting on the body normalized to its mass. If the net force is positive, the object will accelerate. If it is negative, the object will decelerate. If the net force is zero, the object's velocity will remain at a constant level.

Next, we will express the net force in terms of measurable variables and parameters. For a body falling within the vicinity of the earth, the net force is composed of two opposing forces: the downward pull of gravity F_D and the upward force of air resistance F_U (Fig. 1.1):

$$F = F_D + F_U \qquad (1.5)$$

If force in the downward direction is assigned a positive sign, the second law can be used to formulate the force due to gravity as

$$F_D = mg \qquad (1.6)$$

where g is the acceleration due to gravity (9.81 m/s^2).

Air resistance can be formulated in a variety of ways. Knowledge from the science of fluid mechanics suggests that a good first approximation would be to assume that it is proportional to the square of the velocity,

$$F_U = -c_d v^2 \tag{1.7}$$

where c_d is a proportionality constant called the *lumped drag coefficient* (kg/m). Thus, the greater the fall velocity, the greater the upward force due to air resistance. The parameter c_d accounts for properties of the falling object, such as shape or surface roughness, that affect air resistance. For the present case, c_d might be a function of the type of clothing or the orientation used by the jumper during free fall.

The net force is the difference between the downward and upward force. Therefore, Eqs. (1.4) through (1.7) can be combined to yield

$$\frac{dv}{dt} = g - \frac{c_d}{m} v^2 \tag{1.8}$$

Equation (1.8) is a model that relates the acceleration of a falling object to the forces acting on it. It is a *differential equation* because it is written in terms of the differential rate of change (dv/dt) of the variable that we are interested in predicting. However, in contrast to the solution of Newton's second law in Eq. (1.3), the exact solution of Eq. (1.8) for the velocity of the jumper cannot be obtained using simple algebraic manipulation. Rather, more advanced techniques such as those of calculus must be applied to obtain an exact or analytical solution. For example, if the jumper is initially at rest ($v = 0$ at $t = 0$), calculus can be used to solve Eq. (1.8) for

$$v(t) = \sqrt{\frac{gm}{c_d}} \tanh\!\left(\sqrt{\frac{g c_d}{m}} t\right) \tag{1.9}$$

where tanh is the hyperbolic tangent that can be either computed directly[1] or via the more elementary exponential function as in

$$\tanh x = \frac{e^x - e^{-x}}{e^x + e^{-x}} \tag{1.10}$$

Note that Eq. (1.9) is cast in the general form of Eq. (1.1) where $v(t)$ is the dependent variable, t is the independent variable, c_d and m are parameters, and g is the forcing function.

EXAMPLE 1.1   Analytical Solution to the Bungee Jumper Problem

Problem Statement.   A bungee jumper with a mass of 68.1 kg leaps from a stationary hot air balloon. Use Eq. (1.9) to compute velocity for the first 12 s of free fall. Also determine the terminal velocity that will be attained for an infinitely long cord (or alternatively, the jumpmaster is having a particularly bad day!). Use a drag coefficient of 0.25 kg/m.

---

[1] MATLAB allows direct calculation of the hyperbolic tangent via the built-in function tanh(x).

Solution. Inserting the parameters into Eq. (1.9) yields

$$v(t) = \sqrt{\frac{9.81(68.1)}{0.25}} \tanh\left(\sqrt{\frac{9.81(0.25)}{68.1}} t\right) = 51.6938 \tanh(0.18977t)$$

which can be used to compute

t, s	v, m/s
0	0
2	18.7292
4	33.1118
6	42.0762
8	46.9575
10	49.4214
12	50.6175
∞	51.6938

According to the model, the jumper accelerates rapidly (Fig. 1.2). A velocity of 49.4214 m/s (about 110 mi/hr) is attained after 10 s. Note also that after a sufficiently

FIGURE 1.2
The analytical solution for the bungee jumper problem as computed in Example 1.1. Velocity increases with time and asymptotically approaches a terminal velocity.

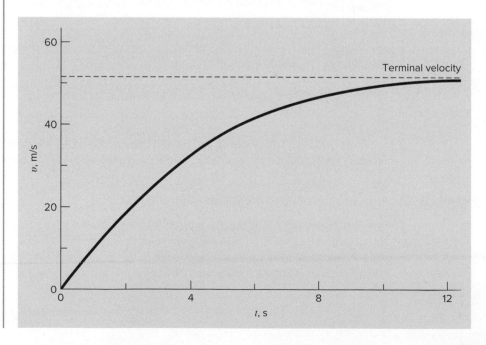

long time, a constant velocity, called the *terminal velocity,* of 51.6983 m/s (115.6 mi/hr) is reached. This velocity is constant because, eventually, the force of gravity will be in balance with the air resistance. Thus, the net force is zero and acceleration has ceased.

Equation (1.9) is called an *analytical* or *closed-form solution* because it exactly satisfies the original differential equation. Unfortunately, there are many mathematical models that cannot be solved exactly. In many of these cases, the only alternative is to develop a numerical solution that approximates the exact solution.

Numerical methods are those in which the mathematical problem is reformulated so it can be solved by arithmetic operations. This can be illustrated for Eq. (1.8) by realizing that the time rate of change of velocity can be approximated by (Fig. 1.3):

$$\frac{dv}{dt} \cong \frac{\Delta v}{\Delta t} = \frac{v(t_{i+1}) - v(t_i)}{t_{i+1} - t_i} \tag{1.11}$$

where Δv and Δt are differences in velocity and time computed over finite intervals, $v(t_i)$ is velocity at an initial time t_i, and $v(t_{i+1})$ is velocity at some later time t_{i+1}. Note that $dv/dt \cong \Delta v/\Delta t$ is approximate because Δt is finite. Remember from calculus that

$$\frac{dv}{dt} = \lim_{\Delta t \to 0} \frac{\Delta v}{\Delta t}$$

Equation (1.11) represents the reverse process.

FIGURE 1.3
The use of a finite difference to approximate the first derivative of v with respect to t.

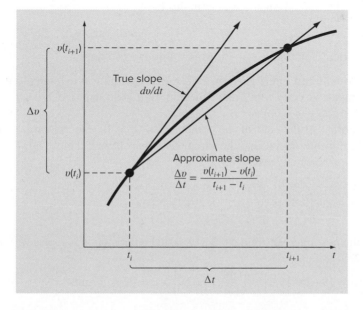

Equation (1.11) is called a *finite-difference approximation* of the derivative at time t_i. It can be substituted into Eq. (1.8) to give

$$\frac{v(t_{i+1}) - v(t_i)}{t_{i+1} - t_i} = g - \frac{c_d}{m} v(t_i)^2$$

This equation can then be rearranged to yield

$$v(t_{i+1}) = v(t_i) + \left[g - \frac{c_d}{m} v(t_i)^2 \right] (t_{i+1} - t_i) \tag{1.12}$$

Notice that the term in brackets is the right-hand side of the differential equation itself [Eq. (1.8)]. That is, it provides a means to compute the rate of change or slope of v. Thus, the equation can be rewritten more concisely as

$$v_{i+1} = v_i + \frac{dv_i}{dt} \Delta t \tag{1.13}$$

where the nomenclature v_i designates velocity at time t_i, and $\Delta t = t_{i+1} - t_i$.

We can now see that the differential equation has been transformed into an equation that can be used to determine the velocity algebraically at t_{i+1} using the slope and previous values of v and t. If you are given an initial value for velocity at some time t_i, you can easily compute velocity at a later time t_{i+1}. This new value of velocity at t_{i+1} can in turn be employed to extend the computation to velocity at t_{i+2} and so on. Thus at any time along the way,

New value = old value + slope × step size

This approach is formally called *Euler's method*. We'll discuss it in more detail when we turn to differential equations later in this book.

EXAMPLE 1.2   Numerical Solution to the Bungee Jumper Problem

Problem Statement.   Perform the same computation as in Example 1.1 but use Eq. (1.12) to compute velocity with Euler's method. Employ a step size of 2 s for the calculation.

Solution.   At the start of the computation ($t_0 = 0$), the velocity of the jumper is zero. Using this information and the parameter values from Example 1.1, Eq. (1.12) can be used to compute velocity at $t_1 = 2$ s:

$$v = 0 + \left[9.81 - \frac{0.25}{68.1} (0)^2 \right] \times 2 = 19.62 \text{ m/s}$$

For the next interval (from $t = 2$ to 4 s), the computation is repeated, with the result

$$v = 19.62 + \left[9.81 - \frac{0.25}{68.1} (19.62)^2 \right] \times 2 = 36.4137 \text{ m/s}$$

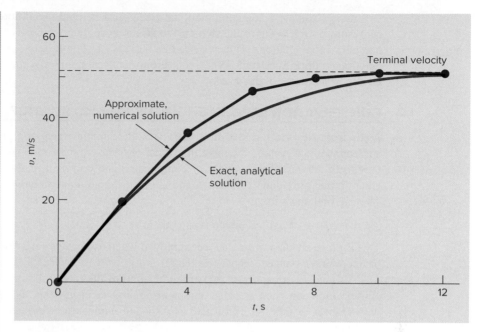

FIGURE 1.4
Comparison of the numerical and analytical solutions for the bungee jumper problem.

The calculation is continued in a similar fashion to obtain additional values:

t, s	v, m/s
0	0
2	19.6200
4	36.4137
6	46.2983
8	50.1802
10	51.3123
12	51.6008
∞	51.6938

The results are plotted in Fig. 1.4 along with the exact solution. We can see that the numerical method captures the essential features of the exact solution. However, because we have employed straight-line segments to approximate a continuously curving function, there is some discrepancy between the two results. One way to minimize such discrepancies is to use a smaller step size. For example, applying Eq. (1.12) at 1-s intervals results in a smaller error, as the straight-line segments track closer to the true solution. Using hand calculations, the effort associated with using smaller and smaller step sizes would make such numerical solutions impractical. However, with the aid of the computer, large numbers of calculations can be performed easily. Thus, you can accurately model the velocity of the jumper without having to solve the differential equation exactly.

As in Example 1.2, a computational price must be paid for a more accurate numerical result. Each halving of the step size to attain more accuracy leads to a doubling of the number of computations. Thus, we see that there is a trade-off between accuracy and computational effort. Such trade-offs figure prominently in numerical methods and constitute an important theme of this book.

1.2   CONSERVATION LAWS IN ENGINEERING AND SCIENCE

Aside from Newton's second law, there are other major organizing principles in science and engineering. Among the most important of these are the *conservation laws*. Although they form the basis for a variety of complicated and powerful mathematical models, the great conservation laws of science and engineering are conceptually easy to understand. They all boil down to

$$\text{Change} = \text{increases} - \text{decreases} \tag{1.14}$$

This is precisely the format that we employed when using Newton's law to develop a force balance for the bungee jumper [Eq. (1.8)].

Although simple, Eq. (1.14) embodies one of the most fundamental ways in which conservation laws are used in engineering and science—that is, to predict changes with respect to time. We will give it a special name—the *time-variable* (or *transient*) computation.

Aside from predicting changes, another way in which conservation laws are applied is for cases where change is nonexistent. If change is zero, Eq. (1.14) becomes

$$\text{Change} = 0 = \text{increases} - \text{decreases}$$

or

$$\text{Increases} = \text{decreases} \tag{1.15}$$

Thus, if no change occurs, the increases and decreases must be in balance. This case, which is also given a special name—the *steady-state* calculation—has many applications in engineering and science. For example, for steady-state incompressible fluid flow in pipes, the flow into a junction must be balanced by flow going out, as in

$$\text{Flow in} = \text{flow out}$$

For the junction in Fig. 1.5, the balance that can be used to compute that the flow out of the fourth pipe must be 60.

For the bungee jumper, the steady-state condition would correspond to the case where the net force was zero or [Eq. (1.8) with $dv/dt = 0$]

$$mg = c_d v^2 \tag{1.16}$$

Thus, at steady state, the downward and upward forces are in balance and Eq. (1.16) can be solved for the terminal velocity

$$v = \sqrt{\frac{gm}{c_d}}$$

Although Eqs. (1.14) and (1.15) might appear trivially simple, they embody the two fundamental ways that conservation laws are employed in engineering and science. As such, they will form an important part of our efforts in subsequent chapters to illustrate the connection between numerical methods and engineering and science.

FIGURE 1.5
A flow balance for steady incompressible fluid flow at the junction of pipes.

Table 1.1 summarizes some models and associated conservation laws that figure prominently in engineering. Many chemical engineering problems involve mass balances for reactors. The mass balance is derived from the conservation of mass. It specifies that the change of mass of a chemical in the reactor depends on the amount of mass flowing in minus the mass flowing out.

Civil and mechanical engineers often focus on models developed from the conservation of momentum. For civil engineering, force balances are utilized to analyze structures such as the simple truss in Table 1.1. The same principles are employed for the mechanical engineering case studies to analyze the transient up-and-down motion or vibrations of an automobile.

Finally, electrical engineering studies employ both current and energy balances to model electric circuits. The current balance, which results from the conservation of charge, is similar in spirit to the flow balance depicted in Fig. 1.5. Just as flow must balance at the junction of pipes, electric current must balance at the junction of electric wires. The energy balance specifies that the changes of voltage around any loop of the circuit must add up to zero.

It should be noted that there are many other branches of engineering beyond chemical, civil, electrical, and mechanical. Many of these are related to the Big Four. For example, chemical engineering skills are used extensively in areas such as environmental, petroleum, and biomedical engineering. Similarly, aerospace engineering has much in common with mechanical engineering. I will endeavor to include examples from these areas in the coming pages.

1.3   NUMERICAL METHODS COVERED IN THIS BOOK

Euler's method was chosen for this introductory chapter because it is typical of many other classes of numerical methods. In essence, most consist of recasting mathematical operations into the simple kind of algebraic and logical operations compatible with digital computers. Figure 1.6 summarizes the major areas covered in this text.

TABLE 1.1 Devices and types of balances that are commonly used in the four major areas of engineering. For each case, the conservation law on which the balance is based is specified.

Field	Device	Organizing Principle	Mathematical Expression
Chemical engineering	Reactors	Conservation of mass	Mass balance: Input → Output Over a unit of time period Δmass = inputs − outputs
Civil engineering	Structure	Conservation of momentum	Force balance: At each node Σ horizontal forces $(F_H) = 0$ Σ vertical forces $(F_V) = 0$
Mechanical engineering	Machine	Conservation of momentum	Force balance: $m\dfrac{d^2x}{dt^2}$ = downward force − upward force
Electrical engineering	Circuit	Conservation of charge	Current balance: For each node Σ current $(i) = 0$
		Conservation of energy	Voltage balance: Around each loop Σ emf's − Σ voltage drops for resistors = 0 $\Sigma \xi - \Sigma iR = 0$

(a) *Part 2*: **Roots and optimization**

Roots: Solve for x so that $f(x) = 0$

Optimization: Solve for x so that $f'(x) = 0$

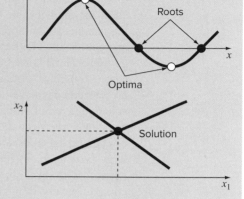

(b) *Part 3*: **Linear algebraic equations**

Given the a's and the b's, solve for the x's
$$a_{11}x_1 + a_{12}x_2 = b_1$$
$$a_{21}x_1 + a_{22}x_2 = b_2$$

(c) *Part 4*: **Curve fitting**

(d) *Part 5*: **Integration and differentiation**

Integration: Find the area under the curve

Differentiation: Find the slope of the curve

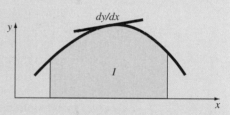

(e) *Part 6*: **Differential equations**

Given
$$\frac{dy}{dt} \approx \frac{\Delta y}{\Delta t} = f(t, y)$$
solve for y as a function of t
$$y_{i+1} = y_i + f(t_i, y_i)\Delta t$$

FIGURE 1.6
Summary of the numerical methods covered in this book.

Part Two deals with two related topics: root finding and optimization. As depicted in Fig. 1.6*a*, *root location* involves searching for the zeros of a function. In contrast, *optimization* involves determining a value or values of an independent variable that correspond to a "best" or optimal value of a function. Thus, as in Fig. 1.6*a*, optimization involves identifying maxima and minima. Although somewhat different approaches are used, root location and optimization both typically arise in design contexts.

Part Three is devoted to solving systems of simultaneous linear algebraic equations (Fig. 1.6*b*). Such systems are similar in spirit to roots of equations in the sense that they are concerned with values that satisfy equations. However, in contrast to satisfying a single equation, a set of values is sought that simultaneously satisfies a set of linear algebraic equations. Such equations arise in a variety of problem contexts and in all disciplines of engineering and science. In particular, they originate in the mathematical modeling of large systems of interconnected elements such as structures, electric circuits, and fluid networks. However, they are also encountered in other areas of numerical methods such as curve fitting and differential equations.

As an engineer or scientist, you will often have occasion to fit curves to data points. The techniques developed for this purpose can be divided into two general categories: regression and interpolation. As described in *Part Four* (Fig. 1.6*c*), *regression* is employed where there is a significant degree of error associated with the data. Experimental results are often of this kind. For these situations, the strategy is to derive a single curve that represents the general trend of the data without necessarily matching any individual points.

In contrast, *interpolation* is used where the objective is to determine intermediate values between relatively error-free data points. Such is usually the case for tabulated information. The strategy in such cases is to fit a curve directly through the data points and use the curve to predict the intermediate values.

As depicted in Fig. 1.6*d*, *Part Five* is devoted to integration and differentiation. A physical interpretation of *numerical integration* is the determination of the area under a curve. Integration has many applications in engineering and science, ranging from the determination of the centroids of oddly shaped objects to the calculation of total quantities based on sets of discrete measurements. In addition, numerical integration formulas play an important role in the solution of differential equations. Part Five also covers methods for *numerical differentiation*. As you know from your study of calculus, this involves the determination of a function's slope or its rate of change.

Finally, *Part Six* focuses on the solution of *ordinary differential equations* (Fig. 1.6*e*). Such equations are of great significance in all areas of engineering and science. This is because many physical laws are couched in terms of the rate of change of a quantity rather than the magnitude of the quantity itself. Examples range from population-forecasting models (rate of change of population) to the acceleration of a falling body (rate of change of velocity). Two types of problems are addressed: initial-value and boundary-value problems.

1.4 CASE STUDY   IT'S A REAL DRAG

Background. In our model of the free-falling bungee jumper, we assumed that drag depends on the square of velocity (Eq. 1.7). A more detailed representation, which was originally formulated by Lord Rayleigh, can be written as

$$F_d = -\frac{1}{2}\rho v^2 A C_d \vec{v} \tag{1.17}$$

where F_d = the drag force (N), ρ = fluid density (kg/m^3), A = the frontal area of the object on a plane perpendicular to the direction of motion (m^2), C_d = a dimensionless drag coefficient, and $\vec{v}$ = a unit vector indicating the direction of velocity.

This relationship, which assumes turbulent conditions (i.e., a high *Reynolds number*), allows us to express the lumped drag coefficient from Eq. (1.7) in more fundamental terms as

$$c_d = \frac{1}{2}\rho A C_d \tag{1.18}$$

Thus, the lumped drag coefficient depends on the object's area, the fluid's density, and a dimensionless drag coefficient. The latter accounts for all the other factors that contribute to air resistance such as the object's "roughness." For example, a jumper wearing a baggy outfit will have a higher C_d than one wearing a sleek jumpsuit.

Note that for cases where velocity is very low, the flow regime around the object will be laminar and the relationship between the drag force and velocity becomes linear. This is referred to as *Stokes drag*.

In developing our bungee jumper model, we assumed that the downward direction was positive. Thus, Eq. (1.7) is an accurate representation of Eq. (1.17), because $\vec{v} = +1$ and the drag force is negative. Hence, drag reduces velocity.

But what happens if the jumper has an upward (i.e., negative) velocity? In this case, $\vec{v} = -1$ and Eq. (1.17) yields a positive drag force. Again, this is physically correct as the positive drag force acts downward against the upward negative velocity.

Unfortunately, for this case, Eq. (1.7) yields a negative drag force because it does not include the unit directional vector. In other words, by squaring the velocity, its sign and hence its direction is lost. Consequently, the model yields the physically unrealistic result that air resistance acts to accelerate an upward velocity!

In this case study, we will modify our model so that it works properly for both downward and upward velocities. We will test the modified model for the same case as Example 1.2, but with an initial value of $v(0) = -40$ m/s. In addition, we will also illustrate how we can extend the numerical analysis to determine the jumper's position.

Solution.   The following simple modification allows the sign to be incorporated into the drag force:

$$F_d = -\frac{1}{2}\rho v|v| A C_d \tag{1.19}$$

or in terms of the lumped drag:

$$F_d = -c_d v|v| \tag{1.20}$$

Thus, the differential equation to be solved is

$$\frac{dv}{dt} = g - \frac{c_d}{m} v|v| \tag{1.21}$$

In order to determine the jumper's position, we recognize that distance traveled, x (m), is related to velocity by

$$\frac{dx}{dt} = -v \tag{1.22}$$

In contrast to velocity, this formulation assumes that upward distance is positive. In the same fashion as Eq. (1.12), this equation can be integrated numerically with Euler's method:

$$x_{i+1} = x_i - v(t_i)\Delta t \tag{1.23}$$

Assuming that the jumper's initial position is defined as $x(0) = 0$, and using the parameter values from Examples 1.1 and 1.2, the velocity and distance at $t = 2$ s can be computed as

$$v(2) = -40 + \left[9.81 - \frac{0.25}{68.1}(-40)(40)\right]2 = -8.6326 \text{ m/s}$$

$$x(2) = 0 - (-40)2 = 80 \text{ m}$$

Note that if we had used the incorrect drag formulation, the results would be -32.1274 m/s and 80 m.

The computation can be repeated for the next interval ($t = 2$ to 4 s):

$$v(4) = -8.6326 + \left[9.81 - \frac{0.25}{68.1}(-8.6326)(8.6326)\right]2 = 11.5346 \text{ m/s}$$

$$x(4) = 80 - (-8.6326)2 = 97.2651 \text{ m}$$

The incorrect drag formulation gives –20.0858 m/s and 144.2549 m.

The calculation is continued and the results shown in Fig. 1.7 along with those obtained with the incorrect drag model. Notice that the correct formulation decelerates more rapidly because drag always diminishes the velocity.

With time, both velocity solutions converge on the same terminal velocity because eventually both are directed downward in which case, Eq. (1.7) is correct. However, the impact on the height prediction is quite dramatic with the incorrect drag case resulting in a much higher trajectory.

This case study demonstrates how important it is to have the correct physical model. In some cases, the solution will yield results that are clearly unrealistic. The current example is more insidious as there is no visual evidence that the incorrect solution is wrong. That is, the incorrect solution "looks" reasonable.

1.4 CASE STUDY   continued

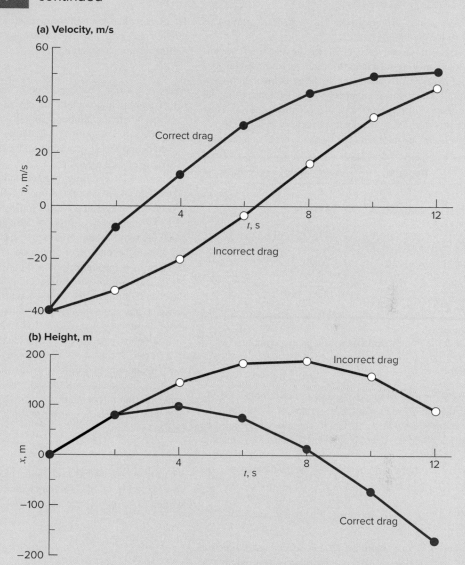

FIGURE 1.7
Plots of (*a*) velocity and (*b*) height for the free-falling bungee jumper with an upward (negative) initial velocity generated with Euler's method. Results for both the correct (Eq. 1.20) and incorrect (Eq. 1.7) drag formulations are displayed.

PROBLEMS

1.1 Use calculus to verify that Eq. (1.9) is a solution of Eq. (1.8) for the initial condition $v(0) = 0$.

1.2 Use calculus to solve Eq. (1.21) for the case where the initial velocity is **(a)** positive and **(b)** negative. **(c)** Based on your results for **(a)** and **(b)**, perform the same computation as in Example 1.1 but with an initial velocity of -40 m/s. Compute values of the velocity from $t = 0$ to 12 s at intervals of 2 s. Note that for this case, the zero velocity occurs at $t = 3.470239$ s.

1.3 The following information is available for a bank account:

Date	Deposits	Withdrawals	Balance
5/1			1512.33
	220.13	327.26	
6/1			
	216.80	378.61	
7/1			
	450.25	106.80	
8/1			
	127.31	350.61	
9/1			

Note that the money earns interest which is computed as

Interest $= iB_i$

where i = the interest rate expressed as a fraction per month, and B_i the initial balance at the beginning of the month.

(a) Use the conservation of cash to compute the balance on 6/1, 7/1, 8/1, and 9/1 if the interest rate is 1% per month ($i = 0.01$/month). Show each step in the computation.

(b) Write a differential equation for the cash balance in the form

$$\frac{dB}{dt} = f[D(t), W(t), i]$$

where t = time (months), $D(t)$ = deposits as a function of time ($/month), $W(t)$ = withdrawals as a function of time ($/month). For this case, assume that interest is compounded continuously; that is, interest $= iB$.

(c) Use Euler's method with a time step of 0.5 month to simulate the balance. Assume that the deposits and withdrawals are applied uniformly over the month.

(d) Develop a plot of balance versus time for **(a)** and **(c)**.

1.4 Repeat Example 1.2. Compute the velocity to $t = 12$ s, with a step size of **(a)** 1 and **(b)** 0.5 s. Can you make any statement regarding the errors of the calculation based on the results?

1.5 Rather than the nonlinear relationship of Eq. (1.7), you might choose to model the upward force on the bungee jumper as a linear relationship:

$$F_U = -c'v$$

where c' = a first-order drag coefficient (kg/s).

(a) Using calculus, obtain the closed-form solution for the case where the jumper is initially at rest ($v = 0$ at $t = 0$).

(b) Repeat the numerical calculation in Example 1.2 with the same initial condition and parameter values. Use a value of 11.5 kg/s for c'.

1.6 For the free-falling bungee jumper with linear drag (Prob. 1.5), assume a first jumper is 70 kg and has a drag coefficient of 12 kg/s. If a second jumper has a drag coefficient of 15 kg/s and a mass of 80 kg, how long will it take her to reach the same velocity jumper 1 reached in 9 s?

1.7 For the second-order drag model (Eq. 1.8), compute the velocity of a free-falling parachutist using Euler's method for the case where $m = 80$ kg and $c_d = 0.25$ kg/m. Perform the calculation from $t = 0$ to 20 s with a step size of 1 s. Use an initial condition that the parachutist has an upward velocity of 20 m/s at $t = 0$. At $t = 10$ s, assume that the chute is instantaneously deployed so that the drag coefficient jumps to 1.5 kg/m.

1.8 The amount of a uniformly distributed radioactive contaminant contained in a closed reactor is measured by its concentration c (becquerel/liter or Bq/L). The contaminant decreases at a decay rate proportional to its concentration; that is

Decay rate $= -kc$

where k is a constant with units of day^{-1}. Therefore, according to Eq. (1.14), a mass balance for the reactor can be written as

$$\frac{dc}{dt} = -kc$$

$$\begin{pmatrix} \text{change} \\ \text{in mass} \end{pmatrix} = \begin{pmatrix} \text{decrease} \\ \text{by decay} \end{pmatrix}$$

(a) Use Euler's method to solve this equation from $t = 0$ to 1 d with $k = 0.175$ d^{-1}. Employ a step size of $\Delta t = 0.1$ d. The concentration at $t = 0$ is 100 Bq/L.

(b) Plot the solution on a semilog graph (i.e., ln c versus t) and determine the slope. Interpret your results.

1.9 A storage tank (Fig. P1.9) contains a liquid at depth y where $y = 0$ when the tank is half full. Liquid is withdrawn at a constant flow rate Q to meet demands. The contents are

FIGURE P1.9

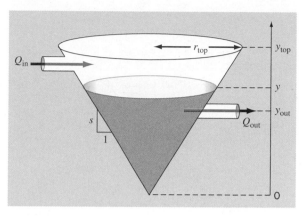

FIGURE P1.11

resupplied at a sinusoidal rate $3Q\sin^2(t)$. Equation (1.14) can be written for this system as

$$\frac{d(Ay)}{dt} = 3Q\sin^2(t) - Q$$

$$\left(\begin{array}{c}\text{change in}\\\text{volume}\end{array}\right) = (\text{inflow}) - (\text{outflow})$$

or, since the surface area A is constant

$$\frac{dy}{dt} = 3\frac{Q}{A}\sin^2(t) - \frac{Q}{A}$$

Use Euler's method to solve for the depth y from $t = 0$ to 10 d with a step size of 0.5 d. The parameter values are $A = 1250\ \text{m}^2$ and $Q = 450\ \text{m}^3/\text{d}$. Assume that the initial condition is $y = 0$.

1.10 For the same storage tank described in Prob. 1.9, suppose that the outflow is not constant but rather depends on the depth. For this case, the differential equation for depth can be written as

$$\frac{dy}{dt} = 3\frac{Q}{A}\sin^2(t) - \frac{\alpha(1+y)^{1.5}}{A}$$

Use Euler's method to solve for the depth y from $t = 0$ to 10 d with a step size of 0.5 d. The parameter values are $A = 1250\ \text{m}^2$, $Q = 450\ \text{m}^3/\text{d}$, and $\alpha = 150$. Assume that the initial condition is $y = 0$.

1.11 Apply the conservation of volume (see Prob. 1.9) to simulate the level of liquid in a conical storage tank (Fig. P1.11).

The liquid flows in at a sinusoidal rate of $Q_{\text{in}} = 3\sin^2(t)$ and flows out according to

$$\begin{array}{ll} Q_{\text{out}} = 3(y - y_{\text{out}})^{1.5} & y > y_{\text{out}}\\ Q_{\text{out}} = 0 & y \le y_{\text{out}}\end{array}$$

where flow has units of m^3/d and $y =$ the elevation of the water surface above the bottom of the tank (m). Use Euler's method to solve for the depth y from $t = 0$ to 10 d with a step size of 0.5 d. The parameter values are $r_{\text{top}} = 2.5\ \text{m}$, $y_{\text{top}} = 4\ \text{m}$, and $y_{\text{out}} = 1\ \text{m}$. Assume that the level is initially below the outlet pipe with $y(0) = 0.8\ \text{m}$.

1.12 A group of 35 students attend a class in an insulated room which measures 11 by 8 by 3 m. Each student takes up about $0.075\ \text{m}^3$ and gives out about 80 W of heat (1 W = 1 J/s). Calculate the air temperature rise during the first 20 minutes of the class if the room is completely sealed and insulated. Assume the heat capacity C_v for air is 0.718 kJ/(kg K). Assume air is an ideal gas at 20 °C and 101.325 kPa. Note that the heat absorbed by the air Q is related to the mass of the air m the heat capacity, and the change in temperature by the following relationship:

$$Q = m\int_{T_1}^{T_2} C_v\,dT = mC_v(T_2 - T_1)$$

The mass of air can be obtained from the ideal gas law:

$$PV = \frac{m}{\text{Mwt}}RT$$

where P is the gas pressure, V is the volume of the gas, Mwt is the molecular weight of the gas (for air, 28.97 kg/kmol), and R is the ideal gas constant [8.314 kPa m^3/(kmol K)].

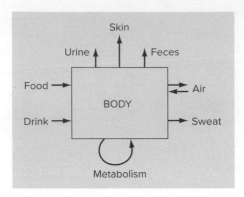

Skin

Urine   Feces

Food →

BODY

Drink →   → Air

→ Sweat

Metabolism

FIGURE P1.13

1.13 Figure P1.13 depicts the various ways in which an average man gains and loses water in one day. One liter is ingested as food, and the body metabolically produces 0.3 liters. In breathing air, the exchange is 0.05 liters while inhaling, and 0.4 liters while exhaling over a one-day period. The body will also lose 0.3, 1.4, 0.2, and 0.35 liters through sweat, urine, feces, and through the skin, respectively. To maintain steady state, how much water must be drunk per day?

1.14 In our example of the free-falling bungee jumper, we assumed that the acceleration due to gravity was a constant value of 9.81 m/s². Although this is a decent approximation when we are examining falling objects near the surface of the earth, the gravitational force decreases as we move above sea level. A more general representation based on Newton's inverse square law of gravitational attraction can be written as

$$g(x) = g(0) \frac{R^2}{(R + x)^2}$$

where $g(x)$ = gravitational acceleration at altitude x (in m) measured upward from the earth's surface (m/s²), $g(0)$ = gravitational acceleration at the earth's surface ($\cong 9.81$ m/s²), and R = the earth's radius ($\cong 6.37 \times 10^6$ m).

(a) In a fashion similar to the derivation of Eq. (1.8), use a force balance to derive a differential equation for velocity as a function of time that utilizes this more complete representation of gravitation. However, for this derivation, assume that upward velocity is positive.

(b) For the case where drag is negligible, use the chain rule to express the differential equation as a function of altitude rather than time. Recall that the chain rule is

$$\frac{dv}{dt} = \frac{dv}{dx} \frac{dx}{dt}$$

(c) Use calculus to obtain the closed form solution where $v = v_0$ at $x = 0$.

(d) Use Euler's method to obtain a numerical solution from $x = 0$ to 100,000 m using a step of 10,000 m where the initial velocity is 1500 m/s upward. Compare your result with the analytical solution.

1.15 Suppose that a spherical droplet of liquid evaporates at a rate that is proportional to its surface area.

$$\frac{dV}{dt} = -kA$$

where V = volume (mm³), t = time (min), k = the evaporation rate (mm/min), and A = surface area (mm²). Use Euler's method to compute the volume of the droplet from $t = 0$ to 10 min using a step size of 0.25 min. Assume that $k = 0.08$ mm/min and that the droplet initially has a radius of 2.5 mm. Assess the validity of your results by determining the radius of your final computed volume and verifying that it is consistent with the evaporation rate.

1.16 A fluid is pumped into the network shown in Fig. P1.16. If $Q_2 = 0.7$, $Q_3 = 0.5$, $Q_7 = 0.1$, and $Q_8 = 0.3$ m³/s, determine the other flows.

1.17 *Newton's law of cooling* says that the temperature of a body changes at a rate proportional to the difference between its temperature and that of the surrounding medium (the ambient temperature),

$$\frac{dT}{dt} = -k(T - T_a)$$

where T = the temperature of the body (°C), t = time (min), k = the proportionality constant (per minute), and T_a = the ambient temperature (°C). Suppose that a cup of coffee originally has a temperature of 70 °C. Use Euler's method to compute the temperature from $t = 0$ to 20 min using a step size of 2 min if $T_a = 20$ °C and $k = 0.019$/min.

Q_1 Q_3 Q_5

Q_2 Q_4 Q_6 Q_7

Q_{10} Q_9 Q_8

FIGURE P1.16

1.18 You are working as a crime scene investigator and must predict the temperature of a homicide victim over a 5-hr period. You know that the room where the victim was found was at 10 °C when the body was discovered.

(a) Use Newton's law of cooling (Prob. 1.17) and Euler's method to compute the victim's body temperature for the 5-hr period using values of $k = 0.12$/hr and $\Delta t = 0.5$ hr. Assume that the victim's body temperature at the time of death was 37 °C, and that the room temperature was at a constant value of 10 °C over the 5-hr period.

(b) Further investigation reveals that the room temperature had actually dropped linearly from 20 to 10 °C over the 5-hr period. Repeat the same calculation as in (a) but incorporate this new information.

(c) Compare the results from (a) and (b) by plotting them on the same graph.

1.19 The velocity is equal to the rate of change of distance, x (m):

$$\frac{dx}{dt} = v(t) \qquad\qquad (P1.19)$$

Use Euler's method to numerically integrate Eqs. (P1.19) and (1.8) in order to determine both the velocity and distance fallen as a function of time for the first 10 seconds of freefall using the same parameters and conditions as in Example 1.2. Develop a plot of your results.

1.20 In addition to the downward force of gravity (weight) and drag, an object falling through a fluid is also subject to a buoyancy force which is proportional to the displaced volume (*Archimedes' principle*). For example, for a sphere with diameter d (m), the sphere's volume is $V = \pi d^3/6$, and its projected area is $A = \pi d^2/4$. The buoyancy force can then be computed as $F_b = -\rho V g$. We neglected buoyancy in our derivation of Eq. (1.8) because it is relatively small for an object like a bungee jumper moving through air. However, for a more dense fluid like water, it becomes more prominent.

(a) Derive a differential equation in the same fashion as Eq. (1.8), but include the buoyancy force and represent the drag force as described in Sec. 1.4.

(b) Rewrite the differential equation from (a) for the special case of a sphere.

(c) Use the equation developed in (b) to compute the terminal velocity (i.e., for the steady-state case). Use the following parameter values for a sphere falling through water: sphere diameter = 1 cm, sphere density = 2700 kg/m³, water density = 1000 kg/m³, and $C_d = 0.47$.

(d) Use Euler's method with a step size of $\Delta t = 0.03125$ s to numerically solve for the velocity from $t = 0$ to 0.25 s with an initial velocity of zero.

1.21 As noted in Sec. 1.4, a fundamental representation of the drag force, which assumes turbulent conditions (i.e., a high Reynolds number), can be formulated as

$$F_d = -\frac{1}{2}\rho A C_d v |v|$$

where F_d = the drag force (N), ρ = fluid density (kg/m³), A = the frontal area of the object on a plane perpendicular to the direction of motion (m²), v = velocity (m/s), and C_d = a dimensionless drag coefficient.

(a) Write the pair of differential equations for velocity and position (see Prob. 1.19) to describe the vertical motion of a sphere with diameter, d (m), and a density of ρ_s (kg/m³). The differential equation for velocity should be written as a function of the sphere's diameter.

(b) Use Euler's method with a step size of $\Delta t = 2$ s to compute the position and velocity of a sphere over the first 14 seconds. Employ the following parameters in your calculation: $d = 120$ cm, $\rho = 1.3$ kg/m³, $\rho_s = 2700$ kg/m³, and $C_d = 0.47$. Assume that the sphere has the initial conditions: $x(0) = 100$ m and $v(0) = -40$ m/s.

(c) Develop a plot of your results (i.e., y and v versus t) and use it to graphically estimate when the sphere would hit the ground.

(d) Compute the value for the bulk second-order drag coefficient, c_d' (kg/m). Note that the bulk second-order drag coefficient is the term in the final differential equation for velocity that multiplies the term $v |v|$.

1.22 As depicted in Fig. P1.22, a spherical particle settling through a quiescent fluid is subject to three forces: the downward force of gravity (F_G), and the upward forces of buoyancy (F_B) and drag (F_D). Both the gravity and buoyancy forces can be computed with Newton's second law with the latter equal to the weight of the displaced fluid. For laminar flow, the drag force can be computed with *Stoke's law*,

$$F_D = 3\pi\mu d v$$

where μ = the dynamic viscosity of the fluid (N s/m²), d = the particle diameter (m), and v = the particle's settling velocity (m/s). The mass of the particle can be expressed as the product of the particle's volume and density, ρ_s (kg/m³), and the mass of the displaced fluid can be computed as the product of the particle's volume and the fluid's density, ρ (kg/m³). The volume of a sphere is $\pi d^3/6$. In addition, laminar flow corresponds to the case where the dimensionless Reynolds number, Re, is less than 1, where Re $= \rho d v/\mu$.

(a) Use a force balance for the particle to develop the differential equation for dv/dt as a function of d, ρ, ρ_s, and μ.

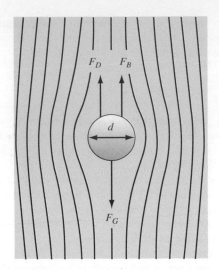

FIGURE P1.22

(b) At steady-state, use this equation to solve for the particle's terminal velocity.

(c) Employ the result of **(b)** to compute the particle's terminal velocity in m/s for a spherical silt particle settling in water: $d = 10\ \mu m$, $\rho = 1$ g/cm^3, $\rho_s = 2.65$ g/cm^3, and $\mu = 0.014$ g/(cm·s).

(d) Check whether flow is laminar.

(e) Use Euler's method to compute the velocity from $t = 0$ to 2^{-15} s with $\Delta t = 2^{-18}$ s given the initial condition: $v(0) = 0$.

1.23 As depicted in Fig. P1.23, the downward deflection, y (m), of a cantilever beam with a uniform load, $w = 10,000$ kg/m, can be computed as

$$y = \frac{w}{24EI}(x^4 - 4Lx^3 + 6L^2x^2)$$

where x = distance (m), E = the modulus of elasticity = 2×10^{11} Pa, I = moment of inertia = 3.25×10^{-4} m^4, and L = length = 4 m. This equation can be differentiated to yield the slope of the downward deflection as a function of x

$$\frac{dy}{dx} = \frac{w}{24EI}(4x^3 - 12Lx^2 + 12L^2x)$$

If $y = 0$ at $x = 0$, use this equation with Euler's method ($\Delta x = 0.125$ m) to compute the deflection from $x = 0$ to L. Develop a plot of your results along with the analytical solution computed with the first equation.

FIGURE P1.23
A cantilever beam.

1.24 Use *Archimedes' principle* to develop a steady-state force balance for a spherical ball of ice floating in seawater. The force balance should be expressed as a third-order polynomial (cubic) in terms of height of the cap above the water line (h), and the seawater's density (ρ_f), the ball's density (ρ_s) and radius (r).

1.25 Beyond fluids, *Archimedes' principle* has proven useful in geology when applied to solids on the earth's crust. Figure P1.25 depicts one such case where a lighter conical granite mountain "floats on" a denser basalt layer at the earth's surface. Note that the part of the cone below the surface is formally referred to as a *frustum*. Develop a steady-state force balance for this case in terms of the following parameters: basalt's density (ρ_b), granite's density (ρ_g), the cone's bottom radius (r), and the height above (h_1) and below (h_2) the earth's surface.

FIGURE P1.24

FIGURE P1.25

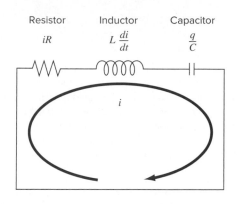

FIGURE P1.26

1.26 As depicted in Fig. P1.26, an *RLC circuit* consists of three elements: a resistor (R), an inductor (L), and a capacitor (C). The flow of current across each element induces a voltage drop. Kirchhoff's second voltage law states that the algebraic sum of these voltage drops around a closed circuit is zero,

$$iR + L\frac{di}{dt} + \frac{q}{C} = 0$$

where i = current, R = resistance, L = inductance, t = time, q = charge, and C = capacitance. In addition, the current is related to charge as in

$$\frac{dq}{dt} = i$$

(a) If the initial values are $i(0) = 0$ and $q(0) = 1$ C, use Euler's method to solve this pair of differential equations from $t = 0$ to 0.1 s using a step size of $\Delta t = 0.01$ s. Employ the following parameters for your calculation: $R = 200\ \Omega$, $L = 5$ H, and $C = 10^{-4}$ F.

(b) Develop a plot of i and q versus t.

1.27 Suppose that a parachutist with linear drag ($m = 70$ kg, $c = 12.5$ kg/s) jumps from an airplane flying at an altitude of 200 m with a horizontal velocity of 180 m/s relative to the ground.

(a) Write a system of four differential equations for x, y, $v_x =$ dx/dt and $v_y = dy/dt$.

(b) If the initial horizontal position is defined as $x = 0$, use Euler's methods with $\Delta t = 1$ s to compute the jumper's position over the first 10 seconds.

(c) Develop plots of y versus t and y versus x. Use the plot to graphically estimate when and where the jumper would hit the ground if the chute failed to open.

1.28 Figure P1.28 shows the forces exerted on a hot air balloon system.

FIGURE P1.28
Forces on a hot air balloon: F_B = buoyancy, F_G = weight of gas, F_P = weight of payload (including the balloon envelope), and F_D = drag. Note that the direction of the drag is downward when the balloon is rising.

Formulate the drag force as

$$F_D = \frac{1}{2}\rho_a v^2 A C_d$$

where ρ_a = air density (kg/m³), v = velocity (m/s), A = projected frontal area (m²), and C_d = the dimensionless drag coefficient ($\cong 0.47$ for a sphere). Note also that the total mass of the balloon consists of two components:

$$m = m_G + m_P$$

where m_G = the mass of the gas inside the expanded balloon (kg), and m_P = the mass of the payload (basket, passengers, and the unexpanded balloon = 265 kg). Assume that the ideal gas law holds ($P = \rho RT$), that the balloon is a perfect sphere with a diameter of 17.3 m, and that the heated air inside the envelope is at roughly the same pressure as the outside air.

Other necessary parameters are:

Normal atmospheric pressure, $P = 101,300$ Pa
The gas constant for dry air, $R = 287$ Joules/(kg K)
The air inside the balloon is heated to an average temperature, $T = 100$ °C
The normal (ambient) air density, $\rho = 1.2$ kg/m³.

(a) Use a force balance to develop the differential equation for dv/dt as a function of the model's fundamental parameters.

(b) At steady-state, calculate the particle's terminal velocity.

(c) Use Euler's method and Excel to compute the velocity from $t = 0$ to 60 s with $\Delta t = 2$ s given the previous parameters along with the initial condition: $v(0) = 0$. Develop a plot of your results.

2

MATLAB Fundamentals

CHAPTER OBJECTIVES

The primary objective of this chapter is to provide an introduction and overview of how MATLAB's calculator mode is used to implement interactive computations. Specific objectives and topics covered are

- Learning how real and complex numbers are assigned to variables.
- Learning how vectors and matrices are assigned values using simple assignment, the colon operator, and the linspace and logspace functions.
- Understanding the priority rules for constructing mathematical expressions.
- Gaining a general understanding of built-in functions and how you can learn more about them with MATLAB's Help facilities.
- Learning how to use vectors to create a simple line plot based on an equation.

YOU'VE GOT A PROBLEM

In Chap. 1, we used a force balance to determine the terminal velocity of a free-falling object like a bungee jumper:

$$v_t = \sqrt{\frac{gm}{c_d}}$$

where v_t = terminal velocity (m/s), g = gravitational acceleration (m/s^2), m = mass (kg), and c_d = a drag coefficient (kg/m). Aside from predicting the terminal velocity, this equation can also be rearranged to compute the drag coefficient

$$c_d = \frac{mg}{v_t^2} \tag{2.1}$$

TABLE 2.1   Data for the mass and associated terminal velocities of a number of jumpers.

m, kg	83.6	60.2	72.1	91.1	92.9	65.3	80.9
v_t, m/s	53.4	48.5	50.9	55.7	54	47.7	51.1

Thus, if we measure the terminal velocity of a number of jumpers of known mass, this equation provides a means to estimate the drag coefficient. The data in Table 2.1 were collected for this purpose.

In this chapter, we will learn how MATLAB can be used to analyze such data. Beyond showing how MATLAB can be employed to compute quantities like drag coefficients, we will also illustrate how its graphical capabilities provide additional insight into such analyses.

2.1   THE MATLAB ENVIRONMENT

MATLAB is a computer program that provides the user with a convenient environment for performing many types of calculations. In particular, it provides a very nice tool to implement numerical methods.

The most common way to operate MATLAB is by entering commands one at a time in the command window. In this chapter, we use this interactive or *calculator mode* to introduce you to common operations such as performing calculations and creating plots. In Chap. 3, we show how such commands can be used to create MATLAB programs.

One further note. This chapter has been written as a hands-on exercise. That is, you should read it while sitting in front of your computer. The most efficient way to become proficient is to actually implement the commands on MATLAB as you proceed through the following material.

MATLAB uses three primary windows:

- Command window. Used to enter commands and data.
- Graphics window. Used to display plots and graphs.
- Edit window. Used to create and edit M-files.

In this chapter, we will make use of the command and graphics windows. In Chap. 3 we will use the edit window to create M-files.

After starting MATLAB, the command window will open with the command prompt being displayed

```
>>
```

The calculator mode of MATLAB operates in a sequential fashion as you type in commands line by line. For each command, you get a result. Thus, you can think of it as operating like a very fancy calculator. For example, if you type in

```
>> 55 - 16
```

MATLAB will display the result[1]

```
ans =
    39
```

---

[1] MATLAB skips a line between the label (ans =) and the number (39). Here, we omit such blank lines for conciseness. You can control whether blank lines are included with the format compact and format loose commands.

Notice that MATLAB has automatically assigned the answer to a variable, ans. Thus, you could now use ans in a subsequent calculation:

```
>> ans + 11
```

with the result

```
ans =
    50
```

MATLAB assigns the result to ans whenever you do not explicitly assign the calculation to a variable of your own choosing.

2.2   ASSIGNMENT

Assignment refers to assigning values to variable names. This results in the storage of the values in the memory location corresponding to the variable name.

2.2.1 Scalars

The assignment of values to scalar variables is similar to other computer languages. Try typing

```
>> a = 4
```

Note how the assignment echo prints to confirm what you have done:

```
a =
    4
```

Echo printing is a characteristic of MATLAB. It can be suppressed by terminating the command line with the semicolon (;) character. Try typing

```
>> A = 6;
```

You can type several commands on the same line by separating them with commas or semicolons. If you separate them with commas, they will be displayed, and if you use the semicolon, they will not. For example,

```
>> a = 4,A = 6;x = 1;

a =
    4
```

MATLAB treats names in a case-sensitive manner—that is, the variable a is not the same as A. To illustrate this, enter

```
>> a
```

and then enter

```
>> A
```

See how their values are distinct. They are distinct names.

We can assign complex values to variables, since MATLAB handles complex arithmetic automatically. The unit imaginary number $\sqrt{-1}$ is preassigned to the variable i. Consequently, a complex value can be assigned simply as in

```
>> x = 2+i*4

x =
   2.0000 + 4.0000i
```

It should be noted that MATLAB allows the symbol j to be used to represent the unit imaginary number for input. However, it always uses an i for display. For example,

```
>> x = 2+j*4

x =
   2.0000 + 4.0000i
```

There are several predefined variables, for example, pi.

```
>> pi

ans =
   3.1416
```

Notice how MATLAB displays four decimal places. If you desire additional precision, enter the following:

```
>> format long
```

Now when pi is entered the result is displayed to 15 significant figures:

```
>> pi

ans =
   3.14159265358979
```

To return to the four decimal version, type

```
>> format short
```

The following is a summary of the format commands you will employ routinely in engineering and scientific calculations. They all have the syntax: format *type*.

type	Result	Example
short	Scaled fixed-point format with 5 digits	3.1416
long	Scaled fixed-point format with 15 digits for double and 7 digits for single	3.14159265358979
short e	Floating-point format with 5 digits	3.1416e+000
long e	Floating-point format with 15 digits for double and 7 digits for single	3.141592653589793e+000
short g	Best of fixed- or floating-point format with 5 digits	3.1416
long g	Best of fixed- or floating-point format with 15 digits for double and 7 digits for single	3.14159265358979
short eng	Engineering format with at least 5 digits and a power that is a multiple of 3	3.1416e+000
long eng	Engineering format with exactly 16 significant digits and a power that is a multiple of 3	3.14159265358979e+000
bank	Fixed dollars and cents	3.14

2.2.2 Arrays, Vectors, and Matrices

An array is a collection of values that are represented by a single variable name. One-dimensional arrays are called *vectors* and two-dimensional arrays are called *matrices*. The scalars used in Sec. 2.2.1 are actually matrices with one row and one column.

Brackets are used to enter arrays in the command mode. For example, a row vector can be assigned as follows:

```
>> a = [1 2 3 4 5]

a =
   1   2   3   4   5
```

Note that this assignment overrides the previous assignment of a = 4.

In practice, row vectors are rarely used to solve mathematical problems. When we speak of vectors, we usually refer to column vectors, which are more commonly used. A column vector can be entered in several ways. Try them.

```
>> b = [2;4;6;8;10]
```

or

```
>> b = [2
4
6
8
10]
```

or, by transposing a row vector with the ' operator,

```
>> b = [2 4 6 8 10]'
```

The result in all three cases will be

```
b =
   2
   4
   6
   8
  10
```

A matrix of values can be assigned as follows:

```
>> A = [1 2 3; 4 5 6; 7 8 9]

A =
   1   2   3
   4   5   6
   7   8   9
```

In addition, the Enter key (carriage return) can be used to separate the rows. For example, in the following case, the Enter key would be struck after the 3, the 6, and the] to assign the matrix:

```
>> A = [1 2 3
4 5 6
7 8 9]
```

Finally, we could construct the same matrix by *concatenating* (i.e., joining) the vectors representing each column:

```
>> A = [[1 4 7]' [2 5 8]' [3 6 9]']
```

At any point in a session, a list of all current variables can be obtained by entering the who command:

```
>> who
```

```
Your variables are:
A    a    ans  b    x
```

or, with more detail, enter the whos command:

```
>> whos
```

```
Name      Size              Bytes  Class
A         3x3                  72  double array
a         1x5                  40  double array
ans       1x1                   8  double array
b         5x1                  40  double array
x         1x1                  16  double array (complex)
Grand total is 21 elements using 176 bytes
```

Note that subscript notation can be used to access an individual element of an array. For example, the fourth element of the column vector b can be displayed as

```
>> b(4)
```

```
ans =
     8
```

For an array, A(m,n) selects the element in mth row and the nth column. For example,

```
>> A(2,3)
```

```
ans =
     6
```

There are several built-in functions that can be used to create matrices. For example, the ones and zeros functions create vectors or matrices filled with ones and zeros, respectively. Both have two arguments, the first for the number of rows and the second for the number of columns. For example, to create a 2 × 3 matrix of zeros:

```
>> E = zeros(2,3)
```

```
E =
     0    0    0
     0    0    0
```

Similarly, the ones function can be used to create a row vector of ones:

```
>> u = ones(1,3)
```

```
u =
     1    1    1
```

2.2.3 The Colon Operator

The colon operator is a powerful tool for creating and manipulating arrays. If a colon is used to separate two numbers, MATLAB generates the numbers between them using an increment of one:

```
>> t = 1:5

t =
     1    2    3    4    5
```

If colons are used to separate three numbers, MATLAB generates the numbers between the first and third numbers using an increment equal to the second number:

```
>> t = 1:0.5:3

t =
     1.0000    1.5000    2.0000    2.5000    3.0000
```

Note that negative increments can also be used

```
>> t = 10:-1:5

t =
    10    9    8    7    6    5
```

Aside from creating series of numbers, the colon can also be used as a wildcard to select the individual rows and columns of a matrix. When a colon is used in place of a specific subscript, the colon represents the entire row or column. For example, the second row of the matrix A can be selected as in

```
>> A(2,:)

ans =
     4    5    6
```

We can also use the colon notation to selectively extract a series of elements from within an array. For example, based on the previous definition of the vector t:

```
>> t(2:4)

ans =
     9    8    7
```

Thus, the second through the fourth elements are returned.

2.2.4 The linspace and logspace Functions

The linspace and logspace functions provide other handy tools to generate vectors of spaced points. The linspace function generates a row vector of equally spaced points. It has the form

```
linspace(x1, x2, n)
```

which generates *n* points between *x1* and *x2*. For example

```
>> linspace(0,1,6)

ans =
        0    0.2000    0.4000    0.6000    0.8000    1.0000
```

If the *n* is omitted, the function automatically generates 100 points.

The logspace function generates a row vector that is logarithmically equally spaced. It has the form

```
logspace(x1, x2, n)
```

which generates *n* logarithmically equally spaced points between decades 10^{x1} and 10^{x2}. For example,

```
>> logspace(-1,2,4)

ans =
     0.1000    1.0000    10.0000   100.0000
```

If *n* is omitted, it automatically generates 50 points.

2.2.5 Character Strings

Aside from numbers, *alphanumeric* information or *character strings* can be represented by enclosing the strings within single quotation marks. For example,

```
>> f = 'Miles ';
>> s = 'Davis';
```

Each character in a string is one element in an array. Thus, we can *concatenate* (i.e., paste together) strings as in

```
>> x = [f s]

x =
Miles Davis
```

Note that very long lines can be continued by placing an *ellipsis* (three consecutive periods) at the end of the line to be continued. For example, a row vector could be entered as

```
>> a = [1 2 3 4 5 ...
6 7 8]

a =
     1    2    3    4    5    6    7    8
```

However, you cannot use an ellipsis within single quotes to continue a string. To enter a string that extends beyond a single line, piece together shorter strings as in

```
>> quote = ['Any fool can make a rule,' ...
' and any fool will mind it']

quote =
Any fool can make a rule, and any fool will mind it
```

TABLE 2.2   Some useful string functions.

Function	Description
n=length(s)	Number of characters, n, in a string, s.
b=strcmp(s1,s2)	Compares two strings, s1 and s2; if equal returns true (**b = 1**). If not equal, returns false (**b = 0**).
n=str2num(s)	Converts a string, s, to a number, n.
s=num2str(n)	Converts a number, n, to a string, s.
s2=strrep(s1,c1,c2)	Replaces characters in a string with different characters.
i=strfind(s1,s2)	Returns the starting indices of any occurrences of the string s2 in the string s1.
S=upper(s)	Converts a string to upper case.
s=lower(S)	Converts a string to lower case.

A number of built-in MATLAB functions are available to operate on strings. Table 2.2 lists a few of the more commonly used ones. For example,

```
>> x1 = 'Canada'; x2 = 'Mexico'; x3 = 'USA'; x4 = '2010'; x5 = 810;

>> strcmp(a1,a2)

ans =

0

>> strcmp(x2,'Mexico')

ans =

1

>> str2num(x4)

ans =

2010

>> num2str(x5)

ans =

810

>> strrep

>> lower

>> upper
```

Note, if you want to display strings in multiple lines, use the sprint function and insert the two-character sequence \n between the strings. For example,

```
>> disp(sprintf('Yo\nAdrian!'))
```

yields

```
Yo

Adrian!
```

2.3 MATHEMATICAL OPERATIONS

Operations with scalar quantities are handled in a straightforward manner, similar to other computer languages. The common operators, in order of priority, are

^	Exponentiation
−	Negation
* /	Multiplication and division
\	Left division[2]
+ −	Addition and subtraction

These operators will work in calculator fashion. Try

```
>> 2*pi

ans =
    6.2832
```

Also, scalar real variables can be included:

```
>> y = pi/4;
>> y ^ 2.45

ans =
    0.5533
```

Results of calculations can be assigned to a variable, as in the next-to-last example, or simply displayed, as in the last example.

As with other computer calculation, the priority order can be overridden with parentheses. For example, because exponentiation has higher priority than negation, the following result would be obtained:

```
>> y = −4 ^ 2

y =
    −16
```

Thus, 4 is first squared and then negated. Parentheses can be used to override the priorities as in

```
>> y = (−4) ^ 2

y =
    16
```

Within each precedence level, operators have equal precedence and are evaluated from left to right. As an example,

```
>> 4^2^3
>> 4^(2^3)
>> (4^2)^3
```

---

[2] Left division applies to matrix algebra. It will be discussed in detail later in this book.

In the first case $4^2 = 16$ is evaluated first, which is then cubed to give 4096. In the second case $2^3 = 8$ is evaluated first and then $4^8 = 65,536$. The third case is the same as the first, but uses parentheses to be clearer.

One potentially confusing operation is negation; that is, when a minus sign is employed with a single argument to indicate a sign change. For example,

```
>> 2*-4
```

The -4 is treated as a number, so you get -8. As this might be unclear, you can use parentheses to clarify the operation

```
>> 2*(-4)
```

Here is a final example where the minus is used for negation

```
>> 2^-4
```

Again -4 is treated as a number, so $2^{\wedge}-4 = 2^{-4} = 1/2^4 = 1/16 = 0.0625$. Parentheses can make the operation clearer

```
>> 2^(-4)
```

Calculations can also involve complex quantities. Here are some examples that use the values of x ($2 + 4i$) and y (16) defined previously:

```
>> 3 * x

ans =
   6.0000 + 12.0000i
>> 1 / x

ans =
   0.1000 - 0.2000i
>> x ^ 2

ans =
 -12.0000 + 16.0000i
>> x + y

ans =
  18.0000 + 4.0000i
```

The real power of MATLAB is illustrated in its ability to carry out vector-matrix calculations. Although we will describe such calculations in detail in Chap. 8, it is worth introducing some examples here.

The *inner product* of two vectors (dot product) can be calculated using the * operator,

```
>> a * b

ans =
   110
```

and likewise, the *outer product*

```
>> b * a

ans =
     2     4     6     8    10
     4     8    12    16    20
     6    12    18    24    30
     8    16    24    32    40
    10    20    30    40    50
```

To further illustrate vector-matrix multiplication, first redefine a and b:

```
>> a = [1 2 3];
```

and

```
>> b = [4 5 6]';
```

Now, try

```
>> a * A

ans =
    30    36    42
```

or

```
>> A * b

ans =
    32
    77
   122
```

Matrices cannot be multiplied if the inner dimensions are unequal. Here is what happens when the dimensions are not those required by the operations. Try

```
>> A * a
```

MATLAB automatically displays the error message:

```
??? Error using ==> mtimes
Inner matrix dimensions must agree.
```

Matrix-matrix multiplication is carried out in likewise fashion:

```
>> A * A

ans =
    30    36    42
    66    81    96
   102   126   150
```

Mixed operations with scalars are also possible:

```
>> A/pi

ans =
    0.3183    0.6366    0.9549
    1.2732    1.5915    1.9099
    2.2282    2.5465    2.8648
```

We must always remember that MATLAB will apply the simple arithmetic operators in vector-matrix fashion if possible. At times, you will want to carry out calculations item by item in a matrix or vector. MATLAB provides for that too. For example,

```
>> A^2

ans =
        30     36     42
        66     81     96
       102    126    150
```

results in matrix multiplication of A with itself.

What if you want to square each element of A? That can be done with

```
>> A.^2

ans =
         1      4      9
        16     25     36
        49     64     81
```

The . preceding the ^ operator signifies that the operation is to be carried out element by element. The MATLAB manual calls these *array operations*. They are also often referred to as *element-by-element operations*.

MATLAB contains a helpful shortcut for performing calculations that you've already done. Press the up-arrow key. You should get back the last line you typed in.

```
>> A.^2
```

Pressing Enter will perform the calculation again. But you can also edit this line. For example, change it to the line below and then press Enter.

```
>> A.^3

ans =
         1      8     27
        64    125    216
       343    512    729
```

Using the up-arrow key, you can go back to any command that you entered. Press the up-arrow until you get back the line

```
>> b * a
```

Alternatively, you can type b and press the up-arrow once and it will automatically bring up the last command beginning with the letter b. The up-arrow shortcut is a quick way to fix errors without having to retype the entire line.

2.4   USE OF BUILT-IN FUNCTIONS

MATLAB and its Toolboxes have a rich collection of built-in functions. You can use online help to find out more about them. For example, if you want to learn about the log function, type in

```
>> help log

LOG    Natural logarithm.
```

```
LOG(X) is the natural logarithm of the elements of X.
Complex results are produced if X is not positive.

See also LOG2, LOG10, EXP, LOGM.
```

For a list of all the elementary functions, type

```
>> help elfun
```

One of their important properties of MATLAB's built-in functions is that they will operate directly on vector and matrix quantities. For example, try

```
>> log(A)

ans =
         0    0.6931    1.0986
    1.3863    1.6094    1.7918
    1.9459    2.0794    2.1972
```

and you will see that the natural logarithm function is applied in array style, element by element, to the matrix A. Most functions, such as sqrt, abs, sin, acos, tanh, and exp, operate in array fashion. Certain functions, such as exponential and square root, have matrix definitions also. MATLAB will evaluate the matrix version when the letter m is appended to the function name. Try

```
>> sqrtm(A)

ans =
    0.4498 + 0.7623i    0.5526 + 0.2068i    0.6555 - 0.3487i
    1.0185 + 0.0842i    1.2515 + 0.0228i    1.4844 - 0.0385i
    1.5873 - 0.5940i    1.9503 - 0.1611i    2.3134 + 0.2717i
```

There are several functions for rounding. For example, suppose that we enter a vector:

```
>> E = [-1.6 -1.5 -1.4 1.4 1.5 1.6];
```

The round function rounds the elements of E to the nearest integers:

```
>> round(E)

ans =
    -2    -2    -1     1     2     2
```

The ceil (short for ceiling) function rounds to the nearest integers toward infinity:

```
>> ceil(E)

ans =
    -1    -1    -1     2     2     2
```

The floor function rounds down to the nearest integers toward minus infinity:

```
>> floor(E)

ans =
    -2    -2    -2     1     1     1
```

There are also functions that perform special actions on the elements of matrices and arrays. For example, the sum function returns the sum of the elements:

```
>> F = [3 5 4 6 1];
>> sum(F)

ans =
    19
```

In a similar way, it should be pretty obvious what's happening with the following commands:

```
>> min(F),max(F),mean(F),prod(F),sort(F)

ans =
    1

ans =
    6

ans =
    3.8000

ans =
    360

ans =
    1   3   4   5   6
```

A common use of functions is to evaluate a formula for a series of arguments. Recall that the velocity of a free-falling bungee jumper can be computed with [Eq. (1.9)]:

$$v = \sqrt{\frac{gm}{c_d}} \tanh\left(\sqrt{\frac{gc_d}{m}} \, t\right)$$

where v is velocity (m/s), g is the acceleration due to gravity (9.81 m/s^2), m is mass (kg), c_d is the drag coefficient (kg/m), and t is time (s).

Create a column vector t that contains values from 0 to 20 in steps of 2:

```
>> t = [0:2:20]'

t =
    0
    2
    4
    6
    8
   10
   12
   14
   16
   18
   20
```

Check the number of items in the t array with the length function:

```
>> length(t)

ans =
   11
```

Assign values to the parameters:

```
>> g = 9.81; m = 68.1; cd = 0.25;
```

MATLAB allows you to evaluate a formula such as $v = f(t)$, where the formula is computed for each value of the t array, and the result is assigned to a corresponding position in the v array. For our case,

```
>> v = sqrt(g*m/cd)*tanh(sqrt(g*cd/m)*t)

v =
         0
   18.7292
   33.1118
   42.0762
   46.9575
   49.4214
   50.6175
   51.1871
   51.4560
   51.5823
   51.6416
```

2.5   GRAPHICS

MATLAB allows graphs to be created quickly and conveniently. For example, to create a graph of the t and v arrays from the data above, enter

```
>> plot(t, v)
```

The graph appears in the graphics window and can be printed or transferred via the clipboard to other programs.

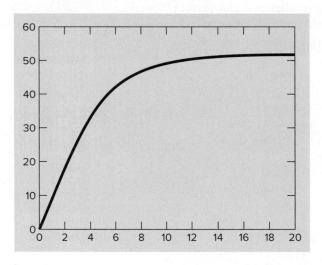

You can customize the graph a bit with commands such as the following:

```
>> title('Plot of v versus t')
>> xlabel('Values of t')
>> ylabel('Values of v')
>> grid
```

The plot command displays a solid thin blue line by default. If you want to plot each point with a symbol, you can include a specifier enclosed in single quotes in the plot function. Table 2.3 lists the available specifiers. For example, if you want to use open circles enter

```
>> plot(t, v, 'o')
```

TABLE 2.3 Specifiers for colors, symbols, and line types.

Colors		Symbols		Line Types	
Blue	b	Point	.	Solid	–
Green	g	Circle	o	Dotted	:
Red	r	X-mark	x	Dashdot	-.
Cyan	c	Plus	+	Dashed	--
Magenta	m	Star	*		
Yellow	y	Square	s		
Black	k	Diamond	d		
White	w	Triangle(down)	v		
		Triangle(up)	^		
		Triangle(left)	<		
		Triangle(right)	>		
		Pentagram	p		
		Hexagram	h		

You can also combine several specifiers. For example, if you want to use square green markers connected by green dashed lines, you could enter

```
>> plot(t, v, 's--g')
```

You can also control the line width as well as the marker's size and its edge and face (i.e., interior) colors. For example, the following command uses a heavier (2-point), dashed, cyan line to connect larger (10-point) diamond-shaped markers with black edges and magenta faces:

```
>> plot(x,y,'--dc','LineWidth', 2,...
    'MarkerSize',10,...
    'MarkerEdgeColor','k',...
    'MarkerFaceColor','m')
```

Note that the default line width is 1 point. For the markers, the default size is 6 point with blue edge color and no face color.

MATLAB allows you to display more than one data set on the same plot. For example, an alternative way to connect each data marker with a straight line would be to type

```
>> plot(t, v, t, v, 'o')
```

It should be mentioned that, by default, previous plots are erased every time the plot command is implemented. The hold on command holds the current plot and all axis properties so that additional graphing commands can be added to the existing plot. The hold off command returns to the default mode. For example, if we had typed the following commands, the final plot would only display symbols:

```
>> plot(t, v)
>> plot(t, v, 'o')
```

In contrast, the following commands would result in both lines and symbols being displayed:

```
>> plot(t, v)
>> hold on
>> plot(t, v, 'o')
>> hold off
```

In addition to hold, another handy function is subplot, which allows you to split the graph window into subwindows or *panes*. It has the syntax

```
subplot(m, n, p)
```

This command breaks the graph window into an *m*-by-*n* matrix of small axes, and selects the *p*-th axes for the current plot.

We can demonstrate subplot by examining MATLAB's capability to generate three-dimensional plots. The simplest manifestation of this capability is the plot3 command which has the syntax

```
plot3(x, y, z)
```

where *x*, *y*, and *z* are three vectors of the same length. The result is a line in three-dimensional space through the points whose coordinates are the elements of *x*, *y*, and *z*.

Plotting a helix provides a nice example to illustrate its utility. First, let's graph a circle with the two-dimensional plot function using the parametric representation: $x = \sin(t)$ and $y = \cos(t)$. We employ the subplot command so we can subsequently add the three-dimensional plot.

```
>> t = 0:pi/50:10*pi;
>> subplot(1,2,1);plot(sin(t),cos(t))
```

```
>> axis square
>> title('(a)')
```

As in Fig. 2.1*a*, the result is a circle. Note that the circle would have been distorted if we had not used the axis square command.

Now, let's add the helix to the graph's right pane. To do this, we again employ a parametric representation: $x = \sin(t)$, $y = \cos(t)$, and $z = t$

```
>> subplot(1,2,2);plot3(sin(t),cos(t),t);
>> title('(b)')
```

The result is shown in Fig. 2.1*b*. Can you visualize what's going on? As time evolves, the x and y coordinates sketch out the circumference of the circle in the x–y plane in the same fashion as the two-dimensional plot. However, simultaneously, the curve rises vertically as the z coordinate increases linearly with time. The net result is the characteristic spring or spiral staircase shape of the helix.

There are other features of graphics that are useful—for example, plotting objects instead of lines, families of curves plots, plotting on the complex plane, log-log or semilog plots, three-dimensional mesh plots, and contour plots. As described next, a variety of resources are available to learn about these as well as other MATLAB capabilities.

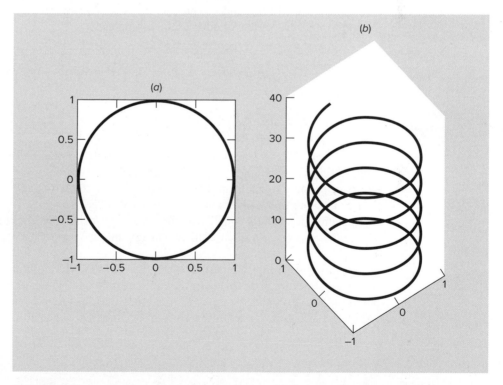

FIGURE 2.1
A two-pane plot of (*a*) a two-dimensional circle and (*b*) a three-dimensional helix.

2.6 OTHER RESOURCES

The foregoing was designed to focus on those features of MATLAB that we will be using in the remainder of this book. As such, it is obviously not a comprehensive overview of all of MATLAB's capabilities. If you are interested in learning more, you should consult one of the excellent books devoted to MATLAB (e.g., Attaway, 2009; Palm, 2007; Hanselman and Littlefield, 2005; and Moore, 2008).

Further, the package itself includes an extensive Help facility that can be accessed by clicking on the Help menu in the command window. This will provide you with a number of different options for exploring and searching through MATLAB's Help material. In addition, it provides access to a number of instructive demos.

As described in this chapter, help is also available in interactive mode by typing the help command followed by the name of a command or function.

If you do not know the name, you can use the lookfor command to search the MATLAB Help files for occurrences of text. For example, suppose that you want to find all the commands and functions that relate to logarithms, you could enter

```
>> lookfor logarithm
```

and MATLAB will display all references that include the word logarithm.

Finally, you can obtain help from The MathWorks, Inc., website at www.mathworks .com. There you will find links to product information, newsgroups, books, and technical support as well as a variety of other useful resources.

2.7 CASE STUDY   EXPLORATORY DATA ANALYSIS

Background.   Your textbooks are filled with formulas developed in the past by renowned scientists and engineers. Although these are of great utility, engineers and scientists often must supplement these relationships by collecting and analyzing their own data. Sometimes this leads to a new formula. However, prior to arriving at a final predictive equation, we usually "play" with the data by performing calculations and developing plots. In most cases, our intent is to gain insight into the patterns and mechanisms hidden in the data.

In this case study, we will illustrate how MATLAB facilitates such exploratory data analysis. We will do this by estimating the drag coefficient of a free-falling human based on Eq. (2.1) and the data from Table 2.1. However, beyond merely computing the drag coefficient, we will use MATLAB's graphical capabilities to discern patterns in the data.

Solution.   The data from Table 2.1 along with gravitational acceleration can be entered as

```
>> m=[83.6 60.2 72.1 91.1 92.9 65.3 80.9];
>> vt=[53.4 48.5 50.9 55.7 54 47.7 51.1];
>> g=9.81;
```

The drag coefficients can then be computed with Eq. (2.1). Because we are performing element-by-element operations on vectors, we must include periods prior to the operators:

```
>> cd=g*m./vt.^2

cd =
      0.2876   0.2511   0.2730   0.2881   0.3125   0.2815   0.3039
```

We can now use some of MATLAB's built-in functions to generate some statistics for the results:

```
>> cdavg=mean(cd),cdmin=min(cd),cdmax=max(cd)
cdavg =
      0.2854
cdmin =
      0.2511
cdmax =
      0.3125
```

Thus, the average value is 0.2854 with a range from 0.2511 to 0.3125 kg/m.

Now, let's start to play with these data by using Eq. (2.1) to make a prediction of the terminal velocity based on the average drag:

```
>> vpred=sqrt(g*m/cdavg)

vpred =
      53.6065   45.4897   49.7831   55.9595   56.5096   47.3774   52.7338
```

Notice that we do not have to use periods prior to the operators in this formula? Do you understand why?

We can plot these values versus the actual measured terminal velocities. We will also superimpose a line indicating exact predictions (the 1:1 line) to help assess the results. Because we are going to eventually generate a second plot, we employ the subplot command:

```
>> subplot(2,1,1);plot(vt,vpred,'o',vt,vt)
>> xlabel('measured')
>> ylabel('predicted')
>> title('Plot of predicted versus measured velocities')
```

As in the top plot of Fig. 2.2, because the predictions generally follow the 1:1 line, you might initially conclude that the average drag coefficient yields decent results. However, notice how the model tends to underpredict the low velocities and overpredict the high. This suggests that rather than being constant, there might be a trend in the drag coefficients. This can be seen by plotting the estimated drag coefficients versus mass:

```
>> subplot(2,1,2);plot(m,cd,'o')
>> xlabel('mass (kg)')
>> ylabel('estimated drag coefficient (kg/m)')
>> title('Plot of drag coefficient versus mass')
```

The resulting plot, which is the bottom graph in Fig. 2.2, suggests that rather than being constant, the drag coefficient seems to be increasing as the mass of the jumper increases. Based on this result, you might conclude that your model needs to be improved. At the

FIGURE 2.2
Two plots created with MATLAB.

least, it might motivate you to conduct further experiments with a larger number of jumpers to confirm your preliminary finding.

In addition, the result might also stimulate you to go to the fluid mechanics literature and learn more about the science of drag. As described previously in Sec. 1.4, you would discover that the parameter c_d is actually a lumped drag coefficient that along with the true drag includes other factors such as the jumper's frontal area and air density:

$$c_d = \frac{C_D \rho A}{2} \tag{2.2}$$

where C_D = a dimensionless drag coefficient, ρ = air density (kg/m^3), and A = frontal area (m^2), which is the area projected on a plane normal to the direction of the velocity.

Assuming that the densities were relatively constant during data collection (a pretty good assumption if the jumpers all took off from the same height on the same day), Eq. (2.2) suggests that heavier jumpers might have larger areas. This hypothesis could be substantiated by measuring the frontal areas of individuals of varying masses.

PROBLEMS

2.1 What is the output when the following commands are implemented?
```
A=[1:3;2:2:6;3:-1:1]
A=A'
A(:,3)=[]
A=[A(:,1) [4 5 7]' A(:,2)]
A=sum(diag(A))
```
2.2 You want to write MATLAB equations to compute a vector of y values using the following equations

(a) $y = \dfrac{6t^3 - 3t - 4}{8 \sin(5t)}$

(b) $y = \dfrac{6t - 4}{8t} - \dfrac{\pi}{2}t$

where t is a vector. Make sure that you use periods only where necessary so the equation handles vector operations properly. Extra periods will be considered incorrect.

2.3 Write a MATLAB expression to compute and display the values of a vector of x values using the following equation

$$x = \frac{y\,(a + bz)^{1.8}}{z(1 - y)}$$

Assume that y and z are vector quantities of equal length and a and b are scalars.

2.4 What is displayed when the following MATLAB statements are executed?

(a) `A=[1 2; 3 4; 5 6]; A(2,:)'`
(b) `y=[0:1.5:7]'`
(c) `a=2; b=8; c=4; a + b / c`

2.5 The MATLAB humps function defines a curve that has 2 maxima (peaks) of unequal height over the interval $0 \le x \le 2$.

$$f(x) = \frac{1}{(x - 0.3)^2 + 0.01} + \frac{1}{(x - 0.9)^2 + 0.04} - 6$$

Use MATLAB to generate a plot of $f(x)$ versus x with

`x = [0:1/256:2];`

Do not use MATLAB's built-in humps function to generate the values of $f(x)$. Also, employ the minimum number of periods to perform the vector operations needed to generate $f(x)$ values for the plot.

2.6 Use the linspace function to create vectors identical to the following created with colon notation:

(a) `t = 4:6:35`
(b) `x = -4:2`

2.7 Use colon notation to create vectors identical to the following created with the linspace function:

(a) `v = linspace(-2,1.5,8)`
(b) `r = linspace(8,4.5,8)`

2.8 The command `linspace(a, b, n)` generates a row vector of n equally spaced points between a and b. Use colon notation to write an alternative one-line command to generate the same vector. Test your formulation for a $= -3$, b $= 5$, n $= 6$.

2.9 The following matrix is entered in MATLAB:

>> `A=[3 2 1;0:0.5:1;linspace(6, 8, 3)]`

(a) Write out the resulting matrix.
(b) Use colon notation to write a single-line MATLAB command to multiply the second row by the third column and assign the result to the variable c.

2.10 The following equation can be used to compute values of y as a function of x:

$$y = be^{-ax} \sin(bx)(0.012x^4 - 0.15x^3 + 0.075x^2 + 2.5x)$$

where a and b are parameters. Write the equation for implementation with MATLAB, where $a = 2$, $b = 5$, and x is a vector holding values from 0 to $\pi/2$ in increments of $\Delta x = \pi/40$. Employ the minimum number of periods (i.e., dot notation) so that your formulation yields a vector for y. In addition, compute the vector $z = y^2$ where each element holds the square of each element of y. Combine x, y, and z into a matrix w, where each column holds one of the variables, and display w using the short g format. In addition, generate a labeled plot of y and z versus x. Include a legend on the plot (use help to understand how to do this). For y, use a 1.5-point, dashdotted red line with 14-point, red-edged, white-faced pentagram-shaped markers. For z, use a standard-sized (i.e., default) solid blue line with standard-sized, blue-edged, green-faced square markers.

2.11 A simple electric circuit consisting of a resistor, a capacitor, and an inductor is depicted in Fig. P2.11. The charge on the capacitor $q(t)$ as a function of time can be computed as

$$q(t) = q_0 e^{-Rt/(2L)} \cos\left[\sqrt{\frac{1}{LC} - \left(\frac{R}{2L}\right)^2}\, t\right]$$

where t = time, q_0 the initial charge, R = the resistance, L = inductance, and C = capacitance. Use MATLAB to generate a plot of this function from $t = 0$ to 0.8, given that $q_0 = 10$, $R = 60$, $L = 9$, and $C = 0.00005$.

2.12 The standard normal probability density function is a bell-shaped curve that can be represented as

$$f(z) = \frac{1}{\sqrt{2\pi}}\, e^{-z^2/2}$$

FIGURE P2.11

Use MATLAB to generate a plot of this function from $z = -5$ to 5. Label the ordinate as frequency and the abscissa as z.

2.13 If a force F (N) is applied to compress a spring, its displacement x (m) can often be modeled by Hooke's law:

$$F = kx$$

where $k =$ the spring constant (N/m). The potential energy stored in the spring U (J) can then be computed as

$$U = \frac{1}{2}kx^2$$

Five springs are tested and the following data compiled:

F, N	14	18	8	9	13
x, m	0.013	0.020	0.009	0.010	0.012

Use MATLAB to store F and x as vectors and then compute vectors of the spring constants and the potential energies. Use the max function to determine the maximum potential energy.

2.14 The density of freshwater can be computed as a function of temperature with the following cubic equation:

$$\rho = 5.5289 \times 10^{-8} T_C^3 - 8.5016 \times 10^{-6} T_C^2$$

$$+ 6.5622 \times 10^{-5} T_C + 0.99987$$

where $\rho =$ density (g/cm^3) and $T_C =$ temperature (°C). Use MATLAB to generate a vector of temperatures ranging from 32 °F to 93.2 °F using increments of 3.6 °F. Convert this vector to degrees Celsius and then compute a vector of densities based on the cubic formula. Create a plot of ρ versus T_C. Recall that $T_C = 5/9(T_F - 32)$.

2.15 Manning's equation can be used to compute the velocity of water in a rectangular open channel:

$$U = \frac{\sqrt{S}}{n} \left(\frac{BH}{B + 2H}\right)^{2/3}$$

where $U =$ velocity (m/s), $S =$ channel slope, $n =$ roughness coefficient, $B =$ width (m), and $H =$ depth (m). The following data are available for five channels:

n	S	B	H
0.035	0.0001	10	2
0.020	0.0002	8	1
0.015	0.0010	20	1.5
0.030	0.0007	24	3
0.022	0.0003	15	2.5

Store these values in a matrix where each row represents one of the channels and each column represents one of the parameters. Write a single-line MATLAB statement to compute a column vector containing the velocities based on the values in the parameter matrix.

2.16 It is general practice in engineering and science that equations be plotted as lines and discrete data as symbols. Here are some data for concentration (c) versus time (t) for the photodegradation of aqueous bromine:

t, min	10	20	30	40	50	60
c, ppm	3.4	2.6	1.6	1.3	1.0	0.5

These data can be described by the following function:

$$c = 4.84 e^{-0.034t}$$

Use MATLAB to create a plot displaying both the data (using diamond-shaped, filled-red symbols) and the function (using a green, dashed line). Plot the function for $t = 0$ to 70 min.

2.17 The semilogy function operates in an identical fashion to the plot function except that a logarithmic (base-10) scale is used for the y axis. Use this function to plot the data and function as described in Prob. 2.16. Explain the results.

2.18 Here are some wind tunnel data for force (F) versus velocity (v):

v, m/s	10	20	30	40	50	60	70	80
F, N	25	70	380	550	610	1220	830	1450

These data can be described by the following function:

$$F = 0.2741 v^{1.9842}$$

Use MATLAB to create a plot displaying both the data (using circular magenta symbols) and the function (using a black dash-dotted line). Plot the function for $v = 0$ to 100 m/s and label the plot's axes.

2.19 The `loglog` function operates in an identical fashion to the `plot` function except that logarithmic scales are used for both the x and y axes. Use this function to plot the data and function as described in Prob. 2.18. Explain the results.

2.20 The Maclaurin series expansion for the cosine is

$$\cos x = 1 - \frac{x^2}{2!} + \frac{x^4}{4!} - \frac{x^6}{6!} + \frac{x^8}{8!} - \cdots$$

Use MATLAB to create a plot of the cosine (solid line) along with a plot of the series expansion (black dashed line) up to and including the term $x^8/8!$. Use the built-in function `factorial` in computing the series expansion. Make the range of the abscissa from $x = 0$ to $3\pi/2$.

2.21 You contact the jumpers used to generate the data in Table 2.1 and measure their frontal areas. The resulting values, which are ordered in the same sequence as the corresponding values in Table 2.1, are

A, m²	0.455	0.402	0.452	0.486	0.531	0.475	0.487

(a) If the air density is $\rho = 1.223$ kg/m³, use MATLAB to compute values of the dimensionless drag coefficient C_D.
(b) Determine the average, minimum, and maximum of the resulting values.
(c) Develop a stacked plot of A versus m (upper) and C_D versus m (lower). Include descriptive axis labels and titles on the plots.

2.22 The following parametric equations generate a *conical helix*.

$$x = t \cos(6t)$$
$$y = t \sin(6t)$$
$$z = t$$

Compute values of x, y, and z for $t = 0$ to 6π with $\Delta t = \pi/64$. Use `subplot` to generate a two-dimensional line plot (red solid line) of (x, y) in the top pane and a three-dimensional line plot (cyan solid line) of (x, y, z) in the bottom pane. Label the axes for both plots.

2.23 Exactly what will be displayed after the following MATLAB commands are typed?
(a)
```
>> x = 5;
>> x ^ 3;
>> y = 8 - x
```

(b)
```
>> q = 4:2:12;
>> r = [7 8 4; 3 6 -5];
>> sum(q) * r(2, 3)
```

2.24 The trajectory of an object can be modeled as

$$y = (\tan \theta_0)x - \frac{g}{2v_0^2 \cos^2 \theta_0} x^2 + y_0$$

where $y =$ height (m), $\theta_0 =$ initial angle (radians), $x =$ horizontal distance (m), $g =$ gravitational acceleration ($= 9.81$ m/s²), $v_0 =$ initial velocity (m/s), and $y_0 =$ initial height. Use MATLAB to find the trajectories for $y_0 = 0$ and $v_0 = 28$ m/s for initial angles ranging from 15 to 75° in increments of 15°. Employ a range of horizontal distances from $x = 0$ to 80 m in increments of 5 m. The results should be assembled in an array where the first dimension (rows) corresponds to the distances, and the second dimension (columns) corresponds to the different initial angles. Use this matrix to generate a single plot of the heights versus horizontal distances for each of the initial angles. Employ a legend to distinguish among the different cases, and scale the plot so that the minimum height is zero using the `axis` command.

2.25 The temperature dependence of chemical reactions can be computed with the *Arrhenius equation:*

$$k = Ae^{-E/(RTa)}$$

where $k =$ reaction rate (s⁻¹), $A =$ the preexponential (or frequency) factor, $E =$ activation energy (J/mol), $R =$ gas constant [8.314 J/(mole · K)], and $T_a =$ absolute temperature (K). A compound has $E = 1 \times 10^5$ J/mol and $A = 7 \times 10^{16}$. Use MATLAB to generate values of reaction rates for temperatures ranging from 253 to 325 K. Use `subplot` to generate a side-by-side graph of (a) k versus T_a (green line) and (b) $\log_{10} k$ (red line) versus $1/T_a$. Employ the `semilogy` function to create (b). Include axis labels and titles for both subplots. Interpret your results.

2.26 Figure P2.26a shows a uniform beam subject to a linearly increasing distributed load. As depicted in Fig. P2.26b, deflection y (m) can be computed with

$$y = \frac{w_0}{120EIL} (-x^5 + 2L^2x^3 - L^4x)$$

where $E =$ the modulus of elasticity and $I =$ the moment of inertia (m⁴). Employ this equation and calculus to generate MATLAB plots of the following quantities versus distance along the beam:
(a) displacement (y),
(b) slope [$\theta(x) = dy/dx$],
(c) moment [$M(x) = EId^2 y/dx^2$],
(d) shear [$V(x) = EId^3 y/dx^3$], and
(e) loading [$w(x) = -EId^4 y/dx^4$].

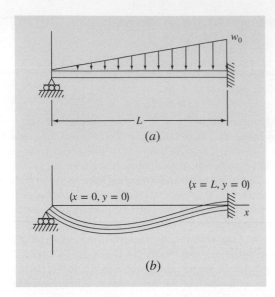

(a)

$(x = 0, y = 0)$

$(x = L, y = 0)$

x

(b)

FIGURE P2.26

Use the following parameters for your computation: $L = 600$ cm, $E = 50{,}000$ kN/cm^2, $I = 30{,}000$ cm^4, $w_0 = 2.5$ kN/cm, and $\Delta x = 10$ cm. Employ the subplot function to display all the plots vertically on the same page in the order **(a)** to **(e)**.

Include labels and use consistent MKS units when developing the plots.

2.27 The *butterfly curve* is given by the following parametric equations:

$$x = \sin(t)\left(e^{\cos t} - 2\cos 4t - \sin^5 \frac{t}{12}\right)$$

$$y = \cos(t)\left(e^{\cos t} - 2\cos 4t - \sin^5 \frac{t}{12}\right)$$

Generate values of x and y for values of t from 0 to 100 with $\Delta t = 1/16$. Construct plots of **(a)** x and y versus t and **(b)** y versus x. Use subplot to stack these plots vertically and make the plot in **(b)** square. Include titles and axis labels on both plots and a legend for **(a)**. For **(a)**, employ a dotted line for y in order to distinguish it from x.

2.28 The butterfly curve from Prob. 2.27 can also be represented in polar coordinates as

$$r = e^{\sin\theta} - 2\cos(4\theta) - \sin^5\left(\frac{2\theta - \pi}{24}\right)$$

Generate values of r for values of θ from 0 to 8π with $\Delta\theta = \pi/32$. Use the MATLAB function polar to generate the polar plot of the butterfly curve with a dashed red line. Employ the MATLAB Help to understand how to generate the plot.

3

Programming with MATLAB

CHAPTER OBJECTIVES

The primary objective of this chapter is to learn how to write M-file programs to implement numerical methods. Specific objectives and topics covered are

- Learning how to create well-documented M-files in the edit window and invoke them from the command window.
- Understanding how script and function files differ.
- Understanding how to incorporate help comments in functions.
- Knowing how to set up M-files so that they interactively prompt users for information and display results in the command window.
- Understanding the role of subfunctions and how they are accessed.
- Knowing how to create and retrieve data files.
- Learning how to write clear and well-documented M-files by employing structured programming constructs to implement logic and repetition.
- Recognizing the difference between if...elseif and switch constructs.
- Recognizing the difference between for...end and while structures.
- Knowing how to animate MATLAB plots.
- Understanding what is meant by vectorization and why it is beneficial.
- Understanding how anonymous functions can be employed to pass functions to function function M-files.

YOU'VE GOT A PROBLEM

In Chap. 1, we used a force balance to develop a mathematical model to predict the fall velocity of a bungee jumper. This model took the form of the following differential equation:

$$\frac{dv}{dt} = g - \frac{c_d}{m} v|v|$$

We also learned that a numerical solution of this equation could be obtained with Euler's method:

$$v_{i+1} = v_i + \frac{dv_i}{dt} \Delta t$$

This equation can be implemented repeatedly to compute velocity as a function of time. However, to obtain good accuracy, many small steps must be taken. This would be extremely laborious and time consuming to implement by hand. However, with the aid of MATLAB, such calculations can be performed easily.

So our problem now is to figure out how to do this. This chapter will introduce you to how MATLAB M-files can be used to obtain such solutions.

3.1 M-FILES

The most common way to operate MATLAB is by entering commands one at a time in the command window. M-files provide an alternative way of performing operations that greatly expand MATLAB's problem-solving capabilities. An *M-file* consists of a series of statements that can be run all at once. Note that the nomenclature "M-file" comes from the fact that such files are stored with a .m extension. M-files come in two flavors: script files and function files.

3.1.1 Script Files

A script file is merely a series of MATLAB commands that are saved on a file. They are useful for retaining a series of commands that you want to execute on more than one occasion. The script can be executed by typing the file name in the command window or by pressing the Run button.

EXAMPLE 3.1   Script File

Problem Statement. Develop a script file to compute the velocity of the free-falling bungee jumper for the case where the initial velocity is zero.

Solution. Open the editor with the selection: **New, Script.** Type in the following statements to compute the velocity of the free-falling bungee jumper at a specific time [recall Eq. (1.9)]:

```
g = 9.81; m = 68.1; t = 12; cd = 0.25;
v = sqrt(g * m / cd) * tanh(sqrt(g * cd / m) * t)
```

Save the file as scriptdemo.m. Return to the command window and type

```
>>scriptdemo
```

The result will be displayed as

```
v =
   50.6175
```

Thus, the script executes just as if you had typed each of its lines in the command window.

As a final step, determine the value of g by typing

```
>> g

g =
    9.8100
```

So you can see that even though g was defined within the script, it retains its value back in the command workspace. As we will see in the following section, this is an important distinction between scripts and functions.

3.1.2 Function Files

Function files are M-files that start with the word function. In contrast to script files, they can accept input arguments and return outputs. Hence they are analogous to user-defined functions in programming languages such as Fortran, Visual Basic or *C*.

The syntax for the function file can be represented generally as

```
function outvar = funcname(arglist)
% helpcomments
statements
outvar = value;
```

where *outvar* = the name of the output variable, *funcname* = the function's name, *arglist* = the function's argument list (i.e., comma-delimited values that are passed into the function), *helpcomments* = text that provides the user with information regarding the function (these can be invoked by typing Help *funcname* in the command window), and *statements* = MATLAB statements that compute the *value* that is assigned to *outvar*.

Beyond its role in describing the function, the first line of the *helpcomments*, called the *H1 line*, is the line that is searched by the lookfor command (recall Sec. 2.6). Thus, you should include key descriptive words related to the file on this line.

The M-file should be saved as *funcname*.m. The function can then be run by typing *funcname* in the command window as illustrated in the following example. Note that even though MATLAB is case-sensitive, your computer's operating system may not be. Whereas MATLAB would treat function names like freefall and FreeFall as two different variables, your operating system might not.

EXAMPLE 3.2   Function File

Problem Statement. As in Example 3.1, compute the velocity of the free-falling bungee jumper but now use a function file for the task.

Solution. Type the following statements in the file editor:

```
function v = freefall(t, m, cd)
% freefall: bungee velocity with second-order drag
% v=freefall(t,m,cd) computes the free-fall velocity
%                    of an object with second-order drag
% input:
```

```
%    t = time (s)
%    m = mass (kg)
%    cd = second-order drag coefficient (kg/m)
% output:
%    v = downward velocity (m/s)

g = 9.81;      % acceleration of gravity
v = sqrt(g * m / cd)*tanh(sqrt(g * cd / m) * t);
```

Save the file as freefall.m. To invoke the function, return to the command window and type in

```
>> freefall(12,68.1,0.25)
```

The result will be displayed as

```
ans =
   50.6175
```

One advantage of a function M-file is that it can be invoked repeatedly for different argument values. Suppose that you wanted to compute the velocity of a 100-kg jumper after 8 s:

```
>> freefall(8,100,0.25)

ans =
   53.1878
```

To invoke the help comments type

```
>> help freefall
```

which results in the comments being displayed

```
freefall: bungee velocity with second-order drag
  v=freefall(t,m,cd) computes the free-fall velocity
                     of an object with second-order drag
input:
  t = time (s)
  m = mass (kg)
  cd = second-order drag coefficient (kg/m)
output:
  v = downward velocity (m/s)
```

If at a later date, you forgot the name of this function, but remembered that it involved bungee jumping, you could enter

```
>> lookfor bungee
```

and the following information would be displayed

```
freefall.m - bungee velocity with second-order drag
```

Note that, at the end of the previous example, if we had typed

```
>> g
```

the following message would have been displayed

```
??? Undefined function or variable 'g'.
```

So even though g had a value of 9.81 within the M-file, it would not have a value in the command workspace. As noted previously at the end of Example 3.1, this is an important distinction between functions and scripts. The variables within a function are said to be *local* and are erased after the function is executed. In contrast, the variables in a script retain their existence after the script is executed.

Function M-files can return more than one result. In such cases, the variables containing the results are comma-delimited and enclosed in brackets. For example, the following function, stats.m, computes the mean and the standard deviation of a vector:

```
function [mean, stdev] = stats(x)
n = length(x);
mean = sum(x)/n;
stdev = sqrt(sum((x-mean).^2/(n-1)));
```

Here is an example of how it can be applied:

```
>> y = [8 5 10 12 6 7.5 4];
>> [m,s] = stats(y)

m =
   7.5000

s =
   2.8137
```

Although we will also make use of script M-files, function M-files will be our primary programming tool for the remainder of this book. Hence, we will often refer to function M-files as simply M-files.

3.1.3 Variable Scope

MATLAB variables have a property known as *scope* that refers to the context of the computing environment in which the variable has a unique identity and value. Typically, a variable's scope is limited either to the MATLAB workspace or within a function. This principle prevents errors when a programmer unintentionally gives the same name to variables in different contexts.

Any variables defined through the command line are within the MATLAB *workspace* and you can readily inspect a workspace variable's value by entering its name at the command line. However, workspace variables are not directly accessible to functions but rather are passed to functions via their arguments. For example, here is a function that adds two numbers

```
function c = adder(a,b)
x = 88
a
c = a + b
```

Suppose in the command window we type

```
>> x = 1; y = 4; c = 8
c =
        8
```

So as expected, the value of c in the workspace is 8. If you type

```
>> d = adder(x,y)
```

the result will be

```
x =
      88
a =
       1
c =
       5
d =
       5
```

But, if you then type

```
>> c, x, a
```

The result is

```
c =
        8
x =
        1

Undefined function or variable 'a'.
Error in ScopeScript (line 6)
c, x, a
```

The point here is that even though x was assigned a new value inside the function, the variable of the same name in the MATLAB workspace is unchanged. Even though they have the same name, the scope of each is limited to their context and does not overlap. In the function, the variables a and b are limited in scope to that function, and only exist while that function is being executed. Such variables are formally called *local variables*. Thus, when we try to display the value of a in the workspace, an error message is generated because the workspace has no access to the a in the function.

Another obvious consequence of limited-scope variables is that any parameter needed by a function must be passed as an input argument or by some other explicit means. A function cannot otherwise access variables in the workspace or in other functions.

3.1.4 Global Variables

As we have just illustrated, the function's argument list is like a window through which information is selectively passed between the workspace and a function, or between two functions. Sometimes, however, it might be convenient to have access to a variable in several contexts without passing it as an argument. In such cases, this can be accomplished by defining the variable as *global*. This is done with the global command, which is defined as

```
global X Y Z
```

where X, Y, and Z are global in scope. If several functions (and possibly the workspace), all declare a particular name as global, then they all share a single value of that variable. Any change to that variable, in any function, is then made to all the other functions that declare it global. Stylistically, MATLAB recommends that global variables use all capital letters, but this is not required.

EXAMPLE 3.3   **Use of Global Variables**

Problem Statement. The *Stefan-Boltzmann* law is used to compute the radiation flux from a black body[1] as in

$$J = \sigma T_a^4$$

where J = radiation flux [W/(m^2 s)], σ = the Stefan-Boltzmann constant (5.670367 × 10^{-8} W m^{-2} K^{-4}), and T_a = absolute temperature (K). In assessing the impact of climate change on water temperature, it is used to compute the radiation terms in a waterbody's heat balance. For example, the long-wave radiation from the atmosphere to the waterbody, J_{an} [W/(m^2 s)], can be calculated as

$$J_{an} = 0.97\sigma(T_{air} + 273.15)^4 \left(0.6 + 0.031\sqrt{e_{air}}\right)$$

where T_{air} = the temperature of the air above the waterbody (°C), and e_{air} = the vapor pressure of the air above the water body (mmHg),

$$e_{air} = 4.596e^{\frac{17.27T_d}{237.3 + T_d}} \tag{E3.3.1}$$

where T_d = the dew-point temperature (°C). The radiation from the water surface back into the atmosphere, J_{br} [W/(m^2 s)], is calculated as

$$J_{br} = 0.97\sigma(T_w + 273.15)^4 \tag{E3.3.2}$$

where T_w = the water temperature (°C). Write a script that utilizes two functions to compute the net long-wave radiation (i.e., the difference between the atmospheric radiation in and the water back radiation out) for a cold lake with a surface temperature of T_w = 15 °C on a hot (T_{air} = 30 °C), humid (T_d = 27.7 °C) summer day. Use global to share the Stefan-Boltzmann constant between the script and functions.

Solution. Here is the script

```
clc, format compact
global SIGMA
SIGMA =5.670367e-8;
Tair = 30; Tw = 15; Td = 27.7;
```

---

[1]A *black body* is an object that absorbs all incident electromagnetic radiation, regardless of frequency or angle of incidence. A *white body* is one that reflects all incident rays completely and uniformly in all directions.

```
Jan = AtmLongWaveRad(Tair, Td)
Jbr = WaterBackRad(Tw)
JnetLongWave = Jan-Jbr
```

Here is a function to compute the incoming long wave radiation from the atmosphere into the lake

```
function Ja = AtmLongWaveRad(Tair, Td)
global SIGMA
eair=4.596*exp(17.27*Td/(237.3+Td));
Ja=0.97*SIGMA*(Tair+273.15)^4*(0.6+0.031*sqrt(eair));
end
```

and here is a function to compute the long wave radiation from the lake back into the atmosphere

```
function Jb = WaterBackRad(Twater)
global SIGMA
Jb=0.97*SIGMA*(Twater+273.15)^4;
end
```

When the script is run, the output is

```
Jan =
  354.8483
Jbr =
  379.1905
JnetLongWave =
  -24.3421
```

Thus, for this case, because the back radiation is larger than the incoming radiation, so the lake loses heat at the rate of 24.3421 $W/(m^2 s)$ due to the two long-wave radiation fluxes.

If you require additional information about global variables, you can always type help global at the command prompt. The help facility can also be invoked to learn about other MATLAB commands dealing with scope such as persistent.

3.1.5 Subfunctions

Functions can call other functions. Although such functions can exist as separate M-files, they may also be contained in a single M-file. For example, the M-file in Example 3.2 (without comments) could have been split into two functions and saved as a single M-file[2]:

```
function v = freefallsubfunc(t, m, cd)
v = vel(t, m, cd);
end

function v = vel(t, m, cd)
g = 9.81;
v = sqrt(g * m / cd)*tanh(sqrt(g * cd / m) * t);
end
```

[2]Note that although end statements are not used to terminate single-function M-files, they are included when subfunctions are involved to demarcate the boundaries between the main function and the subfunctions.

This M-file would be saved as `freefallsubfunc.m`. In such cases, the first function is called the *main* or *primary function*. It is the only function that is accessible to the command window and other functions and scripts. All the other functions (in this case, `vel`) are referred to as *subfunctions*.

A subfunction is only accessible to the main function and other subfunctions within the M-file in which it resides. If we run `freefallsubfunc` from the command window, the result is identical to Example 3.2:

```
>> freefallsubfunc(12,68.1,0.25)

ans =
    50.6175
```

However, if we attempt to run the subfunction `vel`, an error message occurs:

```
>> vel(12,68.1,.25)
??? Undefined function or method 'vel' for input arguments of type 'double'.
```

3.2   INPUT-OUTPUT

As in Sec.3.1, information is passed into the function via the argument list and is output via the function's name. Two other functions provide ways to enter and display information directly using the command window.

The `input` Function. This function allows you to prompt the user for values directly from the command window. Its syntax is

```
n = input('promptstring')
```

The function displays the *promptstring*, waits for keyboard input, and then returns the value from the keyboard. For example,

```
m = input('Mass (kg): ')
```

When this line is executed, the user is prompted with the message

```
Mass (kg):
```

If the user enters a value, it would then be assigned to the variable `m`.

The `input` function can also return user input as a string. To do this, an `'s'` is appended to the function's argument list. For example,

```
name = input('Enter your name: ','s')
```

The `disp` Function. This function provides a handy way to display a value. Its syntax is

```
disp(value)
```

where *value* = the value you would like to display. It can be a numeric constant or variable, or a string message enclosed in hyphens. Its application is illustrated in the following example.

EXAMPLE 3.4   An Interactive M-File Function

Problem Statement. As in Example 3.2, compute the velocity of the free-falling bungee jumper, but now use the input and disp functions for input/output.

Solution. Type the following statements in the file editor:

```
function freefalli
% freefalli: interactive bungee velocity
%   freefalli interactive computation of the
%              free-fall velocity of an object
%              with second-order drag.
g = 9.81;      % acceleration of gravity
m = input('Mass (kg): ');
cd = input('Drag coefficient (kg/m): ');
t = input('Time (s): ');
disp(' ')
disp('Velocity (m/s):')
disp(sqrt(g * m / cd)*tanh(sqrt(g * cd / m) * t))
```

Save the file as freefalli.m. To invoke the function, return to the command window and type

```
>> freefalli
```

```
Mass (kg): 68.1
Drag coefficient (kg/m): 0.25
Time (s): 12

Velocity (m/s):
   50.6175
```

The fprintf Function. This function provides additional control over the display of information. A simple representation of its syntax is

```
fprintf('format', x, ...)
```

where *format* is a string specifying how you want the value of the variable *x* to be displayed. The operation of this function is best illustrated by examples.

A simple example would be to display a value along with a message. For instance, suppose that the variable velocity has a value of 50.6175. To display the value using eight digits with four digits to the right of the decimal point along with a message, the statement along with the resulting output would be

```
>> fprintf('The velocity is %8.4f m/s\n', velocity)
```

```
The velocity is 50.6175 m/s
```

This example should make it clear how the format string works. MATLAB starts at the left end of the string and displays the labels until it detects one of the symbols: % or \. In our example, it first encounters a % and recognizes that the following text is a format code. As in Table 3.1, the *format codes* allow you to specify whether numeric values are displayed in integer, decimal, or scientific format. After displaying the value of velocity, MATLAB continues displaying the character information (in our case the units: m/s) until it detects the symbol \. This tells MATLAB that the following text is a control code. As in Table 3.1, the

TABLE 3.1 Commonly used format and control codes employed with the fprintf function.

Format Code	Description
%d	Integer format
%e	Scientific format with lowercase e
%E	Scientific format with uppercase E
%f	Decimal format
%g	The more compact of %e or %f

Control Code	Description
\n	Start new line
\t	Tab

control codes provide a means to perform actions such as skipping to the next line. If we had omitted the code \n in the previous example, the command prompt would appear at the end of the label m/s rather than on the next line as would typically be desired.

The fprintf function can also be used to display several values per line with different formats. For example,

```
>> fprintf('%5d %10.3f %8.5e\n',100,2*pi,pi);

   100      6.283 3.14159e+000
```

It can also be used to display vectors and matrices. Here is an M-file that enters two sets of values as vectors. These vectors are then combined into a matrix, which is then displayed as a table with headings:

```
function fprintfdemo
x = [1 2 3 4 5];
y = [20.4 12.6 17.8 88.7 120.4];
z = [x;y];
fprintf('     x        y\n');
fprintf('%5d %10.3f\n',z);
```

The result of running this M-file is

```
>> fprintfdemo

   x        y
   1     20.400
   2     12.600
   3     17.800
   4     88.700
   5    120.400
```

3.2.1 Creating and Accessing Files

MATLAB has the capability to both read and write data files. The simplest approach involves a special type of binary file, called a *MAT-file,* which is expressly designed for implementation within MATLAB. Such files are created and accessed with the save and load commands.

The save command can be used to generate a MAT-file holding either the entire workspace or a few selected variables. A simple representation of its syntax is

```
save filename var1 var2 ... varn
```

This command creates a MAT-file named *filename*.mat that holds the variables *var1* through *varn*. If the variables are omitted, all the workspace variables are saved. The load command can subsequently be used to retrieve the file:

```
load filename var1 var2 ... varn
```

which retrieves the variables *var1* through *varn* from *filename*.mat. As was the case with save, if the variables are omitted, all the variables are retrieved.

For example, suppose that you use Eq. (1.9) to generate velocities for a set of drag coefficients:

```
>> g=9.81;m=80;t=5;
>> cd=[.25 .267 .245 .28 .273]';
>> v=sqrt(g*m ./cd).*tanh(sqrt(g*cd/m)*t);
```

You can then create a file holding the values of the drag coefficients and the velocities with

```
>> save veldrag v cd
```

To illustrate how the values can be retrieved at a later time, remove all variables from the workspace with the clear command,

```
>> clear
```

At this point, if you tried to display the velocities you would get the result:

```
>> v
??? Undefined function or variable 'v'.
```

However, you can recover them by entering

```
>> load veldrag
```

Now, the velocities are available as can be verified by typing

```
>> who

Your variables are:
cd  v
```

Although MAT-files are quite useful when working exclusively within the MATLAB environment, a somewhat different approach is required when interfacing MATLAB with other programs. In such cases, a simple approach is to create text files written in ASCII format.

ASCII files can be generated in MATLAB by appending –ascii to the save command. In contrast to MAT-files where you might want to save the entire workspace, you would typically save a single rectangular matrix of values. For example,

```
>> A=[5 7 9 2;3 6 3 9];
>> save simpmatrix.txt –ascii
```

In this case, the save command stores the values in A in 8-digit ASCII form. If you want to store the numbers in double precision, just append –ascii –double. In either case, the file can be accessed by other programs such as spreadsheets or word processors. For example, if you open this file with a text editor, you will see

```
5.0000000e+000      7.0000000e+000      9.0000000e+000      2.0000000e+000
3.0000000e+000      6.0000000e+000      3.0000000e+000      9.0000000e+000
```

Alternatively, you can read the values back into MATLAB with the load command,

```
>> load simpmatrix.txt
```

Because simpmatrix.txt is not a MAT-file, MATLAB creates a double precision array named after the *filename:*

```
>> simpmatrix

simpmatrix =
     5     7     9     2
     3     6     3     9
```

Alternatively, you could use the load command as a function and assign its values to a variable as in

```
>> A = load('simpmatrix.txt')
```

The foregoing material covers but a small portion of MATLAB's file management capabilities. For example, a handy import wizard can be invoked with the menu selections: **File, Import Data.** As an exercise, you can demonstrate the import wizards convenience by using it to open simpmatrix.txt. In addition, you can always consult help to learn more about this and other features.

3.3   STRUCTURED PROGRAMMING

The simplest of all M-files perform instructions sequentially. That is, the program statements are executed line by line starting at the top of the function and moving down to the end. Because a strict sequence is highly limiting, all computer languages include statements allowing programs to take nonsequential paths. These can be classified as

- *Decisions* (or Selection). The branching of flow based on a decision.
- *Loops* (or Repetition). The looping of flow to allow statements to be repeated.

3.3.1 Decisions

The if Structure. This structure allows you to execute a set of statements if a logical condition is true. Its general syntax is

```
if condition
  statements
end
```

where *condition* is a logical expression that is either true or false. For example, here is a simple M-file to evaluate whether a grade is passing:

```
function grader(grade)
% grader(grade):
```

```
%   determines whether grade is passing
% input:
%   grade = numerical value of grade (0–100)
% output:
%   displayed message
if grade >= 60
  disp('passing grade')
end
```

The following illustrates the result

```
>> grader(95.6)
```

```
passing grade
```

For cases where only one statement is executed, it is often convenient to implement the if structure as a single line,

```
if grade > 60, disp('passing grade'), end
```

This structure is called a *single-line if*. For cases where more than one statement is implemented, the multiline if structure is usually preferable because it is easier to read.

Error Function. A nice example of the utility of a single-line if is to employ it for rudimentary error trapping. This involves using the error function which has the syntax,

```
error(msg)
```

When this function is encountered, it displays the text message *msg*, indicates where the error occurred, and causes the M-file to terminate and return to the command window.

An example of its use would be where we might want to terminate an M-file to avoid a division by zero. The following M-file illustrates how this could be done:

```
function f = errortest(x)
if x == 0, error('zero value encountered'), end
f = 1/x;
```

If a nonzero argument is used, the division would be implemented successfully as in

```
>> errortest(10)
```

```
ans =
    0.1000
```

However, for a zero argument, the function would terminate prior to the division and the error message would be displayed in red typeface:

```
>> errortest(0)
```

```
??? Error using ==> errortest at 2
zero value encountered
```

Logical Conditions. The simplest form of the condition is a single relational expression that compares two values as in

$$value_1 \; relation \; value_2$$

where the *values* can be constants, variables, or expressions and the *relation* is one of the relational operators listed in Table 3.2.

TABLE 3.2   Summary of relational operators in MATLAB.

Example	Operator	Relationship
x == 0	==	Equal
unit ~= 'm'	~=	Not equal
a < 0	<	Less than
s > t	>	Greater than
3.9 <= a/3	<=	Less than or equal to
r >= 0	>=	Greater than or equal to

MATLAB also allows testing of more than one logical condition by employing logical operators. We will emphasize the following:

- ~(*Not*). Used to perform logical negation on an expression.

  ~*expression*

  If the *expression* is true, the result is false. Conversely, if the *expression* is false, the result is true.
- & (*And*). Used to perform a logical conjunction on two expressions.

  *expression*₁ & *expression*₂

  If both *expressions* evaluate to true, the result is true. If either or both *expressions* evaluates to false, the result is false.
- || (*Or*). Used to perform a logical disjunction on two expressions.

  *expression*₁ || *expression*₂

  If either or both *expressions* evaluate to true, the result is true.

Table 3.3 summarizes all possible outcomes for each of these operators. Just as for arithmetic operations, there is a priority order for evaluating logical operations. These are from highest to lowest: ~, &, and ||. In choosing between operators of equal priority, MATLAB evaluates them from left to right. Finally, as with arithmetic operators, parentheses can be used to override the priority order.

Let's investigate how the computer employs the priorities to evaluate a logical expression. If a = −1, b = 2, x = 1, and y = 'b', evaluate whether the following is true or false:

    a * b > 0 & b == 2 & x > 7 || ~(y > 'd')

TABLE 3.3   A truth table summarizing the possible outcomes for logical operators employed in MATLAB. The order of priority of the operators is shown at the top of the table.

x	y	Highest ――――――――――――――――――――――→ Lowest		
		~x	x & y	x || y
T	T	F	T	T
T	F	F	F	T
F	T	T	F	T
F	F	T	F	F

To make it easier to evaluate, substitute the values for the variables:

-1 * 2 > 0 & 2 == 2 & 1 > 7 || ~('b' > 'd')

The first thing that MATLAB does is to evaluate any mathematical expressions. In this example, there is only one: -1 * 2,

-2 > 0 & 2 == 2 & 1 > 7 || ~('b' > 'd')

Next, evaluate all the relational expressions

-2 > 0 & 2 == 2 & 1 > 7 || ~('b' > 'd')
  F   &   T   &   F || ~   F

At this point, the logical operators are evaluated in priority order. Since the ~ has highest priority, the last expression (~F) is evaluated first to give

F & T & F || T

The & operator is evaluated next. Since there are two, the left-to-right rule is applied and the first expression (F & T) is evaluated:

F & F || T

The & again has highest priority

F || T

Finally, the || is evaluated as true. The entire process is depicted in Fig. 3.1.

FIGURE 3.1
A step-by-step evaluation of a complex decision.

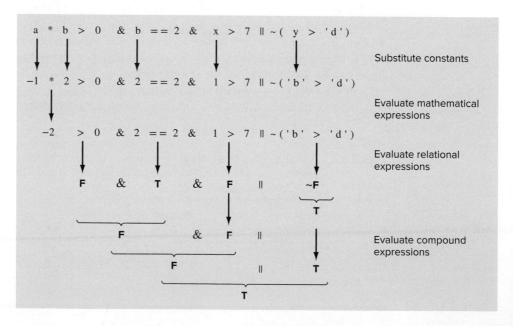

The `if...else` Structure. This structure allows you to execute a set of statements if a logical condition is true and to execute a second set if the condition is false. Its general syntax is

```
if condition
    statements₁
else
    statements₂
end
```

The `if...elseif` Structure. It often happens that the false option of an `if...else` structure is another decision. This type of structure often occurs when we have more than two options for a particular problem setting. For such cases, a special form of decision structure, the `if...elseif` has been developed. It has the general syntax

```
if condition₁
    statements₁
elseif condition₂
    statements₂
elseif condition₃
    statements₃
        .
        .
        .
else
    statementsₑₗₛₑ
end
```

EXAMPLE 3.5   if Structures

Problem Statement. For a scalar, the built-in MATLAB `sign` function returns the sign of its argument (−1, 0, 1). Here's a MATLAB session that illustrates how it works:

```
>> sign(25.6)
ans =
     1
>> sign(-0.776)
ans =
    -1
>> sign(0)
ans =
     0
```

Develop an M-file to perform the same function.

Solution. First, an if structure can be used to return 1 if the argument is positive:

```
function sgn = mysign(x)
% mysign(x) returns 1 if x is greater than zero.
if x > 0
  sgn = 1;
end
```

This function can be run as

```
>> mysign(25.6)
ans =
     1
```

Although the function handles positive numbers correctly, if it is run with a negative or zero argument, nothing is displayed. To partially remedy this shortcoming, an if...else structure can be used to display –1 if the condition is false:

```
function sgn = mysign(x)
% mysign(x) returns 1 if x is greater than zero.
%                   -1 if x is less than or equal to zero.
if x > 0
  sgn = 1;
else
  sgn = -1;
end
```

This function can be run as

```
>> mysign(-0.776)
ans =
    -1
```

Although the positive and negative cases are now handled properly, –1 is erroneously returned if a zero argument is used. An if...elseif structure can be used to incorporate this final case:

```
function sgn = mysign(x)
% mysign(x) returns 1 if x is greater than zero.
%                   -1 if x is less than zero.
%                    0 if x is equal to zero.
if x > 0
  sgn = 1;
elseif x < 0
  sgn = -1;
else
  sgn = 0;
end
```

The function now handles all possible cases. For example,

```
>> mysign(0)
ans =
     0
```

The switch Structure. The switch structure is similar in spirit to the if...elseif structure. However, rather than testing individual conditions, the branching is based on the value of a single test expression. Depending on its value, different blocks of code are implemented. In addition, an optional block is implemented if the expression takes on none of the prescribed values. It has the general syntax

```
switch testexpression
   case value₁
```

```
    statements₁
  case value₂
    statements₂
        .
        .
        .
  otherwise
    statements_otherwise
end
```

As an example, here is function that displays a message depending on the value of the string variable, grade.

```
grade = 'B';
switch grade
  case 'A'
    disp('Excellent')
  case 'B'
    disp('Good')
  case 'C'
    disp('Mediocre')
  case 'D'
    disp('Whoops')
  case 'F'
    disp('Would like fries with your order?')
  otherwise
    disp('Huh!')
end
```

When this code was executed, the message "Good" would be displayed.

Variable Argument List. MATLAB allows a variable number of arguments to be passed to a function. This feature can come in handy for incorporating default values into your functions. A *default value* is a number that is automatically assigned in the event that the user does not pass it to a function.

As an example, recall that earlier in this chapter, we developed a function freefall, which had three arguments:

```
v = freefall(t,m,cd)
```

Although a user would obviously need to specify the time and mass, they might not have a good idea of an appropriate drag coefficient. Therefore, it would be nice to have the program supply a value if they omitted it from the argument list.

MATLAB has a function called nargin that provides the number of input arguments supplied to a function by a user. It can be used in conjunction with decision structures like the if or switch constructs to incorporate default values as well as error messages into your functions. The following code illustrates how this can be done for freefall:

```
function v = freefall2(t, m, cd)
% freefall2: bungee velocity with second-order drag
%   v=freefall2(t,m,cd) computes the free-fall velocity
%                       of an object with second-order drag.
%
% input:
%   t = time (s)
%   m = mass (kg)
```

```
%    cd = drag coefficient (default = 0.27 kg/m)
% output:
%    v = downward velocity (m/s)
switch nargin
  case 0
    error('Must enter time and mass')
  case 1
    error('Must enter mass')
  case 2
    cd = 0.27;
end
g = 9.81;     % acceleration of gravity
v = sqrt(g * m / cd)*tanh(sqrt(g * cd / m) * t);
```

Notice how we have used a `switch` structure to either display error messages or set the default, depending on the number of arguments passed by the user. Here is a command window session showing the results:

```
>> freefall2(12,68.1,0.25)

ans =
    50.6175

>> freefall2(12,68.1)

ans =
    48.8747

>> freefall2(12)

??? Error using ==> freefall2 at 15
Must enter mass

>> freefall2()

??? Error using ==> freefall2 at 13
Must enter time and mass
```

Note that `nargin` behaves a little differently when it is invoked in the command window. In the command window, it must include a string argument specifying the function and it returns the number of arguments in the function. For example,

```
>> nargin('freefall2')

ans =
    3
```

3.3.2 Loops

As the name implies, loops perform operations repetitively. There are two types of loops, depending on how the repetitions are terminated. A *for loop* ends after a specified number of repetitions. A *while loop* ends on the basis of a logical condition.

The for...end Structure. A `for` loop repeats statements a specific number of times. Its general syntax is

```
for index = start:step:finish
  statements
end
```

The for loop operates as follows. The *index* is a variable that is set at an initial value, *start*. The program then compares the *index* with a desired final value, *finish*. If the index is less than or equal to the *finish*, the program executes the *statements*. When it reaches the end line that marks the end of the loop, the *index* variable is increased by the *step* and the program loops back up to the for statement. The process continues until the *index* becomes greater than the *finish* value. At this point, the loop terminates as the program skips down to the line immediately following the end statement.

Note that if an increment of 1 is desired (as is often the case), the *step* can be dropped. For example,

```
for i = 1:5
  disp(i)
end
```

When this executes, MATLAB would display in succession, 1, 2, 3, 4, 5. In other words, the default *step* is 1.

The size of the *step* can be changed from the default of 1 to any other numeric value. It does not have to be an integer, nor does it have to be positive. For example, step sizes of 0.2, –1, or –5, are all acceptable.

If a negative *step* is used, the loop will "countdown" in reverse. For such cases, the loop's logic is reversed. Thus, the *finish* is less than the *start* and the loop terminates when the index is less than the *finish*. For example,

```
for j = 10:-1:1
  disp(j)
end
```

When this executes, MATLAB would display the classic "countdown" sequence: 10, 9, 8, 7, 6, 5, 4, 3, 2, 1.

EXAMPLE 3.6   Using a for Loop to Compute the Factorial

Problem Statement.   Develop an M-file to compute the factorial.[3]

$$0! = 1$$
$$1! = 1$$
$$2! = 1 \times 2 = 2$$
$$3! = 1 \times 2 \times 3 = 6$$
$$4! = 1 \times 2 \times 3 \times 4 = 24$$
$$5! = 1 \times 2 \times 3 \times 4 \times 5 = 120$$
$$\vdots$$

Solution.   A simple function to implement this calculation can be developed as

```
function fout = factor(n)
% factor(n):
%   Computes the product of all the integers from 1 to n.
x = 1;
for i = 1:n
  x = x * i;
end
fout = x;
end
```

[3] Note that MATLAB has a built-in function factorial that performs this computation.

which can be run as

```
>> factor(5)

ans =
    120
```

This loop will execute 5 times (from 1 to 5). At the end of the process, x will hold a value of 5! (meaning 5 factorial or $1 \times 2 \times 3 \times 4 \times 5 = 120$).

Notice what happens if $n = 0$. For this case, the for loop would not execute, and we would get the desired result, $0! = 1$.

Vectorization. The for loop is easy to implement and understand. However, for MATLAB, it is not necessarily the most efficient means to repeat statements a specific number of times. Because of MATLAB's ability to operate directly on arrays, *vectorization* provides a much more efficient option. For example, the following for loop structure:

```
i = 0;
for t = 0:0.02:50
  i = i + 1;
  y(i) = cos(t);
end
```

can be represented in vectorized form as

```
t = 0:0.02:50;
y = cos(t);
```

It should be noted that for more complex code, it may not be obvious how to vectorize the code. That said, wherever possible, vectorization is recommended.

Preallocation of Memory. MATLAB automatically increases the size of arrays every time you add a new element. This can become time consuming when you perform actions such as adding new values one at a time within a loop. For example, here is some code that sets value of elements of y depending on whether or not values of t are greater than one:

```
t = 0:.01:5;
for i = 1:length(t)
  if t(i)>1
    y(i) = 1/t(i);
  else
    y(i) = 1;
  end
end
```

For this case, MATLAB must resize y every time a new value is determined. The following code preallocates the proper amount of memory by using a vectorized statement to assign ones to y prior to entering the loop.

```
t = 0:.01:5;
y = ones(size(t));
for i = 1:length(t)
  if t(i)>1
    y(i) = 1/t(i);
  end
end
```

Thus, the array is only sized once. In addition, preallocation helps reduce memory fragmentation, which also enhances efficiency.

The while Structure. A while loop repeats as long as a logical condition is true. Its general syntax is

```
while condition
   statements
end
```

The *statements* between the while and the end are repeated as long as the *condition* is true. A simple example is

```
x = 8
while x > 0
   x = x - 3;
   disp(x)
end
```

When this code is run, the result is

```
x =
       8
       5
       2
      -1
```

The while...break Structure. Although the while structure is extremely useful, the fact that it always exits at the beginning of the structure on a false result is somewhat constraining. For this reason, languages such as Fortran 90 and Visual Basic have special structures that allow loop termination on a true condition anywhere in the loop. Although such structures are currently not available in MATLAB, their functionality can be mimicked by a special version of the while loop. The syntax of this version, called a *while...break structure,* can be written as

```
while (1)
   statements
   if condition, break, end
   statements
end
```

where break terminates execution of the loop. Thus, a single line if is used to exit the loop if the condition tests true. Note that as shown, the break can be placed in the middle of the loop (i.e., with statements before and after it). Such a structure is called a *midtest loop.*

If the problem required it, we could place the break at the very beginning to create a *pretest loop.* An example is

```
while (1)
   If x < 0, break, end
   x = x - 5;
end
```

Notice how 5 is subtracted from x on each iteration. This represents a mechanism so that the loop eventually terminates. Every decision loop must have such a mechanism. Otherwise it would become a so-called *infinite loop* that would never stop.

Alternatively, we could also place the if...break statement at the very end and create a *posttest loop*,

```
while (1)
  x = x - 5;
  if x < 0, break, end
end
```

It should be clear that, in fact, all three structures are really the same. That is, depending on where we put the exit (beginning, middle, or end) dictates whether we have a pre-, mid- or posttest. It is this simplicity that led the computer scientists who developed Fortran 90 and Visual Basic to favor this structure over other forms of the decision loop such as the conventional while structure.

The pause Command.   There are often times when you might want a program to temporarily halt. The command pause causes a procedure to stop and wait until any key is hit. A nice example involves creating a sequence of plots that a user might want to leisurely peruse before moving on to the next. The following code employs a for loop to create a sequence of interesting plots that can be viewed in this manner:

```
for n = 3:10
  mesh (magic(n))
  pause
end
```

The pause can also be formulated as pause(n), in which case the procedure will halt for n seconds. This feature can be demonstrated by implementing it in conjunction with several other useful MATLAB functions. The beep command causes the computer to emit a beep sound. Two other functions, tic and toc, work together to measure elapsed time. The tic command saves the current time that toc later employs to display the elapsed time. The following code then confirms that pause(n) works as advertised complete with sound effects:

```
tic
beep
pause(5)
beep
toc
```

When this code is run, the computer will beep. Five seconds later it will beep again and display the following message:

```
Elapsed time is 5.006306 seconds.
```

By the way, if you ever have the urge to use the command pause(inf), MATLAB will go into an infinite loop. In such cases, you can return to the command prompt by typing **Ctrl+c** or **Ctrl+Break.**

Although the foregoing examples might seem a tad frivolous, the commands can be quite useful. For instance, tic and toc can be employed to identify the parts of an algorithm that consume the most execution time. Further, the **Ctrl+c** or **Ctrl+Break** key combinations come in real handy in the event that you inadvertently create an infinite loop in one of your M-files.

3.3.3 Animation

There are two simple ways to animate a plot in MATLAB. First, if the computations are sufficiently quick, the standard plot function can be employed in a way that the animation can appear smooth. Here is a code fragment that indicates how a for loop and standard plotting functions can be employed to animate a plot,

```
% create animation with standard plot functions
for j=1:n
  plot commands
end
```

Thus, because we do not include hold on, the plot will refresh on each loop iteration. Through judicious use of axis commands, this can result in a smoothly changing image.

Second, there are special functions, getframe and movie, that allow you to capture a sequence of plots and then play them back. As the name implies, the getframe function captures a snapshot (*pixmap*) of the current axes or figure. It is usually used in a for loop to assemble an array of movie frames for later playback with the movie function, which has the following syntax:

```
movie(m,n,fps)
```

where m = the vector or matrix holding the sequence of frames constituting the movie, n = an optional variable specifying how many times the movie is to be repeated (if it is omitted, the movie plays once), and fps = an optional variable that specifies the movie's *frame rate* (if it is omitted, the default is 12 frames per second). Here is a code fragment that indicates how a for loop along with the two functions can be employed to create a movie,

```
% create animation with getframe and movie
for j=1:n
  plot commands
  M(j) = getframe;
end
movie(M)
```

Each time the loop executes, the *plot commands* create an updated version of a plot, which is then stored in the vector M. After the loop terminates, the n images are then played back by movie.

EXAMPLE 3.7   Animation of Projectile Motion

Problem Statement.   In the absence of air resistance, the Cartesian coordinates of a projectile launched with an initial velocity (v_0) and angle (θ_0) can be computed with

$$x = v_0 \cos(\theta_0)t$$
$$y = v_0 \sin(\theta_0)t - 0.5gt^2$$

where g = 9.81 m/s^2. Develop a script to generate an animated plot of the projectile's trajectory given that v_0 = 5 m/s and θ_0 = 45°.

Solution. A script to generate the animation can be written as

```
clc,clf,clear
g=9.81; theta0=45*pi/180; v0=5;
t(1)=0;x=0;y=0;
plot(x,y,'o','MarkerFaceColor','b','MarkerSize',8)
axis([0 3 0 0.8])
M(1)=getframe;
dt=1/128;
for j = 2:1000
  t(j)=t(j-1)+dt;
  x=v0*cos(theta0)*t(j);
  y=v0*sin(theta0)*t(j)-0.5*g*t(j)^2;
  plot(x,y,'o','MarkerFaceColor','b','MarkerSize',8)
  axis([0 3 0 0.8])
  M(j)=getframe;
  if y<=0, break, end
end
pause
movie(M,1)
```

Several features of this script bear mention. First, notice that we have fixed the ranges for the *x* and *y* axes. If this is not done, the axes will rescale and cause the animation to jump around. Second, we terminate the for loop when the projectile's height *y* falls below zero.

When the script is executed, two animations will be displayed (we've placed a pause between them). The first corresponds to the sequential generation of the frames within the loop, and the second corresponds to the actual movie. Although we cannot show the results here, the trajectory for both cases will look like Fig. 3.2. You should enter and run the foregoing script in MATLAB to see the actual animation.

FIGURE 3.2
Plot of a projectile's trajectory.

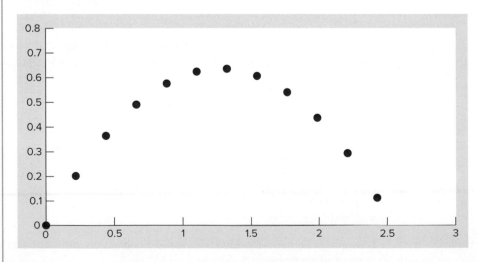

3.4 NESTING AND INDENTATION

We need to understand that structures can be "nested" within each other. *Nesting* refers to placing structures within other structures. The following example illustrates the concept.

EXAMPLE 3.8   Nesting Structures

Problem Statement. The roots of a quadratic equation

$$f(x) = ax^2 + bx + c$$

can be determined with the quadratic formula

$$x = \frac{-b \pm \sqrt{b^2 - 4ac}}{2a}$$

Develop a function to implement this formula given values of the coefficients.

Solution. *Top-down design* provides a nice approach for designing an algorithm to compute the roots. This involves developing the general structure without details and then refining the algorithm. To start, we first recognize that depending on whether the parameter a is zero, we will either have "special" cases (e.g., single roots or trivial values) or conventional cases using the quadratic formula. This "big-picture" version can be programmed as

```
function quadroots(a, b, c)
% quadroots: roots of quadratic equation
%    quadroots(a,b,c): real and complex roots
%                      of quadratic equation
% input:
%    a = second-order coefficient
%    b = first-order coefficient
%    c = zero-order coefficient
% output:
%    r1 = real part of first root
%    i1 = imaginary part of first root
%    r2 = real part of second root
%    i2 = imaginary part of second root
if a == 0
  %special cases
else
  %quadratic formula
end
```

Next, we develop refined code to handle the "special" cases:

```
%special cases
if b ~= 0
  %single root
  r1 = -c / b
else
  %trivial solution
  disp('Trivial solution. Try again')
end
```

And we can develop refined code to handle the quadratic formula cases:

```
%quadratic formula
d = b ^ 2 - 4 * a * c;
```

```
if d >= 0
  %real roots
  r1 = (-b + sqrt(d)) / (2 * a)
  r2 = (-b - sqrt(d)) / (2 * a)
else
  %complex roots
  r1 = -b / (2 * a)
  i1 = sqrt(abs(d)) / (2 * a)
  r2 = r1
  i2 = -i1
end
```

We can then merely substitute these blocks back into the simple "big-picture" framework to give the final result:

```
function quadroots(a, b, c)
% quadroots: roots of quadratic equation
%   quadroots(a,b,c): real and complex roots
%                     of quadratic equation
% input:
%   a = second-order coefficient
%   b = first-order coefficient
%   c = zero-order coefficient
% output:
%   r1 = real part of first root
%   i1 = imaginary part of first root
%   r2 = real part of second root
%   i2 = imaginary part of second root
if a == 0

  %special cases
  if b ~= 0
    %single root
    r1 = -c / b
  else
    %trivial solution
    disp('Trivial solution. Try again')
  end

else

  %quadratic formula
  d = b ^ 2 - 4 * a * c;    %discriminant
  if d >= 0
    %real roots
    r1 = (-b + sqrt(d)) / (2 * a)
    r2 = (-b - sqrt(d)) / (2 * a)
  else
    %complex roots
    r1 = -b / (2 * a)
    i1 = sqrt(abs(d)) / (2 * a)
    r2 = r1
    i2 = -i1
  end

end
```

As highlighted by the shading, notice how indentation helps to make the underlying logical structure clear. Also notice how "modular" the structures are. Here is a command window session illustrating how the function performs:

```
>> quadroots(1,1,1)

r1 =
  -0.5000
i1 =
   0.8660
r2 =
  -0.5000
i2 =
  -0.8660

>> quadroots(1,5,1)

r1 =
  -0.2087
r2 =
  -4.7913

>> quadroots(0,5,1)

r1 =
  -0.2000

>> quadroots(0,0,0)

Trivial solution. Try again
```

3.5   PASSING FUNCTIONS TO M-FILES

Much of the remainder of the book involves developing functions to numerically evaluate other functions. Although a customized function could be developed for every new equation we analyzed, a better alternative is to design a generic function and pass the particular equation we wish to analyze as an argument. In the parlance of MATLAB, these functions are given a special name: *function functions*. Before describing how they work, we will first introduce anonymous functions, which provide a handy means to define simple user-defined functions without developing a full-blown M-file.

3.5.1 Anonymous Functions

Anonymous functions allow you to create a simple function without creating an M-file. They can be defined within the command window with the following syntax:

```
fhandle = @(arglist) expression
```

where *fhandle* = the function handle you can use to invoke the function, *arglist* = a comma separated list of input arguments to be passed to the function, and *expression* = any single valid MATLAB expression. For example,

```
>> f1=@(x,y) x^2 + y^2;
```

Once these functions are defined in the command window, they can be used just as other functions:

```
>> f1(3,4)

ans =
    25
```

Aside from the variables in its argument list, an anonymous function can include variables that exist in the workspace where it is created. For example, we could create an anonymous function $f(x) = 4x^2$ as

```
>> a = 4;
>> b = 2;
>> f2=@(x) a*x^b;
>> f2(3)

ans = 36
```

Note that if subsequently we enter new values for a and b, the anonymous function does not change:

```
>> a = 3;
>> f2(3)

ans = 36
```

Thus, the function handle holds a snapshot of the function at the time it was created. If we want the variables to take on values, we must recreate the function. For example, having changed a to 3,

```
>> f2=@(x) a*x^b;
```

with the result

```
>> f2(3)

ans =
    27
```

It should be noted that prior to MATLAB 7, inline functions performed the same role as anonymous functions. For example, the anonymous function developed above, f1, could be written as

```
>> f1=inline('x^2 + y^2','x','y');
```

Although they are being phased out in favor of anonymous function, some readers might be using earlier versions, and so we thought it would be helpful to mention them. MATLAB help can be consulted to learn more about their use and limitations.

3.5.2 Function Functions

Function functions are functions that operate on other functions which are passed to it as input arguments. The function that is passed to the function function is referred to as the *passed function*. A simple example is the built-in function fplot, which plots the graphs of functions. A simple representation of its syntax is

```
fplot(func,lims)
```

where *func* is the function being plotted between the *x*-axis limits specified by *lims* = [*xmin xmax*]. For this case, *func* is the passed function. This function is "smart" in that it automatically analyzes the function and decides how many values to use so that the plot will exhibit all the function's features.

Here is an example of how fplot can be used to plot the velocity of the free-falling bungee jumper. The function can be created with an anonymous function:

```
>> vel=@(t) ...
sqrt(9.81*68.1/0.25)*tanh(sqrt(9.81*0.25/68.1)*t);
```

We can then generate a plot from *t* = 0 to 12 as

```
>> fplot(vel,[0 12])
```

The result is displayed in Fig. 3.3.

Note that in the remainder of this book, we will have many occasions to use MATLAB's built-in function functions. As in the following example, we will also be developing our own.

FIGURE 3.3
A plot of velocity versus time generated with the **fplot** function.

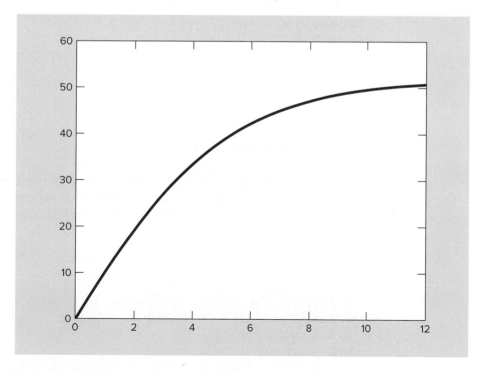

EXAMPLE 3.9   Building and Implementing a Function Function

Problem Statement. Develop an M-file function function to determine the average value of a function over a range. Illustrate its use for the bungee jumper velocity over the range from $t = 0$ to 12 s:

$$v(t) = \sqrt{\frac{gm}{c_d}} \tanh\left(\sqrt{\frac{gc_d}{m}}t\right)$$

where $g = 9.81$, $m = 68.1$, and $c_d = 0.25$.

Solution. The average value of the function can be computed with standard MATLAB commands as

```
>> t=linspace(0,12);
>> v=sqrt(9.81*68.1/0.25)*tanh(sqrt(9.81*0.25/68.1)*t);
>> mean(v)

ans =
   36.0870
```

Inspection of a plot of the function (Fig. 3.3) shows that this result is a reasonable estimate of the curve's average height.

We can write an M-file to perform the same computation:

```
function favg = funcavg(a,b,n)
% funcavg: average function height
%    favg=funcavg(a,b,n): computes average value
%                         of function over a range
% input:
%    a = lower bound of range
%    b = upper bound of range
%    n = number of intervals
% output:
%    favg = average value of function
x = linspace(a,b,n);
y = func(x);
favg = mean(y);
end

function f = func(t)
f=sqrt(9.81*68.1/0.25)*tanh(sqrt(9.81*0.25/68.1)*t);
end
```

The main function first uses linspace to generate equally spaced x values across the range. These values are then passed to a subfunction func in order to generate the corresponding y values. Finally, the average value is computed. The function can be run from the command window as

```
>> funcavg (0,12,60)

ans =
   36.0127
```

Now let's rewrite the M-file so that rather than being specific to func, it evaluates a nonspecific function name f that is passed in as an argument:

```
function favg = funcavg (f,a,b,n)
% funcavg: average function height
```

```
%    favg=funcavg(f,a,b,n): computes average value
%                                  of function over a range
% input:
%    f = function to be evaluated
%    a = lower bound of range
%    b = upper bound of range
%    n = number of intervals
% output:
%    favg = average value of function
x = linspace(a,b,n);
y = f(x);
favg = mean(y);
```

Because we have removed the subfunction func, this version is truly generic. It can be run from the command window as

```
>> vel=@(t) ...
sqrt(9.81*68.1/0.25)*tanh(sqrt(9.81*0.25/68.1)*t);
>> funcavg(vel,0,12,60)

ans =
    36.0127
```

To demonstrate its generic nature, funcavg can easily be applied to another case by merely passing it a different function. For example, it could be used to determine the average value of the built-in sin function between 0 and 2π as

```
>> funcavg(@sin,0,2*pi,180)

ans =
 -6.3001e-017
```

Does this result make sense?

We can see that funcavg is now designed to evaluate any valid MATLAB expression. We will do this on numerous occasions throughout the remainder of this text in a number of contexts ranging from nonlinear equation solving to the solution of differential equations.

3.5.3 Passing Parameters

Recall from Chap. 1 that the terms in mathematical models can be divided into dependent and independent variables, parameters, and forcing functions. For the bungee jumper model, the velocity (v) is the dependent variable, time (t) is the independent variable, the mass (m) and drag coefficient (c_d) are parameters, and the gravitational constant (g) is the forcing function. It is commonplace to investigate the behavior of such models by performing a *sensitivity analysis*. This involves observing how the dependent variable changes as the parameters and forcing functions are varied.

In Example 3.9, we developed a function function, funcavg, and used it to determine the average value of the bungee jumper velocity for the case where the parameters were set at $m = 68.1$ and $c_d = 0.25$. Suppose that we wanted to analyze the same function, but with different parameters. Of course, we could retype the function with new values for each case, but it would be preferable to just change the parameters.

As we learned in Sec. 3.5.1, it is possible to incorporate parameters into anonymous functions. For example, rather than "wiring" the numeric values, we could have done the following:

```
>> m=68.1;cd=0.25;
>> vel=@(t) sqrt(9.81*m/cd)*tanh(sqrt(9.81*cd/m)*t);
>> funcavg(vel,0,12,60)

ans =
   36.0127
```

However, if we want the parameters to take on new values, we must recreate the anonymous function.

MATLAB offers a better alternative by adding the term varargin as the function function's last input argument. In addition, every time the passed function is invoked within the function function, the term varargin{:} should be added to the end of its argument list (note the curly brackets). Here is how both modifications can be implemented for funcavg (omitting comments for conciseness):

```
function favg = funcavg(f,a,b,n,varargin)
x = linspace(a,b,n);
y = f(x,varargin{:});
favg = mean(y);
```

When the passed function is defined, the actual parameters should be added at the end of the argument list. If we used an anonymous function, this can be done as in

```
>> vel=@(t,m,cd) sqrt(9.81*m/cd)*tanh(sqrt(9.81*cd/m)*t);
```

When all these changes have been made, analyzing different parameters becomes easy. To implement the case where $m = 68.1$ and $c_d = 0.25$, we could enter

```
>> funcavg(vel,0,12,60,68.1,0.25)

ans =
   36.0127
```

An alternative case, say $m = 100$ and $c_d = 0.28$, could be rapidly generated by merely changing the arguments:

```
>> funcavg(vel,0,12,60,100,0.28)

ans =
   38.9345
```

A New Approach for Passing Parameters. At the time of this edition's development, MATLAB is going through a transition to a new and better way of passing parameters to function functions. As in the previous example, if the function being passed is

```
>> vel=@(t,m,cd) sqrt(9.81*m/cd)*tanh(sqrt(9.81*cd/m)*t);
```

then you invoke the function as

```
>> funcavg(vel,0,12,60,68.1,0.25)
```

The Mathworks developers thought this approach was cumbersome, so they devised the following alternative

```
>> funcavg(@(t) vel(t,68.1,0.25),0,12,60)
```

Thus, the extra parameters are not strung out at the end making it clear that the parameter list is in the function.

I've described both the "old" and the new way of passing parameters because MATLAB will maintain the old way in functions that support in order to minimize backwards incompatibilities. So if you have old code that has worked in the past, there is no need to go back and convert old code to the new way. For new code, however, I strongly recommend using the new way, because it is easier to read and more versatile.

3.6 CASE STUDY   BUNGEE JUMPER VELOCITY

Background. In this section, we will use MATLAB to solve the free-falling bungee jumper problem we posed at the beginning of this chapter. This involves obtaining a solution of

$$\frac{dv}{dt} = g - \frac{c_d}{m} v|v|$$

Recall that, given an initial condition for time and velocity, the problem involved iteratively solving the formula,

$$v_{i+1} = v_i + \frac{dv_i}{dt} \Delta t$$

Now also remember that to attain good accuracy, we would employ small steps. Therefore, we would probably want to apply the formula repeatedly to step out from our initial time to attain the value at the final time. Consequently, an algorithm to solve the problem would be based on a loop.

Solution. Suppose that we started the computation at $t = 0$ and wanted to predict velocity at $t = 12$ s using a time step of $\Delta t = 0.5$ s. We would therefore need to apply the iterative equation 24 times—that is,

$$n = \frac{12}{0.5} = 24$$

where n = the number of iterations of the loop. Because this result is exact (i.e., the ratio is an integer), we can use a for loop as the basis for the algorithm. Here's an M-file to do this including a subfunction defining the differential equation:

```
function vend = velocity1(dt, ti, tf, vi)
% velocity1: Euler solution for bungee velocity
%    vend = velocity1(dt, ti, tf, vi)
%           Euler method solution of bungee
%           jumper velocity
% input:
%    dt = time step (s)
%    ti = initial time (s)
%    tf = final time (s)
%    vi = initial value of dependent variable (m/s)
```

```
% output:
%    vend = velocity at tf (m/s)
t = ti;
v = vi;
n = (tf - ti) / dt;
for i = 1:n
   dvdt = deriv(v);
   v = v + dvdt * dt;
   t = t + dt;
end
vend = v;
end

function dv = deriv(v)
dv = 9.81 - (0.25 / 68.1) * v*abs(v);
end
```

This function can be invoked from the command window with the result:

```
>> velocity1(0.5,0,12,0)

ans =
   50.9259
```

Note that the true value obtained from the analytical solution is 50.6175 (Example 3.1). We can then try a much smaller value of dt to obtain a more accurate numerical result:

```
>> velocity1(0.001,0,12,0)

ans =
   50.6181
```

Although this function is certainly simple to program, it is not foolproof. In particular, it will not work if the computation interval is not evenly divisible by the time step. To cover such cases, a while . . . break loop can be substituted in place of the shaded area (note that we have omitted the comments for conciseness):

```
function vend = velocity2(dt, ti, tf, vi)
t = ti;
v = vi;
h = dt;
while(1)
   if t + dt > tf, h = tf - t; end
   dvdt = deriv(v);
   v = v + dvdt * h;
   t = t + h;
   if t >= tf, break, end
end
vend = v;
end

function dv = deriv(v)
dv = 9.81 - (0.25 / 68.1) * v*abs(v);
end
```

3.6 CASE STUDY   continued

As soon as we enter the while loop, we use a single line if structure to test whether adding t + dt will take us beyond the end of the interval. If not (which would usually be the case at first), we do nothing. If so, we would shorten up the interval—that is, we set the variable step h to the interval remaining: tf - t. By doing this, we guarantee that the last step falls exactly on tf. After we implement this final step, the loop will terminate because the condition t >= tf will test true.

Notice that before entering the loop, we assign the value of the time step dt to another variable h. We create this *dummy variable* so that our routine does not change the given value of dt if and when we shorten the time step. We do this in anticipation that we might need to use the original value of dt somewhere else in the event that this code were integrated within a larger program.

If we run this new version, the result will be the same as for the version based on the for loop structure:

```
>> velocity2(0.5,0,12,0)

ans =
    50.9259
```

Further, we can use a dt that is not evenly divisible into tf - ti:

```
>> velocity2(0.35,0,12,0)

ans =
    50.8348
```

We should note that the algorithm is still not foolproof. For example, the user could have mistakenly entered a step size greater than the calculation interval (e.g., tf - ti = 5 and dt = 20). Thus, you might want to include error traps in your code to catch such errors and then allow the user to correct the mistake.

As a final note, we should recognize that the foregoing code is not generic. That is, we have designed it to solve the specific problem of the velocity of the bungee jumper. A more generic version can be developed as

```
function yend = odesimp(dydt, dt, ti, tf, yi)
t = ti; y = yi; h = dt;
while (1)
   if t + dt > tf, h = tf - t; end
   y = y + dydt(y) * h;
   t = t + h;
   if t >= tf, break, end
end
yend = y;
```

Notice how we have stripped out the parts of the algorithm that were specific to the bungee example (including the subfunction defining the differential equation) while keeping the essential features of the solution technique. We can then use this routine to solve the bungee jumper example, by specifying the differential equation with an anonymous function and passing its function handle to odesimp to generate the solution

```
>> dvdt=@(v) 9.81-(0.25/68.1)*v*abs(v);
>> odesimp(dvdt,0.5,0,12,0)

ans =
    50.9259
```

We could then analyze a different function without having to go in and modify the M-file. For example, if $y = 10$ at $t = 0$, the differential equation $dy/dt = -0.1y$ has the analytical solution $y = 10e^{-0.1t}$. Therefore, the solution at $t = 5$ would be $y(5) = 10e^{-0.1(5)} = 6.0653$. We can use odesimp to obtain the same result numerically as in

```
>> odesimp(@(y) -0.1*y,0.005,0,5,10)

ans =
     6.0645
```

Finally, we can use varargin and the new way of passing parameters to develop a final and superior version. To do this, the odesimp function is first modified by adding the high-lighted code

```
function yend = odesimp2(dydt, dt, ti, tf, yi, varargin)
t = ti; y = yi; h = dt;
while (1)
  if t + dt > tf, h = tf - t; end
  y = y + dydt(y, varargin{:}) * h;
  t = t + h;
  if t >= tf, break, end
end
yend = y;
```

Then, we can develop a script to perform the computation,

```
clc
format compact
dvdt=@(v,cd,m) 9.81-(cd/m)*v*abs(v);
odesimp2(@(v) dvdt(v,0.25,68.1),0.5,0,12,0)
```

which yields the correct result

```
ans =
    50.9259
```

PROBLEMS

3.1 Figure P3.1 shows a cylindrical tank with a conical base. If the liquid level is quite low, in the conical part, the volume is simply the conical volume of liquid. If the liquid level is midrange in the cylindrical part, the total volume of liquid includes the filled conical part and the partially filled cylindrical part.

Use decisional structures to write an M-file to compute the tank's volume as a function of given values of R and d. Design the function so that it returns the volume for all cases.

FIGURE P3.1

where the depth is less than $3R$. Return an error message ("Overtop") if you overtop the tank—that is, $d > 3R$. Test it with the following data:

| R | 0.9 | 1.5 | 1.3 | 1.3 |
| d | 1 | 1.25 | 3.8 | 4.0 |

Note that the tank's radius is R.

3.2 An amount of money P is invested in an account where interest is compounded at the end of the period. The future worth F yielded at an interest rate i after n periods may be determined from the following formula:

$$F = P(1 + i)^n$$

Write an M-file that will calculate the future worth of an investment for each year from 1 through n. The input to the function should include the initial investment P, the interest rate i (as a decimal), and the number of years n for which the future worth is to be calculated. The output should consist of a table with headings and columns for n and F. Run the program for $P = \$100,000$, $i = 0.05$, and $n = 10$ years.

3.3 Economic formulas are available to compute annual payments for loans. Suppose that you borrow an amount of money P and agree to repay it in n annual payments at an interest rate of i. The formula to compute the annual payment A is

$$A = P \frac{i(1 + i)^n}{(1 + i)^n - 1}$$

Write an M-file to compute A. Test it with $P = \$100,000$ and an interest rate of 3.3% ($i = 0.033$). Compute results for $n = 1, 2, 3, 4$, and 5 and display the results as a table with headings and columns for n and A.

3.4 The average daily temperature for an area can be approximated by the following function:

$$T = T_{mean} + (T_{peak} - T_{mean}) \cos(\omega(t - t_{peak}))$$

where T_{mean} = the average annual temperature, T_{peak} = the peak temperature, ω = the frequency of the annual variation ($= 2\pi/365$), and t_{peak} = day of the peak temperature ($\cong 205$ d). Parameters for some U.S. towns are listed here:

City	T_{mean} (°C)	T_{peak} (°C)
Miami, FL	22.1	28.3
Yuma, AZ	23.1	33.6
Bismarck, ND	5.2	22.1
Seattle, WA	10.6	17.6
Boston, MA	10.7	22.9

Develop an M-file that computes the average temperature between two days of the year for a particular city. Test it for **(a)** January–February in Yuma, AZ ($t = 0$ to 59) and **(b)** July–August temperature in Seattle, WA ($t = 180$ to 242).

3.5 The sine function can be evaluated by the following infinite series:

$$\sin x = x - \frac{x^3}{3!} + \frac{x^5}{5!} - \cdots$$

Create an M-file to implement this formula so that it computes and displays the values of $\sin x$ as each term in the series is added. In other words, compute and display in sequence the values for

$$\sin x = x$$

$$\sin x = x - \frac{x^3}{3!}$$

$$\sin x = x - \frac{x^3}{3!} + \frac{x^5}{5!}$$

$$\vdots$$

up to the order term of your choosing. For each of the preceding, compute and display the percent relative error as

$$\%\text{error} = \frac{\text{true} - \text{series approximation}}{\text{true}} \times 100\%$$

As a test case, employ the program to compute $\sin(0.9)$ for up to and including eight terms—that is, up to the term $x^{15}/15!$.

3.6 Two distances are required to specify the location of a point relative to an origin in two-dimensional space (Fig. P3.6):

- The horizontal and vertical distances (x, y) in Cartesian coordinates.
- The radius and angle (r, θ) in polar coordinates.

It is relatively straightforward to compute Cartesian coordinates (x, y) on the basis of polar coordinates (r, θ). The reverse process is not so simple. The radius can be computed by the following formula:

$$r = \sqrt{x^2 + y^2}$$

If the coordinates lie within the first and fourth coordinates (i.e., $x > 0$), then a simple formula can be used to compute θ:

$$\theta = \tan^{-1}\left(\frac{y}{x}\right)$$

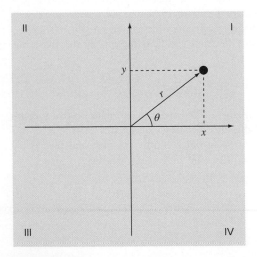

FIGURE P3.6

The difficulty arises for the other cases. The following table summarizes the possibilities:

x	y	θ
<0	>0	$\tan^{-1}(y/x) + \pi$
<0	<0	$\tan^{-1}(y/x) - \pi$
<0	=0	π
=0	>0	$\pi/2$
=0	<0	$-\pi 2$
=0	=0	0

Write a well-structured M-file using `if...elseif` structures to calculate r and θ as a function of x and y. Express the final results for θ in degrees. Test your program by evaluating the following cases:

x	y	r	θ
2	0		
2	1		
0	3		
−3	1		
−2	0		
−1	−2		
0	0		
0	−2		
2	2		

3.7 Develop an M-file to determine polar coordinates as described in Prob. 3.6. However, rather than designing the function to evaluate a single case, pass vectors of x and y. Have the function display the results as a table with columns for x, y, r, and θ. Test the program for the cases outlined in Prob. 3.6.

3.8 Develop an M-file function that is passed a numeric grade from 0 to 100 and returns a letter grade according to the scheme:

Letter	Criteria
A	$90 \leq$ numeric grade ≤ 100
B	$80 \leq$ numeric grade < 90
C	$70 \leq$ numeric grade < 80
D	$60 \leq$ numeric grade < 70
F	numeric grade < 60

The first line of the function should be

```
function grade = lettergrade(score)
```

FIGURE P3.10

Design the function so that it displays an error message and terminates in the event that the user enters a value of score that is less than zero or greater than 100. Test your function with 89.9999, 90, 45, and 120.

3.9 Manning's equation can be used to compute the velocity of water in a rectangular open channel:

$$U = \frac{\sqrt{S}}{n}\left(\frac{BH}{B + 2H}\right)^{2/3}$$

where U = velocity (m/s), S = channel slope, n = roughness coefficient, B = width (m), and H = depth (m). The following data are available for five channels:

n	S	B	H
0.036	0.0001	10	2
0.020	0.0002	8	1
0.015	0.0012	20	1.5
0.030	0.0007	25	3
0.022	0.0003	15	2.6

Write an M-file that computes the velocity for each of these channels. Enter these values into a matrix where each column represents a parameter and each row represents a channel. Have the M-file display the input data along with the computed velocity in tabular form where velocity is the fifth column. Include headings on the table to label the columns.

3.10 A simply supported beam is loaded as shown in Fig. P3.10. Using singularity functions, the displacement along the beam can be expressed by the equation:

$$u_y(x) = \frac{-5}{6}\left[\langle x - 0\rangle^4 - \langle x - 5\rangle^4\right] + \frac{15}{6}\langle x - 8\rangle^3$$

$$+ 75\langle x - 7\rangle^2 + \frac{57}{6}x^3 - 238.25x$$

By definition, the singularity function can be expressed as follows:

$$\langle x - a\rangle^n = \begin{cases} (x - a)^n & \text{when } x > a \\ 0 & \text{when } x \le a \end{cases}$$

Develop an M-file that creates a plot of displacement (dashed line) versus distance along the beam, x. Note that $x = 0$ at the left end of the beam.

3.11 The volume V of liquid in a hollow horizontal cylinder of radius r and length L is related to the depth of the liquid h by

$$V = \left[r^2 \cos^{-1}\left(\frac{r - h}{r}\right) - (r - h)\sqrt{2rh - h^2}\right]L$$

Develop an M-file to create a plot of volume versus depth. Here are the first few lines:

```
function cylinder(r, L, plot_title)
% volume of horizontal cylinder
% inputs:
% r = radius
% L = length
% plot_title = string holding plot title
```

Test your program with

```
>> cylinder(3,5,...
'Volume versus depth for horizontal... cylindrical
tank')
```

3.12 Develop a vectorized version of the following code:

```
tstart=0; tend=20; ni=8;
t(1)=tstart;
y(1)=12 + 6*cos(2*pi*t(1)/(tend-tstart));
for i=2:ni+1
    t(i)=t(i-1)+(tend-tstart)/ni;
    y(i)=12 + 6*cos(2*pi*t(i)/ ...
        (tend-tstart));
end
```

3.13 The "divide and average" method, an old-time method for approximating the square root of any positive number a, can be formulated as

$$x = \frac{x + a/x}{2}$$

Write a well-structured M-file function based on the while...break loop structure to implement this algorithm. Use proper indentation so that the structure is clear. At each step estimate the error in your approximation as

$$\varepsilon = \left| \frac{x_{new} - x_{old}}{x_{new}} \right|$$

Repeat the loop until ε is less than or equal to a specified value. Design your program so that it returns both the result and the error. Make sure that it can evaluate the square root of numbers that are equal to and less than zero. For the latter case, display the result as an imaginary number. For example, the square root of -4 would return $2i$. Test your program by evaluating $a = 0, 2, 10$ and -4 for $\varepsilon = 1 \times 10^{-4}$.

3.14 *Piecewise functions* are sometimes useful when the relationship between a dependent and an independent variable cannot be adequately represented by a single equation. For example, the velocity of a rocket might be described by

$$v(t) = \begin{cases} 10t^2 - 5t & 0 \leq t \leq 8 \\ 624 - 3t & 8 \leq t \leq 16 \\ 36t + 12(t - 16)^2 & 16 \leq t \leq 26 \\ 2136e^{-0.1(t-26)} & t > 26 \\ 0 & \text{otherwise} \end{cases}$$

Develop an M-file function to compute v as a function of t. Then, develop a script that uses this function to generate a plot of v versus t for $t = -5$ to 50.

3.15 Develop an M-file function called rounder to round a number x to a specified number of decimal digits, n. The first line of the function should be set up as

```
function xr = rounder(x, n)
```

Test the program by rounding each of the following to 2 decimal digits: $x = 477.9587, -477.9587, 0.125, 0.135, -0.125,$ and -0.135.

3.16 Develop an M-file function to determine the elapsed days in a year. The first line of the function should be set up as

```
function nd = days(mo, da, leap)
```

where mo = the month (1–12), da = the day (1–31), and leap = (0 for non–leap year and 1 for leap year). Test it for January 1, 1997, February 29, 2004, March 1, 2001, June 21, 2004,

and December 31, 2008. Hint: A nice way to do this combines the for and the switch structures.

3.17 Develop an M-file function to determine the elapsed days in a year. The first line of the function should be set up as

```
function nd = days(mo, da, year)
```

where mo = the month (1–12), da = the day (1–31), and year = the year. Test it for January 1, 1997, February 29, 2004, March 1, 2001, June 21, 2004, and December 31, 2008.

3.18 Develop a function function M-file that returns the difference between the passed function's maximum and minimum value given a range of the independent variable. In addition, have the function generate a plot of the function for the range. Test it for the following cases:
(a) $f(t) = 8e^{-0.25t}\sin(t - 2)$ from $t = 0$ to 6π.
(b) $f(x) = e^{4x}\sin(1/x)$ from $x = 0.01$ to 0.2.
(c) The built-in humps function from $x = 0$ to 2.

3.19 Modify the function function odesimp developed at the end of Sec. 3.6 so that it can be passed the arguments of the passed function. Test it for the following case:

```
>> dvdt=@(v,m,cd) 9.81-(cd/m)*v*abs(v);
>> odesimp(dvdt,0.5,0,12,-10,70,0.23)
```

3.20 A Cartesian vector can be thought of as representing magnitudes along the x-, y-, and z-axes multiplied by a unit vector (i, j, k). For such cases, the dot product of two of these vectors $\{a\}$ and $\{b\}$ corresponds to the product of their magnitudes and the cosine of the angle between their tails as in

$$\{a\}\cdot\{b\} = ab \cos\theta$$

The cross product yields another vector, $\{c\} = \{a\} \times \{b\}$, which is perpendicular to the plane defined by $\{a\}$ and $\{b\}$ such that its direction is specified by the right-hand rule. Develop an M-file function that is passed two such vectors and returns θ, $\{c\}$ and the magnitude of $\{c\}$, and generates a three-dimensional plot of the three vectors $\{a\}$, $\{b\}$, and $\{c\}$ with their origins at zero. Use dashed lines for $\{a\}$ and $\{b\}$ and a solid line for $\{c\}$. Test your function for the following cases:

(a) a = [6 4 2]; b = [2 6 4];
(b) a = [3 2 -6]; b = [4 -3 1];
(c) a = [2 -2 1]; b = [4 2 -4];
(d) a = [-1 0 0]; b = [0 -1 0];

3.21 Based on Example 3.7, develop a script to produce an animation of a bouncing ball where $v_0 = 5$ m/s and $\theta_0 = 50°$. To do this, you must be able to predict exactly when the ball hits the ground. At this point, the direction changes (the new angle will equal the negative of the angle at impact), and the velocity will decrease in magnitude to reflect energy loss due to the collision of the ball with the ground. The change in velocity can be quantified by the *coefficient of restitution* C_R which is equal to the ratio of the velocity after to the velocity before impact. For the present case, use a value of $C_R = 0.8$.

3.22 Develop a function to produce an animation of a particle moving in a circle in Cartesian coordinates based on radial coordinates. Assume a constant radius, r, and allow the angle, θ, to increase from zero to 2π in equal increments. The function's first lines should be

```
function phasor(r, nt, nm)
% function to show the orbit of a phasor
% r = radius
% nt = number of increments for theta
% nm = number of movies
```

Test your function with

```
phasor(1, 256, 10)
```

3.23 Develop a script to produce a movie for the butterfly plot from Prob. 2.22. Use a particle located at the *x-y* coordinates to visualize how the plot evolves in time.

3.24 Develop a MATLAB script to compute the velocity, v, and position, z, of a hot air balloon as described in Prob. 1.28. Perform the calculation from $t = 0$ to 60 s with a step size of 1.6 s. At $z = 200$ m, assume that part of the payload (100 kg) is dropped out of the balloon. Your script should be structured like:

```
% YourFullName
% Hot Air Balloon Script

clear,clc,clf

g=9.81;
global g

% set parameters
r=8.65; % balloon radius
CD=0.47; % dimensionless drag coefficient
mP=265; % mass of payload
P=101300; % atmospheric pressure
Rgas=287; % Universal gas constant for dry air
TC=100; % air temperature
rhoa=1.2; % air density
zd=200; % elevation at which mass is jettisoned
md=100; % mass jettisoned
ti=0; % initial time (s)
tf=60; % final time (s)
vi=0; % initial velocity
zi=0; % initial elevation
dt=1.6; % integration time step
```

```
% precomputations
d = 2 * r; Ta = TC + 273.15; Ab = pi / 4 * d ^ 2;
Vb = pi / 6 * d ^ 3; rhog = P / Rgas / Ta; mG = Vb *
rhog;
FB = Vb * rhoa * g; FG = mG * g; cdp = rhoa * Ab *
CD / 2;

% compute times, velocities and elevations
[t,y] = Balloon(FB, FG, mG, cdp, mP, md, zd,
ti,vi,zi,tf,dt);

% Display results
Your code to display a nice labeled table of times,
velocities, and elevations

% Plot results
Your code to create a nice labeled plot of velocity
and elevation versus time.
```

Your function should be structured like:

```
function [tout,yout]=Balloon(FB, FG, mG, cdp, mP, md,
zd, ti,vi,zi,tf,dt)
global g

% balloon
% function [tout,yout]=Balloon(FB, FG, mG, cdp, mP1,
md, zd, ti,vi,zi,tf,dt)
% Function to generate solutions of vertical
velocity and elevation
% versus time with Euler's method for a hot air
balloon
% Input:
% FB = buoyancy force (N)
% FG = gravity force (N)
% mG = mass (kg)
% cdp=dimensional drag coefficient
% mP= mass of payload (kg)
% md=mass jettisoned (kg)
% zd=elevation at which mass is jettisoned (m)
% ti = initial time (s)
% vi=initial velocity (m/s)
% zi=initial elevation (m)
% tf = final time (s)
% dt=integration time step (s)
% Output:
% tout = vector of times (s)
% yout[:,1] = velocities (m/s)
% yout[:,2] = elevations (m)
% Code to implement Euler's method to compute output
and plot results
```

3.25 A general equation for a sinusoid can be written as

$$y(t) = \bar{y} + \Delta y \sin(2\pi f t - \phi)$$

where y = the dependent variable, $\bar{y}$ = the mean value, Δy = the amplitude, f = the ordinary frequency (i.e., the number of oscillations that occur each unit of time), t = the independent variable (in this case time), and ϕ = phase shift. Develop a MATLAB script to generate a 5 panel vertical plot to illustrate how the function changes as the parameters change. On each plot display the simple sine wave, $y(t) = \sin(2\pi t)$, as a red line. Then, add the following functions to each of the 5 panels as black lines:

subplot	Function	Title
5,1,1	$y(t) = 1 + \sin(2\pi t)$	(a) Effect of mean
5,1,2	$y(t) = 2\sin(2\pi t)$	(b) Effect of amplitude
5,1,3	$y(t) = \sin(4\pi t)$	(c) Effect of frequency
5,1,4	$y(t) = \sin(2\pi t - \pi/4)$	(d) Effect of phase shift
5,1,5	$y(t) = \cos(2\pi t - \pi/2)$	(e) Relationship of sine to cosine

Employ a range of $t = 0$ to 2π and scale each subplot so that the abscissa goes from 0 to 2π and the ordinate goes from -2 to 2. Include the titles with each subplot, label each subplot's ordinate as `'f(t)'`, and the bottom plot's abscissa as `'t'`.

3.26 A *fractal* is a curve or geometric figure, each part of which has the same statistical character as the whole. Fractals are useful in modeling structures (such as eroded coastlines or snowflakes) in which similar patterns recur at progressively smaller scales, and in describing partly random or chaotic phenomena such as crystal growth, fluid turbulence, and galaxy formation. Devaney (1990) has written a nice little book that includes a simple algorithm to create an interesting fractal pattern. Here is a step-by-step description of this algorithm:

Step 1: Assign value to m and n and set hold on.
Step 2: Start a for loop to iterate over i = 1:100000
Step 3: Compute a random number, q = 3*rand(1)
Step 4: If the value of q is less than 1 go to Step 5. Otherwise go to Step 6.
Step 5: Compute new values for m = m/2 and n = n/2 and then go to Step 9.
Step 6: If the value of q is less than 2 go to Step 7. Otherwise go to Step 8.

Step 7: Compute new values for m = m/2 and n = (300 + n)/2, and then go to Step 9.
Step 8: Compute new values for m = (300 + m)/2 and n = (300 + n)/2.
Step 9: If i is less than 100000 then go to Step 10. Otherwise, go to Step 11.
Step 10: Plot a point at the coordinate,(m,n).
Step 11: Terminate ii loop.
Step 12: Set hold off.

Develop a MATLAB script for this algorithm using for and if structures. Run it for the following two cases **(a)** m = 2 and n = 1 and **(b)** m = 100 and n = 200.

3.27 Write a well-structured MATLAB function procedure named Fnorm to calculate the Frobenius norm of an $m \times n$ matrix,

$$\|A\|_f = \sqrt{\sum_{i=1}^{m} \sum_{j=1}^{n} a_{i,j}^2}$$

Here is a script that uses the function

```
A = [5 7 9; 1 8 4; 7 6 2];
Fn = Fnorm(A)
```

Here is the first line of the function

```
function Norm = Fnorm(x)
```

Develop two versions of the function: **(a)** using nested for loops and **(b)** using sum functions.

3.28 The pressure and temperature of the atmosphere are constantly changing depending on a number of factors including altitude, latitude/longitude, time of day, and season. To take all these variations into account when considering the design and performance of flight vehicles is impractical. Therefore, a *standard atmosphere* is frequently used to provide engineers and scientists with a common reference for their research and development. The *International Standard Atmosphere* is one such model of how conditions of the earth's atmosphere change over a wide range of altitudes or elevations. The following table shows values of temperature and pressure at selected altitudes.

Layer Index, i	Layer Name	Base Geopotential Altitude Above MSL, h (km)	Lapse Rate (°C/km)	Base Temperature T (°C)	Base Pressure, p (Pa)
1	Troposphere	0	−6.5	15	101,325
2	Tropopause	11	0	−56.5	22,632
3	Stratosphere	20	1	−56.5	5474.9
4	Stratosphere	32	2.8	−44.5	868.02
5	Stratopause	47	0	−2.5	110.91
6	Mesosphere	51	−2.8	−2.5	66.939
7	Mesosphere	71	−2.0	−58.5	3.9564
8	Mesopause	84.852	−	−86.28	0.3734

The temperature at each altitude can then be computed as

$$T(h) = T_i + \gamma_i (h - h_i) \qquad\qquad h_i < h \leq h_{i+1}$$

where $T(h)$ = temperature at altitude h (°C), T_i = the base temperature for layer i (°C), γ_i = lapse rate or the rate at which atmospheric temperature decreases linearly with increase in altitude for layer i (°C/km), and h_i = base geopotential altitude above mean sea level (MSL) for layer i. The pressure at each altitude can then be computed as

$$p(h) = p_i + \frac{p_{i+1} - p_i}{h_{i+1} - h_i} (h - h_i)$$

where $p(h)$ = pressure at altitude h (Pa ≡ N/m²), p_i = the base pressure for layer i (Pa). The density, ρ (kg/m³), can then be calculated according to a molar form of the *ideal gas law*:

$$\rho = \frac{pM}{RT_a}$$

where M = molar mass (≅ 0.0289644 kg/mol), R = the universal gas constant (8.3144621 J/(mol·K), and T_a = absolute temperature (K) = $T + 273.15$.

Develop a MATLAB function, StdAtm, to determine values of the three properties for a given altitude. If the user requests a value outside the range of altitudes, have the function display an error message and terminate the application. Use the following script as the starting point to create a 3-panel plot of altitude versus the properties.

```
% Script to generate a plot of temperature, pressure
and density
% for the International Standard Atmosphere
clc,clf
h=[0 11 20 32 47 51 71 84.852];
gamma=[-6.5 0 1 2.8 0 -2.8 -2];
T=[15 -56.5 -56.5 -44.5 -2.5 -2.5 -58.5 -86.28];
p=[101325 22632 5474.9 868.02 110.91 66.939 3.9564
0.3734];
hint=[0:0.1:84.852];
for i=1:length(hint)
   [Tint(i),pint(i),rint(i)]=StdAtm(h,T,p,gamma,hint
(i));
end

% Create plot
% Function call to test error trap
[Tint(i),pint(i),rint(i)]=StdAtm(h,T,p,gamma,85);
```

3.29 Develop a MATLAB function to convert a vector of temperatures from Celsius to Fahrenheit and vice versa. Test it with the following data for the average monthly temperatures at Death Volley, CA and at the South Pole.

Day	Death Valley °F	South Pole °C
15	54	−27
45	60	−40
75	69	−53
105	77	−56
135	87	−57
165	96	−57
195	102	−59
225	101	−59
255	92	−59
285	78	−50
315	63	−38
345	52	−27

Use the following script as the starting point to generate a 2-panel stacked plot of the temperatures versus day for both of the sites with the Celsius time series at the top and the Fahrenheit at the bottom. If the user requests a unit other than 'C' or 'F', have the function display an error message and terminate the application.

```
% Script to generate stacked plots of temperatures
versus time
% for Death Valley and the South Pole with Celsius
time series
% on the top plot and Fahrenheit on the bottom.

clc,clf
t=[15 45 75 105 135 165 195 225 255 285 315 345];
TFDV=[54 60 69 77 87 96 102 101 92 78 63 52];
TCSP=[-27 -40 -53 -56 -57 -57 -59 -59 -59 -50
-38 -27];
TCDV=TempConv(TFDV,'C');
TFSP=TempConv(TCSP,'F');

% Create plot

% Test of error trap
TKSP=TempConv(TCSP,'K');
```

3.30 Because there are only two possibilities, as in Prob. 3.29, it's relatively easy to convert between Celsius and Fahrenheit temperature units. Because there are many more units in common use, converting between pressure units is more challenging. Here are some of the possibilities along with the number of Pascals represented by each:

Index, i	Unit, U_i	Description or Usage	# of Pa, C_i
1	psi	Tire pressure, above ambient pressure	6894.76
2	atm	Used for high pressure experiments	101,325
3	inHg	Atmospheric pressure given by weatherperson	3376.85
4	kg/cm^2	Europe metric unit in situations where psi used in U.S.	98,066.5
5	inH$_2$O	Used in heating/ventilating systems in buildings	248.843
6	Pa	Standard SI (metric) unit of pressure, 1 N/m^2	1
7	bar	Frequented used by meteorologists	100,000
8	dyne/cm^2	Older scientific pressure unit from the CGS system	0.1
9	ftH$_2$O	American and English low value pressure unit.	2988.98
10	mmHg	Used in laboratory pressure measurements	133.322
11	torr	Same as 1 mmHg but used in vacuum measurement	133.322
12	ksi	Used in structural engineering	6,894,760

The information in the table can be used to implement a conversion calculation. One way to do this is to store both the units and the corresponding number of the Pascals in individual arrays with the subscripts corresponding to each entry. For example,

```
U(1)='psi'  C(1)=6894.76
U(2)='atm'  C(2)=101325.
    .          .
    .          .
    .          .
```

The conversion from one unit to another could then be computed with the following general formula:

$$P_d = \frac{C_j}{C_i} P_g$$

where P_g = given pressure, P_d = desired pressure, j = the index of the desired unit, and i = the index of the given unit. As an example, to convert a tire pressure of say 28.6 psi to atm, we would use

$$P_d = \frac{C_2}{C_1} P_g = \frac{101325. \text{ atm/Pa}}{6894.76 \text{ psi/Pa}} \ 28.6 \text{ psi} = 420.304 \text{ atm}$$

So we see that the conversion from one unit to another involves first determining the indices corresponding to the given and the desired units, and then implementing the conversion equation. Here is a step-by-step algorithm to do this:

1. Assign the values of the unit, U, and conversion, C, arrays.
2. Have the user select the input units by entering the value of i.

    If the user enters a correct value within the range, 1–12, continue to step 3.

    If the user enters a value outside the range, display an error message and repeat step 2.

3. Have the user enter the given value of pressure, P_i.
4. Have the user select the desired units by entering the value of j.

    If the user enters a correct value within the range, 1–12, continue to step 5.

    If the user enters a value outside the range, display an error message and repeat step 4.

5. Use the formula to convert the quantity in input units to its value in the desired output units.
6. Display the original quantity and units and the output quantity and units.
7. Ask if another output result, for the same input, is desired.

    If yes, go back to step 4 and continue from there.
    If no, go on to step 8.

8. Ask if another conversion is desired.

    If yes, go back to step 2 and continue from there.
    If no, end of algorithm.

Develop a well-structured MATLAB script using loop and if structures to implement this algorithm. Test it with the following:

(a) Duplicate the hand calculation example to ensure that you get about 420.304 atm for an input of 28.6 psi.
(b) Try to enter a choice code of $i = 13$. Does the program trap that error and allow you to correct it? If not, it should. Now, try a choice code of the letter Q. What happens?

4

Roundoff and Truncation Errors

CHAPTER OBJECTIVES

The primary objective of this chapter is to acquaint you with the major sources of errors involved in numerical methods. Specific objectives and topics covered are

- Understanding the distinction between accuracy and precision.
- Learning how to quantify error.
- Learning how error estimates can be used to decide when to terminate an iterative calculation.
- Understanding how roundoff errors occur because digital computers have a limited ability to represent numbers.
- Understanding why floating-point numbers have limits on their range and precision.
- Recognizing that truncation errors occur when exact mathematical formulations are represented by approximations.
- Knowing how to use the Taylor series to estimate truncation errors.
- Understanding how to write forward, backward, and centered finite-difference approximations of first and second derivatives.
- Recognizing that efforts to minimize truncation errors can sometimes increase roundoff errors.

YOU'VE GOT A PROBLEM

I n Chap. 1, you developed a numerical model for the velocity of a bungee jumper. To solve the problem with a computer, you had to approximate the derivative of velocity with a finite difference:

$$\frac{dv}{dt} \cong \frac{\Delta v}{\Delta t} = \frac{v(t_{i+1}) - v(t_i)}{t_{i+1} - t_i}$$

Thus, the resulting solution is not exact—that is, it has error.

In addition, the computer you use to obtain the solution is also an imperfect tool. Because it is a digital device, the computer is limited in its ability to represent the magnitudes and precision of numbers. Consequently, the machine itself yields results that contain error.

So both your mathematical approximation and your digital computer cause your resulting model prediction to be uncertain. Your problem is: How do you deal with such uncertainty? In particular, is it possible to understand, quantify, and control such errors in order to obtain acceptable results? This chapter introduces you to some approaches and concepts that engineers and scientists use to deal with this dilemma.

4.1 ERRORS

Engineers and scientists constantly find themselves having to accomplish objectives based on uncertain information. Although perfection is a laudable goal, it is rarely if ever attained. For example, despite the fact that the model developed from Newton's second law is an excellent approximation, it would never in practice exactly predict the jumper's fall. A variety of factors such as winds and slight variations in air resistance would result in deviations from the prediction. If these deviations are systematically high or low, then we might need to develop a new model. However, if they are randomly distributed and tightly grouped around the prediction, then the deviations might be considered negligible and the model deemed adequate. Numerical approximations also introduce similar discrepancies into the analysis.

This chapter covers basic topics related to the identification, quantification, and minimization of these errors. General information concerned with the quantification of error is reviewed in this section. This is followed by Sections 4.2 and 4.3, dealing with the two major forms of numerical error: roundoff error (due to computer approximations) and truncation error (due to mathematical approximations). We also describe how strategies to reduce truncation error sometimes increase roundoff. Finally, we briefly discuss errors not directly connected with the numerical methods themselves. These include blunders, model errors, and data uncertainty.

4.1.1 Accuracy and Precision

The errors associated with both calculations and measurements can be characterized with regard to their accuracy and precision. *Accuracy* refers to how closely a computed or measured value agrees with the true value. *Precision* refers to how closely individual computed or measured values agree with each other.

These concepts can be illustrated graphically using an analogy from target practice. The bullet holes on each target in Fig. 4.1 can be thought of as the predictions of a numerical technique, whereas the bull's-eye represents the truth. *Inaccuracy* (also called *bias*) is defined as systematic deviation from the truth. Thus, although the shots in Fig. 4.1c are more tightly grouped than in Fig. 4.1a, the two cases are equally biased because they are both centered on the upper left quadrant of the target. *Imprecision* (also called *uncertainty*), on the other hand, refers to the magnitude of the scatter. Therefore, although Fig. 4.1b and d are equally accurate (i.e., centered on the bull's-eye), the latter is more precise because the shots are tightly grouped.

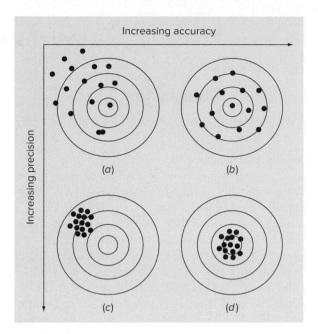

FIGURE 4.1
An example from marksmanship illustrating the concepts of accuracy and precision:
(*a*) inaccurate and imprecise, (*b*) accurate and imprecise, (*c*) inaccurate and precise,
and (*d*) accurate and precise.

Numerical methods should be sufficiently accurate or unbiased to meet the requirements of a particular problem. They also should be precise enough for adequate design.
In this book, we will use the collective term *error* to represent both the inaccuracy and
imprecision of our predictions.

4.1.2 Error Definitions

Numerical errors arise from the use of approximations to represent exact mathematical
operations and quantities. For such errors, the relationship between the exact, or true, result
and the approximation can be formulated as

$$\text{True value} = \text{approximation} + \text{error} \tag{4.1}$$

By rearranging Eq. (4.1), we find that the numerical error is equal to the discrepancy
between the truth and the approximation, as in

$$E_t = \text{true value} - \text{approximation} \tag{4.2}$$

where E_t is used to designate the exact value of the error. The subscript t is included to
designate that this is the "true" error. This is in contrast to other cases, as described shortly,
where an "approximate" estimate of the error must be employed. Note that the true error is
commonly expressed as an absolute value and referred to as the *absolute error*.

A shortcoming of this definition is that it takes no account of the order of magnitude of
the value under examination. For example, an error of a centimeter is much more significant

if we are measuring a rivet than a bridge. One way to account for the magnitudes of the quantities being evaluated is to normalize the error to the true value, as in

$$\text{True fractional relative error} = \frac{\text{true value} - \text{approximation}}{\text{true value}}$$

The relative error can also be multiplied by 100% to express it as

$$\varepsilon_t = \frac{\text{true value} - \text{approximation}}{\text{true value}} 100\% \tag{4.3}$$

where ε_t designates the true percent relative error.

For example, suppose that you have the task of measuring the lengths of a bridge and a rivet and come up with 9999 and 9 cm, respectively. If the true values are 10,000 and 10 cm, respectively, the error in both cases is 1 cm. However, their percent relative errors can be computed using Eq. (4.3) as 0.01% and 10%, respectively. Thus, although both measurements have an absolute error of 1 cm, the relative error for the rivet is much greater. We would probably conclude that we have done an adequate job of measuring the bridge, whereas our estimate for the rivet leaves something to be desired.

Notice that for Eqs. (4.2) and (4.3), E and ε are subscripted with a t to signify that the error is based on the true value. For the example of the rivet and the bridge, we were provided with this value. However, in actual situations such information is rarely available. For numerical methods, the true value will only be known when we deal with functions that can be solved analytically. Such will typically be the case when we investigate the theoretical behavior of a particular technique for simple systems. However, in real-world applications, we will obviously not know the true answer *a priori*. For these situations, an alternative is to normalize the error using the best available estimate of the true value—that is, to the approximation itself, as in

$$\varepsilon_a = \frac{\text{approximate error}}{\text{approximation}} 100\% \tag{4.4}$$

where the subscript a signifies that the error is normalized to an approximate value. Note also that for real-world applications, Eq. (4.2) cannot be used to calculate the error term in the numerator of Eq. (4.4). One of the challenges of numerical methods is to determine error estimates in the absence of knowledge regarding the true value. For example, certain numerical methods use *iteration* to compute answers. In such cases, a present approximation is made on the basis of a previous approximation. This process is performed repeatedly, or iteratively, to successively compute (hopefully) better and better approximations. For such cases, the error is often estimated as the difference between the previous and present approximations. Thus, percent relative error is determined according to

$$\varepsilon_a = \frac{\text{present approximation} - \text{previous approximation}}{\text{present approximation}} 100\% \tag{4.5}$$

This and other approaches for expressing errors is elaborated on in subsequent chapters.

The signs of Eqs. (4.2) through (4.5) may be either positive or negative. If the approximation is greater than the true value (or the previous approximation is greater than the current approximation), the error is negative; if the approximation is less than the true value, the error is positive. Also, for Eqs. (4.3) to (4.5), the denominator may be less than zero, which can also lead to a negative error. Often, when performing computations, we may not be concerned with the sign of the error but are interested in whether the absolute value of the percent relative error is lower than a prespecified tolerance ε_s. Therefore, it is often useful to employ the absolute value of Eq. (4.5). For such cases, the computation is repeated until

$$|\varepsilon_a| < \varepsilon_s \tag{4.6}$$

This relationship is referred to as a *stopping criterion*. If it is satisfied, our result is assumed to be within the prespecified acceptable level ε_s. Note that for the remainder of this text, we almost always employ absolute values when using relative errors.

It is also convenient to relate these errors to the number of significant figures in the approximation. It can be shown (Scarborough, 1966) that if the following criterion is met, we can be assured that the result is correct to *at least n* significant figures.

$$\varepsilon_s = (0.5 \times 10^{2-n})\% \tag{4.7}$$

EXAMPLE 4.1   Error Estimates for Iterative Methods

Problem Statement. In mathematics, functions can often be represented by infinite series. For example, the exponential function can be computed using

$$e^x = 1 + x + \frac{x^2}{2} + \frac{x^3}{3!} + \cdots + \frac{x^n}{n!} \tag{E4.1.1}$$

Thus, as more terms are added in sequence, the approximation becomes a better and better estimate of the true value of e^x. Equation (E4.1.1) is called a *Maclaurin series expansion*.

Starting with the simplest version, $e^x = 1$, add terms one at a time in order to estimate $e^{0.5}$. After each new term is added, compute the true and approximate percent relative errors with Eqs. (4.3) and (4.5), respectively. Note that the true value is $e^{0.5} = 1.648721 \ldots$. Add terms until the absolute value of the approximate error estimate ε_a falls below a prespecified error criterion ε_s conforming to three significant figures.

Solution. First, Eq. (4.7) can be employed to determine the error criterion that ensures a result that is correct to at least three significant figures:

$$\varepsilon_s = (0.5 \times 10^{2-3})\% = 0.05\%$$

Thus, we will add terms to the series until ε_a falls below this level.

The first estimate is simply equal to Eq. (E4.1.1) with a single term. Thus, the first estimate is equal to 1. The second estimate is then generated by adding the second term as in

$$e^x = 1 + x$$

or for $x = 0.5$

$$e^{0.5} = 1 + 0.5 = 1.5$$

This represents a true percent relative error of [Eq. (4.3)]

$$\varepsilon_t = \left| \frac{1.648721 - 1.5}{1.648721} \right| \times 100\% = 9.02\%$$

Equation (4.5) can be used to determine an approximate estimate of the error, as in

$$\varepsilon_a = \left| \frac{1.5 - 1}{1.5} \right| \times 100\% = 33.3\%$$

Because ε_a is not less than the required value of ε_s, we would continue the computation by adding another term, $x^2/2!$, and repeating the error calculations. The process is continued until $|\varepsilon_a| < \varepsilon_s$. The entire computation can be summarized as

Terms	Result	ε_t, %	ε_a, %
1	1	39.3	
2	1.5	9.02	33.3
3	1.625	1.44	7.69
4	1.645833333	0.175	1.27
5	1.648437500	0.0172	0.158
6	1.648697917	0.00142	0.0158

Thus, after six terms are included, the approximate error falls below $\varepsilon_s = 0.05\%$, and the computation is terminated. However, notice that, rather than three significant figures, the result is accurate to five! This is because, for this case, both Eqs. (4.5) and (4.7) are conservative. That is, they ensure that the result is at least as good as they specify. Although, this is not always the case for Eq. (4.5), it is true most of the time.

4.1.3 Computer Algorithm for Iterative Calculations

Many of the numerical methods described in the remainder of this text involve iterative calculations of the sort illustrated in Example 4.1. These all entail solving a mathematical problem by computing successive approximations to the solution starting from an initial guess.

The computer implementation of such iterative solutions involves loops. As we saw in Sec. 3.3.2, these come in two basic flavors: count-controlled and decision loops. Most iterative solutions use decision loops. Thus, rather than employing a prespecified number of iterations, the process typically is repeated until an approximate error estimate falls below a stopping criterion as in Example 4.1.

To do this for the same problem as Example 4.1, the series expansion can be expressed as

$$e^x \cong \sum_{i=0}^{n} \frac{x^n}{n!}$$

An M-file to implement this formula is shown in Fig. 4.2. The function is passed the value to be evaluated (x) along with a stopping error criterion (es) and a maximum allowable number of iterations (maxit). If the user omits either of the latter two parameters, the function assigns default values.

```
function [fx,ea,iter] = IterMeth(x,es,maxit)
% Maclaurin series of exponential function
%    [fx,ea,iter] = IterMeth(x,es,maxit)
% input:
%    x = value at which series evaluated
%    es = stopping criterion (default = 0.0001)
%    maxit = maximum iterations (default = 50)
% output:
%    fx = estimated value
%    ea = approximate relative error (%)
%    iter = number of iterations

% defaults:
if nargin<2|isempty(es),es=0.0001;end
if nargin<3|isempty(maxit),maxit=50;end
% initialization
iter = 1; sol = 1; ea = 100;
% iterative calculation
while (1)
  solold = sol;
  sol = sol + x ^ iter / factorial(iter);
  iter = iter + 1;
  if sol~=0
    ea=abs((sol - solold)/sol)*100;
  end
  if ea<=es | iter>=maxit,break,end
end
fx = sol;
end
```

FIGURE 4.2
An M-file to solve an iterative calculation. This example is set up to evaluate the Maclaurin series expansion for e^x as described in Example 4.1.

The function then initializes three variables: (a) iter, which keeps track of the number of iterations, (b) sol, which holds the current estimate of the solution, and (c) a variable, ea, which holds the approximate percent relative error. Note that ea is initially set to a value of 100 to ensure that the loop executes at least once.

These initializations are followed by a decision loop that actually implements the iterative calculation. Prior to generating a new solution, the previous value, sol, is first assigned to solold. Then a new value of sol is computed and the iteration counter is incremented. If the new value of sol is nonzero, the percent relative error, ea, is determined. The stopping criteria are then tested. If both are false, the loop repeats. If either is true, the loop terminates and the final solution is sent back to the function call.

When the M-file is implemented, it generates an estimate for the exponential function which is returned along with the approximate error and the number of iterations. For example, e^1 can be evaluated as

```
>> format long
>> [approxval, ea, iter] = IterMeth(1,1e-6,100)
```

```
approxval =
    2.718281826198493
ea =
     9.216155641522974e-007
iter =
    12
```

We can see that after 12 iterations, we obtain a result of 2.7182818 with an approximate error estimate of $= 9.2162 \times 10^{-7}\%$. The result can be verified by using the built-in exp function to directly calculate the exact value and the true percent relative error,

```
>> trueval=exp(1)

trueval =
    2.718281828459046

>> et=abs((trueval- approxval)/trueval)*100

et =
    8.316108397236229e-008
```

As was the case with Example 4.1, we obtain the desirable outcome that the true error is less than the approximate error.

4.2   ROUNDOFF ERRORS

Roundoff errors arise because digital computers cannot represent some quantities exactly. They are important to engineering and scientific problem solving because they can lead to erroneous results. In certain cases, they can actually lead to a calculation going unstable and yielding obviously erroneous results. Such calculations are said to be *ill-conditioned*. Worse still, they can lead to subtler discrepancies that are difficult to detect.

There are two major facets of roundoff errors involved in numerical calculations:

1. Digital computers have magnitude and precision limits on their ability to represent numbers.
2. Certain numerical manipulations are highly sensitive to roundoff errors. This can result from both mathematical considerations as well as from the way in which computers perform arithmetic operations.

4.2.1 Computer Number Representation

Numerical roundoff errors are directly related to the manner in which numbers are stored in a computer. The fundamental unit whereby information is represented is called a *word*. This is an entity that consists of a string of *binary digits*, or *bits*. Numbers are typically stored in one or more words. To understand how this is accomplished, we must first review some material related to number systems.

A *number system* is merely a convention for representing quantities. Because we have 10 fingers and 10 toes, the number system that we are most familiar with is the *decimal,* or *base-10,* number system. A base is the number used as the reference for constructing the system. The base-10 system uses the 10 digits—0, 1, 2, 3, 4, 5, 6, 7, 8, and 9—to represent numbers. By themselves, these digits are satisfactory for counting from 0 to 9.

For larger quantities, combinations of these basic digits are used, with the position or *place value* specifying the magnitude. The rightmost digit in a whole number represents a number from 0 to 9. The second digit from the right represents a multiple of 10. The third digit from the right represents a multiple of 100 and so on. For example, if we have the number 8642.9, then we have eight groups of 1000, six groups of 100, four groups of 10, two groups of 1, and nine groups of 0.1, or

$$(8 \times 10^3) + (6 \times 10^2) + (4 \times 10^1) + (2 \times 10^0) + (9 \times 10^{-1}) = 8642.9$$

This type of representation is called *positional notation*.

Now, because the decimal system is so familiar, it is not commonly realized that there are alternatives. For example, if human beings happened to have eight fingers and toes we would undoubtedly have developed an *octal,* or *base-8,* representation. In the same sense, our friend the computer is like a two-fingered animal who is limited to two states—either 0 or 1. This relates to the fact that the primary logic units of digital computers are on/off electronic components. Hence, numbers on the computer are represented with a *binary,* or *base-2,* system. Just as with the decimal system, quantities can be represented using positional notation. For example, the binary number 101.1 is equivalent to $(1 \times 2^2) + (0 \times 2^1) + (1 \times 2^0) + (1 \times 2^{-1}) = 4 + 0 + 1 + 0.5 = 5.5$ in the decimal system.

Integer Representation. Now that we have reviewed how base-10 numbers can be represented in binary form, it is simple to conceive of how integers are represented on a computer. The most straightforward approach, called the *signed magnitude method,* employs the first bit of a word to indicate the sign, with a 0 for positive and a 1 for negative. The remaining bits are used to store the number. For example, the integer value of 173 is represented in binary as 10101101:

$$(10101101)_2 = 2^7 + 2^5 + 2^3 + 2^2 + 2^0 = 128 + 32 + 8 + 4 + 1 = (173)_{10}$$

Therefore, the binary equivalent of -173 would be stored on a 16-bit computer, as depicted in Fig. 4.3.

If such a scheme is employed, there clearly is a limited range of integers that can be represented. Again assuming a 16-bit word size, if one bit is used for the sign, the 15 remaining bits can represent binary integers from 0 to 111111111111111. The upper limit can be converted to a decimal integer, as in $(1 \times 2^{14}) + (1 \times 2^{13}) + \cdots + (1 \times 2^1) + (1 \times 2^0) = 32{,}767$.

FIGURE 4.3
The binary representation of the decimal integer –173 on a 16-bit computer using the signed magnitude method.

Note that this value can be simply evaluated as $2^{15} - 1$. Thus, a 16-bit computer word can store decimal integers ranging from $-32,767$ to $32,767$.

In addition, because zero is already defined as 0000000000000000, it is redundant to use the number 1000000000000000 to define a "minus zero." Therefore, it is conventionally employed to represent an additional negative number: $-32,768$, and the range is from $-32,768$ to $32,767$. For an n-bit word, the range would be from -2^{n-1} to $2^{n-1} - 1$. Thus, 32-bit integers would range from $-2,147,483,648$ to $+2,147,483,647$.

Note that, although it provides a nice way to illustrate our point, the signed magnitude method is not actually used to represent integers for conventional computers. A preferred approach called the *2s complement* technique directly incorporates the sign into the number's magnitude rather than providing a separate bit to represent plus or minus. Regardless, the range of numbers is still the same as for the signed magnitude method described above.

The foregoing serves to illustrate how all digital computers are limited in their capability to represent integers. That is, numbers above or below the range cannot be represented. A more serious limitation is encountered in the storage and manipulation of fractional quantities as described next.

Floating-Point Representation. Fractional quantities are typically represented in computers using *floating-point format*. In this approach, which is very much like scientific notation, the number is expressed as

$$\pm s \times b^e$$

where s = the *significand* (or *mantissa*), b = the base of the number system being used, and e = the exponent.

Prior to being expressed in this form, the number is *normalized* by moving the decimal place over so that only one significant digit is to the left of the decimal point. This is done so computer memory is not wasted on storing useless nonsignificant zeros. For example, a value like 0.005678 could be represented in a wasteful manner as 0.005678×10^0. However, normalization would yield 5.678×10^{-3} which eliminates the useless zeroes.

Before describing the base-2 implementation used on computers, we will first explore the fundamental implications of such floating-point representation. In particular, what are the ramifications of the fact that in order to be stored in the computer, both the mantissa and the exponent must be limited to a finite number of bits? As in the next example, a nice way to do this is within the context of our more familiar base-10 decimal world.

EXAMPLE 4.2   Implications of Floating-Point Representation

Problem Statement. Suppose that we had a hypothetical base-10 computer with a 5-digit word size. Assume that one digit is used for the sign, two for the exponent, and two for the mantissa. For simplicity, assume that one of the exponent digits is used for its sign, leaving a single digit for its magnitude.

Solution. A general representation of the number following normalization would be

$$s_1 d_1.d_2 \times 10^{s_0 d_0}$$

where s_0 and s_1 = the signs, d_0 = the magnitude of the exponent, and d_1 and d_2 = the magnitude of the significand digits.

Now, let's play with this system. First, what is the largest possible positive quantity that can be represented? Clearly, it would correspond to both signs being positive and all magnitude digits set to the largest possible value in base-10, that is, 9:

$$\text{Largest value} = +9.9 \times 10^{+9}$$

So the largest possible number would be a little less than 10 billion. Although this might seem like a big number, it's really not that big. For example, this computer would be incapable of representing a commonly used constant like Avogadro's number (6.022×10^{23}).

In the same sense, the smallest possible positive number would be

$$\text{Smallest value} = +1.0 \times 10^{-9}$$

Again, although this value might seem pretty small, you could not use it to represent a quantity like Planck's constant $(6.626 \times 10^{-34}\ \text{J} \cdot \text{s})$.

Similar negative values could also be developed. The resulting ranges are displayed in Fig. 4.4. Large positive and negative numbers that fall outside the range would cause an *overflow error*. In a similar sense, for very small quantities there is a "hole" at zero, and very small quantities would usually be converted to zero.

Recognize that the exponent overwhelmingly determines these range limitations. For example, if we increase the mantissa by one digit, the maximum value increases slightly to 9.99×10^9. In contrast, a one-digit increase in the exponent raises the maximum by 90 orders of magnitude to 9.9×10^{99}!

When it comes to precision, however, the situation is reversed. Whereas the significand plays a minor role in defining the range, it has a profound effect on specifying the precision. This is dramatically illustrated for this example where we have limited the significand to only 2 digits. As in Fig. 4.5, just as there is a "hole" at zero, there are also "holes" between values.

For example, a simple rational number with a finite number of digits like $2^{-5} = 0.03125$ would have to be stored as 3.1×10^{-2} or 0.031. Thus, a *roundoff error* is introduced. For this case, it represents a relative error of

$$\frac{0.03125 - 0.031}{0.03125} = 0.008$$

FIGURE 4.4

The number line showing the possible ranges corresponding to the hypothetical base-10 floating-point scheme described in Example 4.2.

FIGURE 4.5
A small portion of the number line corresponding to the hypothetical base-10 floating-point scheme described in Example 4.2. The numbers indicate values that can be represented exactly. All other quantities falling in the "holes" between these values would exhibit some roundoff error.

While we could store a number like 0.03125 exactly by expanding the digits of the significand, quantities with infinite digits must always be approximated. For example, a commonly used constant such as $\pi \, (= 3.14159\ldots)$ would have to be represented as 3.1×10^0 or 3.1. For this case, the relative error is

$$\frac{3.14159 - 3.1}{3.14159} = 0.0132$$

Although adding significand digits can improve the approximation, such quantities will always have some roundoff error when stored in a computer.

Another more subtle effect of floating-point representation is illustrated by Fig. 4.5. Notice how the interval between numbers increases as we move between orders of magnitude. For numbers with an exponent of -1 (i.e., between 0.1 and 1), the spacing is 0.01. Once we cross over into the range from 1 to 10, the spacing increases to 0.1. This means that the roundoff error of a number will be proportional to its magnitude. In addition, it means that the relative error will have an upper bound. For this example, the maximum relative error would be 0.05. This value is called the *machine epsilon* (or machine precision).

As illustrated in Example 4.2, the fact that both the exponent and significand are finite means that there are both range and precision limits on floating-point representation. Now, let us examine how floating-point quantities are actually represented in a real computer using base-2 or binary numbers.

First, let's look at normalization. Since binary numbers consist exclusively of 0s and 1s, a bonus occurs when they are normalized. That is, the bit to the left of the binary point will always be one! This means that this leading bit does not have to be stored. Hence, nonzero binary floating-point numbers can be expressed as

$$\pm(1 + f) \times 2^e$$

where $f =$ the *mantissa* (i.e., the fractional part of the significand). For example, if we normalized the binary number 1101.1, the result would be $1.1011 \times (2)^{-3}$ or $(1 + 0.1011) \times 2^{-3}$.

FIGURE 4.6
The manner in which a floating-point number is stored in an 8-byte word in IEEE double-precision format.

Thus, although the original number has five significant bits, we only have to store the four fractional bits: 0.1011.

By default, MATLAB has adopted the *IEEE double-precision format* in which eight bytes (64 bits) are used to represent floating-point numbers. As in Fig. 4.6, one bit is reserved for the number's sign. In a similar spirit to the way in which integers are stored, the exponent and its sign are stored in 11 bits. Finally, 52 bits are set aside for the mantissa. However, because of normalization, 53 bits can be stored.

Now, just as in Example 4.2, this means that the numbers will have a limited range and precision. However, because the IEEE format uses many more bits, the resulting number system can be used for practical purposes.

Range. In a fashion similar to the way in which integers are stored, the 11 bits used for the exponent translates into a range from -1022 to 1023. The largest positive number can be represented in binary as

$$\text{Largest value} = +1.1111 \ldots 1111 \times 2^{+1023}$$

where the 52 bits in the mantissa are all 1. Since the significand is approximately 2 (it is actually $2 - 2^{-52}$), the largest value is therefore $2^{1024} = 1.7977 \times 10^{308}$. In a similar fashion, the smallest positive number can be represented as

$$\text{Smallest value} = +1.0000 \ldots 0000 \times 2^{-1022}$$

This value can be translated into a base-10 value of $2^{-1022} = 2.2251 \times 10^{-308}$.

Precision. The 52 bits used for the mantissa correspond to about 15 to 16 base-10 digits. Thus, π would be expressed as

```
>> format long
>> pi

ans =
   3.14159265358979
```

Note that the machine epsilon is $2^{-52} = 2.2204 \times 10^{-16}$.

MATLAB has a number of built-in functions related to its internal number representation. For example, the *realmax* function displays the largest positive real number:

```
>> format long
>> realmax

ans =
    1.797693134862316e+308
```

Numbers occurring in computations that exceed this value create an overflow. In MATLAB they are set to infinity, inf. The *realmin* function displays the smallest positive real number:

```
>> realmin

ans =
    2.225073858507201e-308
```

Numbers that are smaller than this value create an *underflow* and, in MATLAB, are set to zero. Finally, the *eps* function displays the machine epsilon:

```
>> eps

ans =
    2.220446049250313e-016
```

4.2.2 Arithmetic Manipulations of Computer Numbers

Aside from the limitations of a computer's number system, the actual arithmetic manipulations involving these numbers can also result in roundoff error. To understand how this occurs, let's look at how the computer performs simple addition and subtraction.

Because of their familiarity, normalized base-10 numbers will be employed to illustrate the effect of roundoff errors on simple addition and subtraction. Other number bases would behave in a similar fashion. To simplify the discussion, we will employ a hypothetical decimal computer with a 4-digit mantissa and a 1-digit exponent.

When two floating-point numbers are added, the numbers are first expressed so that they have the same exponents. For example, if we want to add $1.557 + 0.04341$, the computer would express the numbers as $0.1557 \times 10^1 + 0.004341 \times 10^1$. Then the mantissas are added to give 0.160041×10^1. Now, because this hypothetical computer only carries a 4-digit mantissa, the excess number of digits get chopped off and the result is 0.1600×10^1. Notice how the last two digits of the second number (41) that were shifted to the right have essentially been lost from the computation.

Subtraction is performed identically to addition except that the sign of the subtrahend is reversed. For example, suppose that we are subtracting 26.86 from 36.41. That is,

$$\begin{array}{r} 0.3641 \times 10^2 \\ -0.2686 \times 10^2 \\ \hline 0.0955 \times 10^2 \end{array}$$

For this case the result must be normalized because the leading zero is unnecessary. So we must shift the decimal one place to the right to give $0.9550 \times 10^1 = 9.550$.

Notice that the zero added to the end of the mantissa is not significant but is merely appended to fill the empty space created by the shift. Even more dramatic results would be obtained when the numbers are very close as in

$$0.7642 \times 10^3$$
$$\underline{-0.7641 \times 10^3}$$
$$0.0001 \times 10^3$$

which would be converted to $0.1000 \times 10^0 = 0.1000$. Thus, for this case, three nonsignificant zeros are appended.

The subtracting of two nearly equal numbers is called *subtractive cancellation*. It is the classic example of how the manner in which computers handle mathematics can lead to numerical problems. Other calculations that can cause problems include:

Large Computations. Certain methods require extremely large numbers of arithmetic manipulations to arrive at their final results. In addition, these computations are often interdependent. That is, the later calculations are dependent on the results of earlier ones. Consequently, even though an individual roundoff error could be small, the cumulative effect over the course of a large computation can be significant. A very simple case involves summing a round base-10 number that is not round in base-2. Suppose that the following M-file is constructed:

```
function sout = sumdemo()
s = 0;
for i = 1:10000
  s = s + 0.0001;
end
sout = s;
```

When this function is executed, the result is

```
>> format long
>> sumdemo

ans =
   0.99999999999991
```

The `format long` command lets us see the 15 significant-digit representation used by MATLAB. You would expect that sum would be equal to 1. However, although 0.0001 is a nice round number in base-10, it cannot be expressed exactly in base-2. Thus, the sum comes out to be slightly different than 1. We should note that MATLAB has features that are designed to minimize such errors. For example, suppose that you form a vector as in

```
>> format long
>> s = [0:0.0001:1];
```

For this case, rather than being equal to 0.99999999999991, the last entry will be exactly one as verified by

```
>> s(10001)

ans =
   1
```

Adding a Large and a Small Number. Suppose we add a small number, 0.0010, to a large number, 4000, using a hypothetical computer with the 4-digit mantissa and the 1-digit exponent. After modifying the smaller number so that its exponent matches the larger,

$$
\begin{array}{r}
0.4000 \times 10^4 \\
0.0000001 \times 10^4 \\
\hline
0.4000001 \times 10^4
\end{array}
$$

which is chopped to 0.4000×10^4. Thus, we might as well have not performed the addition! This type of error can occur in the computation of an infinite series. The initial terms in such series are often relatively large in comparison with the later terms. Thus, after a few terms have been added, we are in the situation of adding a small quantity to a large quantity. One way to mitigate this type of error is to sum the series in reverse order. In this way, each new term will be of comparable magnitude to the accumulated sum.

Smearing. Smearing occurs whenever the individual terms in a summation are larger than the summation itself. One case where this occurs is in a series of mixed signs.

Inner Products. As should be clear from the last sections, some infinite series are particularly prone to roundoff error. Fortunately, the calculation of series is not one of the more common operations in numerical methods. A far more ubiquitous manipulation is the calculation of inner products as in

$$
\sum_{i=1}^{n} x_i \, y_i = x_1 \, y_1 + x_2 \, y_2 + \cdots + x_n \, y_n
$$

This operation is very common, particularly in the solution of simultaneous linear algebraic equations. Such summations are prone to roundoff error. Consequently, it is often desirable to compute such summations in double precision as is done automatically in MATLAB.

4.3 TRUNCATION ERRORS

Truncation errors are those that result from using an approximation in place of an exact mathematical procedure. For example, in Chap. 1 we approximated the derivative of velocity of a bungee jumper by a finite-difference equation of the form [Eq. (1.11)]

$$
\frac{dv}{dt} \cong \frac{\Delta v}{\Delta t} = \frac{v(t_{i+1}) - v(t_i)}{t_{i+1} - t_i}
\tag{4.8}
$$

A truncation error was introduced into the numerical solution because the difference equation only approximates the true value of the derivative (recall Fig. 1.3). To gain insight into the properties of such errors, we now turn to a mathematical formulation that is used widely in numerical methods to express functions in an approximate fashion—the Taylor series.

4.3.1 The Taylor Series

Taylor's theorem and its associated formula, the Taylor series, is of great value in the study of numerical methods. In essence, the *Taylor theorem* states that any smooth function can be approximated as a polynomial. The *Taylor series* then provides a means to express this idea mathematically in a form that can be used to generate practical results.

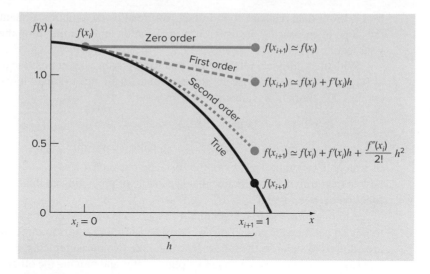

FIGURE 4.7

The approximation of $f(x) = -0.1x^4 - 0.15x^3 - 0.5x^2 - 0.25x + 1.2$ at $x = 1$ by zero-order, first-order, and second-order Taylor series expansions.

A useful way to gain insight into the Taylor series is to build it term by term. A good problem context for this exercise is to predict a function value at one point in terms of the function value and its derivatives at another point.

Suppose that you are blindfolded and taken to a location on the side of a hill facing downslope (Fig. 4.7). We'll call your horizontal location x_i and your vertical distance with respect to the base of the hill $f(x_i)$. You are given the task of predicting the height at a position x_{i+1}, which is a distance h away from you.

At first, you are placed on a platform that is completely horizontal so that you have no idea that the hill is sloping down away from you. At this point, what would be your best guess at the height at x_{i+1}? If you think about it (remember you have no idea whatsoever what's in front of you), the best guess would be the same height as where you're standing now! You could express this prediction mathematically as

$$f(x_{i+1}) \cong f(x_i) \tag{4.9}$$

This relationship, which is called the *zero-order approximation,* indicates that the value of f at the new point is the same as the value at the old point. This result makes intuitive sense because if x_i and x_{i+1} are close to each other, it is likely that the new value is probably similar to the old value.

Equation (4.9) provides a perfect estimate if the function being approximated is, in fact, a constant. For our problem, you would be right only if you happened to be standing on a perfectly flat plateau. However, if the function changes at all over the interval, additional terms of the Taylor series are required to provide a better estimate.

So now you are allowed to get off the platform and stand on the hill surface with one leg positioned in front of you and the other behind. You immediately sense that the front foot is lower than the back foot. In fact, you're allowed to obtain a quantitative estimate of the slope by measuring the difference in elevation and dividing it by the distance between your feet.

With this additional information, you're clearly in a better position to predict the height at $f(x_{i+1})$. In essence, you use the slope estimate to project a straight line out to x_{i+1}. You can express this prediction mathematically by

$$f(x_{i+1}) \cong f(x_i) + f'(x_i)h \tag{4.10}$$

This is called a *first-order approximation* because the additional first-order term consists of a slope $f'(x_i)$ multiplied by h, the distance between x_i and x_{i+1}. Thus, the expression is now in the form of a straight line that is capable of predicting an increase or decrease of the function between x_i and x_{i+1}.

Although Eq. (4.10) can predict a change, it is only exact for a straight-line, or *linear*, trend. To get a better prediction, we need to add more terms to our equation. So now you are allowed to stand on the hill surface and take two measurements. First, you measure the slope behind you by keeping one foot planted at x_i and moving the other one back a distance Δx. Let's call this slope $f'_b(x_i)$. Then you measure the slope in front of you by keeping one foot planted at x_i and moving the other one forward Δx. Let's call this slope $f'_f(x_i)$. You immediately recognize that the slope behind is milder than the one in front. Clearly the drop in height is "accelerating" downward in front of you. Thus, the odds are that $f(x_i)$ is even lower than your previous linear prediction.

As you might expect, you're now going to add a second-order term to your equation and make it into a parabola. The Taylor series provides the correct way to do this as in

$$f(x_{i+1}) \cong f(x_i) + f'(x_i)h + \frac{f''(x_i)}{2!}h^2 \tag{4.11}$$

To make use of this formula, you need an estimate of the second derivative. You can use the last two slopes you determined to estimate it as

$$f''(x_{i+1}) \cong \frac{f'_f(x_i) - f'_b(x_i)}{\Delta x} \tag{4.12}$$

Thus, the second derivative is merely a derivative of a derivative; in this case, the rate of change of the slope.

Before proceeding, let's look carefully at Eq. (4.11). Recognize that all the values subscripted i represent values that you have estimated. That is, they are numbers. Consequently, the only unknowns are the values at the prediction position x_{i+1}. Thus, it is a quadratic equation of the form

$$f(h) \cong a_2 h^2 + a_1 h + a_0$$

Thus, we can see that the second-order Taylor series approximates the function with a second-order polynomial.

Clearly, we could keep adding more derivatives to capture more of the function's curvature. Thus, we arrive at the complete Taylor series expansion

$$f(x_{i+1}) \cong f(x_i) + f'(x_i)h + \frac{f''(x_i)}{2!}h^2 + \frac{f^{(3)}(x_i)}{3!}h^3 + \cdots + \frac{f^{(n)}(x_i)}{n!}h^n + R_n \tag{4.13}$$

Note that because Eq. (4.13) is an infinite series, an equal sign replaces the approximate sign that was used in Eqs. (4.9) through (4.11). A remainder term is also included to account for all terms from $n + 1$ to infinity:

$$R_n = \frac{f^{(n+1)}(\xi)}{(n + 1)!} h^{n+1} \tag{4.14}$$

where the subscript n connotes that this is the remainder for the nth-order approximation and ξ is a value of x that lies somewhere between x_i and x_{i+1}.

We can now see why the Taylor theorem states that any smooth function can be approximated as a polynomial and that the Taylor series provides a means to express this idea mathematically.

In general, the nth-order Taylor series expansion will be exact for an nth-order polynomial. For other differentiable and continuous functions, such as exponentials and sinusoids, a finite number of terms will not yield an exact estimate. Each additional term will contribute some improvement, however slight, to the approximation. This behavior will be demonstrated in Example 4.3. Only if an infinite number of terms are added will the series yield an exact result.

Although the foregoing is true, the practical value of Taylor series expansions is that, in most cases, the inclusion of only a few terms will result in an approximation that is close enough to the true value for practical purposes. The assessment of how many terms are required to get "close enough" is based on the remainder term of the expansion (Eq. 4.14). This relationship has two major drawbacks. First, ξ is not known exactly but merely lies somewhere between x_i and x_{i+1}. Second, to evaluate Eq. (4.14), we need to determine the $(n + 1)$th derivative of $f(x)$. To do this, we need to know $f(x)$. However, if we knew $f(x)$, there would be no need to perform the Taylor series expansion in the present context!

Despite this dilemma, Eq. (4.14) is still useful for gaining insight into truncation errors. This is because we *do* have control over the term h in the equation. In other words, we can choose how far away from x we want to evaluate $f(x)$, and we can control the number of terms we include in the expansion. Consequently, Eq. (4.14) is often expressed as

$$R_n = O(h^{n+1})$$

where the nomenclature $O(h^{n+1})$ means that the truncation error is of the order of h^{n+1}. That is, the error is proportional to the step size h raised to the $(n + 1)$th power. Although this approximation implies nothing regarding the magnitude of the derivatives that multiply h^{n+1}, it is extremely useful in judging the comparative error of numerical methods based on Taylor series expansions. For example, if the error is $O(h)$, halving the step size will halve the error. On the other hand, if the error is $O(h^2)$, halving the step size will quarter the error.

In general, we can usually assume that the truncation error is decreased by the addition of terms to the Taylor series. In many cases, if h is sufficiently small, the first- and other lower-order terms usually account for a disproportionately high percent of the error. Thus, only a few terms are required to obtain an adequate approximation. This property is illustrated by the following example.

EXAMPLE 4.3   Approximation of a Function with a Taylor Series Expansion

Problem Statement. Use Taylor series expansions with $n = 0$ to 6 to approximate $f(x) = \cos x$ at $x_{i+1} = \pi/3$ on the basis of the value of $f(x)$ and its derivatives at $x_i = \pi/4$. Note that this means that $h = \pi/3 - \pi/4 = \pi/12$.

Solution. Our knowledge of the true function allows us to determine the correct value $f(\pi/3) = 0.5$. The zero-order approximation is [Eq. (4.9)]

$$f\left(\frac{\pi}{3}\right) \cong \cos\left(\frac{\pi}{4}\right) = 0.707106781$$

which represents a percent relative error of

$$\varepsilon_t = \left|\frac{0.5 - 0.707106781}{0.5}\right| 100\% = 41.4\%$$

For the first-order approximation, we add the first derivative term where $f'(x) = -\sin x$:

$$f\left(\frac{\pi}{3}\right) \cong \cos\left(\frac{\pi}{4}\right) - \sin\left(\frac{\pi}{4}\right)\left(\frac{\pi}{12}\right) = 0.521986659$$

which has $|\varepsilon_t| = 0.40\%$. For the second-order approximation, we add the second derivative term where $f''(x) = -\cos x$:

$$f\left(\frac{\pi}{3}\right) \cong \cos\left(\frac{\pi}{4}\right) - \sin\left(\frac{\pi}{4}\right)\left(\frac{\pi}{12}\right) - \frac{\cos(\pi/4)}{2}\left(\frac{\pi}{12}\right)^2 = 0.497754491$$

with $|\varepsilon_t| = 0.449\%$. Thus, the inclusion of additional terms results in an improved estimate. The process can be continued and the results listed as in

Order n	$f^{(n)}(x)$	$f(\pi/3)$	$|\varepsilon_t|$
0	cos x	0.707106781	41.4
1	−sin x	0.521986659	4.40
2	−cos x	0.497754491	0.449
3	sin x	0.499869147	2.62×10^{-2}
4	cos x	0.500007551	1.51×10^{-3}
5	−sin x	0.500000304	6.08×10^{-5}
6	−cos x	0.499999988	2.44×10^{-6}

Notice that the derivatives never go to zero as would be the case for a polynomial. Therefore, each additional term results in some improvement in the estimate. However, also notice how most of the improvement comes with the initial terms. For this case, by the time we have added the third-order term, the error is reduced to 0.026%, which means that we have attained 99.974% of the true value. Consequently, although the addition of more terms will reduce the error further, the improvement becomes negligible.

4.3.2 The Remainder for the Taylor Series Expansion

Before demonstrating how the Taylor series is actually used to estimate numerical errors, we must explain why we included the argument ξ in Eq. (4.14). To do this, we will use a simple, visually based explanation.

Suppose that we truncated the Taylor series expansion [Eq. (4.13)] after the zero-order term to yield

$$f(x_{i+1}) \cong f(x_i)$$

A visual depiction of this zero-order prediction is shown in Fig. 4.8. The remainder, or error, of this prediction, which is also shown in the illustration, consists of the infinite series of terms that were truncated

$$R_0 = f'(x_i)h + \frac{f''(x_i)}{2!} h^2 + \frac{f^{(3)}(x_i)}{3!} h^3 + \cdots$$

It is obviously inconvenient to deal with the remainder in this infinite series format. One simplification might be to truncate the remainder itself, as in

$$R_0 \cong f'(x_i)h \tag{4.15}$$

Although, as stated in the previous section, lower-order derivatives usually account for a greater share of the remainder than the higher-order terms, this result is still inexact because of the neglected second- and higher-order terms. This "inexactness" is implied by the approximate equality symbol ($\cong$) employed in Eq. (4.15).

An alternative simplification that transforms the approximation into an equivalence is based on a graphical insight. As in Fig. 4.9, the *derivative mean-value theorem* states that if a function $f(x)$ and its first derivative are continuous over an interval from x_i to x_{i+1}, then

---

FIGURE 4.8
Graphical depiction of a zero-order Taylor series prediction and remainder.

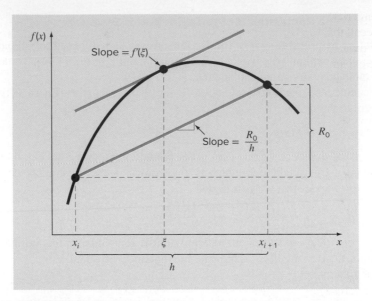

FIGURE 4.9
Graphical depiction of the derivative mean-value theorem.

there exists at least one point on the function that has a slope, designated by $f'(\xi)$, that is parallel to the line joining $f(x_i)$ and $f(x_{i+1})$. The parameter ξ marks the x value where this slope occurs (Fig. 4.9). A physical illustration of this theorem is that, if you travel between two points with an average velocity, there will be at least one moment during the course of the trip when you will be moving at that average velocity.

By invoking this theorem, it is simple to realize that, as illustrated in Fig. 4.9, the slope $f'(\xi)$ is equal to the rise R_0 divided by the run h, or

$$f'(\xi) = \frac{R_0}{h}$$

which can be rearranged to give

$$R_0 = f'(\xi)h \tag{4.16}$$

Thus, we have derived the zero-order version of Eq. (4.14). The higher-order versions are merely a logical extension of the reasoning used to derive Eq. (4.16). The first-order version is

$$R_1 = \frac{f''(\xi)}{2!} h^2 \tag{4.17}$$

For this case, the value of ξ conforms to the x value corresponding to the second derivative that makes Eq. (4.17) exact. Similar higher-order versions can be developed from Eq. (4.14).

4.3.3 Using the Taylor Series to Estimate Truncation Errors

Although the Taylor series will be extremely useful in estimating truncation errors throughout this book, it may not be clear to you how the expansion can actually be applied to numerical methods. In fact, we have already done so in our example of the bungee jumper. Recall that the objective of both Examples 1.1 and 1.2 was to predict velocity as a function of time. That is, we were interested in determining $v(t)$. As specified by Eq. (4.13), $v(t)$ can be expanded in a Taylor series:

$$v(t_{i+1}) = v(t_i) + v'(t_i)(t_{i+1} - t_i) + \frac{v''(t_i)}{2!}(t_{i+1} - t_i)^2 + \cdots + R_n$$

Now let us truncate the series after the first derivative term:

$$v(t_{i+1}) = v(t_i) + v'(t_i)(t_{i+1} - t_i) + R_1 \tag{4.18}$$

Equation (4.18) can be solved for

$$v'(t_i) = \underbrace{\frac{v(t_{i+1}) - v(t_i)}{t_{i+1} - t_i}}_{\substack{\text{First-order} \\ \text{approximation}}} - \underbrace{\frac{R_1}{t_{i+1} - t_i}}_{\substack{\text{Truncation} \\ \text{error}}} \tag{4.19}$$

The first part of Eq. (4.19) is exactly the same relationship that was used to approximate the derivative in Example 1.2 [Eq. (1.11)]. However, because of the Taylor series approach, we have now obtained an estimate of the truncation error associated with this approximation of the derivative. Using Eqs. (4.14) and (4.19) yields

$$\frac{R_1}{t_{i+1} - t_i} = \frac{v''(\xi)}{2!}(t_{i+1} - t_i)$$

or

$$\frac{R_1}{t_{i+1} - t_i} = O(t_{i+1} - t_i)$$

Thus, the estimate of the derivative [Eq. (1.11) or the first part of Eq. (4.19)] has a truncation error of order $t_{i+1} - t_i$. In other words, the error of our derivative approximation should be proportional to the step size. Consequently, if we halve the step size, we would expect to halve the error of the derivative.

4.3.4 Numerical Differentiation

Equation (4.19) is given a formal label in numerical methods—it is called a *finite difference*. It can be represented generally as

$$f'(x_i) = \frac{f(x_{i+1}) - f(x_i)}{x_{i+1} - x_i} + O(x_{i+1} - x_i) \tag{4.20}$$

or

$$f'(x_i) = \frac{f(x_{i+1}) - f(x_i)}{h} + O(h) \tag{4.21}$$

where h is called the step size—that is, the length of the interval over which the approximation is made, $x_{i+1} - x_i$. It is termed a "forward" difference because it utilizes data at i and $i + 1$ to estimate the derivative (Fig. 4.10a).

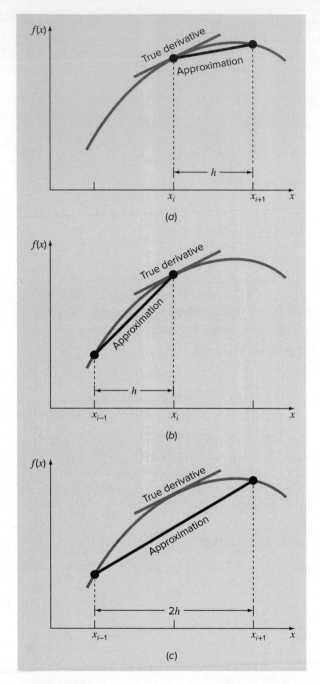

FIGURE 4.10
Graphical depiction of (*a*) forward, (*b*) backward, and (*c*) centered finite-difference approximations of the first derivative.

This forward difference is but one of many that can be developed from the Taylor series to approximate derivatives numerically. For example, backward and centered difference approximations of the first derivative can be developed in a fashion similar to the derivation of Eq. (4.19). The former utilizes values at x_{i-1} and x_i (Fig. 4.10b), whereas the latter uses values that are equally spaced around the point at which the derivative is estimated (Fig. 4.10c). More accurate approximations of the first derivative can be developed by including higher-order terms of the Taylor series. Finally, all the foregoing versions can also be developed for second, third, and higher derivatives. The following sections provide brief summaries illustrating how some of these cases are derived.

Backward Difference Approximation of the First Derivative. The Taylor series can be expanded backward to calculate a previous value on the basis of a present value, as in

$$f(x_{i-1}) = f(x_i) - f'(x_i)h + \frac{f''(x_i)}{2!} h^2 - \cdots \qquad (4.22)$$

Truncating this equation after the first derivative and rearranging yields

$$f'(x_i) \cong \frac{f(x_i) - f(x_{i-1})}{h} \qquad (4.23)$$

where the error is $O(h)$.

Centered Difference Approximation of the First Derivative. A third way to approximate the first derivative is to subtract Eq. (4.22) from the forward Taylor series expansion:

$$f(x_{i+1}) = f(x_i) + f'(x_i)h + \frac{f''(x_i)}{2!} h^2 + \cdots \qquad (4.24)$$

to yield

$$f(x_{i+1}) = f(x_{i-1}) + 2f'(x_i)h + 2\frac{f^{(3)}(x_i)}{3!} h^3 + \cdots$$

which can be solved for

$$f'(x_i) = \frac{f(x_{i+1}) - f(x_{i-1})}{2h} - \frac{f^{(3)}(x_i)}{6} h^2 + \cdots$$

or

$$f'(x_i) = \frac{f(x_{i+1}) - f(x_{i-1})}{2h} - O(h^2) \qquad (4.25)$$

Equation (4.25) is a *centered finite difference* representation of the first derivative. Notice that the truncation error is of the order of h^2 in contrast to the forward and backward approximations that were of the order of h. Consequently, the Taylor series analysis yields the practical information that the centered difference is a more accurate representation of the derivative (Fig. 4.10c). For example, if we halve the step size using a forward or backward difference, we would approximately halve the truncation error, whereas for the central difference, the error would be quartered.

EXAMPLE 4.4   Finite-Difference Approximations of Derivatives

Problem Statement. Use forward and backward difference approximations of $O(h)$ and a centered difference approximation of $O(h^2)$ to estimate the first derivative of

$$f(x) = -0.1x^4 - 0.15x^3 - 0.5x^2 - 0.25x + 1.2$$

at $x = 0.5$ using a step size $h = 0.5$. Repeat the computation using $h = 0.25$. Note that the derivative can be calculated directly as

$$f'(x) = -0.4x^3 - 0.45x^2 - 1.0x - 0.25$$

and can be used to compute the true value as $f'(0.5) = -0.9125$.

Solution. For $h = 0.5$, the function can be employed to determine

$$x_{i-1} = 0 \qquad f(x_{i-1}) = 1.2$$
$$x_i = 0.5 \qquad f(x_i) = 0.925$$
$$x_{i+1} = 1.0 \qquad f(x_{i+1}) = 0.2$$

These values can be used to compute the forward difference [Eq. (4.21)],

$$f'(0.5) \cong \frac{0.2 - 0.925}{0.5} = -1.45 \qquad |\varepsilon_t| = 58.9\%$$

the backward difference [Eq. (4.23)],

$$f'(0.5) \cong \frac{0.925 - 1.2}{0.5} = -0.55 \qquad |\varepsilon_t| = 39.7\%$$

and the centered difference [Eq. (4.25)],

$$f'(0.5) \cong \frac{0.2 - 1.2}{1.0} = -1.0 \qquad |\varepsilon_t| = 9.6\%$$

For $h = 0.25$,

$$x_{i-1} = 0.25 \qquad f(x_{i-1}) = 1.10351563$$
$$x_i = 0.5 \qquad f(x_i) = 0.925$$
$$x_{i+1} = 0.75 \qquad f(x_{i+1}) = 0.63632813$$

which can be used to compute the forward difference,

$$f'(0.5) \cong \frac{0.63632813 - 0.925}{0.25} = -1.155 \qquad |\varepsilon_t| = 26.5\%$$

the backward difference,

$$f'(0.5) \cong \frac{0.925 - 1.10351563}{0.25} = 0.714 \qquad |\varepsilon_t| = 21.7\%$$

and the centered difference,

$$f'(0.5) \cong \frac{0.63632813 - 1.10351563}{0.5} = -0.934 \qquad |\varepsilon_t| = 2.4\%$$

For both step sizes, the centered difference approximation is more accurate than forward or backward differences. Also, as predicted by the Taylor series analysis, halving the step size approximately halves the error of the backward and forward differences and quarters the error of the centered difference.

Finite-Difference Approximations of Higher Derivatives. Besides first derivatives, the Taylor series expansion can be used to derive numerical estimates of higher derivatives. To do this, we write a forward Taylor series expansion for $f(x_{i+2})$ in terms of $f(x_i)$:

$$f(x_{i+2}) = f(x_i) + f'(x_i)(2h) + \frac{f''(x_i)}{2!}(2h)^2 + \cdots \tag{4.26}$$

Equation (4.24) can be multiplied by 2 and subtracted from Eq. (4.26) to give

$$f(x_{i+2}) - 2f(x_{i+1}) = -f(x_i) + f''(x_i)h^2 + \cdots$$

which can be solved for

$$f''(x_i) = \frac{f(x_{i+2}) - 2f(x_{i+1}) + f(x_i)}{h^2} + O(h) \tag{4.27}$$

This relationship is called the *second forward finite difference*. Similar manipulations can be employed to derive a backward version

$$f''(x_i) = \frac{f(x_i) - 2f(x_{i-1}) + f(x_{i-2})}{h^2} + O(h)$$

A centered difference approximation for the second derivative can be derived by adding Eqs. (4.22) and (4.24) and rearranging the result to give

$$f''(x_i) = \frac{f(x_{i+1}) - 2f(x_i) + f(x_{i-1})}{h^2} + O(h^2)$$

As was the case with the first-derivative approximations, the centered case is more accurate. Notice also that the centered version can be alternatively expressed as

$$f''(x_i) \cong \frac{\dfrac{f(x_{i+1}) - f(x_i)}{h} - \dfrac{f(x_i) - f(x_{i-1})}{h}}{h}$$

Thus, just as the second derivative is a derivative of a derivative, the second finite difference approximation is a difference of two first finite differences [recall Eq. (4.12)].

4.4   TOTAL NUMERICAL ERROR

The *total numerical error* is the summation of the truncation and roundoff errors. In general, the only way to minimize roundoff errors is to increase the number of significant figures of the computer. Further, we have noted that roundoff error may *increase* due to subtractive cancellation or due to an increase in the number of computations in an analysis. In contrast, Example 4.4 demonstrated that the truncation error can be reduced by decreasing the step size. Because a decrease in step size can lead to subtractive cancellation or to an increase in computations, the truncation errors are *decreased* as the roundoff errors are *increased*.

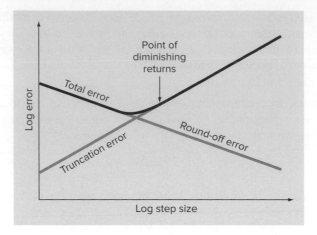

FIGURE 4.11
A graphical depiction of the trade-off between roundoff and truncation error that sometimes comes into play in the course of a numerical method. The point of diminishing returns is shown, where roundoff error begins to negate the benefits of step-size reduction.

Therefore, we are faced by the following dilemma: The strategy for decreasing one component of the total error leads to an increase of the other component. In a computation, we could conceivably decrease the step size to minimize truncation errors only to discover that in doing so, the roundoff error begins to dominate the solution and the total error grows! Thus, our remedy becomes our problem (Fig. 4.11). One challenge that we face is to determine an appropriate step size for a particular computation. We would like to choose a large step size to decrease the amount of calculations and roundoff errors without incurring the penalty of a large truncation error. If the total error is as shown in Fig. 4.11, the challenge is to identify the point of diminishing returns where roundoff error begins to negate the benefits of step-size reduction.

When using MATLAB, such situations are relatively uncommon because of its 15- to 16-digit precision. Nevertheless, they sometimes do occur and suggest a sort of "numerical uncertainty principle" that places an absolute limit on the accuracy that may be obtained using certain computerized numerical methods. We explore such a case in the following section.

4.4.1 Error Analysis of Numerical Differentiation

As described in Sec. 4.3.4, a centered difference approximation of the first derivative can be written as [Eq. (4.25)]

$$f'(x_i) = \frac{f(x_{i+1}) - f(x_{i-1})}{2h} - \frac{f^{(3)}(\xi)}{6} h^2$$

(4.28)

True value	Finite-difference approximation	Truncation error

Thus, if the two function values in the numerator of the finite-difference approximation have no roundoff error, the only error is due to truncation.

However, because we are using digital computers, the function values do include roundoff error as in

$$f(x_{i-1}) = \tilde{f}(x_{i-1}) + e_{i-1}$$
$$f(x_{i+1}) = \tilde{f}(x_{i+1}) + e_{i+1}$$

where the $\tilde{f}$'s are the rounded function values and the e's are the associated roundoff errors. Substituting these values into Eq. (4.28) gives

$$f'(x_i) = \frac{\tilde{f}(x_{i+1}) - \tilde{f}(x_{i-1})}{2h} + \frac{e_{i+1} - e_{i-1}}{2h} - \frac{f^{(3)}(\xi)}{6}h^2$$

| True value | Finite-difference approximation | Roundoff error | Truncation error |

We can see that the total error of the finite-difference approximation consists of a roundoff error that decreases with step size and a truncation error that increases with step size.

Assuming that the absolute value of each component of the roundoff error has an upper bound of ε, the maximum possible value of the difference $e_{i+1} - e_{i-1}$ will be 2ε. Further, assume that the third derivative has a maximum absolute value of M. An upper bound on the absolute value of the total error can therefore be represented as

$$\text{Total error} = \left| f'(x_i) - \frac{\tilde{f}(x_{i+1}) - \tilde{f}(x_{i-1})}{2h} \right| \leq \frac{\varepsilon}{h} + \frac{h^2 M}{6} \tag{4.29}$$

An optimal step size can be determined by differentiating Eq. (4.29), setting the result equal to zero and solving for

$$h_{opt} = \sqrt[3]{\frac{3\varepsilon}{M}} \tag{4.30}$$

EXAMPLE 4.5   Roundoff and Truncation Errors in Numerical Differentiation

Problem Statement. In Example 4.4, we used a centered difference approximation of $O(h^2)$ to estimate the first derivative of the following function at $x = 0.5$,

$$f(x) = -0.1x^4 - 0.15x^3 - 0.5x^2 - 0.25x + 1.2$$

Perform the same computation starting with $h = 1$. Then progressively divide the step size by a factor of 10 to demonstrate how roundoff becomes dominant as the step size is reduced. Relate your results to Eq. (4.30). Recall that the true value of the derivative is -0.9125.

Solution. We can develop the following M-file to perform the computations and plot the results. Notice that we pass both the function and its analytical derivative as arguments:

```
function diffex(func,dfunc,x,n)
format long
dftrue=dfunc(x);
h=1;
H(1)=h;
D(1)=(func(x+h)-func(x-h))/(2*h);
F(1)=abs(dftrue-D(1));
```

```
for i=2:n
  h=h/10;
  H(i)=h;
  D(i)=(func(x+h)-func(x-h))/(2*h);
  E(i)=abs(dftrue-D(i));
end
L=[H' D' E']';
fprintf('   step size    finite difference    true error\n');
fprintf('%14.10f %16.14f %16.13f\n',L);
loglog(H,E),xlabel('Step Size'),ylabel('Error')
title('Plot of Error Versus Step Size')
format short
```

The M-file can then be run using the following commands:

```
>> ff=@(x) -0.1*x^4-0.15*x^3-0.5*x^2-0.25*x+1.2;
>> df=@(x) -0.4*x^3-0.45*x^2-x-0.25;
>> diffex(ff,df,0.5,11)
```

step size	finite difference	true error
1.0000000000	-1.26250000000000	0.3500000000000
0.1000000000	-0.91600000000000	0.0035000000000
0.0100000000	-0.91253500000000	0.0000350000000
0.0010000000	-0.91250035000001	0.0000003500000
0.0001000000	-0.91250000349985	0.0000000034998
0.0000100000	-0.91250000003318	0.0000000000332
0.0000010000	-0.91250000000542	0.0000000000054
0.0000001000	-0.91249999945031	0.0000000005497
0.0000000100	-0.91250000333609	0.0000000033361
0.0000000010	-0.91250001998944	0.0000000199894
0.0000000001	-0.91250007550059	0.0000000755006

As depicted in Fig. 4.12, the results are as expected. At first, roundoff is minimal and the estimate is dominated by truncation error. Hence, as in Eq. (4.29), the total error drops by a factor of 100 each time we divide the step by 10. However, starting at about $h = 0.0001$, we see roundoff error begin to creep in and erode the rate at which the error diminishes. A minimum error is reached at $h = 10^{-6}$. Beyond this point, the error increases as roundoff dominates.

Because we are dealing with an easily differentiable function, we can also investigate whether these results are consistent with Eq. (4.30). First, we can estimate M by evaluating the function's third derivative as

$$M = |f^{(3)}(0.5)| = |-2.4(0.5) - 0.9| = 2.1$$

Because MATLAB has a precision of about 15 to 16 base-10 digits, a rough estimate of the upper bound on roundoff would be about $\varepsilon = 0.5 \times 10^{-16}$. Substituting these values into Eq. (4.30) gives

$$h_{opt} = \sqrt[3]{\frac{3(0.5 \times 10^{-16})}{2.1}} = 4.3 \times 10^{-6}$$

which is on the same order as the result of 1×10^{-6} obtained with MATLAB.

FIGURE 4.12

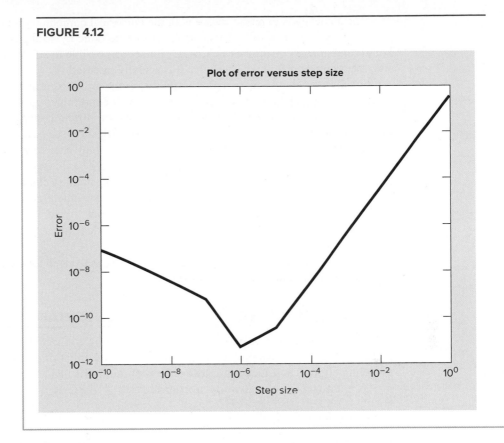

4.4.2 Control of Numerical Errors

For most practical cases, we do not know the exact error associated with numerical methods. The exception, of course, is when we know the exact solution, which makes our numerical approximations unnecessary. Therefore, for most engineering and scientific applications we must settle for some estimate of the error in our calculations.

There are no systematic and general approaches to evaluating numerical errors for all problems. In many cases error estimates are based on the experience and judgment of the engineer or scientist.

Although error analysis is to a certain extent an art, there are several practical programming guidelines we can suggest. First and foremost, avoid subtracting two nearly equal numbers. Loss of significance almost always occurs when this is done. Sometimes you can rearrange or reformulate the problem to avoid subtractive cancellation. If this is not possible, you may want to use extended-precision arithmetic. Furthermore, when adding and subtracting numbers, it is best to sort the numbers and work with the smallest numbers first. This avoids loss of significance.

Beyond these computational hints, one can attempt to predict total numerical errors using theoretical formulations. The Taylor series is our primary tool for analysis of such errors. Prediction of total numerical error is very complicated for even moderately sized problems and tends to be pessimistic. Therefore, it is usually attempted for only small-scale tasks.

The tendency is to push forward with the numerical computations and try to estimate the accuracy of your results. This can sometimes be done by seeing if the results satisfy some condition or equation as a check. Or it may be possible to substitute the results back into the original equation to check that it is actually satisfied.

Finally you should be prepared to perform numerical experiments to increase your awareness of computational errors and possible ill-conditioned problems. Such experiments may involve repeating the computations with a different step size or method and comparing the results. We may employ sensitivity analysis to see how our solution changes when we change model parameters or input values. We may want to try different numerical algorithms that have different theoretical foundations, arc based on different computational strategies, or have different convergence properties and stability characteristics.

When the results of numerical computations are extremely critical and may involve loss of human life or have severe economic ramifications, it is appropriate to take special precautions. This may involve the use of two or more independent groups to solve the same problem so that their results can be compared.

The roles of errors will be a topic of concern and analysis in all sections of this book. We will leave these investigations to specific sections.

4.5   BLUNDERS, MODEL ERRORS, AND DATA UNCERTAINTY

Although the following sources of error are not directly connected with most of the numerical methods in this book, they can sometimes have great impact on the success of a modeling effort. Thus, they must always be kept in mind when applying numerical techniques in the context of real-world problems.

4.5.1 Blunders

We are all familiar with gross errors, or blunders. In the early years of computers, erroneous numerical results could sometimes be attributed to malfunctions of the computer itself. Today, this source of error is highly unlikely, and most blunders must be attributed to human imperfection.

Blunders can occur at any stage of the mathematical modeling process and can contribute to all the other components of error. They can be avoided only by sound knowledge of fundamental principles and by the care with which you approach and design your solution to a problem.

Blunders are usually disregarded in discussions of numerical methods. This is no doubt due to the fact that, try as we may, mistakes are to a certain extent unavoidable. However, we believe that there are a number of ways in which their occurrence can be minimized. In particular, the good programming habits that were outlined in Chap. 3 are extremely useful for mitigating programming blunders. In addition, there are usually simple ways to check whether a particular numerical method is working properly. Throughout this book, we discuss ways to check the results of numerical calculations.

4.5.2 Model Errors

Model errors relate to bias that can be ascribed to incomplete mathematical models. An example of a negligible model error is the fact that Newton's second law does not account for relativistic effects. This does not detract from the adequacy of the solution in Example 1.1

because these errors are minimal on the time and space scales associated with the bungee jumper problem.

However, suppose that air resistance is not proportional to the square of the fall velocity, as in Eq. (1.7), but is related to velocity and other factors in a different way. If such were the case, both the analytical and numerical solutions obtained in Chap. 1 would be erroneous because of model error. You should be cognizant of this type of error and realize that, if you are working with a poorly conceived model, no numerical method will provide adequate results.

4.5.3 Data Uncertainty

Errors sometimes enter into an analysis because of uncertainty in the physical data on which a model is based. For instance, suppose we wanted to test the bungee jumper model by having an individual make repeated jumps and then measuring his or her velocity after a specified time interval. Uncertainty would undoubtedly be associated with these measurements, as the parachutist would fall faster during some jumps than during others. These errors can exhibit both inaccuracy and imprecision. If our instruments consistently underestimate or overestimate the velocity, we are dealing with an inaccurate, or biased, device. On the other hand, if the measurements are randomly high and low, we are dealing with a question of precision.

Measurement errors can be quantified by summarizing the data with one or more well-chosen statistics that convey as much information as possible regarding specific characteristics of the data. These descriptive statistics are most often selected to represent (1) the location of the center of the distribution of the data and (2) the degree of spread of the data. As such, they provide a measure of the bias and imprecision, respectively. We will return to the topic of characterizing data uncertainty when we discuss regression in Part Four.

Although you must be cognizant of blunders, model errors, and uncertain data, the numerical methods used for building models can be studied, for the most part, independently of these errors. Therefore, for most of this book, we will assume that we have not made gross errors, we have a sound model, and we are dealing with error-free measurements. Under these conditions, we can study numerical errors without complicating factors.

PROBLEMS

4.1 The "divide and average" method, an old-time method for approximating the square root of any positive number a, can be formulated as

$$x = \frac{x + a/x}{2}$$

Write a well-structured function to implement this algorithm based on the algorithm outlined in Fig. 4.2.

4.2 Convert the following base-2 numbers to base 10: **(a)** 1011001, **(b)** 0.01011, and **(c)** 110.01001.

4.3 Convert the following base-8 numbers to base 10: 61,565 and 2.71.

4.4 For computers, the machine epsilon ε can also be thought of as the smallest number that when added to one

gives a number greater than 1. An algorithm based on this idea can be developed as

Step 1: Set $\varepsilon = 1$.
Step 2: If $1 + \varepsilon$ is less than or equal to 1, then go to Step 5. Otherwise go to Step 3.
Step 3: $\varepsilon = \varepsilon/2$
Step 4: Return to Step 2
Step 5: $\varepsilon = 2 \times \varepsilon$

Write your own M-file based on this algorithm to determine the machine epsilon. Validate the result by comparing it with the value computed with the built-in function eps.

4.5 In a fashion similar to Prob. 4.4, develop your own M-file to determine the smallest positive real number used in MATLAB. Base your algorithm on the notion that your computer will be unable to reliably distinguish between zero and a quantity that is smaller than this number. Note that the result you obtain will differ from the value computed with realmin. Challenge question: Investigate the results by taking the base-2 logarithm of the number generated by your code and those obtained with realmin.

4.6 Although it is not commonly used, MATLAB allows numbers to be expressed in single precision. Each value is stored in 4 bytes with 1 bit for the sign, 23 bits for the mantissa, and 8 bits for the signed exponent. Determine the smallest and largest positive floating-point numbers as well as the machine epsilon for single precision representation. Note that the exponents range from -126 to 127.

4.7 For the hypothetical base-10 computer in Example 4.2, prove that the machine epsilon is 0.05.

4.8 The derivative of $f(x) = 1/(1 - 3x^2)$ is given by

$$\frac{6x}{(1 - 3x^2)^2}$$

Do you expect to have difficulties evaluating this function at $x = 0.577$? Try it using 3- and 4-digit arithmetic with chopping.

4.9 (a) Evaluate the polynomial

$$y = x^3 - 7x^2 + 8x - 0.35$$

at $x = 1.37$. Use 3-digit arithmetic with chopping. Evaluate the percent relative error.

(b) Repeat **(a)** but express y as

$$y = ((x - 7)x + 8)x - 0.35$$

Evaluate the error and compare with part **(a)**.

4.10 The following infinite series can be used to approximate e^x:

$$e^x = 1 + x + \frac{x^2}{2} + \frac{x^3}{3!} + \cdots + \frac{x^n}{n!}$$

(a) Prove that this Maclaurin series expansion is a special case of the Taylor series expansion (Eq. 4.13) with $x_i = 0$ and $h = x$.

(b) Use the Taylor series to estimate $f(x) = e^{-x}$ at $x_{i+1} = 1$ for $x_i = 0.25$. Employ the zero-, first-, second-, and third-order versions and compute the $|\varepsilon_t|$ for each case.

4.11 The Maclaurin series expansion for $\cos x$ is

$$\cos x = 1 - \frac{x^2}{2} - \frac{x^4}{4!} + \frac{x^6}{6!} - \frac{x^8}{8!} - \cdots$$

Starting with the simplest version, $\cos x = 1$, add terms one at a time to estimate $\cos(\pi/3)$. After each new term is

added, compute the true and approximate percent relative errors. Use your calculator or MATLAB to determine the true value. Add terms until the absolute value of the approximate error estimate falls below an error criterion conforming to two significant figures.

4.12 Perform the same computation as in Prob. 4.11, but use the Maclaurin series expansion for the sin x to estimate $\sin(\pi/3)$.

$$\sin x = x - \frac{x^3}{3!} + \frac{x^5}{5!} - \frac{x^7}{7!} + \cdots$$

4.13 Use zero- through third-order Taylor series expansions to predict $f(3)$ for

$$f(x) = 25x^3 - 6x^2 + 7x - 88$$

using a base point at $x = 1$. Compute the true percent relative error for each approximation.

4.14 Prove that Eq. (4.11) is exact for all values of x if $f(x) = ax^2 + bx + c$.

4.15 Use zero- through fourth-order Taylor series expansions to predict $f(2)$ for $f(x) = \ln x$ using a base point at $x = 1$. Compute the true percent relative error ε_t for each approximation. Discuss the meaning of the results.

4.16 Use forward and backward difference approximations of $O(h)$ and a centered difference approximation of $O(h^2)$ to estimate the first derivative of the function examined in Prob. 4.13. Evaluate the derivative at $x = 2$ using a step size of $h = 0.25$. Compare your results with the true value of the derivative. Interpret your results on the basis of the remainder term of the Taylor series expansion.

4.17 Use a centered difference approximation of $O(h^2)$ to estimate the second derivative of the function examined in Prob. 4.13. Perform the evaluation at $x = 2$ using step sizes of $h = 0.2$ and 0.1. Compare your estimates with the true value of the second derivative. Interpret your results on the basis of the remainder term of the Taylor series expansion.

4.18 If $|x| < 1$ it is known that

$$\frac{1}{1 - x} = 1 + x + x^2 + x^3 + \cdots$$

Repeat Prob. 4.11 for this series for $x = 0.1$.

4.19 To calculate a planet's space coordinates, we have to solve the function

$$f(x) = x - 1 - 0.5 \sin x$$

Let the base point be $a = x_i = \pi/2$ on the interval $[0, \pi]$. Determine the highest-order Taylor series expansion resulting in a maximum error of 0.015 on the specified interval. The error is equal to the absolute value of the difference between the given function and the specific Taylor series expansion. (Hint: Solve graphically.)

4.20 Consider the function $f(x) = x^3 - 2x + 4$ on the interval $[-2, 2]$ with $h = 0.25$. Use the forward, backward, and centered finite difference approximations for the first and second derivatives so as to graphically illustrate which approximation is most accurate. Graph all three first-derivative finite difference approximations along with the theoretical, and do the same for the second derivative as well.

4.21 Derive Eq. (4.30).

4.22 Repeat Example 4.5, but for $f(x) = \cos(x)$ at $x = \pi/6$.

4.23 Repeat Example 4.5, but for the forward divided difference (Eq. 4.21).

4.24 One common instance where subtractive cancellation occurs involves finding the roots of a parabola, $ax^2 + bx + c$, with the quadratic formula:

$$x = \frac{-b \pm \sqrt{b^2 - 4ac}}{2a}$$

For cases where $b^2 \gg 4ac$, the difference in the numerator can be very small and roundoff errors can occur. In such cases, an alternative formulation can be used to minimize subtractive cancellation:

$$x = \frac{-2c}{b \pm \sqrt{b^2 - 4ac}}$$

Use 5-digit arithmetic with chopping to determine the roots of the following equation with both versions of the quadratic formula.

$$x^2 - 5000.002x + 10$$

4.25 Develop a well-structured MATLAB function to compute the Maclaurin series expansion for the cosine function

as described in Prob. 4.11. Pattern your function after the one for the exponential function in Fig. 4.2. Test your program for $\theta = \pi/3$ (60°) and $\theta = 2\pi + \pi/3 = 7\pi/3$ (420°). Explain the difference in the number of iterations required to obtain the correct result with the desired approximate absolute error (ε_a).

4.26 Develop a well-structured MATLAB function to compute the Maclaurin series expansion for the sine function as described in Prob. 4.12. Pattern your function after the one for the exponential function in Fig. 4.2. Test your program for $\theta = \pi/3$ (60°) and $\theta = 2\pi + \pi/3 = 7\pi/3$ (420°). Explain the difference in the number of iterations required to obtain the correct result with the desired approximate absolute error (ε_a).

4.27 Recall from your calculus class that the *Maclaurin series,* named after the Scottish mathematician Colin Maclaurin (1698–1746), is a Taylor series expansion of a function about 0. Use the Taylor series to derive the first four terms of the Maclaurin series expansion for the cosine employed in Probs. 4.11 and 4.25.

4.28 The Maclaurin series expansion for the arctangent of x is defined for $|x| \le 1$ as

$$\arctan x = \sum_{n=0}^{\infty} \frac{(-1)^n}{2n+1} x^{2n+1}$$

(a) Write out the first 4 terms ($n = 0, ..., 3$).

(b) Starting with the simplest version, $\arctan x = x$, add terms one at a time to estimate $\arctan(\pi/6)$. After each new term is added, compute the true and approximate percent relative errors. Use your calculator to determine the true value. Add terms until the absolute value of the approximate error estimate falls below an error criterion conforming to two significant figures.

PART TWO

Roots and Optimization

2.1 OVERVIEW

Years ago, you learned to use the quadratic formula

$$x = \frac{-b \pm \sqrt{b^2 - 4ac}}{2a} \tag{PT2.1}$$

to solve

$$f(x) = ax^2 + bx + c = 0 \tag{PT2.2}$$

The values calculated with Eq. (PT2.1) are called the "roots" of Eq. (PT2.2). They represent the values of x that make Eq. (PT2.2) equal to zero. For this reason, roots are sometimes called the *zeros* of the equation.

Although the quadratic formula is handy for solving Eq. (PT2.2), there are many other functions for which the root cannot be determined so easily. Before the advent of digital computers, there were a number of ways to solve for the roots of such equations. For some cases, the roots could be obtained by direct methods, as with Eq. (PT2.1). Although

there were equations like this that could be solved directly, there were many more that could not. In such instances, the only alternative is an approximate solution technique.

One method to obtain an approximate solution is to plot the function and determine where it crosses the x axis. This point, which represents the x value for which $f(x) = 0$, is the root. Although graphical methods are useful for obtaining rough estimates of roots, they are limited because of their lack of precision. An alternative approach is to use *trial and error*. This "technique" consists of guessing a value of x and evaluating whether $f(x)$ is zero. If not (as is almost always the case), another guess is made, and $f(x)$ is

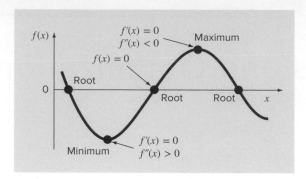

FIGURE PT2.1
A function of a single variable illustrating the difference between roots and optima.

again evaluated to determine whether the new value provides a better estimate of the root. The process is repeated until a guess results in an $f(x)$ that is close to zero.

Such haphazard methods are obviously inefficient and inadequate for the requirements of engineering and science practice. Numerical methods represent alternatives that are also approximate but employ systematic strategies to home in on the true root. As elaborated in the following pages, the combination of these systematic methods and computers makes the solution of most applied roots-of-equations problems a simple and efficient task.

Besides roots, another feature of interest to engineers and scientists are a function's minimum and maximum values. The determination of such optimal values is referred to as *optimization*. As you learned in calculus, such solutions can be obtained analytically by determining the value at which the function is flat; that is, where its derivative is zero. Although such analytical solutions are sometimes feasible, most practical optimization problems require numerical, computer solutions. From a numerical standpoint, such optimization methods are similar in spirit to the root-location methods we just discussed. That is, both involve guessing and searching for a location on a function. The fundamental difference between the two types of problems is illustrated in Fig. PT2.1. Root location involves searching for the location where the function equals zero. In contrast, optimization involves searching for the function's extreme points.

2.2 PART ORGANIZATION

The first two chapters in this part are devoted to root location. *Chapter 5* focuses on *bracketing methods* for finding roots. These methods start with guesses that bracket, or contain, the root and then systematically reduce the width of the bracket. Two specific methods are covered: *bisection* and *false position*. Graphical methods are used to provide visual insight into the techniques. Error formulations are developed to help you determine how much computational effort is required to estimate the root to a prespecified level of precision.

Chapter 6 covers *open methods*. These methods also involve systematic trial-and-error iterations but do not require that the initial guesses bracket the root. We will discover that these methods are usually more computationally efficient than bracketing methods but that they do not always work. We illustrate several open methods including the *fixed-point iteration, Newton-Raphson,* and *secant* methods.

Following the description of these individual open methods, we then discuss a hybrid approach called *Brent's root-finding method* that exhibits the reliability of the bracketing methods while exploiting the speed of the open methods. As such, it forms the basis for MATLAB's root-finding function, fzero. After illustrating how fzero can be used for engineering and scientific problems solving, Chap. 6 ends with a brief discussion of special methods devoted to finding the roots of *polynomials*. In particular, we describe MATLAB's excellent built-in capabilities for this task.

Chapter 7 deals with *optimization*. First, we describe two bracketing methods, *golden-section search* and *parabolic interpolation,* for finding the optima of a function of a single variable. Then, we discuss a robust, hybrid approach that combines golden-section search and quadratic interpolation. This approach, which again is attributed to Brent, forms the basis for MATLAB's one-dimensional root-finding function: fminbnd. After describing and illustrating fminbnd, the last part of the chapter provides a brief description of optimization of multidimensional functions. The emphasis is on describing and illustrating the use of MATLAB's capability in this area: the fminsearch function. Finally, the chapter ends with an example of how MATLAB can be employed to solve optimization problems in engineering and science.

5

Roots: Bracketing Methods

CHAPTER OBJECTIVES

The primary objective of this chapter is to acquaint you with bracketing methods for finding the root of a single nonlinear equation. Specific objectives and topics covered are

- Understanding what roots problems are and where they occur in engineering and science.
- Knowing how to determine a root graphically.
- Understanding the incremental search method and its shortcomings.
- Knowing how to solve a roots problem with the bisection method.
- Knowing how to estimate the error of bisection and why it differs from error estimates for other types of root-location algorithms.
- Understanding false position and how it differs from bisection.

YOU'VE GOT A PROBLEM

Medical studies have established that a bungee jumper's chances of sustaining a significant vertebrae injury increase significantly if the free-fall velocity exceeds 36 m/s after 4 s of free fall. Your boss at the bungee-jumping company wants you to determine the mass at which this criterion is exceeded given a drag coefficient of 0.25 kg/m.

You know from your previous studies that the following analytical solution can be used to predict fall velocity as a function of time:

$$v(t) = \sqrt{\frac{gm}{c_d}} \tanh\left(\sqrt{\frac{gc_d}{m}}\, t\right) \tag{5.1}$$

Try as you might, you cannot manipulate this equation to explicitly solve for m—that is, you cannot isolate the mass on the left side of the equation.

An alternative way of looking at the problem involves subtracting $v(t)$ from both sides to give a new function:

$$f(m) = \sqrt{\frac{gm}{c_d}} \tanh\left(\sqrt{\frac{gc_d}{m}}\, t\right) - v(t) \tag{5.2}$$

Now we can see that the answer to the problem is the value of m that makes the function equal to zero. Hence, we call this a "roots" problem. This chapter will introduce you to how the computer is used as a tool to obtain such solutions.

5.1   ROOTS IN ENGINEERING AND SCIENCE

Although they arise in other problem contexts, roots of equations frequently occur in the area of design. Table 5.1 lists a number of fundamental principles that are routinely used in design work. As introduced in Chap. 1, mathematical equations or models derived from these principles are employed to predict dependent variables as a function of independent variables, forcing functions, and parameters. Note that in each case, the dependent variables reflect the state or performance of the system, whereas the parameters represent its properties or composition.

An example of such a model is the equation for the bungee jumper's velocity. If the parameters are known, Eq. (5.1) can be used to predict the jumper's velocity. Such computations can be performed directly because v is expressed *explicitly* as a function of the model parameters. That is, it is isolated on one side of the equal sign.

However, as posed at the start of the chapter, suppose that we had to determine the mass for a jumper with a given drag coefficient to attain a prescribed velocity in a set time period. Although Eq. (5.1) provides a mathematical representation of the interrelationship among the model variables and parameters, it cannot be solved explicitly for mass. In such cases, m is said to be *implicit*.

TABLE 5.1   Fundamental principles used in design problems.

Fundamental Principle	Dependent Variable	Independent Variable	Parameters
Heat balance	Temperature	Time and position	Thermal properties of material, system geometry
Mass balance	Concentration or quantity of mass	Time and position	Chemical behavior of material, mass transfer, system geometry
Force balance	Magnitude and direction of forces	Time and position	Strength of material, structural properties, system geometry
Energy balance	Changes in kinetic and potential energy	Time and position	Thermal properties, mass of material, system geometry
Newton's laws of motion	Acceleration, velocity, or location	Time and position	Mass of material, system geometry, dissipative parameters
Kirchhoff's laws	Currents and voltages	Time	Electrical properties (resistance, capacitance, inductance)

This represents a real dilemma, because many design problems involve specifying the properties or composition of a system (as represented by its parameters) to ensure that it performs in a desired manner (as represented by its variables). Thus, these problems often require the determination of implicit parameters.

The solution to the dilemma is provided by numerical methods for roots of equations. To solve the problem using numerical methods, it is conventional to reexpress Eq. (5.1) by subtracting the dependent variable v from both sides of the equation to give Eq. (5.2). The value of m that makes $f(m) = 0$ is, therefore, the root of the equation. This value also represents the mass that solves the design problem.

The following pages deal with a variety of numerical and graphical methods for determining roots of relationships such as Eq. (5.2). These techniques can be applied to many other problems confronted routinely in engineering and science.

5.2 GRAPHICAL METHODS

A simple method for obtaining an estimate of the root of the equation $f(x) = 0$ is to make a plot of the function and observe where it crosses the x axis. This point, which represents the x value for which $f(x) = 0$, provides a rough approximation of the root.

EXAMPLE 5.1 The Graphical Approach

Problem Statement. Use the graphical approach to determine the mass of the bungee jumper with a drag coefficient of 0.25 kg/m to have a velocity of 36 m/s after 4 s of free fall. Note: The acceleration of gravity is 9.81 m/s^2.

Solution. The following MATLAB session sets up a plot of Eq. (5.2) versus mass:

```
>> cd = 0.25; g = 9.81; v = 36; t = 4;
>> mp = linspace(50,200);
>> fp = sqrt(g*mp/cd).*tanh(sqrt(g*cd./mp)*t)-v;
>> plot(mp,fp),grid
```

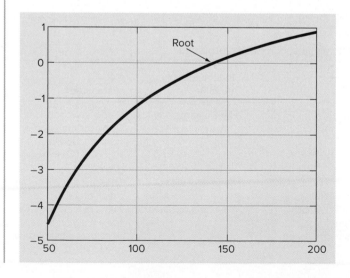

The function crosses the m axis between 140 and 150 kg. Visual inspection of the plot provides a rough estimate of the root of 145 kg (about 320 lb). The validity of the graphical estimate can be checked by substituting it into Eq. (5.2) to yield

```
>> sqrt(g*145/cd)*tanh(sqrt(g*cd/145)*t)-v
```

```
ans =
   0.0456
```

which is close to zero. It can also be checked by substituting it into Eq. (5.1) along with the parameter values from this example to give

```
>> sqrt(g*145/cd)*tanh(sqrt(g*cd/145)*t)
```

```
ans =
   36.0456
```

which is close to the desired fall velocity of 36 m/s.

Graphical techniques are of limited practical value because they are not very precise. However, graphical methods can be utilized to obtain rough estimates of roots. These estimates can be employed as starting guesses for numerical methods discussed in this chapter.

Aside from providing rough estimates of the root, graphical interpretations are useful for understanding the properties of the functions and anticipating the pitfalls of the numerical methods. For example, Fig. 5.1 shows a number of ways in which roots can occur (or be absent) in an interval prescribed by a lower bound x_l and an upper bound x_u. Figure 5.1b depicts the case where a single root is bracketed by negative and positive values of $f(x)$. However, Fig. 5.1d, where $f(x_l)$ and $f(x_u)$ are also on opposite sides of the x axis, shows three roots occurring within the interval. In general, if $f(x_l)$ and $f(x_u)$ have opposite signs, there are an odd number of roots in the interval. As indicated by Fig. 5.1a and c, if $f(x_l)$ and $f(x_u)$ have the same sign, there are either no roots or an even number of roots between the values.

Although these generalizations are usually true, there are cases where they do not hold. For example, functions that are tangential to the x axis (Fig. 5.2a) and discontinuous functions (Fig. 5.2b) can violate these principles. An example of a function that is tangential to the axis is the cubic equation $f(x) = (x - 2)(x - 2)(x - 4)$. Notice that $x = 2$ makes two terms in this polynomial equal to zero. Mathematically, $x = 2$ is called a *multiple root*. Although they are beyond the scope of this book, there are special techniques that are expressly designed to locate multiple roots (Chapra and Canale, 2010).

The existence of cases of the type depicted in Fig. 5.2 makes it difficult to develop foolproof computer algorithms guaranteed to locate all the roots in an interval. However, when used in conjunction with graphical approaches, the methods described in the following sections are extremely useful for solving many problems confronted routinely by engineers, scientists, and applied mathematicians.

5.3   BRACKETING METHODS AND INITIAL GUESSES

If you had a roots problem in the days before computing, you'd often be told to use "trial and error" to come up with the root. That is, you'd repeatedly make guesses until the function was sufficiently close to zero. The process was greatly facilitated by the advent of software

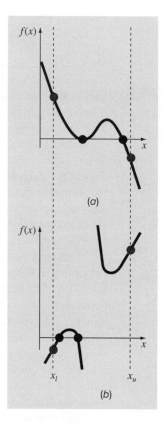

FIGURE 5.1

Illustration of a number of general ways that a root may occur in an interval prescribed by a lower bound x_l and an upper bound x_u. Parts (a) and (c) indicate that if both $f(x_l)$ and $f(x_u)$ have the same sign, either there will be no roots or there will be an even number of roots within the interval. Parts (b) and (d) indicate that if the function has different signs at the end points, there will be an odd number of roots in the interval.

FIGURE 5.2

Illustration of some exceptions to the general cases depicted in Fig. 5.1. (a) Multiple roots that occur when the function is tangential to the x axis. For this case, although the end points are of opposite signs, there are an even number of axis interceptions for the interval. (b) Discontinuous functions where end points of opposite sign bracket an even number of roots. Special strategies are required for determining the roots for these cases.

tools such as spreadsheets. By allowing you to make many guesses rapidly, such tools can actually make the trial-and-error approach attractive for some problems.

But, for many other problems, it is preferable to have methods that come up with the correct answer automatically. Interestingly, as with trial and error, these approaches require an initial "guess" to get started. Then they systematically home in on the root in an iterative fashion.

The two major classes of methods available are distinguished by the type of initial guess. They are

- *Bracketing methods*. As the name implies, these are based on two initial guesses that "bracket" the root—that is, are on either side of the root.
- *Open methods*. These methods can involve one or more initial guesses, but there is no need for them to bracket the root.

For well-posed problems, the bracketing methods always work but converge slowly (i.e., they typically take more iterations to home in on the answer). In contrast, the open methods do not always work (i.e., they can diverge), but when they do they usually converge quicker.

In both cases, initial guesses are required. These may naturally arise from the physical context you are analyzing. However, in other cases, good initial guesses may not be obvious. In such cases, automated approaches to obtain guesses would be useful. The following section describes one such approach, the incremental search.

5.3.1 Incremental Search

When applying the graphical technique in Example 5.1, you observed that $f(x)$ changed sign on opposite sides of the root. In general, if $f(x)$ is real and continuous in the interval from x_l to x_u and $f(x_l)$ and $f(x_u)$ have opposite signs, that is,

$$f(x_l)f(x_u) < 0 \tag{5.3}$$

then there is at least one real root between x_l and x_u.

Incremental search methods capitalize on this observation by locating an interval where the function changes sign. A potential problem with an incremental search is the choice of the increment length. If the length is too small, the search can be very time consuming. On the other hand, if the length is too great, there is a possibility that closely spaced roots might be missed (Fig. 5.3). The problem is compounded by the possible existence of multiple roots.

An M-file can be developed[1] that implements an incremental search to locate the roots of a function func within the range from xmin to xmax (Fig. 5.4). An optional argument ns allows the user to specify the number of intervals within the range. If ns is omitted, it is automatically set to 50. A for loop is used to step through each interval. In the event that a sign change occurs, the upper and lower bounds are stored in an array xb.

---

[1] This function is a modified version of an M-file originally presented by Recktenwald (2000).

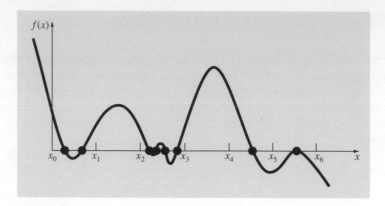

FIGURE 5.3
Cases where roots could be missed because the incremental length of the search procedure is too large. Note that the last root on the right is multiple and would be missed regardless of the increment length.

```
function xb = incsearch(func,xmin,xmax,ns)
% incsearch: incremental search root locator
%   xb = incsearch(func,xmin,xmax,ns):
%       finds brackets of x that contain sign changes
%       of a function on an interval
% input:
%   func = name of function
%   xmin, xmax = endpoints of interval
%   ns = number of subintervals (default = 50)
% output:
%   xb(k,1) is the lower bound of the kth sign change
%   xb(k,2) is the upper bound of the kth sign change
%   If no brackets found, xb = [].

if nargin < 3, error('at least 3 arguments required'), end
if nargin < 4, ns = 50; end %if ns blank set to 50

% Incremental search
x = linspace(xmin,xmax,ns);
f = func(x);
nb = 0; xb = []; %xb is null unless sign change detected
for k = 1:length(x)-1
  if sign(f(k)) ~= sign(f(k+1)) %check for sign change
    nb = nb + 1;
    xb(nb,1) = x(k);
    xb(nb,2) = x(k+1);
  end
end
if isempty(xb)     %display that no brackets were found
  disp('no brackets found')
  disp('check interval or increase ns')
else
  disp('number of brackets:') %display number of brackets
  disp(nb)
end
```

FIGURE 5.4
An M-file to implement an incremental search.

EXAMPLE 5.2   Incremental Search

Problem Statement.   Use the M-file incsearch (Fig. 5.4) to identify brackets within the interval [3, 6] for the function:

$$f(x) = \sin(10x) + \cos(3x) \tag{5.4}$$

Solution.   The MATLAB session using the default number of intervals (50) is

```
>> incsearch(@(x) sin(10*x)+cos(3*x),3,6)
number of brackets:
     5

ans =

     3.2449    3.3061
     3.3061    3.3673
     3.7347    3.7959
     4.6531    4.7143
     5.6327    5.6939
```

A plot of Eq. (5.4) along with the root locations is shown here.

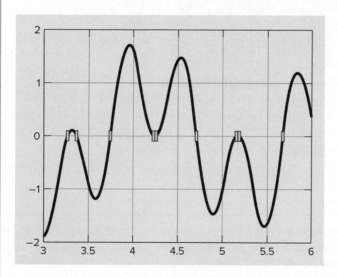

Although five sign changes are detected, because the subintervals are too wide, the function misses possible roots at $x \cong 4.25$ and 5.2. These possible roots look like they might be double roots. However, by using the zoom in tool, it is clear that each represents two real roots that are very close together. The function can be run again with more subintervals with the result that all nine sign changes are located

```
>> incsearch(@(x) sin(10*x)+cos(3*x),3,6,100)
number of brackets:
     9

ans =
     3.2424    3.2727
     3.3636    3.3939
     3.7273    3.7576
```

4.2121	4.2424
4.2424	4.2727
4.6970	4.7273
5.1515	5.1818
5.1818	5.2121
5.6667	5.6970

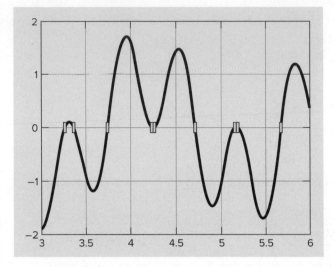

The foregoing example illustrates that brute-force methods such as incremental search are not foolproof. You would be wise to supplement such automatic techniques with any other information that provides insight into the location of the roots. Such information can be found by plotting the function and through understanding the physical problem from which the equation originated.

5.4 BISECTION

The *bisection method* is a variation of the incremental search method in which the interval is always divided in half. If a function changes sign over an interval, the function value at the midpoint is evaluated. The location of the root is then determined as lying within the subinterval where the sign change occurs. The subinterval then becomes the interval for the next iteration. The process is repeated until the root is known to the required precision. A graphical depiction of the method is provided in Fig. 5.5. The following example goes through the actual computations involved in the method.

EXAMPLE 5.3   The Bisection Method

Problem Statement. Use bisection to solve the same problem approached graphically in Example 5.1.

Solution. The first step in bisection is to guess two values of the unknown (in the present problem, m) that give values for $f(m)$ with different signs. From the graphical solution in

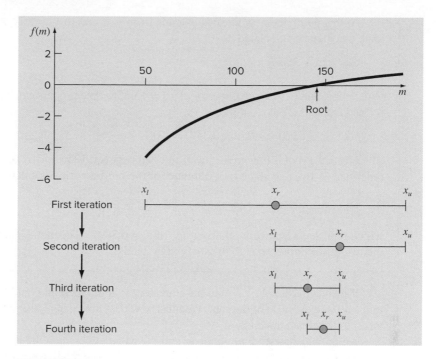

FIGURE 5.5
A graphical depiction of the bisection method. This plot corresponds to the first four iterations from Example 5.3.

Example 5.1, we can see that the function changes sign between values of 50 and 200. The plot obviously suggests better initial guesses, say 140 and 150, but for illustrative purposes let's assume we don't have the benefit of the plot and have made conservative guesses. Therefore, the initial estimate of the root x_r lies at the midpoint of the interval

$$x_r = \frac{50 + 200}{2} = 125$$

Note that the exact value of the root is 142.7376. This means that the value of 125 calculated here has a true percent relative error of

$$|\varepsilon_t| = \left| \frac{142.7376 - 125}{142.7376} \right| \times 100\% = 12.43\%$$

Next we compute the product of the function value at the lower bound and at the midpoint:

$$f(50)f(125) = -4.579(-0.409) = 1.871$$

which is greater than zero, and hence no sign change occurs between the lower bound and the midpoint. Consequently, the root must be located in the upper interval between 125 and 200. Therefore, we create a new interval by redefining the lower bound as 125.

At this point, the new interval extends from $x_l = 125$ to $x_u = 200$. A revised root estimate can then be calculated as

$$x_r = \frac{125 + 200}{2} = 162.5$$

which represents a true percent error of $|\varepsilon_t| = 13.85\%$. The process can be repeated to obtain refined estimates. For example,

$$f(125)\,f(162.5) = -0.409(0.359) = -0.147$$

Therefore, the root is now in the lower interval between 125 and 162.5. The upper bound is redefined as 162.5, and the root estimate for the third iteration is calculated as

$$x_r = \frac{125 + 162.5}{2} = 143.75$$

which represents a percent relative error of $\varepsilon_t = 0.709\%$. The method can be repeated until the result is accurate enough to satisfy your needs.

We ended Example 5.3 with the statement that the method could be continued to obtain a refined estimate of the root. We must now develop an objective criterion for deciding when to terminate the method.

An initial suggestion might be to end the calculation when the error falls below some prespecified level. For instance, in Example 5.3, the true relative error dropped from 12.43 to 0.709% during the course of the computation. We might decide that we should terminate when the error drops below, say, 0.5%. This strategy is flawed because the error estimates in the example were based on knowledge of the true root of the function. This would not be the case in an actual situation because there would be no point in using the method if we already knew the root.

Therefore, we require an error estimate that is not contingent on foreknowledge of the root. One way to do this is by estimating an approximate percent relative error as in [recall Eq. (4.5)]

$$|\varepsilon_a| = \left| \frac{x_r^{\text{new}} - x_r^{\text{old}}}{x_r^{\text{new}}} \right| 100\% \qquad (5.5)$$

where x_r^{new} is the root for the present iteration and x_r^{old} is the root from the previous iteration. When ε_a becomes less than a prespecified stopping criterion ε_s, the computation is terminated.

EXAMPLE 5.4   Error Estimates for Bisection

Problem Statement. Continue Example 5.3 until the approximate error falls below a stopping criterion of $\varepsilon_s = 0.5\%$. Use Eq. (5.5) to compute the errors.

Solution. The results of the first two iterations for Example 5.3 were 125 and 162.5. Substituting these values into Eq. (5.5) yields

$$|\varepsilon_a| = \left| \frac{162.5 - 125}{162.5} \right| 100\% = 23.08\%$$

Recall that the true percent relative error for the root estimate of 162.5 was 13.85%. Therefore, $|\varepsilon_a|$ is greater than $|\varepsilon_t|$. This behavior is manifested for the other iterations:

Iteration	x_l	x_u	x_r	$|\varepsilon_a|$ (%)	$|\varepsilon_t|$ (%)
1	50	200	125		12.43
2	125	200	162.5	23.08	13.85
3	125	162.5	143.75	13.04	0.71
4	125	143.75	134.375	6.98	5.86
5	134.375	143.75	139.0625	3.37	2.58
6	139.0625	143.75	141.4063	1.66	0.93
7	141.4063	143.75	142.5781	0.82	0.11
8	142.5781	143.75	143.1641	0.41	0.30

Thus after eight iterations $|\varepsilon_a|$ finally falls below $\varepsilon_s = 0.5\%$, and the computation can be terminated.

These results are summarized in Fig. 5.6. The "ragged" nature of the true error is due to the fact that, for bisection, the true root can lie anywhere within the bracketing interval. The true and approximate errors are far apart when the interval happens to be centered on the true root. They are close when the true root falls at either end of the interval.

FIGURE 5.6
Errors for the bisection method. True and approximate errors are plotted versus the number of iterations.

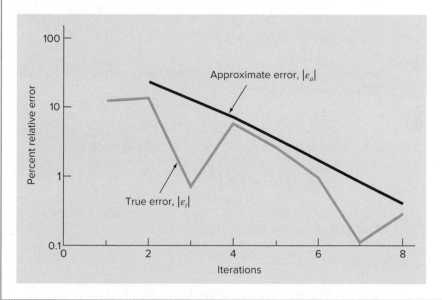

Although the approximate error does not provide an exact estimate of the true error, Fig. 5.6 suggests that $|\varepsilon_a|$ captures the general downward trend of $|\varepsilon_t|$. In addition, the plot exhibits the extremely attractive characteristic that $|\varepsilon_a|$ is always greater than $|\varepsilon_t|$. Thus, when $|\varepsilon_a|$ falls below ε_s, the computation could be terminated with confidence that the root is known to be at least as accurate as the prespecified acceptable level.

While it is dangerous to draw general conclusions from a single example, it can be demonstrated that $|\varepsilon_a|$ will always be greater than $|\varepsilon_t|$ for bisection. This is due to the fact that each time an approximate root is located using bisection as $x_r = (x_l + x_u)/2$, we know that the true root lies somewhere within an interval of $\Delta x = x_u - x_l$. Therefore, the root must lie within $\pm \Delta x/2$ of our estimate. For instance, when Example 5.4 was terminated, we could make the definitive statement that

$$x_r = 143.1641 \pm \frac{143.7500 - 142.5781}{2} = 143.1641 \pm 0.5859$$

In essence, Eq. (5.5) provides an upper bound on the true error. For this bound to be exceeded, the true root would have to fall outside the bracketing interval, which by definition could never occur for bisection. Other root-locating techniques do not always behave as nicely. Although bisection is generally slower than other methods, the neatness of its error analysis is a positive feature that makes it attractive for certain engineering and scientific applications.

Another benefit of the bisection method is that the number of iterations required to attain an absolute error can be computed *a priori*—that is, before starting the computation. This can be seen by recognizing that before starting the technique, the absolute error is

$$E_a^0 = x_u^0 - x_l^0 = \Delta x^0$$

where the superscript designates the iteration. Hence, before starting the method we are at the "zero iteration." After the first iteration, the error becomes

$$E_a^1 = \frac{\Delta x^0}{2}$$

Because each succeeding iteration halves the error, a general formula relating the error and the number of iterations n is

$$E_a^n = \frac{\Delta x^0}{2^n}$$

If $E_{a,d}$ is the desired error, this equation can be solved for[2]

$$n = \frac{\log(\Delta x^0 / E_{a,d})}{\log 2} = \log_2 \left(\frac{\Delta x^0}{E_{a,d}} \right) \tag{5.6}$$

Let's test the formula. For Example 5.4, the initial interval was $\Delta x_0 = 200 - 50 = 150$. After eight iterations, the absolute error was

$$E_a = \frac{|143.7500 - 142.5781|}{2} = 0.5859$$

We can substitute these values into Eq. (5.6) to give

$$n = \log_2(150/0.5859) = 8$$

---

[2] MATLAB provides the log2 function to evaluate the base-2 logarithm directly. If the pocket calculator or computer language you are using does not include the base-2 logarithm as an intrinsic function, this equation shows a handy way to compute it. In general, $\log_b(x) = \log(x)/\log(b)$.

Thus, if we knew beforehand that an error of less than 0.5859 was acceptable, the formula tells us that eight iterations would yield the desired result.

Although we have emphasized the use of relative errors for obvious reasons, there will be cases where (usually through knowledge of the problem context) you will be able to specify an absolute error. For these cases, bisection along with Eq. (5.6) can provide a useful root-location algorithm.

5.4.1 MATLAB M-file: bisect

An M-file to implement bisection is displayed in Fig. 5.7. It is passed the function (func) along with lower (xl) and upper (xu) guesses. In addition, an optional stopping criterion (es)

FIGURE 5.7
An M-file to implement the bisection method.

```
function [root,fx,ea,iter]=bisect(func,xl,xu,es,maxit,varargin)
% bisect: root location zeroes
%   [root,fx,ea,iter]=bisect(func,xl,xu,es,maxit,p1,p2,...):
%       uses bisection method to find the root of func
% input:
%   func = name of function
%   xl, xu = lower and upper guesses
%   es = desired relative error (default = 0.0001%)
%   maxit = maximum allowable iterations (default = 50)
%   p1,p2,... = additional parameters used by func
% output:
%   root = real root
%   fx = function value at root
%   ea = approximate relative error (%)
%   iter = number of iterations

if nargin<3,error('at least 3 input arguments required'),end
test = func(xl,varargin{:})*func(xu,varargin{:});
if test>0,error('no sign change'),end
if nargin<4|isempty(es), es=0.0001;end
if nargin<5|isempty(maxit), maxit=50;end
iter = 0; xr = xl; ea = 100;
while (1)
  xrold = xr;
  xr = (xl + xu)/2;
  iter = iter + 1;
  if xr ~= 0,ea = abs((xr - xrold)/xr) * 100;end
  test = func(xl,varargin{:})*func(xr,varargin{:});
  if test < 0
    xu = xr;
  elseif test > 0
    xl = xr;
  else
    ea = 0;
  end
  if ea <= es | iter >= maxit,break,end
end
root = xr; fx = func(xr, varargin{:});
```

and maximum iterations (maxit) can be entered. The function first checks whether there are sufficient arguments and if the initial guesses bracket a sign change. If not, an error message is displayed and the function is terminated. It also assigns default values if maxit and es are not supplied. Then a while...break loop is employed to implement the bisection algorithm until the approximate error falls below es or the iterations exceed maxit.

We can employ this function to solve the problem posed at the beginning of the chapter. Recall that you need to determine the mass at which a bungee jumper's free-fall velocity exceeds 36 m/s after 4 s of free fall given a drag coefficient of 0.25 kg/m. Thus, you have to find the root of

$$f(m) = \sqrt{\frac{9.81m}{0.25}} \tanh\left(\sqrt{\frac{9.81(0.25)}{m}}\, 4\right) - 36$$

In Example 5.1 we generated a plot of this function versus mass and estimated that the root fell between 140 and 150 kg. The bisect function from Fig. 5.7 can be used to determine the root with the following script

```
fm=@(m,cd,t,v) sqrt(9.81*m/cd)*tanh(sqrt(9.81*cd/m)*t)-v;
[mass fx ea iter]=bisect(@(m) fm(m,0.25,4,36),40,200)

mass =
        142.7377
fx =
    4.6089e-007
ea =
     5.345e-005
iter =
     21
```

Thus, a result of $m = 142.74$ kg is obtained after 21 iterations with an approximate relative error of $\varepsilon_a = 0.00005345\%$, and a function value close to zero.

5.5 FALSE POSITION

False position (also called the linear interpolation method) is another well-known bracketing method. It is very similar to bisection with the exception that it uses a different strategy to come up with its new root estimate. Rather than bisecting the interval, it locates the root by joining $f(x_l)$ and $f(x_u)$ with a straight line (Fig. 5.8). The intersection of this line with the x axis represents an improved estimate of the root. Thus, the shape of the function influences the new root estimate. Using similar triangles, the intersection of the straight line with the x axis can be estimated as (see Chapra and Canale, 2010, for details),

$$x_r = x_u - \frac{f(x_u)(x_l - x_u)}{f(x_l) - f(x_u)} \tag{5.7}$$

This is the *false-position formula*. The value of x_r computed with Eq. (5.7) then replaces whichever of the two initial guesses, x_l or x_u, yields a function value with the same sign as $f(x_r)$. In this way the values of x_l and x_u always bracket the true root. The process

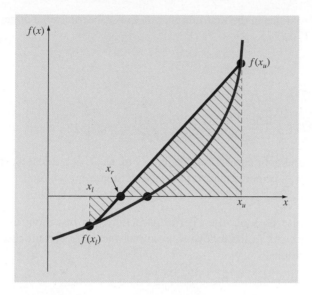

FIGURE 5.8
False position.

is repeated until the root is estimated adequately. The algorithm is identical to the one for bisection (Fig. 5.7) with the exception that Eq. (5.7) is used.

EXAMPLE 5.5 The False-Position Method

Problem Statement. Use false position to solve the same problem approached graphically and with bisection in Examples 5.1 and 5.3.

Solution. As in Example 5.3, initiate the computation with guesses of $x_l = 50$ and $x_u = 200$.

First iteration:

$$x_l = 50 \qquad f(x_l) = -4.579387$$

$$x_u = 200 \qquad f(x_u) = 0.860291$$

$$x_r = 200 - \frac{0.860291(50 - 200)}{-4.579387 - 0.860291} = 176.2773$$

which has a true relative error of 23.5%.

Second iteration:

$$f(x_l)\,f(x_r) = -2.592732$$

Therefore, the root lies in the first subinterval, and x_r becomes the upper limit for the next iteration, $x_u = 176.2773$.

$$x_l = 50 \qquad\qquad f(x_l) = -4.579387$$

$$x_u = 176.2773 \qquad f(x_u) = 0.566174$$

$$x_r = 176.2773 - \frac{0.566174(50 - 176.2773)}{-4.579387 - 0.566174} = 162.3828$$

which has true and approximate relative errors of 13.76% and 8.56%, respectively. Additional iterations can be performed to refine the estimates of the root.

Although false position often performs better than bisection, there are other cases where it does not. As in the following example, there are certain cases where bisection yields superior results.

EXAMPLE 5.6   A Case Where Bisection Is Preferable to False Position

Problem Statement. Use bisection and false position to locate the root of

$$f(x) = x^{10} - 1$$

between $x = 0$ and 1.3.

Solution. Using bisection, the results can be summarized as

Iteration	x_l	x_u	x_r	ε_a (%)	ε_t (%)
1	0	1.3	0.65	100.0	35
2	0.65	1.3	0.975	33.3	2.5
3	0.975	1.3	1.1375	14.3	13.8
4	0.975	1.1375	1.05625	7.7	5.6
5	0.975	1.05625	1.015625	4.0	1.6

Thus, after five iterations, the true error is reduced to less than 2%. For false position, a very different outcome is obtained:

Iteration	x_l	x_u	x_r	ε_a (%)	ε_t (%)
1	0	1.3	0.09430		90.6
2	0.09430	1.3	0.18176	48.1	81.8
3	0.18176	1.3	0.26287	30.9	73.7
4	0.26287	1.3	0.33811	22.3	66.2
5	0.33811	1.3	0.40788	17.1	59.2

After five iterations, the true error has only been reduced to about 59%. Insight into these results can be gained by examining a plot of the function. As in Fig. 5.9, the curve violates the premise on which false position was based—that is, if $f(x_l)$ is much closer to

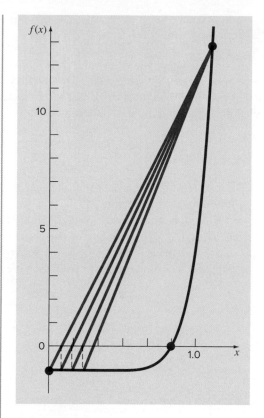

FIGURE 5.9
Plot of $f(x) = x^{10} - 1$, illustrating slow convergence of the false-position method.

zero than $f(x_u)$, then the root should be much closer to x_l than to x_u (recall Fig. 5.8). Because of the shape of the present function, the opposite is true.

The foregoing example illustrates that blanket generalizations regarding root-location methods are usually not possible. Although a method such as false position is often superior to bisection, there are invariably cases that violate this general conclusion. Therefore, in addition to using Eq. (5.5), the results should always be checked by substituting the root estimate into the original equation and determining whether the result is close to zero.

The example also illustrates a major weakness of the false-position method: its one-sidedness. That is, as iterations are proceeding, one of the bracketing points will tend to stay fixed. This can lead to poor convergence, particularly for functions with significant curvature. Possible remedies for this shortcoming are available elsewhere (Chapra and Canale, 2010).

5.6 CASE STUDY   GREENHOUSE GASES AND RAINWATER

Background. It is well documented that the atmospheric levels of several so-called "greenhouse" gases have been increasing over the past 50 years. For example, Fig. 5.10 shows data for the partial pressure of carbon dioxide (CO_2) collected at Mauna Loa, Hawaii from 1958 through 2008. The trend in these data can be nicely fit with a quadratic polynomial,[3]

$$p_{CO_2} = 0.012226(t - 1983)^2 + 1.418542(t - 1983) + 342.38309$$

where p_{CO_2} = CO_2 partial pressure (ppm). These data indicate that levels have increased a little over 22% over the period from 315 to 386 ppm.

One question that we can address is how this trend is affecting the pH of rainwater. Outside of urban and industrial areas, it is well documented that carbon dioxide is the primary determinant of the pH of the rain. pH is the measure of the activity of hydrogen ions and, therefore, its acidity or alkalinity. For dilute aqueous solutions, it can be computed as

$$pH = -\log_{10}[H^+] \tag{5.8}$$

where $[H^+]$ is the molar concentration of hydrogen ions.

The following five equations govern the chemistry of rainwater:

$$K_1 = 10^6 \frac{[H^+][HCO_3^-]}{K_H p_{CO_2}} \tag{5.9}$$

FIGURE 5.10
Average annual partial pressures of atmospheric carbon dioxide (ppm) measured at Mauna Loa, Hawaii.

---

[3] In Part Four, we will learn how to determine such polynomials.

continued

$$K_2 = \frac{[H^+][CO_3^{-2}]}{[HCO_3^-]} \tag{5.10}$$

$$K_w = [H^+][OH^-] \tag{5.11}$$

$$c_T = \frac{K_H p_{CO_2}}{10^6} + [HCO_3^-] + [CO_3^{-2}] \tag{5.12}$$

$$0 = [HCO_3^-] + 2[CO_3^{-2}] + [OH^-] - [H^+] \tag{5.13}$$

where K_H = Henry's constant, and K_1, K_2, and K_w are equilibrium coefficients. The five unknowns are c_T = total inorganic carbon, $[HCO_3^-]$ = bicarbonate, $[CO_3^{-2}]$ = carbonate, $[H^+]$ = hydrogen ion, and $[OH^-]$ = hydroxyl ion. Notice how the partial pressure of CO_2 shows up in Eqs. (5.9) and (5.12).

Use these equations to compute the pH of rainwater given that $K_H = 10^{-1.46}$, $K_1 = 10^{-6.3}$, $K_2 = 10^{-10.3}$, and $K_w = 10^{-14}$. Compare the results in 1958 when the p_{CO_2} was 315 and in 2008 when it was 386 ppm. When selecting a numerical method for your computation, consider the following:

- You know with certainty that the pH of rain in pristine areas always falls between 2 and 12.
- You also know that pH can only be measured to two places of decimal precision.

Solution. There are a variety of ways to solve this system of five equations. One way is to eliminate unknowns by combining them to produce a single function that only depends on $[H^+]$. To do this, first solve Eqs. (5.9) and (5.10) for

$$[HCO_3^-] = \frac{K_1}{10^6[H^+]} K_H p_{CO_2} \tag{5.14}$$

$$[CO_3^{-2}] = \frac{K_2[HCO_3^-]}{[H^+]} \tag{5.15}$$

Substitute Eq. (5.14) into (5.15)

$$[CO_3^{-2}] = \frac{K_2 K_1}{10^6[H^+]^2} K_H p_{CO_2} \tag{5.16}$$

Equations (5.14) and (5.16) can be substituted along with Eq. (5.11) into Eq. (5.13) to give

$$0 = \frac{K_1}{10^6[H^+]} K_H p_{CO_2} + 2\frac{K_2 K_1}{10^6[H^+]^2} K_H p_{CO_2} + \frac{K_w}{[H^+]} - [H^+] \tag{5.17}$$

Although it might not be immediately apparent, this result is a third-order polynomial in $[H^+]$. Thus, its root can be used to compute the pH of the rainwater.

Now we must decide which numerical method to employ to obtain the solution. There are two reasons why bisection would be a good choice. First, the fact that the pH always falls within the range from 2 to 12, provides us with two good initial guesses. Second, because the pH can only be measured to two decimal places of precision, we will be satisfied with an absolute error of $E_{a,d} = \pm0.005$. Remember that given an initial bracket and the desired error, we can compute the number of iteration *a priori*. Substituting the present values into Eq. (5.6) gives

```
>> dx=12-2;
>> Ead=0.005;
>> n=log2(dx/Ead)

n =
    10.9658
```

Eleven iterations of bisection will produce the desired precision.

Before implementing bisection, we must first express Eq. (5.17) as a function. Because it is relatively complicated, we will store it as an M-file:

```
function f = fpH(pH,pCO2)
K1=10^-6.3;K2=10^-10.3;Kw=10^-14;
KH=10^-1.46;
H=10^-pH;
f=K1/(1e6*H)*KH*pCO2+2*K2*K1/(1e6*H)*KH*pCO2+Kw/H-H;
```

We can then use the M-file from Fig. 5.7 to obtain the solution. Notice how we have set the value of the desired relative error ($\varepsilon_a = 1 \times 10^{-8}$) at a very low level so that the iteration limit (maxit) is reached first so that exactly 11 iterations are implemented

```
>> [pH1958 fx ea iter]=bisect(@(pH) fpH(pH,315),2,12,1e-8,11)

pH1958 =
    5.6279
fx =
 -2.7163e-008
ea =
    0.0868
iter =
    11
```

Thus, the pH is computed as 5.6279 with a relative error of 0.0868%. We can be confident that the rounded result of 5.63 is correct to two decimal places. This can be verified by performing another run with more iterations. For example, setting maxit to 50 yields

```
>> [pH1958 fx ea iter]=bisect(@(pH) fpH(pH,315),2,12,1e-8,50)

pH1958 =
    5.6304
fx =
    1.615e-015
ea =
    5.1690e-009
iter =
    35
```

For 2008, the result is

```
>> [pH2008 fx ea iter]=bisect(@(pH) fpH(pH,386),2,12,1e-8,50)
```

5.6 CASE STUDY continued

```
pH2008 =
    5.5864
fx =
    3.2926e-015
ea =
    5.2098e-009
iter =
    35
```

Interestingly, the results indicate that the 22.5% rise in atmospheric CO_2 levels has produced only a 0.78% drop in pH. Although this is certainly true, remember that the pH represents a logarithmic scale as defined by Eq. (5.8). Consequently, a unit drop in pH represents an order-of-magnitude (i.e., a 10-fold) increase in the hydrogen ion. The concentration can be computed as $[H^+] = 10^{-pH}$ and its percent change can be calculated as

```
>> ((10^-pH2008-10^-pH1958)/10^-pH1958)*100

ans =
    10.6791
```

Therefore, the hydrogen ion concentration has increased about 10.7%.

There is quite a lot of controversy related to the meaning of the greenhouse gas trends. Most of this debate focuses on whether the increases are contributing to global warming. However, regardless of the ultimate implications, it is sobering to realize that something as large as our atmosphere has changed so much over a relatively short time period. This case study illustrates how numerical methods and MATLAB can be employed to analyze and interpret such trends. Over the coming years, engineers and scientists can hopefully use such tools to gain increased understanding of such phenomena and help rationalize the debate over their ramifications.

PROBLEMS

5.1 Use bisection to determine the drag coefficient needed so that an 95-kg bungee jumper has a velocity of 46 m/s after 9 s of free fall. Note: The acceleration of gravity is 9.81 m/s². Start with initial guesses of $x_l = 0.2$ and $x_u = 0.5$ and iterate until the approximate relative error falls below 5%.

5.2 Develop your own M-file for bisection in a similar fashion to Fig. 5.7. However, rather than using the maximum iterations and Eq. (5.5), employ Eq. (5.6) as your stopping criterion. Make sure to round the result of Eq. (5.6) up to the next highest integer (Hint: the ceil function provides a handy way to do this). The first line of your function should be

```
function [root,Ea,ea,n] = bisectnew(func,xl,xu,Ead,
varargin)
```

Note that for the output, Ea = the approximate absolute error and ea = the approximate percent relative error. Then develop your own script, called **LastNameHmwk04Script** to solve Prob. 5.1. Note that you **MUST** pass the parameters via the argument. In addition, set up the function so that it uses a default value for Ead = 0.000001.

5.3 Figure P5.3 shows a pinned-fixed beam subject to a uniform load. The equation for the resulting deflections is

$$y = -\frac{w}{48EI}(2x^4 - 3Lx^3 + L^3x)$$

Develop a MATLAB script that

(a) plots the function, dy/dx versus x (with appropriate labels) and

FIGURE P5.3

(b) uses `LastNameBisect` to determine the point of maximum deflection (i.e., the value of x where $dy/dx = 0$). Then substitute this value into the deflection equation to determine the value of the maximum deflection. Employ initial guesses of $x_l = 0$ and $x_u = 0.9L$. Use the following parameter values in your computation (making sure that you use consistent units): $L = 400$ cm, $E = 52,000$ kN/cm^2, $I = 32,000$ cm^4, and $w = 4$ kN/cm. In addition, use Ead = 0.0000001 m. Also, set `format long` in your script so you display 15 significant digits for your results.

5.4 As shown in Fig. P5.4, the velocity of water, v (m/s), discharged from a cylindrical tank through a long pipe can be computed as

$$v = \sqrt{2gH} \, \tanh\left(\sqrt{\frac{2gH}{2L}} \, t\right)$$

where $g = 9.81$ m/s^2, H = initial head (m), L = pipe length (m), and t = elapsed time (s). Develop a MATLAB script that
(a) plots the function $f(H)$ versus H for $H = 0$ to 4 m (make sure to label the plot) and
(b) uses `LastNameBisect` with initial guesses of $x_l = 0$ and $x_u = 4$ m to determine the initial head needed to achieve $v = 5$ m/s in 2.5 s for a 4-m long pipe. In addition, use Ead = 0.0000001. Also, set `format long` in your script so you display 15 significant digits for your results.

5.5 Repeat Prob. 5.1, but use the false-position method to obtain your solution.

5.6 Develop an M-file for the false-position method. Test it by solving Prob. 5.1.

5.7 (a) Determine the roots of $f(x) = -12 - 21x + 18x^2 - 2.75x^3$ graphically. In addition, determine the first root of the function with
(b) bisection and (c) false position. For (b) and
(c) use initial guesses of $x_l = -1$ and $x_u = 0$ and a stopping criterion of 1%.

5.8 Locate the first nontrivial root of $\sin(x) = x^2$ where x is in radians. Use a graphical technique and bisection with the initial interval from 0.5 to 1. Perform the computation until ε_a is less than $\varepsilon_s = 2\%$.

5.9 Determine the positive real root of $\ln(x^2) = 0.7$ (a) graphically, (b) using three iterations of the bisection method, with initial guesses of $x_l = 0.5$ and $x_u = 2$, and (c) using three iterations of the false-position method, with the same initial guesses as in (b).

5.10 The saturation concentration of dissolved oxygen in freshwater can be calculated with the equation

$$\ln o_{sf} = -139.34411 + \frac{1.575701 \times 10^5}{T_a}$$
$$- \frac{6.642308 \times 10^7}{T_a^2} + \frac{1.243800 \times 10^{10}}{T_a^3}$$
$$- \frac{8.621949 \times 10^{11}}{T_a^4}$$

where o_{sf} = the saturation concentration of dissolved oxygen in freshwater at 1 atm (mg L^{-1}); and T_a = absolute temperature (K). Remember that $T_a = T + 273.15$, where T = temperature (°C). According to this equation, saturation decreases with increasing temperature. For typical natural waters in temperate climates, the equation can be used to determine that oxygen concentration ranges from 14.621 mg/L at 0 °C to 6.949 mg/L at 35 °C. Given a value of oxygen concentration, this formula and the bisection method can be used to solve for temperature in °C.
(a) If the initial guesses are set as 0 and 35 °C, how many bisection iterations would be required to determine temperature to an absolute error of 0.05 °C?
(b) Based on (a), develop and test a bisection M-file function to determine T as a function of a given oxygen concentration. Test your function for $o_{sf} = 8$, 10, and 14 mg/L. Check your results.

5.11 A beam is loaded as shown in Fig. P5.11. Use the bisection method to solve for the position inside the beam where there is no moment.

5.12 Water is flowing in a trapezoidal channel at a rate of $Q = 20$ m^3/s. The critical depth y for such a channel must satisfy the equation

$$0 = 1 - \frac{Q^2}{gA_c^3} B$$

FIGURE P5.4

100 lb/ft   100 lb

├── 3' ──┤── 3' ──┤── 4' ──┤─ 2' ─┤

FIGURE P5.11

where $g = 9.81$ m/s^2, A_c = the cross-sectional area (m^2), and B = the width of the channel at the surface (m). For this case, the width and the cross-sectional area can be related to depth y by

$$B = 3 + y$$

and

$$A_c = 3y + \frac{y^2}{2}$$

Solve for the critical depth using (a) the graphical method, (b) bisection, and (c) false position. For (b) and (c) use initial guesses of $x_l = 0.5$ and $x_u = 2.5$, and iterate until the approximate error falls below 1% or the number of iterations exceeds 10. Discuss your results.

5.13 The Michaelis-Menten model describes the kinetics of enzyme mediated reactions:

$$\frac{dS}{dt} = -v_m \frac{S}{k_s + S}$$

where S = substrate concentration (moles/L), v_m = maximum uptake rate (moles/L/d), and k_s = the half-saturation constant, which is the substrate level at which uptake is half of the maximum [moles/L]. If the initial substrate level at $t = 0$ is S_0, this differential equation can be solved for

$$S = S_0 - v_m t + k_s \ln(S_0/S)$$

Develop an M-file to generate a plot of S versus t for the case where $S_0 = 8$ moles/L, $v_m = 0.7$ moles/L/d, and $k_s = 2.5$ moles/L.

5.14 A reversible chemical reaction

$$2A + B \underset{\leftarrow}{\rightarrow} C$$

can be characterized by the equilibrium relationship

$$K = \frac{c_c}{c_a^2 c_b}$$

where the nomenclature c_i represents the concentration of constituent i. Suppose that we define a variable x as representing the number of moles of C that are produced. Conservation of mass can be used to reformulate the equilibrium relationship as

$$K = \frac{(c_{c,0} + x)}{(c_{a,0} - 2x)^2 (c_{b,0} - x)}$$

where the subscript 0 designates the initial concentration of each constituent. If $K = 0.016$, $c_{a,0} = 42$, $c_{b,0} = 28$, and $c_{c,0} = 4$, determine the value of x.

(a) Obtain the solution graphically.

(b) On the basis of (a), solve for the root with initial guesses of $x_l = 0$ and $x_u = 20$ to $\varepsilon_s = 0.5\%$. Choose either bisection or false position to obtain your solution. Justify your choice.

5.15 Figure P5.15a shows a uniform beam subject to a linearly increasing distributed load. The equation for the resulting elastic curve is (see Fig. P5.15b)

$$y = \frac{w_0}{120 E I L} (-x^5 + 2L^2 x^3 - L^4 x) \tag{P5.15}$$

Use bisection to determine the point of maximum deflection (i.e., the value of x where $dy/dx = 0$). Then substitute this value into Eq. (P5.15) to determine the value of the maximum deflection. Use the following parameter values in your computation: $L = 600$ cm, $E = 50{,}000$ kN/cm^2, $I = 30{,}000$ cm^4, and $w_0 = 2.5$ kN/cm.

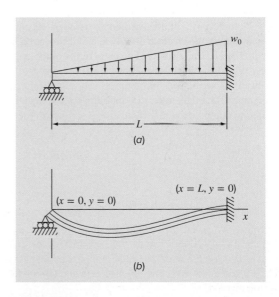

w_0

├─────── L ───────┤

(a)

$(x = 0, y = 0)$ $(x = L, y = 0)$

x

(b)

FIGURE P5.15

5.16 You buy a $35,000 vehicle for nothing down at $8500 per year for 7 years. Use the bisect function from Fig. 5.7 to determine the interest rate that you are paying. Employ initial guesses for the interest rate of 0.01 and 0.3 and a stopping criterion of 0.00005. The formula relating present worth P, annual payments A, number of years n, and interest rate i is

$$A = P \frac{i(1 + i)^n}{(1 + i)^n - 1}$$

5.17 Many fields of engineering require accurate population estimates. For example, transportation engineers might find it necessary to determine separately the population growth trends of a city and adjacent suburb. The population of the urban area is declining with time according to

$$P_u(t) = P_{u,max} e^{-k_u t} + P_{u,min}$$

while the suburban population is growing, as in

$$P_s(t) = \frac{P_{s,max}}{1 + [P_{s,max}/P_0 - 1]e^{-k_s t}}$$

where $P_{u,max}$, k_u, $P_{s,max}$, P_0, and k_s = empirically derived parameters. Determine the time and corresponding values of $P_u(t)$ and $P_s(t)$ when the suburbs are 20% larger than the city. The parameter values are $P_{u,max} = 80,000$, $k_u = 0.05/yr$, $P_{u,min} = 110,000$ people, $P_{s,max} = 320,000$ people, $P_0 = 10,000$ people, and $k_s = 0.09/yr$. To obtain your solutions, use (a) graphical and (b) false-position methods.

5.18 The resistivity ρ of doped silicon is based on the charge q on an electron, the electron density n, and the electron mobility μ. The electron density is given in terms of the doping density N and the intrinsic carrier density n_i. The electron mobility is described by the temperature T, the reference temperature T_0, and the reference mobility μ_0. The equations required to compute the resistivity are

$$\rho = \frac{1}{qn\mu}$$

where

$$n = \frac{1}{2}\left(N + \sqrt{N^2 + 4n_i^2}\right) \quad \text{and} \quad \mu = \mu_0 \left(\frac{T}{T_0}\right)^{-2.42}$$

Determine N, given $T_0 = 300$ K, $T = 1000$ K, $\mu_0 = 1360$ cm^2 (V s)$^{-1}$, $q = 1.7 \times 10^{-19}$ C, $n_i = 6.21 \times 10^9$ cm^{-3}, and a desired $\rho = 6.5 \times 10^6$ V s cm/C. Employ initial guesses of $N = 0$ and 2.5×10^{10}. Use (a) bisection and (b) the false position method.

FIGURE P5.19

5.19 A total charge Q is uniformly distributed around a ring-shaped conductor with radius a. A charge q is located at a distance x from the center of the ring (Fig. P5.19). The force exerted on the charge by the ring is given by

$$F = \frac{1}{4\pi e_0} \frac{qQx}{(x^2 + a^2)^{3/2}}$$

where $e_0 = 8.9 \times 10^{-12}$ C^2/(N m^2). Find the distance x where the force is 1.25 N if q and Q are 2×10^{-5} C for a ring with a radius of 0.85 m.

5.20 For fluid flow in pipes, friction is described by a dimensionless number, the *Fanning friction factor f*. The Fanning friction factor is dependent on a number of parameters related to the size of the pipe and the fluid, which can all be represented by another dimensionless quantity, the *Reynolds number* Re. A formula that predicts f given Re is the *von Karman equation*:

$$\frac{1}{\sqrt{f}} = 4 \log_{10}\left(\text{Re} \sqrt{f}\right) - 0.4$$

Typical values for the Reynolds number for turbulent flow are 10,000 to 500,000 and for the Fanning friction factor are 0.001 to 0.01. Develop a function that uses bisection to solve for f given a user-supplied value of Re between 2500 and 1,000,000. Design the function so that it ensures that the absolute error in the result is $E_{a,d} < 0.000005$.

5.21 Mechanical engineers, as well as most other engineers, use thermodynamics extensively in their work. The following polynomial can be used to relate the zero-pressure specific heat of dry air c_p kJ/(kg K) to temperature (K):

$$c_p = 0.99403 + 1.671 \times 10^{-4}T + 9.7215 \times 10^{-8}T^2 \\ -9.5838 \times 10^{-11}T^3 + 1.9520 \times 10^{-14}T^4$$

Develop a plot of c_p versus a range of $T = 0$ to 1200 K, and then use bisection to determine the temperature that corresponds to a specific heat of 1.1 kJ/(kg K).

5.22 The upward velocity of a rocket can be computed by the following formula:

$$v = u \ln \frac{m_0}{m_0 - qt} - gt$$

where v = upward velocity, u = the velocity at which fuel is expelled relative to the rocket, m_0 = the initial mass of the rocket at time $t = 0$, q = the fuel consumption rate, and g = the downward acceleration of gravity (assumed constant = 9.81 m/s^2). If u = 1800 m/s, m_0 = 160,000 kg, and q = 2600 kg/s, compute the time at which v = 750 m/s. (Hint: t is somewhere between 10 and 50 s.) Determine your result so that it is within 1% of the true value. Check your answer.

5.23 Although we did not mention it in Sec. 5.6, Eq. (5.13) is an expression of *electroneutrality*—that is, that positive and negative charges must balance. This can be seen more clearly by expressing it as

$$[H^+] = [HCO_3^-] + 2[CO_3^{2-}] + [OH^-]$$

In other words, the positive charges must equal the negative charges. Thus, when you compute the pH of a natural water body such as a lake, you must also account for other ions that may be present. For the case where these ions originate from nonreactive salts, the net negative minus positive charges due to these ions are lumped together in a quantity called *alkalinity,* and the equation is reformulated as

$$Alk + [H^+] = [HCO_3^-] + 2[CO_3^{2-}] + [OH^-] \qquad (P5.23)$$

where Alk = alkalinity (eq/L). For example, the alkalinity of Lake Superior is approximately 0.4×10^{-3} eq/L. Perform the same calculations as in Sec. 5.6 to compute the pH of Lake Superior in 2008. Assume that just like the raindrops, the lake is in equilibrium with atmospheric CO_2 but account for the alkalinity as in Eq. (P5.23).

5.24 According to *Archimedes' principle,* the *buoyancy* force is equal to the weight of fluid displaced by the submerged portion of the object. For the sphere depicted in Fig. P5.24, use bisection to determine the height, h, of the portion that is above water. Employ the following values for your computation: r = 1 m, ρ_s = density of sphere = 200 kg/m^3, and ρ_w = density of water = 1000 kg/m^3. Note that the volume of the above-water portion of the sphere can be computed with

$$V = \frac{\pi h^2}{3} (3r - h)$$

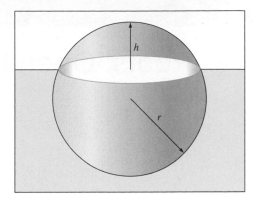

FIGURE P5.24

5.25 Perform the same computation as in Prob. 5.24, but for the *frustum* of a cone as depicted in Fig. P5.25. Employ the following values for your computation: r_1 = 0.5 m, r_2 = 1 m, h = 1 m, ρ_f = frustum density = 200 kg/m^3, and ρ_w = water density = 1000 kg/m^3. Note that the volume of a frustum is given by

$$V = \frac{\pi h}{3} (r_1^2 + r_2^2 + r_1 r_2)$$

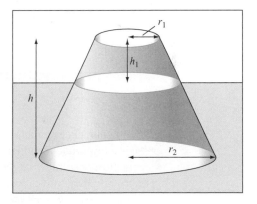

FIGURE P5.25

6

Roots: Open Methods

CHAPTER OBJECTIVES

The primary objective of this chapter is to acquaint you with open methods for finding the root of a single nonlinear equation. Specific objectives and topics covered are

- Recognizing the difference between bracketing and open methods for root location.
- Understanding the fixed-point iteration method and how you can evaluate its convergence characteristics.
- Knowing how to solve a roots problem with the Newton-Raphson method and appreciating the concept of quadratic convergence.
- Knowing how to implement both the secant and the modified secant methods.
- Understanding how Brent's method combines reliable bracketing methods with fast open methods to locate roots in a robust and efficient manner.
- Knowing how to use MATLAB's fzero function to estimate roots.
- Learning how to manipulate and determine the roots of polynomials with MATLAB.

For the bracketing methods in Chap. 5, the root is located within an interval prescribed by a lower and an upper bound. Repeated application of these methods always results in closer estimates of the true value of the root. Such methods are said to be *convergent* because they move closer to the truth as the computation progresses (Fig. 6.1*a*).

In contrast, the *open methods* described in this chapter require only a single starting value or two starting values that do not necessarily bracket the root. As such, they sometimes *diverge* or move away from the true root as the computation progresses (Fig. 6.1*b*). However, when the open methods converge (Fig. 6.1*c*) they usually do so much more quickly than the bracketing methods. We will begin our discussion of open techniques with a simple approach that is useful for illustrating their general form and also for demonstrating the concept of convergence.

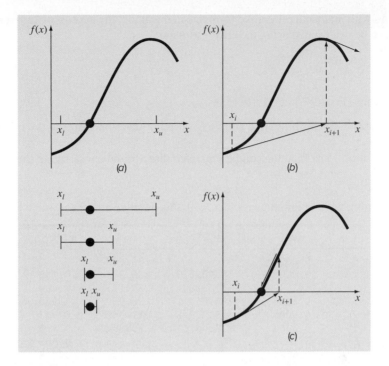

FIGURE 6.1
Graphical depiction of the fundamental difference between the (*a*) bracketing and (*b*) and (*c*) open methods for root location. In (*a*), which is bisection, the root is constrained within the interval prescribed by x_l and x_u. In contrast, for the open method depicted in (*b*) and (*c*), which is Newton-Raphson, a formula is used to project from x_i to x_{i+1} in an iterative fashion. Thus the method can either (*b*) diverge or (*c*) converge rapidly, depending on the shape of the function and the value of the initial guess.

6.1   SIMPLE FIXED-POINT ITERATION

As just mentioned, open methods employ a formula to predict the root. Such a formula can be developed for simple *fixed-point iteration* (or, as it is also called, *one-point iteration* or *successive substitution*) by rearranging the function $f(x) = 0$ so that x is on the left-hand side of the equation:

$$x = g(x) \tag{6.1}$$

This transformation can be accomplished either by algebraic manipulation or by simply adding x to both sides of the original equation.

The utility of Eq. (6.1) is that it provides a formula to predict a new value of x as a function of an old value of x. Thus, given an initial guess at the root x_i, Eq. (6.1) can be used to compute a new estimate x_{i+1} as expressed by the iterative formula

$$x_{i+1} = g(x_i) \tag{6.2}$$

As with many other iterative formulas in this book, the approximate error for this equation can be determined using the error estimator:

$$\varepsilon_a = \left| \frac{x_{i+1} - x_i}{x_{i+1}} \right| 100\% \tag{6.3}$$

EXAMPLE 6.1   Simple Fixed-Point Iteration

Problem Statement. Use simple fixed-point iteration to locate the root of $f(x) = e^{-x} - x$.

Solution. The function can be separated directly and expressed in the form of Eq. (6.2) as

$$x_{i+1} = e^{-x_i}$$

Starting with an initial guess of $x_0 = 0$, this iterative equation can be applied to compute:

i	x_i	$|\varepsilon_a|$, %	$|\varepsilon_t|$, %	$|\varepsilon_t|_i/|\varepsilon_t|_{i-1}$
0	0.0000		100.000	
1	1.0000	100.000	76.322	0.763
2	0.3679	171.828	35.135	0.460
3	0.6922	46.854	22.050	0.628
4	0.5005	38.309	11.755	0.533
5	0.6062	17.447	6.894	0.586
6	0.5454	11.157	3.835	0.556
7	0.5796	5.903	2.199	0.573
8	0.5601	3.481	1.239	0.564
9	0.5711	1.931	0.705	0.569
10	0.5649	1.109	0.399	0.566

Thus, each iteration brings the estimate closer to the true value of the root: 0.56714329.

Notice that the true percent relative error for each iteration of Example 6.1 is roughly proportional (for this case, by a factor of about 0.5 to 0.6) to the error from the previous iteration. This property, called *linear convergence,* is characteristic of fixed-point iteration.

Aside from the "rate" of convergence, we must comment at this point about the "possibility" of convergence. The concepts of convergence and divergence can be depicted graphically. Recall that in Sec. 5.2, we graphed a function to visualize its structure and behavior. Such an approach is employed in Fig. 6.2a for the function $f(x) = e^{-x} - x$. An alternative graphical approach is to separate the equation into two component parts, as in

$$f_1(x) = f_2(x)$$

Then the two equations

$$y_1 = f_1(x) \tag{6.4}$$

and

$$y_2 = f_2(x) \tag{6.5}$$

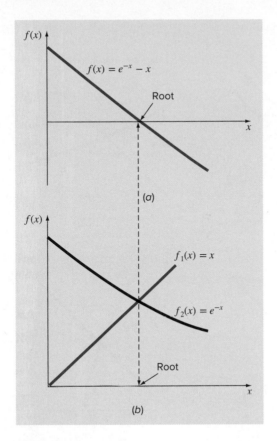

FIGURE 6.2
Two alternative graphical methods for determining the root of $f(x) = e^{-x} - x$. (a) Root at the
point where it crosses the x axis; (b) root at the intersection of the component functions.

can be plotted separately (Fig. 6.2b). The x values corresponding to the intersections of
these functions represent the roots of $f(x) = 0$.

The two-curve method can now be used to illustrate the convergence and divergence
of fixed-point iteration. First, Eq. (6.1) can be reexpressed as a pair of equations $y_1 = x$ and
$y_2 = g(x)$. These two equations can then be plotted separately. As was the case with Eqs.
(6.4) and (6.5), the roots of $f(x) = 0$ correspond to the abscissa value at the intersection of
the two curves. The function $y_1 = x$ and four different shapes for $y_2 = g(x)$ are plotted in
Fig. 6.3.

For the first case (Fig. 6.3a), the initial guess of x_0 is used to determine the corresponding point on the y_2 curve $[x_0, g(x_0)]$. The point $[x_1, x_1]$ is located by moving left
horizontally to the y_1 curve. These movements are equivalent to the first iteration of
the fixed-point method:

$$x_1 = g(x_0)$$

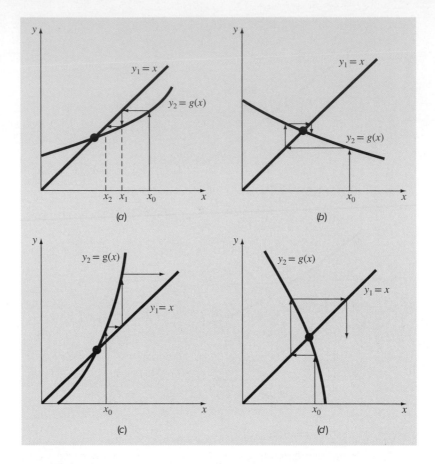

FIGURE 6.3
"*Cobweb plots*" depicting convergence (*a* and *b*) and divergence (*c* and *d*). Graphs (*a*) and (*c*) are called monotone patterns whereas (*b*) and (*c*) are called oscillating or spiral patterns. Note that convergence occurs when $|g'(x)| < 1$.

Thus, in both the equation and in the plot, a starting value of x_0 is used to obtain an estimate of x_1. The next iteration consists of moving to $[x_1, g(x_1)]$ and then to $[x_2, x_2]$. This iteration is equivalent to the equation

$$x_2 = g(x_1)$$

The solution in Fig. 6.3*a* is *convergent* because the estimates of x move closer to the root with each iteration. The same is true for Fig. 6.3*b*. However, this is not the case for Fig. 6.3*c* and *d*, where the iterations diverge from the root.

A theoretical derivation can be used to gain insight into the process. As described in Chapra and Canale (2010), it can be shown that the error for any iteration is linearly proportional to the error from the previous iteration multiplied by the absolute value of the slope of g:

$$E_{i+1} = g'(\xi)E_i$$

Consequently, if $|g'| < 1$, the errors decrease with each iteration. For $|g'| > 1$ the errors grow. Notice also that if the derivative is positive, the errors will be positive, and hence the errors will have the same sign (Fig. 6.3*a* and *c*). If the derivative is negative, the errors will change sign on each iteration (Fig. 6.3*b* and *d*).

6.2   NEWTON-RAPHSON

Perhaps the most widely used of all root-locating formulas is the *Newton-Raphson method* (Fig. 6.4). If the initial guess at the root is x_i, a tangent can be extended from the point $[x_i, f(x_i)]$. The point where this tangent crosses the x axis usually represents an improved estimate of the root.

The Newton-Raphson method can be derived on the basis of this geometrical interpretation. As in Fig. 6.4, the first derivative at x is equivalent to the slope:

$$f'(x_i) = \frac{f(x_i) - 0}{x_i - x_{i+1}}$$

which can be rearranged to yield

$$x_{i+1} = x_i - \frac{f(x_i)}{f'(x_i)} \tag{6.6}$$

which is called the *Newton-Raphson formula*.

EXAMPLE 6.2   Newton-Raphson Method

Problem Statement.   Use the Newton-Raphson method to estimate the root of $f(x) = e^{-x} - x$ employing an initial guess of $x_0 = 0$.

Solution.   The first derivative of the function can be evaluated as

$$f'(x) = -e^{-x} - 1$$

which can be substituted along with the original function into Eq. (6.6) to give

$$x_{i+1} = x_i - \frac{e^{-x_i} - x_i}{-e^{-x_i} - 1}$$

Starting with an initial guess of $x_0 = 0$, this iterative equation can be applied to compute

i	x_i	$|\varepsilon_t|$, %
0	0	100
1	0.500000000	11.8
2	0.566311003	0.147
3	0.567143165	0.0000220
4	0.567143290	$<10^{-8}$

Thus, the approach rapidly converges on the true root. Notice that the true percent relative error at each iteration decreases much faster than it does in simple fixed-point iteration (compare with Example 6.1).

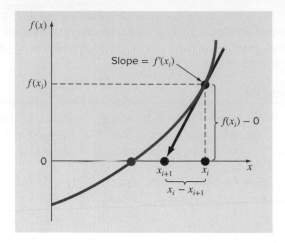

FIGURE 6.4
Graphical depiction of the Newton-Raphson method. A tangent to the function of x_i [i.e., $f'(x)$] is extrapolated down to the x axis to provide an estimate of the root at x_{i+1}.

As with other root-location methods, Eq. (6.3) can be used as a termination criterion. In addition, a theoretical analysis (Chapra and Canale, 2010) provides insight regarding the rate of convergence as expressed by

$$E_{t,i+1} = \frac{-f''(x_r)}{2f'(x_r)} E_{t,i}^2 \qquad (6.7)$$

Thus, the error should be roughly proportional to the square of the previous error. In other words, the number of significant figures of accuracy approximately doubles with each iteration. This behavior is called *quadratic convergence* and is one of the major reasons for the popularity of the method.

Although the Newton-Raphson method is often very efficient, there are situations where it performs poorly. A special case—multiple roots—is discussed elsewhere (Chapra and Canale, 2010). However, even when dealing with simple roots, difficulties can also arise, as in the following example.

EXAMPLE 6.3   A Slowly Converging Function with Newton-Raphson

Problem Statement. Determine the positive root of $f(x) = x^{10} - 1$ using the Newton-Raphson method and an initial guess of $x = 0.5$.

Solution. The Newton-Raphson formula for this case is

$$x_{i+1} = x_i - \frac{x_i^{10} - 1}{10x_i^9}$$

which can be used to compute

i	x_i	$|\varepsilon_a|$, %
0	0.5	
1	51.65	99.032
2	46.485	11.111
3	41.8365	11.111
4	37.65285	11.111
.		
.		
.		
40	1.002316	2.130
41	1.000024	0.229
42	1	0.002

Thus, after the first poor prediction, the technique is converging on the true root of 1, but at a very slow rate.

Why does this happen? As shown in Fig. 6.5, a simple plot of the first few iterations is helpful in providing insight. Notice how the first guess is in a region where the slope is near zero. Thus, the first iteration flings the solution far away from the initial guess to a new value ($x = 51.65$) where $f(x)$ has an extremely high value. The solution then plods along for over 40 iterations until converging on the root with adequate accuracy.

FIGURE 6.5
Graphical depiction of the Newton-Raphson method for a case with slow convergence. The inset shows how a near-zero slope initially shoots the solution far from the root. Thereafter, the solution very slowly converges on the root.

Aside from slow convergence due to the nature of the function, other difficulties can arise, as illustrated in Fig. 6.6. For example, Fig. 6.6a depicts the case where an inflection

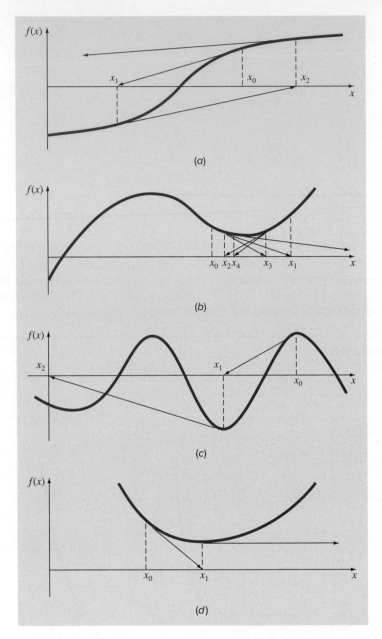

FIGURE 6.6
Four cases where the Newton-Raphson method exhibits poor convergence.

point (i.e., $f''(x) = 0$) occurs in the vicinity of a root. Notice that iterations beginning at x_0 progressively diverge from the root. Fig. 6.6b illustrates the tendency of the Newton-Raphson technique to oscillate around a local maximum or minimum. Such oscillations may persist, or, as in Fig. 6.6b, a near-zero slope is reached whereupon the solution is sent far from the area of interest. Figure 6.6c shows how an initial guess that is close to one root can jump to a location several roots away. This tendency to move away from the area of interest is due to the fact that near-zero slopes are encountered. Obviously, a zero slope $[f'(x) = 0]$ is a real disaster because it causes division by zero in the Newton-Raphson formula [Eq. (6.6)]. As in Fig. 6.6d, it means that the solution shoots off horizontally and never hits the x axis.

Thus, there is no general convergence criterion for Newton-Raphson. Its convergence depends on the nature of the function and on the accuracy of the initial guess. The only remedy is to have an initial guess that is "sufficiently" close to the root. And for some functions, no guess will work! Good guesses are usually predicated on knowledge of the physical problem setting or on devices such as graphs that provide insight into the behavior of the solution. It also suggests that good computer software should be designed to recognize slow convergence or divergence.

6.2.1 MATLAB M-file: newtraph

An algorithm for the Newton-Raphson method can be easily developed (Fig. 6.7). Note that the program must have access to the function (func) and its first derivative (dfunc). These can be simply accomplished by the inclusion of user-defined functions to compute these quantities. Alternatively, as in the algorithm in Fig. 6.7, they can be passed to the function as arguments.

After the M-file is entered and saved, it can be invoked to solve for root. For example, for the simple function $x^2 - 9$, the root can be determined as in

```
>> newtraph(@(x) x^2-9,@(x) 2*x,5)

ans =
     3
```

EXAMPLE 6.4   Newton-Raphson Bungee Jumper Problem

Problem Statement. Use the M-file function from Fig. 6.7 to determine the mass of the bungee jumper with a drag coefficient of 0.25 kg/m to have a velocity of 36 m/s after 4 s of free fall. The acceleration of gravity is 9.81 m/s².

Solution. The function to be evaluated is

$$f(m) = \sqrt{\frac{gm}{c_d}} \tanh\left(\sqrt{\frac{gc_d}{m}} t\right) - v(t) \tag{E6.4.1}$$

To apply the Newton-Raphson method, the derivative of this function must be evaluated with respect to the unknown, m:

$$\frac{df(m)}{dm} = \frac{1}{2}\sqrt{\frac{g}{mc_d}} \tanh\left(\sqrt{\frac{gc_d}{m}}t\right) - \frac{g}{2m} t \operatorname{sech}^2\left(\sqrt{\frac{gc_d}{m}}t\right) \tag{E6.4.2}$$

```
function [root,ea,iter]=newtraph(func,dfunc,xr,es,maxit,varargin)
% newtraph: Newton-Raphson root location zeroes
%   [root,ea,iter]=newtraph(func,dfunc,xr,es,maxit,p1,p2,...):
%        uses Newton-Raphson method to find the root of func
% input:
%   func = name of function
%   dfunc = name of derivative of function
%   xr = initial guess
%   es = desired relative error (default = 0.0001%)
%   maxit = maximum allowable iterations (default = 50)
%   p1,p2,... = additional parameters used by function
% output:
%   root = real root
%   ea = approximate relative error (%)
%   iter = number of iterations

if nargin<3,error('at least 3 input arguments required'),end
if nargin<4|isempty(es),es=0.0001;end
if nargin<5|isempty(maxit),maxit=50;end
iter = 0;
while (1)
  xrold = xr;
  xr = xr - func(xr)/dfunc(xr);
  iter = iter + 1;
  if xr ~= 0, ea = abs((xr - xrold)/xr) * 100; end
  if ea <= es | iter >= maxit, break, end
end
root = xr;
```

FIGURE 6.7
An M-file to implement the Newton-Raphson method.

We should mention that although this derivative is not difficult to evaluate in principle, it involves a bit of concentration and effort to arrive at the final result.

The two formulas can now be used in conjunction with the function newtraph to evaluate the root:

```
>> y = @(m) sqrt(9.81*m/0.25)*tanh(sqrt(9.81*0.25/m)*4)-36;
>> dy = @(m) 1/2*sqrt(9.81/(m*0.25))*tanh((9.81*0.25/m) ...
       ^(1/2)*4)-9.81/(2*m)*sech(sqrt(9.81*0.25/m)*4)^2;

>> newtraph(y,dy,140,0.00001)

ans =
  142.7376
```

6.3 SECANT METHODS

As in Example 6.4, a potential problem in implementing the Newton-Raphson method is the evaluation of the derivative. Although this is not inconvenient for polynomials and many other functions, there are certain functions whose derivatives may be difficult or inconvenient to evaluate. For these cases, the derivative can be approximated by a backward finite divided difference:

$$f'(x_i) \cong \frac{f(x_{i-1}) - f(x_i)}{x_{i-1} - x_i}$$

This approximation can be substituted into Eq. (6.6) to yield the following iterative equation:

$$x_{i+1} = x_i - \frac{f(x_i)(x_{i-1} - x_i)}{f(x_{i-1}) - f(x_i)} \tag{6.8}$$

Equation (6.8) is the formula for the *secant method*. Notice that the approach requires two initial estimates of x. However, because $f(x)$ is not required to change signs between the estimates, it is not classified as a bracketing method.

Rather than using two arbitrary values to estimate the derivative, an alternative approach involves a fractional perturbation of the independent variable to estimate $f'(x)$,

$$f'(x_i) \cong \frac{f(x_i + \delta x_i) - f(x_i)}{\delta x_i}$$

where δ = a small perturbation fraction. This approximation can be substituted into Eq. (6.6) to yield the following iterative equation:

$$x_{i+1} = x_i - \frac{\delta x_i f(x_i)}{f(x_i + \delta x_i) - f(x_i)} \tag{6.9}$$

We call this the *modified secant method*. As in the following example, it provides a nice means to attain the efficiency of Newton-Raphson without having to compute derivatives.

EXAMPLE 6.5   Modified Secant Method

Problem Statement. Use the modified secant method to determine the mass of the bungee jumper with a drag coefficient of 0.25 kg/m to have a velocity of 36 m/s after 4 s of free fall. Note: The acceleration of gravity is 9.81 m/s². Use an initial guess of 50 kg and a value of 10^{-6} for the perturbation fraction.

Solution. Inserting the parameters into Eq. (6.9) yields

First iteration:

$x_0 = 50$ $f(x_0) = -4.57938708$

$x_0 + \delta x_0 = 50.00005$ $f(x_0 + \delta x_0) = -4.579381118$

$$x_1 = 50 - \frac{10^{-6}(50)(-4.57938708)}{-4.579381118 - (-4.57938708)}$$

$$= 88.39931(|\varepsilon_t| = 38.1\%; |\varepsilon_a| = 43.4\%)$$

Second iteration:

$$x_1 = 88.39931 \qquad\qquad f(x_1) = -1.69220771$$

$$x_1 + \delta x_1 = 88.39940 \qquad\qquad f(x_1 + \delta x_1) = -1.692203516$$

$$x_2 = 88.39931 - \frac{10^{-6}(88.39931)(-1.69220771)}{-1.692203516 - (-1.69220771)}$$

$$= 124.08970 (|\varepsilon_t| = 13.1\%; |\varepsilon_a| = 28.76\%)$$

The calculation can be continued to yield

i	x_i	$|\varepsilon_t|$, %	$|\varepsilon_a|$, %
0	50.0000	64.971	
1	88.3993	38.069	43.438
2	124.0897	13.064	28.762
3	140.5417	1.538	11.706
4	142.7072	0.021	1.517
5	142.7376	4.1×10^{-6}	0.021
6	142.7376	3.4×10^{-12}	4.1×10^{-6}

The choice of a proper value for δ is not automatic. If δ is too small, the method can be swamped by roundoff error caused by subtractive cancellation in the denominator of Eq. (6.9). If it is too big, the technique can become inefficient and even divergent. However, if chosen correctly, it provides a nice alternative for cases where evaluating the derivative is difficult and developing two initial guesses is inconvenient.

Further, in its most general sense, a univariate function is merely an entity that returns a single value in return for values sent to it. Perceived in this sense, functions are not always simple formulas like the one-line equations solved in the preceding examples in this chapter. For example, a function might consist of many lines of code that could take a significant amount of execution time to evaluate. In some cases, the function might even represent an independent computer program. For such cases, the secant and modified secant methods are valuable.

6.4 BRENT'S METHOD

Wouldn't it be nice to have a hybrid approach that combined the reliability of bracketing with the speed of the open methods? *Brent's root-location method* is a clever algorithm that does just that by applying a speedy open method wherever possible, but reverting to a reliable bracketing method if necessary. The approach was developed by Richard Brent (1973) based on an earlier algorithm of Theodorus Dekker (1969).

The bracketing technique is the trusty bisection method (Sec. 5.4), whereas two different open methods are employed. The first is the secant method described in Sec. 6.3. As explained next, the second is inverse quadratic interpolation.

6.4.1 Inverse Quadratic Interpolation

Inverse quadratic interpolation is similar in spirit to the secant method. As in Fig. 6.8*a*, the secant method is based on computing a straight line that goes through two guesses. The intersection of this straight line with the *x* axis represents the new root estimate. For this reason, it is sometimes referred to as a *linear interpolation method.*

Now suppose that we had three points. In that case, we could determine a quadratic function of *x* that goes through the three points (Fig. 6.8*b*). Just as with the linear secant method, the intersection of this parabola with the *x* axis would represent the new root estimate. And as illustrated in Fig. 6.8*b*, using a curve rather than a straight line often yields a better estimate.

Although this would seem to represent a great improvement, the approach has a fundamental flaw: it is possible that the parabola might not intersect the *x* axis! Such would be the case when the resulting parabola had complex roots. This is illustrated by the parabola, $y = f(x)$, in Fig. 6.9.

The difficulty can be rectified by employing inverse quadratic interpolation. That is, rather than using a parabola in *x*, we can fit the points with a parabola in *y*. This amounts to reversing the axes and creating a "sideways" parabola [the curve, $x = f(y)$, in Fig. 6.9].

If the three points are designated as (x_{i-2}, y_{i-2}), (x_{i-1}, y_{i-1}), and (x_i, y_i), a quadratic function of *y* that passes through the points can be generated as

$$g(y) = \frac{(y - y_{i-1})(y - y_i)}{(y_{i-2} - y_{i-1})(y_{i-2} - y_i)}x_{i-2} + \frac{(y - y_{i-2})(y - y_i)}{(y_{i-1} - y_{i-2})(y_{i-1} - y_i)}x_{i-1}$$

$$+ \frac{(y - y_{i-2})(y - y_{i-1})}{(y_i - y_{i-2})(y_i - y_{i-1})}x_i \tag{6.10}$$

FIGURE 6.8
Comparison of (*a*) the secant method and (*b*) inverse quadratic interpolation. Note that the approach in (*b*) is called "inverse" because the quadratic function is written in *y* rather than in *x*.

(*a*)   (*b*)

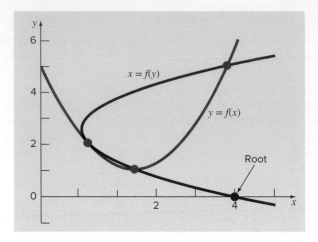

FIGURE 6.9
Two parabolas fit to three points. The parabola written as a function of x, $y = f(x)$, has complex roots and hence does not intersect the x axis. In contrast, if the variables are reversed, and the parabola developed as $x = f(y)$, the function does intersect the x axis.

As we will learn in Sec. 18.2, this form is called a *Lagrange polynomial.* The root, x_{i+1}, corresponds to $y = 0$, which when substituted into Eq. (6.10) yields

$$x_{i+1} = \frac{y_{i-1}\,y_i}{(y_{i-2} - y_{i-1})(y_{i-2} - y_i)}x_{i-2} + \frac{y_{i-2}y_i}{(y_{i-1} - y_{i-2})(y_{i-1} - y_i)}x_{i-1}$$

$$+ \frac{y_{i-2}\,y_{i-1}}{(y_i - y_{i-2})(y_i - y_{i-1})}x_i \tag{6.11}$$

As shown in Fig. 6.9, such a "sideways" parabola always intersects the x axis.

EXAMPLE 6.6   Inverse Quadratic Interpolation

Problem Statement. Develop quadratic equations in both x and y for the data points depicted in Fig. 6.9: (1, 2), (2, 1), and (4, 5). For the first, $y = f(x)$, employ the quadratic formula to illustrate that the roots are complex. For the latter, $x = g(y)$, use inverse quadratic interpolation (Eq. 6.11) to determine the root estimate.

Solution.   By reversing the x's and y's, Eq. (6.10) can be used to generate a quadratic in x as

$$f(x) = \frac{(x-2)(x-4)}{(1-2)(1-4)}\,2 + \frac{(x-1)(x-4)}{(2-1)(2-4)}\,1 + \frac{(x-1)(x-2)}{(4-1)(4-2)}\,5$$

or collecting terms

$$f(x) = x^2 - 4x + 5$$

This equation was used to generate the parabola, $y = f(x)$, in Fig. 6.9. The quadratic formula can be used to determine that the roots for this case are complex,

$$x = \frac{4 \pm \sqrt{(-4)^2 - 4(1)(5)}}{2} = 2 \pm i$$

Equation (6.10) can be used to generate the quadratic in y as

$$g(y) = \frac{(y-1)(y-5)}{(2-1)(2-5)}1 + \frac{(y-2)(y-5)}{(1-2)(1-5)}2 + \frac{(y-2)(y-1)}{(5-2)(5-1)}4$$

or collecting terms:

$$g(y) = 0.5x^2 - 2.5x + 4$$

Finally, Eq. (6.11) can be used to determine the root as

$$x_{i+1} = \frac{-1(-5)}{(2-1)(2-5)}1 + \frac{-2(-5)}{(1-2)(1-5)}2 + \frac{-2(-1)}{(5-2)(5-1)}4 = 4$$

Before proceeding to Brent's algorithm, we need to mention one more case where inverse quadratic interpolation does not work. If the three y values are not distinct (i.e., $y_{i-2} = y_{i-1}$ or $y_{i-1} = y_i$), an inverse quadratic function does not exist. So this is where the secant method comes into play. If we arrive at a situation where the y values are not distinct, we can always revert to the less efficient secant method to generate a root using two of the points. If $y_{i-2} = y_{i-1}$, we use the secant method with x_{i-1} and x_i. If $y_{i-1} = y_i$, we use x_{i-2} and x_{i-1}.

6.4.2 Brent's Method Algorithm

The general idea behind the *Brent's root-finding method* is whenever possible to use one of the quick open methods. In the event that these generate an unacceptable result (i.e., a root estimate that falls outside the bracket), the algorithm reverts to the more conservative bisection method. Although bisection may be slower, it generates an estimate guaranteed to fall within the bracket. This process is then repeated until the root is located to within an acceptable tolerance. As might be expected, bisection typically dominates at first but as the root is approached, the technique shifts to the faster open methods.

Figure 6.10 presents a function based on a MATLAB M-file developed by Cleve Moler (2004). It represents a stripped down version of the fzero function which is the professional root-location function employed in MATLAB. For that reason, we call the simplified version: fzerosimp. Note that it requires another function f that holds the equation for which the root is being evaluated.

The fzerosimp function is passed two initial guesses that must bracket the root. Then, the three variables defining the search interval (a, b, c) are initialized, and f is evaluated at the endpoints.

A main loop is then implemented. If necessary, the three points are rearranged to satisfy the conditions required for the algorithm to work effectively. At this point, if the

```
function b = fzerosimp(xl,xu)
a = xl; b = xu; fa = f(a); fb = f(b);
c = a; fc = fa; d = b - c; e = d;
while (1)
  if fb == 0, break, end
  if sign(fa) == sign(fb) %If needed, rearrange points
    a = c; fa = fc; d = b - c; e = d;
  end
  if abs(fa) < abs(fb)
    c = b; b = a; a = c;
    fc = fb; fb = fa; fa = fc;
  end
  m = 0.5*(a - b); %Termination test and possible exit
  tol = 2 * eps * max(abs(b), 1);
  if abs(m) <= tol | fb == 0.
    break
  end
  %Choose open methods or bisection
  if abs(e) >= tol & abs(fc) > abs(fb)
    s = fb/fc;
    if a == c                    %Secant method
      p = 2*m*s;
      q = 1 - s;
    else             %Inverse quadratic interpolation
      q = fc/fa; r = fb/fa;
      p = s * (2*m*q * (q - r) - (b - c)*(r - 1));
      q = (q - 1)*(r - 1)*(s - 1);
    end
    if p > 0, q = -q; else p = -p; end;
    if 2*p < 3*m*q - abs(tol*q) & p < abs(0.5*e*q)
      e = d; d = p/q;
    else
      d = m; e = m;
    end
  else                              %Bisection
    d = m; e = m;
  end
  c = b; fc = fb;
  if abs(d) > tol, b=b+d; else b=b-sign(b-a)*tol; end
  fb = f(b);
end
```

FIGURE 6.10
Function for Brent's root-finding algorithm based on a MATLAB M-file developed by Cleve Moler (2004).

stopping criteria are met, the loop is terminated. Otherwise, a decision structure chooses among the three methods and checks whether the outcome is acceptable. A final section then evaluates f at the new point and the loop is repeated. Once the stopping criteria are met, the loop terminates and the final root estimate is returned.

6.5   MATLAB FUNCTION: fzero

The fzero function is designed to find the real root of a single equation. A simple representation of its syntax is

```
fzero(function,x0)
```

where *function* is the name of the function being evaluated, and *x0* is the initial guess. Note that two guesses that bracket the root can be passed as a vector:

```
fzero(function,[x0 x1])
```

where *x0* and *x1* are guesses that bracket a sign change.

Here is a MATLAB session that solves for the root of a simple quadratic: $x^2 - 9$. Clearly two roots exist at -3 and 3. To find the negative root:

```
>> x = fzero(@(x) x^2-9,-4)

x =
    -3
```

If we want to find the positive root, use a guess that is near it:

```
>> x = fzero(@(x) x^2-9,4)

x =
    3
```

If we put in an initial guess of zero, it finds the negative root:

```
>> x = fzero(@(x) x^2-9,0)

x =
    -3
```

If we wanted to ensure that we found the positive root, we could enter two guesses as in

```
>> x = fzero(@(x) x^2-9,[0 4])

x =
    3
```

Also, if a sign change does not occur between the two guesses, an error message is displayed

```
>> x = fzero(@(x) x^2-9,[-4 4])

??? Error using ==> fzero
The function values at the interval endpoints must ...
differ in sign.
```

The fzero function works as follows. If a single initial guess is passed, it first performs a search to identify a sign change. This search differs from the incremental search described in Sec. 5.3.1, in that the search starts at the single initial guess and then takes increasingly bigger steps in both the positive and negative directions until a sign change is detected.

Thereafter, the fast methods (secant and inverse quadratic interpolation) are used unless an unacceptable result occurs (e.g., the root estimate falls outside the bracket). If an unacceptable result happens, bisection is implemented until an acceptable root is obtained with one of the fast methods. As might be expected, bisection typically dominates at first but as the root is approached, the technique shifts to the faster methods.

A more complete representation of the fzero syntax can be written as

$$[x,fx] = fzero(function,x0,options,p1,p2,...)$$

where $[x,fx]$ = a vector containing the root x and the function evaluated at the root fx, *options* is a data structure created by the optimset function, and $p1$, $p2$... are any parameters that the function requires. Note that if you desire to pass in parameters but not use the options, pass an empty vector [] in its place.

The optimset function has the syntax

$$options = optimset('par_1',val_1,'par_2',val_2,...)$$

where the parameter par_i has the value val_i. A complete listing of all the possible parameters can be obtained by merely entering optimset at the command prompt. The parameters that are commonly used with the fzero function are

display: When set to 'iter' displays a detailed record of all the iterations.
tolx: A positive scalar that sets a termination tolerance on x.

EXAMPLE 6.7   The fzero and optimset Functions

Problem Statement. Recall that in Example 6.3, we found the positive root of $f(x) = x^{10} - 1$ using the Newton-Raphson method with an initial guess of 0.5. Solve the same problem with optimset and fzero.

Solution. An interactive MATLAB session can be implemented as follows:

```
>> options = optimset('display','iter');
>> [x,fx] = fzero(@(x) x^10-1,0.5,options)
 Func-count        x            f(x)           Procedure
      1           0.5         -0.999023        initial
      2        0.485858       -0.999267        search
      3        0.514142       -0.998709        search
      4          0.48         -0.999351        search
      5          0.52         -0.998554        search
      6        0.471716       -0.999454        search
      .
      .
      .
     23        0.952548       -0.385007        search
     24         -0.14            -1            search
     25          1.14          2.70722         search
```

```
Looking for a zero in the interval [-0.14, 1.14]
26      0.205272           -1         interpolation
27      0.672636    -0.981042         bisection
28      0.906318    -0.626056         bisection
29      1.02316      0.257278         bisection
30      0.989128    -0.103551         interpolation
31      0.998894    -0.0110017        interpolation
32      1.00001   7.68385e-005        interpolation
33           1   -3.83061e-007        interpolation
34           1   -1.3245e-011         interpolation
35           1            0           interpolation
Zero found in the interval: [-0.14, 1.14].

x =
    1

fx =
    0
```

Thus, after 25 iterations of searching, fzero finds a sign change. It then uses interpolation and bisection until it gets close enough to the root so that interpolation takes over and rapidly converges on the root.

Suppose that we would like to use a less stringent tolerance. We can use the optimset function to set a low maximum tolerance and a less accurate estimate of the root results:

```
>> options = optimset ('tolx', 1e-3);
>> [x,fx] = fzero(@(x) x^10-1,0.5,options)

x =
    1.0009

fx =
    0.0090
```

6.6   POLYNOMIALS

Polynomials are a special type of nonlinear algebraic equation of the general form

$$f_n(x) = a_1 x^n + a_2 x^{n-1} + \cdots + a_{n-1} x^2 + a_n x + a_{n+1} \tag{6.12}$$

where n is the order of the polynomial, and the a's are constant coefficients. In many (but not all) cases, the coefficients will be real. For such cases, the roots can be real and/or complex. In general, an nth order polynomial will have n roots.

Polynomials have many applications in engineering and science. For example, they are used extensively in curve fitting. However, one of their most interesting and powerful applications is in characterizing dynamic systems—and, in particular, linear systems. Examples include reactors, mechanical devices, structures, and electrical circuits.

6.6.1 MATLAB Function: roots

If you are dealing with a problem where you must determine a single real root of a polynomial, the techniques such as bisection and the Newton-Raphson method can have utility. However, in many cases, engineers desire to determine all the roots, both real and complex. Unfortunately, simple techniques like bisection and Newton-Raphson are not available for determining all the roots of higher-order polynomials. However, MATLAB has an excellent built-in capability, the roots function, for this task.

The roots function has the syntax,

```
x = roots(c)
```

where x is a column vector containing the roots and c is a row vector containing the polynomial's coefficients.

So how does the roots function work? MATLAB is very good at finding the eigenvalues of a matrix. Consequently, the approach is to recast the root evaluation task as an eigenvalue problem. Because we will be describing eigenvalue problems later in the book, we will merely provide an overview here.

Suppose we have a polynomial

$$a_1 x^5 + a_2 x^4 + a_3 x^3 + a_4 x^2 + a_5 x + a_6 = 0 \tag{6.13}$$

Dividing by a_1 and rearranging yields

$$x^5 = -\frac{a_2}{a_1} x^4 - \frac{a_3}{a_1} x^3 - \frac{a_4}{a_1} x^2 - \frac{a_5}{a_1} x - \frac{a_6}{a_1}$$

A special matrix can be constructed by using the coefficients from the right-hand side as the first row and with 1's and 0's written for the other rows as shown:

$$\begin{bmatrix} -a_2/a_1 & -a_3/a_1 & -a_4/a_1 & -a_5/a_1 & -a_6/a_1 \\ 1 & 0 & 0 & 0 & 0 \\ 0 & 1 & 0 & 0 & 0 \\ 0 & 0 & 1 & 0 & 0 \\ 0 & 0 & 0 & 1 & 0 \end{bmatrix} \tag{6.14}$$

Equation (6.14) is called the polynomial's *companion matrix*. It has the useful property that its eigenvalues are the roots of the polynomial. Thus, the algorithm underlying the roots function consists of merely setting up the companion matrix and then using MATLAB's powerful eigenvalue evaluation function to determine the roots. Its application, along with some other related polynomial manipulation functions, are described in the following example.

We should note that roots has an inverse function called poly, which when passed the values of the roots, will return the polynomial's coefficients. Its syntax is

```
c = poly(r)
```

where r is a column vector containing the roots and c is a row vector containing the polynomial's coefficients.

EXAMPLE 6.8   Using MATLAB to Manipulate Polynomials and Determine Their Roots

Problem Statement. Use the following equation to explore how MATLAB can be employed to manipulate polynomials:

$$f_5(x) = x^5 - 3.5x^4 + 2.75x^3 + 2.125x^2 - 3.875x + 1.25 \qquad \text{(E6.8.1)}$$

Note that this polynomial has three real roots: 0.5, −1.0, and 2; and one pair of complex roots: $1 \pm 0.5i$.

Solution. Polynomials are entered into MATLAB by storing the coefficients as a row vector. For example, entering the following line stores the coefficients in the vector a:

```
>> a = [1 -3.5 2.75 2.125 -3.875 1.25];
```

We can then proceed to manipulate the polynomial. For example we can evaluate it at $x = 1$, by typing

```
>> polyval(a,1)
```

with the result, $1(1)^5 - 3.5(1)^4 + 2.75(1)^3 + 2.125(1)^2 - 3.875(1) + 1.25 = -0.25$:

```
ans =

    -0.2500
```

We can create a quadratic polynomial that has roots corresponding to two of the original roots of Eq. (E6.8.1): 0.5 and −1. This quadratic is $(x - 0.5)(x + 1) = x^2 + 0.5x - 0.5$. It can be entered into MATLAB as the vector b:

```
>> b = [1 .5 -.5]

b =
   1.0000    0.5000   -0.5000
```

Note that the poly function can be used to perform the same task as in

```
>> b = poly([0.5 -1])

b =
   1.0000    0.5000   -0.5000
```

We can divide this polynomial into the original polynomial by

```
>> [q,r] = deconv(a,b)
```

with the result being a quotient (a third-order polynomial, q) and a remainder (r)

```
q =
   1.0000   -4.0000    5.2500   -2.5000
r =
      0    0    0    0    0    0
```

Because the polynomial is a perfect divisor, the remainder polynomial has zero coefficients. Now, the roots of the quotient polynomial can be determined as

```
>> x = roots(q)
```

with the expected result that the remaining roots of the original polynomial Eq. (E6.8.1) are found:

```
x =
2.0000
1.0000 + 0.5000i
1.0000 - 0.5000i
```

We can now multiply q by b to come up with the original polynomial:

```
>> a = conv(q,b)

a =
    1.0000   -3.5000    2.7500    2.1250   -3.8750    1.2500
```

We can then determine all the roots of the original polynomial by

```
>> x = roots(a)

x =
    2.0000
   -1.0000
    1.0000 + 0.5000i
    1.0000 - 0.5000i
    0.5000
```

Finally, we can return to the original polynomial again by using the poly function:

```
>> a = poly(x)

a =
    1.0000   -3.5000    2.7500    2.1250   -3.8750    1.2500
```

6.7 CASE STUDY   PIPE FRICTION

Background. Determining fluid flow through pipes and tubes has great relevance in many areas of engineering and science. In engineering, typical applications include the flow of liquids and gases through pipelines and cooling systems. Scientists are interested in topics ranging from flow in blood vessels to nutrient transmission through a plant's vascular system.

The resistance to flow in such conduits is parameterized by a dimensionless number called the *friction factor*. For turbulent flow, the *Colebrook equation* provides a means to calculate the friction factor:

$$0 = \frac{1}{\sqrt{f}} + 2.0 \log\left(\frac{\varepsilon}{3.7D} + \frac{2.51}{\text{Re } \sqrt{f}}\right) \tag{6.15}$$

6.7 CASE STUDY   continued

where ε = the roughness (m), D = diameter (m), and Re = the *Reynolds number:*

$$\text{Re} = \frac{\rho V D}{\mu}$$

where ρ = the fluid's density (kg/m^3), V = its velocity (m/s), and μ = dynamic viscosity (N·s/m^2). In addition to appearing in Eq. (6.15), the Reynolds number also serves as the criterion for whether flow is turbulent (Re > 4000).

In this case study, we will illustrate how the numerical methods covered in this part of the book can be employed to determine f for air flow through a smooth, thin tube. For this case, the parameters are ρ = 1.23 kg/m^3, μ = 1.79 × 10^{-5} N·s/m^2, D = 0.005 m, V = 40 m/s, and ε = 0.0015 mm. Note that friction factors range from about 0.008 to 0.08. In addition, an explicit formulation called the *Swamee-Jain equation* provides an approximate estimate:

$$f = \frac{1.325}{\left[\ln\left(\dfrac{\varepsilon}{3.7\,D} + \dfrac{5.74}{\text{Re}^{0.9}}\right)\right]^2} \tag{6.16}$$

Solution. The Reynolds number can be computed as

$$\text{Re} = \frac{\rho V D}{\mu} = \frac{1.23(40)0.005}{1.79 \times 10^{-5}} = 13{,}743$$

This value along with the other parameters can be substituted into Eq. (6.15) to give

$$g(f) = \frac{1}{\sqrt{f}} + 2.0 \log\left(\frac{0.0000015}{3.7(0.005)} + \frac{2.51}{13{,}743\sqrt{f}}\right)$$

Before determining the root, it is advisable to plot the function to estimate initial guesses and to anticipate possible difficulties. This can be done easily with MATLAB:

```
>> rho=1.23;mu=1.79e-5;D=0.005;V=40;e=0.0015/1000;
>> Re=rho*V*D/mu;
>> g=@(f) 1/sqrt(f)+2*log10(e/(3.7*D)+2.51/(Re*sqrt(f)));
>> fplot(g,[0.008 0.08]),grid,xlabel('f'),ylabel('g(f)')
```

As in Fig. 6.11, the root is located at about 0.03.

Because we are supplied initial guesses (x_l = 0.008 and x_u = 0.08), either of the bracketing methods from Chap. 5 could be used. For example, the bisect function developed in Fig. 5.7 gives a value of f = 0.0289678 with a percent relative error of error of 5.926 × 10^{-5} in 22 iterations. False position yields a result of similar precision in 26 iterations. Thus, although they produce the correct result, they are somewhat inefficient. This would not be important for a single application, but could become prohibitive if many evaluations were made.

6.7 CASE STUDY   continued

FIGURE 6.11

We could try to attain improved performance by turning to an open method. Because Eq. (6.15) is relatively straightforward to differentiate, the Newton-Raphson method is a good candidate. For example, using an initial guess at the lower end of the range ($x_0 = 0.008$), the newtraph function developed in Fig. 6.7 converges quickly:

```
>> dg=@(f) -2/log(10)*1.255/Re*f^(-3/2)/(e/D/3.7 ...
                            +2.51/Re/sqrt(f))-0.5/f^(3/2);
>> [f ea iter]=newtraph(g,dg,0.008)

f =
    0.02896781017144
ea =
     6.870124190058040e-006
iter =
     6
```

However, when the initial guess is set at the upper end of the range ($x_0 = 0.08$), the routine diverges,

```
>> [f ea iter]=newtraph(g,dg,0.08)

f =
                NaN +             NaNi
```

As can be seen by inspecting Fig. 6.11, this occurs because the function's slope at the initial guess causes the first iteration to jump to a negative value. Further runs demonstrate that for this case, convergence only occurs when the initial guess is below about 0.066.

6.7 CASE STUDY   continued

So we can see that although the Newton-Raphson is very efficient, it requires good initial guesses. For the Colebrook equation, a good strategy might be to employ the Swamee-Jain equation (Eq. 6.16) to provide the initial guess as in

```
>> fSJ=1.325/log(e/(3.7*D)+5.74/Re^0.9)^2

fSJ =
    0.02903099711265
>> [f ea iter]=newtraph(g,dg,fSJ)

f =
    0.02896781017144
ea =
     8.510189472800060e-010
iter =
     3
```

Aside from our homemade functions, we can also use MATLAB's built-in fzero function. However, just as with the Newton-Raphson method, divergence also occurs when fzero function is used with a single guess. However, in this case, guesses at the lower end of the range cause problems. For example,

```
>> fzero(g,0.008)

Exiting fzero: aborting search for an interval containing a sign
change because complex function value encountered ...
                                              during search.
(Function value at -0.0028 is -4.92028-20.2423i.)
Check function or try again with a different starting value.
ans =
    NaN
```

If the iterations are displayed using optimset (recall Example 6.7), it is revealed that a negative value occurs during the search phase before a sign change is detected and the routine aborts. However, for single initial guesses above about 0.016, the routine works nicely. For example, for the guess of 0.08 that caused problems for Newton-Raphson, fzero does just fine:

```
>> fzero(g,0.08)

ans =
    0.02896781017144
```

As a final note, let's see whether convergence is possible for simple fixed-point iteration. The easiest and most straightforward version involves solving for the first f in Eq. (6.15):

$$f_{i+1} = \frac{0.25}{\left(\log \left(\frac{\varepsilon}{3.7\,D} + \frac{2.51}{\mathrm{Re}\sqrt{f_i}} \right) \right)^2} \tag{6.17}$$

6.7 CASE STUDY continued

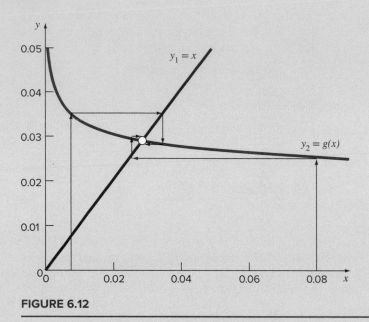

FIGURE 6.12

The two-curve display of this function depicted indicates a surprising result (Fig. 6.12). Recall that fixed-point iteration converges when the y_2 curve has a relatively flat slope (i.e., $|g'(\xi)| < 1$). As indicated by Fig. 6.12, the fact that the y_2 curve is quite flat in the range from $f = 0.008$ to 0.08 means that not only does fixed-point iteration converge, but it converges fairly rapidly! In fact, for initial guesses anywhere between 0.008 and 0.08, fixed-point iteration yields predictions with percent relative errors less than 0.008% in six or fewer iterations! Thus, this simple approach that requires only one guess and no derivative estimates performs really well for this particular case.

The take-home message from this case study is that even great, professionally developed software like MATLAB is not always foolproof. Further, there is usually no single method that works best for all problems. Sophisticated users understand the strengths and weaknesses of the available numerical techniques. In addition, they understand enough of the underlying theory so that they can effectively deal with situations where a method breaks down.

PROBLEMS

6.1 Employ fixed-point iteration to locate the root of

$$f(x) = \sin(\sqrt{x}) - x$$

Use an initial guess of $x_0 = 0.5$ and iterate until $\varepsilon_a \leq 0.01\%$. Verify that the process is linearly convergent as described at the end of Sec. 6.1.

6.2 Use (**a**) fixed-point iteration and (**b**) the Newton-Raphson method to determine a root of $f(x) = -0.9x^2 + 1.7x + 2.5$ using $x_0 = 5$. Perform the computation until ε_a is less than $\varepsilon_s = 0.01\%$. Also check your final answer.

6.3 Determine the highest real root of $f(x) = x^3 - 6x^2 + 11x - 6.1$:

(**a**) Graphically.

(**b**) Using the Newton-Raphson method (three iterations, $x_0 = 3.5$

(**c**) Using the secant method (three iterations, $x_{-1} = 2.5$ and $x_0 = 3.5$).

(**d**) Using the modified secant method (three iterations, $x_0 = 3.5$, $\delta = 0.01$).

(**e**) Determine all the roots with MATLAB.

6.4 Determine the lowest positive root of $f(x) = 7\sin(x)e^{-x} - 1$:

(**a**) Graphically.

(**b**) Using the Newton-Raphson method (three iterations, $x_0 = 0.3$).

(**c**) Using the secant method (three iterations, $x_{-1} = 0.5$ and $x_0 = 0.4$.

(**d**) Using the modified secant method (five iterations, $x_0 = 0.3$, $\delta = 0.01$).

6.5 Use (**a**) the Newton-Raphson method and (**b**) the modified secant method ($\delta = 0.05$) to determine a root of $f(x) = x^5 - 16.05x^4 + 88.75x^3 - 192.0375x^2 + 116.35x + 31.6875$ using an initial guess of $x = 0.5825$ and $\varepsilon_s = 0.01\%$. Explain your results.

6.6 Develop an M-file for the secant method. Along with the two initial guesses, pass the function as an argument. Test it by solving Prob. 6.3.

6.7 Develop an M-file for the modified secant method. Along with the initial guess and the perturbation fraction, pass the function as an argument. Test it by solving Prob. 6.3.

6.8 Differentiate Eq. (E6.4.1) to get Eq. (E6.4.2).

6.9 Employ the Newton-Raphson method to determine a real root for $f(x) = -2 + 6x - 4x^2 + 0.5x^3$, using an initial guess of (**a**) 4.5 and (**b**) 4.43. Discuss and use graphical and analytical methods to explain any peculiarities in your results.

6.10 The "divide and average" method, an old-time method for approximating the square root of any positive number a, can be formulated as

$$x_{i+1} = \frac{x_i + a/x_i}{2}$$

Prove that this formula is based on the Newton-Raphson algorithm.

6.11 (**a**) Apply the Newton-Raphson method to the function $f(x) = \tanh(x^2 - 9)$ to evaluate its known real root at $x = 3$. Use an initial guess of $x_0 = 3.2$ and take a minimum of three iterations. (**b**) Did the method exhibit convergence onto its real root? Sketch the plot with the results for each iteration labeled.

6.12 The polynomial $f(x) = 0.0074x^4 - 0.284x^3 + 3.355x^2 - 12.183x + 5$ has a real root between 15 and 20. Apply the Newton-Raphson method to this function using an initial guess of $x_0 = 16.15$. Explain your results.

6.13 Mechanical engineers, as well as most other engineers, use thermodynamics extensively in their work. The following polynomial can be used to relate the zero-pressure specific heat of dry air c_p in kJ/(kg K) to temperature in K:

$$c_p = 0.99403 + 1.671 \times 10^{-4}T + 9.7215 \times 10^{-8}T^2$$

$$-9.5838 \times 10^{-11}T^3 + 1.9520 \times 10^{-14}T^4$$

Write a MATLAB script (**a**) to plot c_p versus a range of $T = 0$ to 1200 K and (**b**) to determine the temperature that corresponds to a specific heat of 1.1 kJ/(kg K) with MATLAB polynomial functions.

6.14 In a chemical engineering process, water vapor (H_2O) is heated to sufficiently high temperatures that a significant portion of the water dissociates, or splits apart, to form oxygen (O_2) and hydrogen (H_2):

$$H_2O \rightleftharpoons H_2 + \tfrac{1}{2}O_2$$

If it is assumed that this is the only reaction involved, the mole fraction x of H_2O that dissociates can be represented by

$$K = \frac{x}{1-x}\sqrt{\frac{2p_t}{2+x}} \qquad (P6.14.1)$$

where K is the reaction's equilibrium constant and p_t is the total pressure of the mixture. If $p_t = 3$ atm and $K = 0.05$, determine the value of x that satisfies Eq. (P6.14.1).

6.15 The *Redlich-Kwong* equation of state is given by

$$p = \frac{RT}{v - b} - \frac{a}{v(v + b)\sqrt{T}}$$

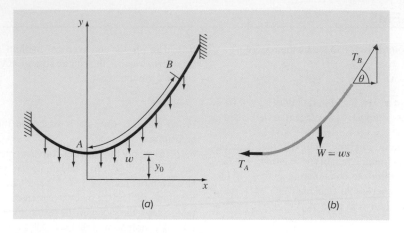

FIGURE P6.17

where R = the universal gas constant [= 0.518 kJ/(kg K)], T = absolute temperature (K), p = absolute pressure (kPa), and v = the volume of a kg of gas (m³/kg). The parameters a and b are calculated by

$$a = 0.427 \frac{R^2 T_c^{2.5}}{p_c} \qquad b = 0.0866R \frac{T_c}{p_c}$$

where p_c = 4600 kPa and T_c = 191 K. As a chemical engineer, you are asked to determine the amount of methane fuel that can be held in a 3-m³ tank at a temperature of −40 °C with a pressure of 65,000 kPa. Use a root-locating method of your choice to calculate v and then determine the mass of methane contained in the tank.

6.16 The volume of liquid V in a hollow horizontal cylinder of radius r and length L is related to the depth of the liquid h by

$$V = \left[r^2 \cos^{-1}\left(\frac{r-h}{r} \right) - (r-h)\sqrt{2rh - h^2} \right]L$$

Determine h given r = 2 m, L = 5 m, and V = 8 m³.

6.17 A catenary cable is one which is hung between two points not in the same vertical line. As depicted in Fig. P6.17a, it is subject to no loads other than its own weight. Thus, its weight acts as a uniform load per unit length along the cable w (N/m). A free-body diagram of a section AB is depicted in Fig. P6.17b, where T_A and T_B are the tension forces at the end. Based on horizontal and vertical force balances, the following differential equation model of the cable can be derived:

$$\frac{d^2 y}{dx^2} = \frac{w}{T_A}\sqrt{1 + \left(\frac{dy}{dx}\right)^2}$$

Calculus can be employed to solve this equation for the height of the cable y as a function of distance x:

$$y = \frac{T_A}{w}\cosh\left(\frac{w}{T_A}x\right) + y_0 - \frac{T_A}{w}$$

(a) Use a numerical method to calculate a value for the parameter T_A given values for the parameters w = 10 and y_0 = 5, such that the cable has a height of y = 15 at x = 50.

(b) Develop a plot of y versus x for x = −50 to 100.

6.18 An oscillating current in an electric circuit is described by $I = 9e^{-t}\sin(2\pi t)$, where t is in seconds. Determine all values of t such that I = 3.5

6.19 Figure P6.19 shows a circuit with a resistor, an inductor, and a capacitor in parallel. Kirchhoff's rules can be used to express the impedance of the system as

$$\frac{1}{Z} = \sqrt{\frac{1}{R^2} + \left(\omega C - \frac{1}{\omega L}\right)^2}$$

where Z = impedance (Ω), and ω is the angular frequency. Find the ω that results in an impedance of 100 Ω using the fzero function with initial guesses of 1 and 1000 for the following parameters: R = 225 Ω, C = 0.6 × 10⁻⁶ F, and L = 0.5 H.

FIGURE P6.19

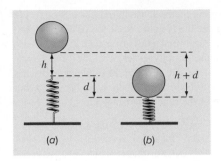

FIGURE P6.20

6.20 Real mechanical systems may involve the deflection of nonlinear springs. In Fig. P6.20, a block of mass m is released a distance h above a nonlinear spring. The resistance force F of the spring is given by

$$F = -(k_1 d + k_2 d^{3/2})$$

Conservation of energy can be used to show that

$$0 = \frac{2k_2 d^{5/2}}{5} + \frac{1}{2} k_1 d^2 - mgd - mgh$$

Solve for d, given the following parameter values: $k_1 = 40{,}000$ g/s^2, $k_2 = 40$ g/(s^2 m$^{0.5}$), $m = 95$ g, $g = 9.81$ m/s^2, and $h = 0.43$ m.

6.21 Aerospace engineers sometimes compute the trajectories of projectiles such as rockets. A related problem deals with the trajectory of a thrown ball. The trajectory of a ball thrown by a right fielder is defined by the (x, y) coordinates as displayed in Fig. P6.21. The trajectory can be modeled as

$$y = (\tan \theta_0)x - \frac{g}{2v_0^2 \cos^2 \theta_0} x^2 + y_0$$

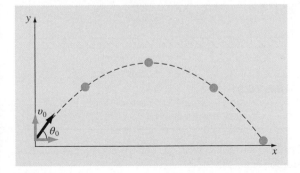

FIGURE P6.21

Find the appropriate initial angle θ_0, if $v_0 = 30$ m/s, and the distance to the catcher is 90 m. Note that the throw leaves the right fielder's hand at an elevation of 1.8 m and the catcher receives it at 1 m.

6.22 You are designing a spherical tank (Fig. P6.22) to hold water for a small village in a developing country. The volume of liquid it can hold can be computed as

$$V = \pi h^2 \frac{[3R - h]}{3}$$

where V = volume [m^3], h = depth of water in tank [m], and R = the tank radius [m].

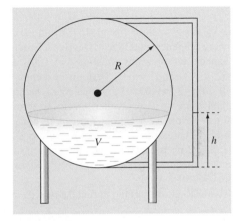

FIGURE P6.22

If $R = 3$ m, what depth must the tank be filled to so that it holds 30 m^3? Use three iterations of the most efficient numerical method possible to determine your answer. Determine the approximate relative error after each iteration. Also, provide justification for your choice of method. Extra information: (**a**) For bracketing methods, initial guesses of 0 and R will bracket a single root for this example. (**b**) For open methods, an initial guess of R will always converge.

6.23 Perform the identical MATLAB operations as those in Example 6.8 to manipulate and find all the roots of the polynomial

$$f_5(x) = (x + 2)(x + 5)(x - 6)(x - 4)(x - 8)$$

6.24 In control systems analysis, transfer functions are developed that mathematically relate the dynamics of a system's input to its output. A transfer function for a robotic positioning system is given by

$$G(s) = \frac{C(s)}{N(s)} = \frac{s^3 + 9s^2 + 26s + 24}{s^4 + 15s^3 + 77s^2 + 153s + 90}$$

where $G(s)$ = system gain, $C(s)$ = system output, $N(s)$ = system input, and s = Laplace transform complex frequency. Use MATLAB to find the roots of the numerator and denominator and factor these into the form

$$G(s) = \frac{(s + a_1)(s + a_2)(s + a_3)}{(s + b_1)(s + b_2)(s + b_3)(s + b_4)}$$

where a_i and b_i = the roots of the numerator and denominator, respectively.

6.25 The Manning equation can be written for a rectangular open channel as

$$Q = \frac{\sqrt{S}\,(BH)^{5/3}}{n(B + 2H)^{2/3}}$$

where Q = flow (m³/s), S = slope (m/m), H = depth (m), and n = the Manning roughness coefficient. Develop a fixed-point iteration scheme to solve this equation for H given $Q = 5$, $S = 0.0002$, $B = 20$, and $n = 0.03$. Perform the computation until ε_a is less than $\varepsilon_s = 0.05\%$. Prove that your scheme converges for all initial guesses greater than or equal to zero.

6.26 See if you can develop a foolproof function to compute the friction factor based on the Colebrook equation as described in Sec. 6.7. Your function should return a precise result for Reynolds number ranging from 4000 to 10^7 and for ε/D ranging from 0.00001 to 0.05.

6.27 Use the Newton-Raphson method to find the root of

$$f(x) = e^{-0.5x}(4 - x) - 2$$

Employ initial guesses of **(a)** 2, **(b)** 6, and **(c)** 8. Explain your results.

6.28 Given

$$f(x) = -2x^6 - 1.5x^4 + 10x + 2$$

Use a root-location technique to determine the maximum of this function. Perform iterations until the approximate relative error falls below 5%. If you use a bracketing method, use initial guesses of $x_l = 0$ and $x_u = 1$. If you use the Newton-Raphson or the modified secant method, use an initial guess of $x_i = 1$. If you use the secant method, use initial guesses of $x_{i-1} = 0$ and $x_i = 1$. Assuming that convergence is not an issue, choose the technique that is best suited to this problem. Justify your choice.

6.29 You must determine the root of the following easily differentiable function:

$$e^{0.5x} = 5 - 5x$$

Pick the best numerical technique, justify your choice, and then use that technique to determine the root. Note that it is known that for positive initial guesses, all techniques except fixed-point iteration will eventually converge. Perform iterations until the approximate relative error falls below 2%.

If you use a bracketing method, use initial guesses of $x_l = 0$. and $x_u = 2$. If you use the Newton-Raphson or the modified secant method, use an initial guess of $x_i = 0.7$. If you use the secant method, use initial guesses of $x_{i-1} = 0$ and $x_i = 2$.

6.30 (a) Develop an M-file function to implement Brent's root-location method. Base your function on Fig. 6.10, but with the beginning of the function changed to

```
function [b,fb] = fzeronew(f,xl,xu,varargin)
% fzeronew: Brent root location zeroes
% [b,fb] = fzeronew(f,xl,xu,p1,p2,...):
%   uses Brent's method to find the root of f
% input:
%   f = name of function
%   xl, xu = lower and upper guesses
%   p1,p2,... = additional parameters used by f
% output:
%   b = real root
%   fb = function value at root
```

Make the appropriate modifications so that the function performs as outlined in the documentation statements. In addition, include error traps to ensure that the function's three required arguments (f,xl,xu) are prescribed, and that the initial guesses bracket a root.

(b) Test your function by using it to solve for the root of the function from Example 5.6 using

```
>> [x,fx] = fzeronew(@(x,n) x^n-1,0,1.3,10)
```

6.31 Figure P6.31 shows a side view of a broad crested weir. The symbols shown in Fig. P6.31 are defined as: H_w = the height of the weir (m), H_h = the head above the weir (m), and $H = H_w + H_h$ = the depth of the river upstream of the weir (m).

FIGURE P6.31
A broad-crested weir used to control depth and velocity of rivers and streams.

The flow across the weir, Q_w (m³/s), can be computed as (Munson et al., 2009)

$$Q_w = C_w B_w \sqrt{g}\left(\frac{2}{3}\right)^{3/2} H_h^{3/2} \qquad (P6.31.1)$$

where C_w = a weir coefficient (dimensionless), B_w = the weir width (m), and g = the gravitational constant (m/s²). C_w can be determined using the weir height (H_w) as in

$$C_w = 1.125 \sqrt{\frac{1 + H_h/H_w}{2 + H_h/H_w}} \qquad (P6.31.2)$$

Given $g = 9.81$ m/s², $H_w = 0.8$ m, $B_w = 8$ m, and $Q_w = 1.3$ m³/s, determine the upstream depth, H, using **(a)** the modified secant method with $\delta = 10^{-5}$, **(b)** fixed-point iteration, and **(c)** the MATLAB function fzero. For all cases employ an initial guess of $0.5H_w$, which for this case is 0.4. For **(b)** also prove that your result will be convergent for positive initial guesses.

6.32 The following reversible chemical reaction describes how gaseous phases of methane and water react to form carbon dioxide and hydrogen in a closed reactor,

$$CH_4 + 2H_2O \Leftrightarrow CO_2 + 4H_2$$

with the equilibrium relationship

$$K = \frac{[CO_2][H_2]^4}{[CH_4][H_2O]^2}$$

where K = the equilibrium coefficient and the brackets [] designate molar concentrations (mole/L). Conservation of mass can be used to reformulate the equilibrium relationship as

$$K = \frac{\left(\frac{x}{V}\right)\left(\frac{4x}{V}\right)^4}{\left(\frac{M_{CH_4} - x}{V}\right)\left(\frac{M_{H_2O} - 2x}{V}\right)_2}$$

where x = the number of moles created in the forward direction (mole), V = the volume of the reactor (L), and M_i = the initial number of moles of constituent i (mole). Given that $K = 7 \times 10^{-3}$, $V = 20$ L, and $M_{CH_4} = M_{H_2O} = 1$ moles, determine x using **(a)** fixed-point iteration and **(b)** fzero.

6.33 The concentration of pollutant bacteria c in a lake decreases according to

$$c = 77e^{-1.5t} + 20e^{-0.08t}$$

Determine the time required for the bacteria concentration to be reduced to 15 using the Newton-Raphson method with an initial guess of $t = 6$ and a stopping criterion of 1%. Check your result with fzero.

6.34 You are asked to solve for the root of the following equation with fixed-point iteration:

$$x^4 = 5x + 10$$

Determine the solution approach that converges for initial guesses in the range of $0 < x < 7$. Use either a graphical or analytical approach to prove that your formulation always converges in the given range.

6.35 A circular pipe made out of new cast iron is used to convey water at a volume flow rate of $Q = 0.3$ m³/s. Assume that the flow is steady and fully developed and the water is incompressible. The head loss, friction, and diameter are related by the *Darcy-Weisbach equation*,

$$h_L = f \frac{LV^2}{D2g} \qquad (P6.35.1)$$

where f = the friction factor (dimensionless), L = length (m), D = pipe inner diameter (m), V = velocity (m/s), and g = the gravitational constant (= 9.81 m/s²). The velocity can be related to flow by

$$Q = A_c V \qquad (P6.35.2)$$

where A_c = the pipe's cross-sectional area (m²) = $\pi D^2/4$ and the friction factor can be determined by the *Colebrook equation*. If you want the head loss to be less than 0.006 m per meter of pipe, develop a MATLAB function to determine the smallest diameter pipe to achieve this objective. Use the following parameter values: $v = 1.16 \times 10^{-6}$ m²/s and $\varepsilon = 0.4$ mm.

6.36 Figure P6.36 shows an asymmetric diamond-shaped supersonic airfoil. The orientation of the airfoil relative to the airflow is represented by a number of angles: α = the angle of attack, β = the shock angle, θ = the deflection angle, with the subscripts "l" and "u" designating the lower and upper surfaces of the airfoil. The following formula relates the deflection angle to the oblique shock angle and speed,

$$\tan \theta = \frac{2\cot \beta(M^2 \sin^2 \beta - 1)}{M^2(k + \cos 2\beta + 2)}$$

FIGURE P6.36
A diamond-shaped airfoil.

where M = the *Mach number* which is the ratio of the jet's speed, v (m/s), to the speed of sound, c (m/s), where

$$c = \sqrt{kRT_a}$$

where k = the ratio of specific heats which for air is c_p/c_v (= 1.4), R = the air gas constant (= 287 N m/(kg K)), and T_a = the air's absolute temperature (K). Given estimates of M, k, and θ, the shock angle can be determined as the root of

$$f(\beta) = \frac{2\cot\beta(M^2\sin^2\beta - 1)}{M^2(k + \cos2\beta + 2)} - \tan\theta$$

The pressure on the airfoil surface, p_a (kPa), can then be computed as

$$p_a = p\left(\frac{2k}{k+1}(M\sin\beta)^2 - \frac{k-1}{k+1}\right)$$

Suppose that the airfoil is attached to a jet traveling at a speed $v = 625$ m/s through air with a temperature $T = 4\,^\circ$C, pressure $p = 110$ kPa, and $\theta_u = 4^\circ$. Develop a MATLAB script to (a) generate a plot of $f(\beta_u)$ versus $\beta_u = 2^\circ$ to 88°, and (b) compute the pressure on the upper surface of the airfoil.

6.37 As described in Sec. 1.4, for objects falling through fluids at very low speeds, the flow regime around the object will be laminar and the relationship between the drag force and velocity is linear. In addition, in such cases, the buoyancy force must also be included. For such cases, a force balance can be written as

$$\frac{dv}{dt} = g \quad - \quad \frac{\rho_f V}{m}g \quad - \quad \frac{c_d}{m}v \qquad \text{(P6.37)}$$
$$\text{(gravity)} \quad \text{(buoyancy)} \quad \text{(drag)}$$

where v = velocity (m/s), t = time (s), m = the mass of the particle (kg), g = the gravitational constant (= 9.81 m/s²), ρ_f = fluid density (kg/m³), V = particle volume (m³), and c_d = the linear drag coefficient (kg/m). Note that the mass of the particle can be computed as $V\rho_s$, where ρ_s = the density of the particle (kg/m³). For a small sphere, Stokes developed the following formula for the drag coefficient, $c_d = 6\pi\mu r$, where μ = the fluid's dynamic viscosity (N s/m²), and r = the sphere's radius (m).

You release an iron sphere at the surface of a container ($x = 0$) filled with honey (Fig. P6.37) and then measure how long it takes to settle to the bottom ($x = L$). Use this information to estimate the honey's viscosity based on the following parameter values: $\rho_f = 1420$ kg/m³, $\rho_s = 7850$ kg/m³, $r = 0.02$ m, $L = 0.5$ m, and $t(x = 0.5) = 3.6$ s. Check the Reynolds number (Re = $\rho_f v d/\mu$, where d = diameter) to confirm that laminar conditions occurred during the experiment.

[Hint: The problem can be solved by integrating Eq. (P6.37) two times to yield an equation for x as a function of t.]

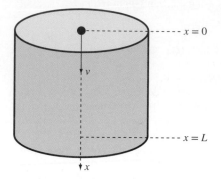

FIGURE P6.37
A sphere settling in a cylinder filled with viscous honey.

6.38 As depicted in Fig. P6.38a, a scoreboard is suspended above a sports arena by two cables, pinned at A, B, and C. The cables are initially horizontal and of length L. After the scoreboard is hung, a free-body diagram at node B can be developed as shown in Fig. P6.38b. Assuming that the weight of each cable is negligible, determine the deflection, d (m), that results if the scoreboard weighs $W = 9000$ N. Also, compute how much each cable is elongated. Note that each cable obeys Hooke's law such that the axial elongation is represented by $L' - L = FL/(A_c E)$ where F = the axial force (N), A_c = the cable's cross-sectional area (m²), and E = the modulus of elasticity (N/m²). Use the following parameters for your calculations: $L = 45$ m, $A_c = 6.362 \times 10^{-4}$ m², and $E = 1.5 \times 10^{11}$ N/m².

FIGURE P6.38
(a) Two thin cables pinned at A, B, and C with a scoreboard suspended from B. (b) Free-body diagram of the pin at B after the scoreboard is hung.

6.39 A water tower is connected to a pipe with a valve at its end as depicted in Fig. P6.39. Under a number of simplifying assumptions (e.g., minor friction losses neglected), the following energy balance can be written

$$gh - \frac{v^2}{2} = f\left(\frac{L+h}{d} + \frac{L_{e,e}}{d} + \frac{L_{e,v}}{d}\right)\frac{v^2}{2} + K\frac{v^2}{2}$$

where g = gravitational acceleration (= 9.81 m/s²), h = tower height (m), v = mean water velocity in pipe (m/s), f = the pipe's friction factor, L = horizontal pipe length (m),

FIGURE P6.39
A water tower connected to a pipe with a valve at its end.

d = pipe diameter (m), $L_{e,e}$ = equivalent length for the elbow (m), $L_{e,v}$ = equivalent length for the valve (m), and K = loss coefficient for the contraction at the bottom of the tank. Write a MATLAB script to determine the flow exiting the valve, Q (m³/s), using the following parameter values: h = 24 m, L = 65 m, d = 100 mm, $L_{e,e}/d$ = 30, $L_{e,v}/d$ = 8, and K = 0.5. In addition, the kinematic viscosity of water is $v = \mu/\rho = 1.2 \times 10^{-6}$ m²/s.

6.40 Modify the fzerosimp function (Fig. 6.10) so that it can be passed any function with a single unknown and uses varargin to pass the function's parameters. Then test it with the following script to obtain a solution for pipe friction based on Case Study 6.3,

```
clc
format long, format compact
rho=1.23;mu=1.79e-5;D=0.005;V=40;e=0.0015/1000;
Re=rho*V*D/mu;
g=@(f,e,D) 1/sqrt(f)+2*log10(e/(3.7*D)+2.51/
(Re*sqrt(f)));
f=fzerosimp(@(x) g(x,e,D),0.008,0.08)
```

7

Optimization

YOU'VE GOT A PROBLEM

An object like a bungee jumper can be projected upward at a specified velocity. If it is subject to linear drag, its altitude as a function of time can be computed as

$$z = z_0 + \frac{m}{c}\left(v_0 + \frac{mg}{c}\right)\left(1 - e^{-(c/m)t}\right) - \frac{mg}{c}t \tag{7.1}$$

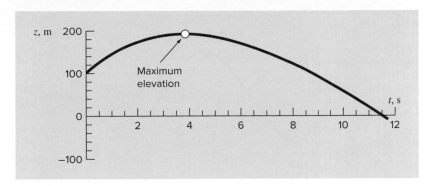

FIGURE 7.1
Elevation as a function of time for an object initially projected upward with an initial velocity.

where z = altitude (m) above the earth's surface (defined as $z = 0$), z_0 = the initial altitude (m), m = mass (kg), c = a linear drag coefficient (kg/s), v_0 = initial velocity (m/s), and t = time (s). Note that for this formulation, positive velocity is considered to be in the upward direction. Given the following parameter values: $g = 9.81$ m/s^2, $z_0 = 100$ m, $v_0 = 55$ m/s, $m = 80$ kg, and $c = 15$ kg/s, Eq. (7.1) can be used to calculate the jumper's altitude. As displayed in Fig. 7.1, the jumper rises to a peak elevation of about 190 m at about $t = 4$ s.

Suppose that you are given the job of determining the exact time of the peak elevation. The determination of such extreme values is referred to as optimization. This chapter will introduce you to how the computer is used to make such determinations.

7.1 INTRODUCTION AND BACKGROUND

In the most general sense, optimization is the process of creating something that is as effective as possible. As engineers, we must continuously design devices and products that perform tasks in an efficient fashion for the least cost. Thus, engineers are always confronting optimization problems that attempt to balance performance and limitations. In addition, scientists have interest in optimal phenomena ranging from the peak elevation of projectiles to the minimum free energy.

From a mathematical perspective, optimization deals with finding the maxima and minima of a function that depends on one or more variables. The goal is to determine the values of the variables that yield maxima or minima for the function. These can then be substituted back into the function to compute its optimal values.

Although these solutions can sometimes be obtained analytically, most practical optimization problems require numerical, computer solutions. From a numerical standpoint, optimization is similar in spirit to the root-location methods we just covered in Chaps. 5 and 6. That is, both involve guessing and searching for a point on a function. The fundamental difference between the two types of problems is illustrated in Fig. 7.2. Root location involves searching for the location where the function equals zero. In contrast, optimization involves searching for the function's extreme points.

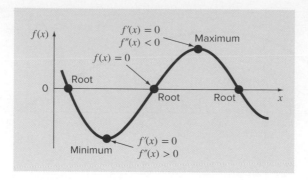

FIGURE 7.2
A function of a single variable illustrating the difference between roots and optima.

As can be seen in Fig. 7.2, the optimums are the points where the curve is flat. In mathematical terms, this corresponds to the x value where the derivative $f'(x)$ is equal to zero. Additionally, the second derivative, $f''(x)$, indicates whether the optimum is a minimum or a maximum: if $f''(x) < 0$, the point is a maximum; if $f''(x) > 0$, the point is a minimum.

Now, understanding the relationship between roots and optima would suggest a possible strategy for finding the latter. That is, you can differentiate the function and locate the root (i.e., the zero) of the new function. In fact, some optimization methods do just this by solving the root problem: $f'(x) = 0$.

EXAMPLE 7.1   Determining the Optimum Analytically by Root Location

Problem Statement. Determine the time and magnitude of the peak elevation based on Eq. (7.1). Use the following parameter values for your calculation: $g = 9.81$ m/s^2, $z_0 = 100$ m, $v_0 = 55$ m/s, $m = 80$ kg, and $c = 15$ kg/s.

Solution. Equation (7.1) can be differentiated to give

$$\frac{dz}{dt} = v_0 e^{-(c/m)t} - \frac{mg}{c}(1 - e^{-(c/m)t}) \tag{E7.1.1}$$

Note that because $v = dz/dt$, this is actually the equation for the velocity. The maximum elevation occurs at the value of t that drives this equation to zero. Thus, the problem amounts to determining the root. For this case, this can be accomplished by setting the derivative to zero and solving Eq. (E7.1.1) analytically for

$$t = \frac{m}{c} \ln\left(1 + \frac{cv_0}{mg}\right)$$

Substituting the parameters gives

$$t = \frac{80}{15} \ln\left(1 + \frac{15(55)}{80(9.81)}\right) = 3.83166 \text{ s}$$

This value along with the parameters can then be substituted into Eq. (7.1) to compute the maximum elevation as

$$z = 100 + \frac{80}{15}\left(50 + \frac{80(9.81)}{15}\right)\left(1 - e^{-(15/80)3.83166}\right) - \frac{80(9.81)}{15}(3.83166) = 192.8609 \text{ m}$$

We can verify that the result is a maximum by differentiating Eq. (E7.1.1) to obtain the second derivative

$$\frac{d^2z}{dt^2} = -\frac{c}{m}v_0 e^{-(c/m)t} - g e^{-(c/m)t} = -9.81 \frac{\text{m}}{\text{s}^2}$$

The fact that the second derivative is negative tells us that we have a maximum. Further, the result makes physical sense since the acceleration should be solely equal to the force of gravity at the maximum when the vertical velocity (and hence drag) is zero.

Although an analytical solution was possible for this case, we could have obtained the same result using the root-location methods described in Chaps. 5 and 6. This will be left as a homework exercise.

Although it is certainly possible to approach optimization as a roots problem, a variety of direct numerical optimization methods are available. These methods are available for both one-dimensional and multidimensional problems. As the name implies, one-dimensional problems involve functions that depend on a single dependent variable. As in Fig. 7.3a, the search then consists of climbing or descending one-dimensional peaks and valleys. Multidimensional problems involve functions that depend on two or more dependent variables.

FIGURE 7.3
(a) One-dimensional optimization. This figure also illustrates how minimization of $f(x)$ is equivalent to the maximization of $-f(x)$. (b) Two-dimensional optimization. Note that this figure can be taken to represent either a maximization (contours increase in elevation up to the maximum like a mountain) or a minimization (contours decrease in elevation down to the minimum like a valley).

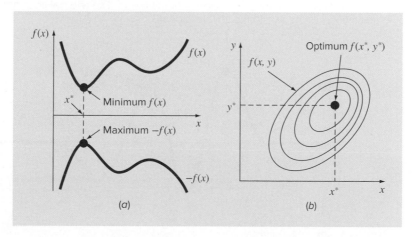

In the same spirit, a two-dimensional optimization can again be visualized as searching out peaks and valleys (Fig. 7.3b). However, just as in real hiking, we are not constrained to walk a single direction; instead the topography is examined to efficiently reach the goal.

Finally, the process of finding a maximum versus finding a minimum is essentially identical because the same value x^* both minimizes $f(x)$ and maximizes $-f(x)$. This equivalence is illustrated graphically for a one-dimensional function in Fig. 7.3a.

In the next section, we will describe some of the more common approaches for one-dimensional optimization. Then we will provide a brief description of how MATLAB can be employed to determine optima for multidimensional functions.

7.2 ONE-DIMENSIONAL OPTIMIZATION

This section will describe techniques to find the minimum or maximum of a function of a single variable $f(x)$. A useful image in this regard is the one-dimensional "roller coaster"–like function depicted in Fig. 7.4. Recall from Chaps. 5 and 6 that root location was complicated by the fact that several roots can occur for a single function. Similarly, both local and global optima can occur in optimization.

A *global optimum* represents the very best solution. A *local optimum,* though not the very best, is better than its immediate neighbors. Cases that include local optima are called *multimodal.* In such cases, we will almost always be interested in finding the global optimum. In addition, we must be concerned about mistaking a local result for the global optimum.

Just as in root location, optimization in one dimension can be divided into bracketing and open methods. As described in the next section, the golden-section search is an example of a bracketing method that is very similar in spirit to the bisection method for root location. This is followed by a somewhat more sophisticated bracketing approach—parabolic interpolation. We will then show how these two methods are combined and implemented with MATLAB's fminbnd function.

FIGURE 7.4
A function that asymptotically approaches zero at plus and minus ∞ and has two maximum and two minimum points in the vicinity of the origin. The two points to the right are local optima, whereas the two to the left are global.

7.2.1 Golden-Section Search

In many cultures, certain numbers are ascribed magical qualities. For example, we in the West are all familiar with "lucky 7" and "Friday the 13th." Beyond such superstitious quantities, there are several well-known numbers that have such interesting and powerful mathematical properties that they could truly be called "magical." The most common of these are the ratio of a circle's circumference to its diameter π and the base of the natural logarithm e.

Although not as widely known, the *golden ratio* should surely be included in the pantheon of remarkable numbers. This quantity, which is typically represented by the Greek letter ϕ (pronounced: fee), was originally defined by Euclid (ca. 300 BCE) because of its role in the construction of the pentagram or five-pointed star. As depicted in Fig. 7.5, Euclid's definition reads: "A straight line is said to have been cut in extreme and mean ratio when, as the whole line is to the greater segment, so is the greater to the lesser."

The actual value of the golden ratio can be derived by expressing Euclid's definition as

$$\frac{\ell_1 + \ell_2}{\ell_1} = \frac{\ell_1}{\ell_2} \tag{7.2}$$

Multiplying by ℓ_1/ℓ_2 and collecting terms yields

$$\phi^2 - \phi - 1 = 0 \tag{7.3}$$

where $\phi = \ell_1/\ell_2$. The positive root of this equation is the golden ratio:

$$\phi = \frac{1 + \sqrt{5}}{2} = 1.61803398874989 \ldots \tag{7.4}$$

The golden ratio has long been considered aesthetically pleasing in Western cultures. In addition, it arises in a variety of other contexts including biology. For our purposes, it provides the basis for the golden-section search, a simple, general-purpose method for determining the optimum of a single-variable function.

The golden-section search is similar in spirit to the bisection approach for locating roots in Chap. 5. Recall that bisection hinged on defining an interval, specified by a lower guess (x_l) and an upper guess (x_u) that bracketed a single root. The presence of a root between these bounds was verified by determining that $f(x_l)$ and $f(x_u)$ had different signs. The root was then estimated as the midpoint of this interval:

$$x_r = \frac{x_l + x_u}{2} \tag{7.5}$$

FIGURE 7.5
Euclid's definition of the golden ratio is based on dividing a line into two segments so that the ratio of the whole line to the larger segment is equal to the ratio of the larger segment to the smaller segment. This ratio is called the golden ratio.

The final step in a bisection iteration involved determining a new smaller bracket. This was done by replacing whichever of the bounds x_l or x_u had a function value with the same sign as $f(x_r)$. A key advantage of this approach was that the new value x_r replaced one of the old bounds.

Now suppose that instead of a root, we were interested in determining the minimum of a one-dimensional function. As with bisection, we can start by defining an interval that contains a single answer. That is, the interval should contain a single minimum, and hence is called *unimodal*. We can adopt the same nomenclature as for bisection, where x_l and x_u defined the lower and upper bounds, respectively, of such an interval. However, in contrast to bisection, we need a new strategy for finding a minimum within the interval. Rather than using a single intermediate value (which is sufficient to detect a sign change, and hence a zero), we would need two intermediate function values to detect whether a minimum occurred.

The key to making this approach efficient is the wise choice of the intermediate points. As in bisection, the goal is to minimize function evaluations by replacing old values with new values. For bisection, this was accomplished by choosing the midpoint. For the golden-section search, the two intermediate points are chosen according to the golden ratio:

$$x_1 = x_l + d \tag{7.6}$$
$$x_2 = x_u - d \tag{7.7}$$

where

$$d = (\phi - 1)(x_u - x_l) \tag{7.8}$$

The function is evaluated at these two interior points. Two results can occur:

1. If, as in Fig. 7.6a, $f(x_1) < f(x_2)$, then $f(x_1)$ is the minimum, and the domain of x to the left of x_2, from x_l to x_2, can be eliminated because it does not contain the minimum. For this case, x_2 becomes the new x_l for the next round.
2. If $f(x_2) < f(x_1)$, then $f(x_2)$ is the minimum and the domain of x to the right of x_1, from x_1 to x_u would be eliminated. For this case, x_1 becomes the new x_u for the next round.

Now, here is the real benefit from the use of the golden ratio. Because the original x_1 and x_2 were chosen using the golden ratio, we do not have to recalculate all the function values for the next iteration. For example, for the case illustrated in Fig. 7.6, the old x_1 becomes the new x_2. This means that we already have the value for the new $f(x_2)$, since it is the same as the function value at the old x_1.

To complete the algorithm, we need only determine the new x_1. This is done with Eq. (7.6) with d computed with Eq. (7.8) based on the new values of x_l and x_u. A similar approach would be used for the alternate case where the optimum fell in the left subinterval. For this case, the new x_2 would be computed with Eq. (7.7).

As the iterations are repeated, the interval containing the extremum is reduced rapidly. In fact, each round the interval is reduced by a factor of $\phi - 1$ (about 61.8%). That means that after 10 rounds, the interval is shrunk to about 0.618^{10} or 0.008 or 0.8% of its initial length. After 20 rounds, it is about 0.0066%. This is not quite as good as the reduction achieved with bisection (50%), but optimization is a harder problem than root location.

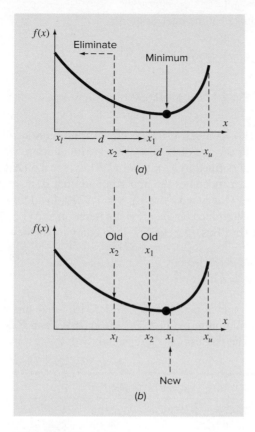

FIGURE 7.6
(*a*) The initial step of the golden-section search algorithm involves choosing two interior points according to the golden ratio. (*b*) The second step involves defining a new interval that encompasses the optimum.

EXAMPLE 7.2   Golden-Section Search

Problem Statement.   Use the golden-section search to find the minimum of

$$f(x) = \frac{x^2}{10} - 2 \sin x$$

within the interval from $x_l = 0$ to $x_u = 4$.

Solution.   First, the golden ratio is used to create the two interior points:

$$d = 0.61803(4 - 0) = 2.4721$$
$$x_1 = 0 + 2.4721 = 2.4721$$
$$x_2 = 4 - 2.4721 = 1.5279$$

The function can be evaluated at the interior points:

$$f(x_2) = \frac{1.5279^2}{10} - 2\sin(1.5279) = -1.7647$$

$$f(x_1) = \frac{2.4721^2}{10} - 2\sin(2.4721) = -0.6300$$

Because $f(x_2) < f(x_1)$, our best estimate of the minimum at this point is that it is located at $x = 1.5279$ with a value of $f(x) = -1.7647$. In addition, we also know that the minimum is in the interval defined by x_l, x_2, and x_1. Thus, for the next iteration, the lower bound remains $x_l = 0$, and x_1 becomes the upper bound, that is, $x_u = 2.4721$. In addition, the former x_2 value becomes the new x_1, that is, $x_1 = 1.5279$. In addition, we do not have to recalculate $f(x_1)$, it was determined on the previous iteration as $f(1.5279) = -1.7647$.

All that remains is to use Eqs. (7.8) and (7.7) to compute the new value of d and x_2:

$$d = 0.61803(2.4721 - 0) = 1.5279$$

$$x_2 = 2.4721 - 1.5279 = 0.9443$$

The function evaluation at x_2 is $f(0.9943) = -1.5310$. Since this value is less than the function value at x_1, the minimum is $f(1.5279) = -1.7647$, and it is in the interval prescribed by x_2, x_1, and x_u. The process can be repeated, with the results tabulated here:

i	x_l	$f(x_l)$	x_2	$f(x_2)$	x_1	$f(x_1)$	x_u	$f(x_u)$	d
1	0	0	1.5279	−1.7647	2.4721	−0.6300	4.0000	3.1136	2.4721
2	0	0	0.9443	−1.5310	1.5279	−1.7647	2.4721	−0.6300	1.5279
3	0.9443	−1.5310	1.5279	−1.7647	1.8885	−1.5432	2.4721	−0.6300	0.9443
4	0.9443	−1.5310	1.3050	−1.7595	1.5279	−1.7647	1.8885	−1.5432	0.5836
5	1.3050	−1.7595	1.5279	−1.7647	1.6656	−1.7136	1.8885	−1.5432	0.3607
6	1.3050	−1.7595	1.4427	−1.7755	1.5279	−1.7647	1.6656	−1.7136	0.2229
7	1.3050	−1.7595	1.3901	−1.7742	1.4427	−1.7755	1.5279	−1.7647	0.1378
8	1.3901	−1.7742	1.4427	−1.7755	1.4752	−1.7732	1.5279	−1.7647	0.0851

Note that the current minimum is highlighted for every iteration. After the eighth iteration, the minimum occurs at $x = 1.4427$ with a function value of -1.7755. Thus, the result is converging on the true value of -1.7757 at $x = 1.4276$.

Recall that for bisection (Sec. 5.4), an exact upper bound for the error can be calculated at each iteration. Using similar reasoning, an upper bound for golden-section search can be derived as follows: Once an iteration is complete, the optimum will either fall in one of two intervals. If the optimum function value is at x_2, it will be in the lower interval (x_l, x_2, x_1). If the optimum function value is at x_1, it will be in the upper interval (x_2, x_1, x_u). Because the interior points are symmetrical, either case can be used to define the error.

Looking at the upper interval (x_2, x_1, x_u), if the true value were at the far left, the maximum distance from the estimate would be

$$\Delta x_a = x_1 - x_2$$
$$= x_1 + (\phi - 1)(x_u - x_1) - x_u + (\phi - 1)(x_u - x_l)$$
$$= (x_l - x_u) + 2(\phi - 1)(x_u - x_l)$$
$$= (2\phi - 3)(x_u - x_l)$$

or $0.2361 (x_u - x_l)$. If the true value were at the far right, the maximum distance from the estimate would be

$$\Delta x_b = x_u - x_1$$
$$= x_u - x_l - (\phi - 1)(x_u - x_l)$$
$$= (x_u - x_l) - (\phi - 1)(x_u - x_l)$$
$$= (2 - \phi)(x_u - x_l)$$

or $0.3820 (x_u - x_l)$. Therefore, this case would represent the maximum error. This result can then be normalized to the optimal value for that iteration x_{opt} to yield

$$\varepsilon_a = (2 - \phi) \left| \frac{x_u - x_l}{x_{\text{opt}}} \right| \times 100\% \tag{7.9}$$

This estimate provides a basis for terminating the iterations.

An M-file function for the golden-section search for minimization is presented in Fig. 7.7. The function returns the location of the minimum, the value of the function, the approximate error, and the number of iterations.

The M-file can be used to solve the problem from Example 7.1.

```
>> g=9.81;v0=55;m=80;c=15;z0=100;
>> z=@(t) -(z0+m/c*(v0+m*g/c)*(1-exp(-c/m*t))-m*g/c*t);
>> [xmin,fmin,ea]=goldmin(z,0,8)

xmin =
     3.8317
fmin =
 -192.8609
ea =
  6.9356e-005
```

Notice how because this is a maximization, we have entered the negative of Eq. (7.1). Consequently, fmin corresponds to a maximum height of 192.8609.

You may be wondering why we have stressed the reduced function evaluations of the golden-section search. Of course, for solving a single optimization, the speed savings would be negligible. However, there are two important contexts where minimizing the number of function evaluations can be important. These are

1. Many evaluations. There are cases where the golden-section search algorithm may be a part of a much larger calculation. In such cases, it may be called many times. Therefore, keeping function evaluations to a minimum could pay great dividends for such cases.

```
function [x,fx,ea,iter]=goldmin(f,xl,xu,es,maxit,varargin)
% goldmin: minimization golden section search
% [x,fx,ea,iter]=goldmin(f,xl,xu,es,maxit,p1,p2,...):
%    uses golden section search to find the minimum of f
% input:
%   f = name of function
%   xl, xu = lower and upper guesses
%   es = desired relative error (default = 0.0001%)
%   maxit = maximum allowable iterations (default = 50)
%   p1,p2,... = additional parameters used by f
% output:
%   x = location of minimum
%   fx = minimum function value
%   ea = approximate relative error (%)
%   iter = number of iterations

if nargin<3,error('at least 3 input arguments required'),end
if nargin<4|isempty(es), es=0.0001;end
if nargin<5|isempty(maxit), maxit=50;end
phi=(1+sqrt(5))/2; iter = 0;
d = (phi-1)*(xu - xl);
x1 = xl + d; x2 = xu - d;
f1 = f(x1,varargin{:}); f2 = f(x2,varargin{:});
while(1)
  xint = xu - xl;
  if f1 < f2
    xopt = x1; xl = x2; x2 = x1; f2 = f1;
    x1 = xl + (phi-1)*(xu-xl); f1 = f(x1,varargin{:});
  else
    xopt = x2; xu = x1; x1 = x2; f1 = f2;
    x2 = xu - (phi-1)*(xu-xl); f2 = f(x2,varargin{:});
  end
  iter=iter+1;
  if xopt~=0, ea = (2 - phi) * abs(xint / xopt) * 100;end
  if ea <= es | iter >= maxit,break,end
end
x=xopt; fx=f(xopt,varargin{:});
```

FIGURE 7.7
An M-file to determine the minimum of a function with the golden-section search.

2. Time-consuming evaluation. For pedagogical reasons, we use simple functions in most of our examples. You should understand that a function can be very complex and time-consuming to evaluate. For example, optimization can be used to estimate the parameters of a model consisting of a system of differential equations. For such cases, the "function" involves time-consuming model integration. Any method that minimizes such evaluations would be advantageous.

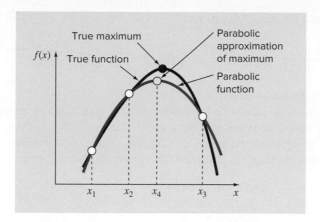

FIGURE 7.8
Graphical depiction of parabolic interpolation.

7.2.2 Parabolic Interpolation

Parabolic interpolation takes advantage of the fact that a second-order polynomial often provides a good approximation to the shape of $f(x)$ near an optimum (Fig. 7.8).

Just as there is only one straight line connecting two points, there is only one parabola connecting three points. Thus, if we have three points that jointly bracket an optimum, we can fit a parabola to the points. Then we can differentiate it, set the result equal to zero, and solve for an estimate of the optimal x. It can be shown through some algebraic manipulations that the result is

$$x_4 = x_2 - \frac{1}{2} \frac{(x_2 - x_1)^2 [f(x_2) - f(x_3)] - (x_2 - x_3)^2 [f(x_2) - f(x_1)]}{(x_2 - x_1) [f(x_2) - f(x_3)] - (x_2 - x_3) [f(x_2) - f(x_1)]} \tag{7.10}$$

where x_1, x_2, and x_3 are the initial guesses, and x_4 is the value of x that corresponds to the optimum value of the parabolic fit to the guesses.

EXAMPLE 7.3   Parabolic Interpolation

Problem Statement. Use parabolic interpolation to approximate the minimum of

$$f(x) = \frac{x^2}{10} - 2 \sin x$$

with initial guesses of $x_1 = 0$, $x_2 = 1$, and $x_3 = 4$.

Solution. The function values at the three guesses can be evaluated:

$$
\begin{aligned}
x_1 &= 0 & f(x_1) &= 0 \\
x_2 &= 1 & f(x_2) &= -1.5829 \\
x_3 &= 4 & f(x_3) &= 3.1136
\end{aligned}
$$

and substituted into Eq. (7.10) to give

$$x_4 = 1 - \frac{1}{2} \frac{(1-0)^2 [-1.5829 - 3.1136] - (1-4)^2 [-1.5829 - 0]}{(1-0) [-1.5829 - 3.1136] - (1-4) [-1.5829 - 0]} = 1.5055$$

which has a function value of $f(1.5055) = -1.7691$.

Next, a strategy similar to the golden-section search can be employed to determine which point should be discarded. Because the function value for the new point is lower than for the intermediate point (x_2) and the new x value is to the right of the intermediate point, the lower guess (x_1) is discarded. Therefore, for the next iteration:

$$x_1 = 1 \qquad f(x_1) = -1.5829$$
$$x_2 = 1.5055 \qquad f(x_2) = -1.7691$$
$$x_3 = 4 \qquad f(x_3) = 3.1136$$

which can be substituted into Eq. (7.10) to give

$$x_4 = 1.5055 - \frac{1}{2} \frac{(1.5055-1)^2 [-1.7691 - 3.1136] - (1.5055-4)^2 [-1.7691 - (-1.5829)]}{(1.5055-1) [-1.7691 - 3.1136] - (1.5055-4) [-1.7691 - (-1.5829)]}$$

$$= 1.4903$$

which has a function value of $f(1.4903) = -1.7714$. The process can be repeated, with the results tabulated here:

i	x_1	$f(x_1)$	x_2	$f(x_2)$	x_3	$f(x_3)$	x_4	$f(x_4)$
1	0.0000	0.0000	1.0000	-1.5829	4.0000	3.1136	1.5055	-1.7691
2	1.0000	-1.5829	1.5055	-1.7691	4.0000	3.1136	1.4903	-1.7714
3	1.0000	-1.5829	1.4903	-1.7714	1.5055	-1.7691	1.4256	-1.7757
4	1.0000	-1.5829	1.4256	-1.7757	1.4903	-1.7714	1.4266	-1.7757
5	1.4256	-1.7757	1.4266	-1.7757	1.4903	-1.7714	1.4275	-1.7757

Thus, within five iterations, the result is converging rapidly on the true value of -1.7757 at $x = 1.4276$.

7.2.3 MATLAB Function: fminbnd

Recall that in Sec. 6.4 we described Brent's method for root location, which combined several root-finding methods into a single algorithm that balanced reliability with efficiency. Because of these qualities, it forms the basis for the built-in MATLAB function fzero.

Brent also developed a similar approach for one-dimensional minimization which forms the basis for the MATLAB fminbnd function. It combines the slow, dependable golden-section search with the faster, but possibly unreliable, parabolic interpolation. It first attempts parabolic interpolation and keeps applying it as long as acceptable results are obtained. If not, it uses the golden-section search to get matters in hand.

A simple expression of its syntax is

[*xmin, fval*] = fminbnd(*function,x1,x2*)

where x and fval are the location and value of the minimum, *function* is the name of the function being evaluated, and *x1* and *x2* are the bounds of the interval being searched.

Here is a simple MATLAB session that uses fminbnd to solve the problem from Example 7.1.

```
>> g=9.81;v0=55;m=80;c=15;z0=100;
>> z=@(t) -(z0+m/c*(v0+m*g/c)*(1-exp(-c/m*t))-m*g/c*t);
>> [x,f]=fminbnd(z,0,8)

x =
    3.8317
f =
  -192.8609
```

As with fzero, optional parameters can be specified using optimset. For example, we can display calculation details:

```
>> options = optimset('display','iter');
>> fminbnd(z,0,8,options)
```

Func-count	x	f(x)	Procedure
1	3.05573	-189.759	initial
2	4.94427	-187.19	golden
3	1.88854	-171.871	golden
4	3.87544	-192.851	parabolic
5	3.85836	-192.857	parabolic
6	3.83332	-192.861	parabolic
7	3.83162	-192.861	parabolic
8	3.83166	-192.861	parabolic
9	3.83169	-192.861	parabolic

```
Optimization terminated:
 the current x satisfies the termination criteria using
OPTIONS.TolX of 1.000000e-004

ans =
    3.8317
```

Thus, after three iterations, the method switches from golden to parabolic, and after eight iterations, the minimum is determined to a tolerance of 0.0001.

7.3 MULTIDIMENSIONAL OPTIMIZATION

Aside from one-dimensional functions, optimization also deals with multidimensional functions. Recall from Fig. 7.3a that our visual image of a one-dimensional search was like a roller coaster. For two-dimensional cases, the image becomes that of mountains and valleys (Fig. 7.3b). As in the following example, MATLAB's graphic capabilities provide a handy means to visualize such functions.

EXAMPLE 7.4   Visualizing a Two-Dimensional Function

Problem Statement. Use MATLAB's graphical capabilities to display the following function and visually estimate its minimum in the range $-2 \leq x_1 \leq 0$ and $0 \leq x_2 \leq 3$:

$$f(x_1, x_2) = 2 + x_1 - x_2 + 2x_1^2 + 2x_1 x_2 + x_2^2$$

Solution. The following script generates contour and mesh plots of the function:

```
x=linspace(-2,0,40);y=linspace(0,3,40);
[X,Y] = meshgrid(x,y);
Z=2+X-Y+2*X.^2+2*X.*Y+Y.^2;
subplot(1,2,1);
cs=contour(X,Y,Z);clabel(cs);
xlabel('x_1');ylabel('x_2');
title('(a) Contour plot');grid;
subplot(1,2,2);
cs=surfc(X,Y,Z);
zmin=floor(min(Z));
zmax=ceil(max(Z));
xlabel('x_1');ylabel('x_2');zlabel('f(x_1,x_2)');
title('(b) Mesh plot');
```

As displayed in Fig. 7.9, both plots indicate that function has a minimum value of about $f(x_1, x_2) = 0$ to 1 located at about $x_1 = -1$ and $x_2 = 1.5$.

FIGURE 7.9
(a) Contour and (b) mesh plots of a two-dimensional function.

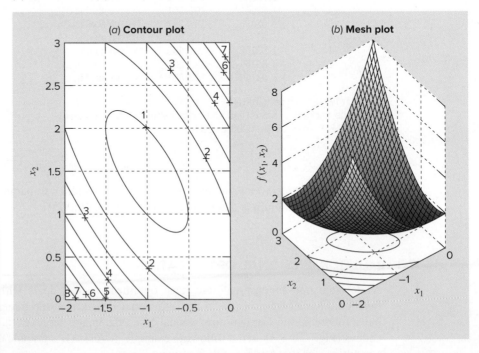

Techniques for multidimensional unconstrained optimization can be classified in a number of ways. For purposes of the present discussion, we will divide them depending on whether they require derivative evaluation. Those that require derivatives are called *gradient,* or *descent* (or *ascent*), methods. The approaches that do not require derivative evaluation are called *nongradient,* or *direct,* methods. As described next, the built-in MATLAB function fminsearch is a direct method.

7.3.1 MATLAB Function: fminsearch

Standard MATLAB has a function fminsearch that can be used to determine the minimum of a multidimensional function. It is based on the Nelder-Mead method, which is a direct-search method that uses only function values (does not require derivatives) and handles non-smooth objective functions. A simple expression of its syntax is

[*xmin, fval*] = fminsearch(*function,x0*)

where *xmin* and *fval* are the location and value of the minimum, *function* is the name of the function being evaluated, and *x0* is the initial guess. Note that *x0* can be a scalar, vector, or a matrix.

Here is a simple MATLAB session that uses fminsearch to determine minimum for the function we just graphed in Example 7.4:

```
>> f=@(x) 2+x(1)-x(2)+2*x(1)^2+2*x(1)*x(2)+x(2)^2;
>> [x,fval]=fminsearch(f,[-0.5,0.5])

x =
    -1.0000    1.5000
fval =
     0.7500
```

7.4 CASE STUDY   EQUILIBRIUM AND MINIMUM POTENTIAL ENERGY

Background. As in Fig. 7.10*a*, an unloaded spring can be attached to a wall mount. When a horizontal force is applied, the spring stretches. The displacement is related to the force by *Hookes law, F = kx.* The *potential energy* of the deformed state consists of the difference between the strain energy of the spring and the work done by the force:

$$PE(x) = 0.5kx^2 - Fx \qquad (7.11)$$

FIGURE 7.10
(*a*) An unloaded spring attached to a wall mount. (*b*) Application of a horizontal force stretches the spring where the relationship between force and displacement is described by Hooke's law.

Equation (7.11) defines a parabola. Since the potential energy will be at a minimum at equilibrium, the solution for displacement can be viewed as a one-dimensional optimization problem. Because this equation is so easy to differentiate, we can solve for the displacement as $x = F/k$. For example, if $k = 2$ N/cm and $F = 5$ N, $x = 5\text{N}/(2 \text{ N/cm}) = 2.5$ cm.

A more interesting two-dimensional case is shown in Fig. 7.11. In this system, there are two degrees of freedom in that the system can move both horizontally and vertically. In the same way that we approached the one-dimensional system, the equilibrium deformations are the values of x_1 and x_2 that minimize the potential energy:

$$PE(x_1, x_2) = 0.5k_a \left(\sqrt{x_1^2 + (L_a - x_2)^2} - L_a \right)^2$$
$$+ 0.5k_b \left(\sqrt{x_1^2 + (L_b + x_2)^2} - L_b \right)^2 - F_1 x_1 - F_2 x_2 \tag{7.12}$$

If the parameters are $k_a = 9$ N/cm, $k_b = 2$ N/cm, $L_a = 10$ cm, $L_b = 10$ cm, $F_1 = 2$ N, and $F_2 = 4$ N, use MATLAB to solve for the displacements and the potential energy.

FIGURE 7.11
A two-spring system: (*a*) unloaded and (*b*) loaded.

7.4 CASE STUDY continued

Solution. An M-file can be developed to hold the potential energy function:

```
function p=PE(x,ka,kb,La,Lb,F1,F2)
PEa=0.5*ka*(sqrt(x(1)^2+(La-x(2))^2)-La)^2;
PEb=0.5*kb*(sqrt(x(1)^2+(Lb+x(2))^2)-Lb)^2;
W=F1*x(1)+F2*x(2);
p=PEa+PEb-W;
```

The solution can be obtained with the fminsearch function:

```
>> ka=9;kb=2;La=10;Lb=10;F1=2;F2=4;
>> [x,f]=fminsearch(@(x)PE(x,ka,kb,La,Lb,F1,F2),[-0.5,0.5])

x =
     4.9523    1.2769
f =
    -9.6422
```

Thus, at equilibrium, the potential energy is -9.6422 N·cm. The connecting point is located 4.9523 cm to the right and 1.2759 cm above its original position.

PROBLEMS

7.1 Perform three iterations of the Newton-Raphson method to determine the root of Eq. (E7.1.1). Use the parameter values from Example 7.1 along with an initial guess of $t = 3$ s.

7.2 Given the formula

$$f(x) = -x^2 + 8x - 12$$

(a) Determine the maximum and the corresponding value of x for this function analytically (i.e., using differentiation).

(b) Verify that Eq. (7.10) yields the same results based on initial guesses of $x_1 = 0$, $x_2 = 2$, and $x_3 = 6$.

7.3 Consider the following function:

$$f(x) = 3 + 6x + 5x^2 + 3x^3 + 4x^4$$

Locate the minimum by finding the root of the derivative of this function. Use bisection with initial guesses of $x_l = -2$ and $x_u = 1$.

7.4 Given

$$f(x) = -1.5x^6 - 2x^4 + 12x$$

(a) Plot the function.

(b) Use analytical methods to prove that the function is concave for all values of x.

(c) Differentiate the function and then use a root-location method to solve for the maximum $f(x)$ and the corresponding value of x.

7.5 Solve for the value of x that maximizes $f(x)$ in Prob. 7.4 using the golden-section search. Employ initial guesses of $x_l = 0$ and $x_u = 2$, and perform three iterations.

7.6 Repeat Prob. 7.5, except use parabolic interpolation. Employ initial guesses of $x_1 = 0$, $x_2 = 1$, and $x_3 = 2$, and perform three iterations.

7.7 Employ the following methods to find the maximum of

$$f(x) = 4x - 1.8x^2 + 1.2x^3 - 0.3x^4$$

(a) Golden-section search ($x_l = -2$, $x_u = 4$, $\varepsilon_s = 1\%$).

(b) Parabolic interpolation ($x_1 = 1.75$, $x_2 = 2$, $x_3 = 2.5$, iterations = 5).

7.8 Consider the following function:

$$f(x) = x^4 + 2x^3 + 8x^2 + 5x$$

Use analytical and graphical methods to show the function has a minimum for some value of x in the range $-2 \leq x \leq 1$.

7.9 Employ the following methods to find the minimum of the function from Prob. 7.8:
(a) Golden-section search ($x_l = -2, x_u = 1, \varepsilon_s = 1\%$).
(b) Parabolic interpolation ($x_1 = -2, x_2 = -1, x_3 = 1$, iterations = 5).
7.10 Consider the following function:

$$f(x) = 2x + \frac{3}{x}$$

Perform 10 iterations of parabolic interpolation to locate the minimum. Comment on the convergence of your results ($x_1 = 0.1, x_2 = 0.5, x_3 = 5$)
7.11 The following function defines a curve with several unequal minima over the interval: $2 \le x \le 20$,

$$f(x) = \sin(x) + \sin\left(\frac{2}{3}x\right)$$

Develop a MATLAB script to **(a)** plot the function over the interval. Determine the minimum **(b)** with `fminbnd` and **(c)** by hand with golden-section search with a stopping criterion corresponding to three significant figures. For **(b)** and **(c)**, use initial guesses of [4, 8].
7.12 Use the golden-section search to determine the location, x_{max}, and maximum, $f(x_{max})$, of the following function by hand,

$$f(x) = -0.8x^4 + 2.2x^2 + 0.6$$

Use initial guesses of $x_l = 0.7$ and $x_u = 1.4$ and perform sufficient iterations so that $\varepsilon_s = 10\%$. Determine the approximate relative error of your final result.
7.13 Develop a single script to **(a)** generate contour and mesh subplots of the following temperature field in a similar fashion to Example 7.4:

$$T(x, y) = 2x^2 + 3y^2 - 4xy - y - 3x$$

and **(b)** determine the minimum with `fminsearch`.
7.14 The head of a groundwater aquifer is described in Cartesian coordinates by

$$h(x, y) = \frac{1}{1 + x^2 + y^2 + x + xy}$$

Develop a single script to **(a)** generate contour and mesh subplots of the function in a similar fashion to Example 7.4, and **(b)** determine the maximum with `fminsearch`.
7.15 Recent interest in competitive and recreational cycling has meant that engineers have directed their skills toward the design and testing of mountain bikes (Fig. P7.15a). Suppose that you are given the task of predicting the horizontal

FIGURE P7.15
(a) A mountain bike along with (b) a free-body diagram for a part of the frame.

and vertical displacement of a bike bracketing system in response to a force. Assume the forces you must analyze can be simplified as depicted in Fig. P7.15b. You are interested in testing the response of the truss to a force exerted in any number of directions designated by the angle θ. The parameters for the problem are $E =$ Young's modulus $= 2 \times 10^{11}$ Pa, $A =$ cross-sectional area $= 0.0001$ m^2, $w =$ width $= 0.44$ m, $\ell =$ length $= 0.56$ m, and $h =$ height $= 0.5$ m. The displacements x and y can be solved by determining the values that yield a minimum potential energy. Determine the displacements for a force of 10,000 N and a range of θ's from $0°$ (horizontal) to $90°$ (vertical).
7.16 As electric current moves through a wire (Fig. P7.16), heat generated by resistance is conducted through a layer of insulation and then convected to the surrounding air. The steady-state temperature of the wire can be computed as

$$T = T_{air} + \frac{q}{2\pi}\left[\frac{1}{k}\ln\left(\frac{r_w + r_i}{r_w}\right) + \frac{1}{h}\frac{1}{r_w + r_i}\right]$$

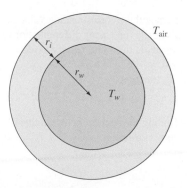

FIGURE P7.16
Cross-section of an insulated wire.

Determine the thickness of insulation r_i(m) that minimizes the wire's temperature given the following parameters: q = heat generation rate = 75 W/m, r_w = wire radius = 6 mm, k = thermal conductivity of insulation = 0.17 W/(m K), h = convective heat transfer coefficient = 12 W/(m^2 K), and T_{air} = air temperature = 293 K.

7.17 Develop an M-file that is expressly designed to locate a maximum with the golden-section search. In other words, set it up so that it directly finds the maximum rather than finding the minimum of $-f(x)$. The function should have the following features:

- Iterate until the relative error falls below a stopping criterion or exceeds a maximum number of iterations.
- Return both the optimal x and $f(x)$.

Test your program with the same problem as Example 7.1.

7.18 Develop an M-file to locate a minimum with the golden-section search. Rather than using the maximum iterations and Eq. (7.9) as the stopping criteria, determine the number of iterations needed to attain a desired tolerance. Test your function by solving Example 7.2 using $E_{a,d}$ = 0.0001.

7.19 Develop an M-file to implement parabolic interpolation to locate a minimum. The function should have the following features:

- Base it on two initial guesses, and have the program generate the third initial value at the midpoint of the interval.
- Check whether the guesses bracket a maximum. If not, the function should not implement the algorithm, but should return an error message.
- Iterate until the relative error falls below a stopping criterion or exceeds a maximum number of iterations.
- Return both the optimal x and $f(x)$.

Test your program with the same problem as Example 7.3.

7.20 Pressure measurements are taken at certain points behind an airfoil over time. These data best fit the curve $y = 6 \cos x - 1.5 \sin x$ from $x = 0$ to 6 s. Use four iterations of the golden-search method to find the minimum pressure. Set $x_l = 2$ and $x_u = 4$.

7.21 The trajectory of a ball can be computed with

$$y = (\tan \theta_0)x - \frac{g}{2v_0^2 \cos^2 \theta_0} x^2 + y_0$$

where y = the height (m), θ_0 = the initial angle (radians), v_0 = the initial velocity (m/s), g = the gravitational constant = 9.81 m/s^2, and y_0 = the initial height (m). Use the golden-section search to determine the maximum height

given y_0 = 2 m, v_0 = 20 m/s, and θ_0 = 45°. Iterate until the approximate error falls below ε_s = 10% using initial guesses of x_l = 10 and x_u = 30 m.

7.22 The deflection of a uniform beam subject to a linearly increasing distributed load can be computed as

$$y = \frac{w_0}{120E\,I\,L}(-x^5 + 2L^2x^3 - L^4x)$$

Given that L = 600 cm, E = 50,000 kN/cm^2, I = 30,000 cm^4, and w_0 = 2.5 kN/cm, determine the point of maximum deflection **(a)** graphically, **(b)** using the golden-section search until the approximate error falls below ε_s = 1% with initial guesses of x_l = 0 and x_u = L.

7.23 A object with a mass of 90 kg is projected upward from the surface of the earth at a velocity of 60 m/s. If the object is subject to linear drag (c = 15 kg/s), use the golden-section search to determine the maximum height the object attains.

7.24 The normal distribution is a bell-shaped curve defined by

$$y = e^{-x^2}$$

Use the golden-section search to determine the location of the inflection point of this curve for positive x.

7.25 Use the fminsearch function to determine the minimum of

$$f(x, y) = 2y^2 - 2.25xy - 1.75y + 1.5x^2$$

7.26 Use the fminsearch function to determine the maximum of

$$f(x, y) = 4x + 2y + x^2 - 2x^4 + 2xy - 3y^2$$

7.27 Given the following function:

$$f(x, y) = -8x + x^2 + 12y + 4y^2 - 2xy$$

Determine the minimum **(a)** graphically, **(b)** numerically with the fminsearch function, and **(c)** substitute the result of **(b)** back into the function to determine the minimum $f(x, y)$.

7.28 The specific growth rate of a yeast that produces an antibiotic is a function of the food concentration c:

$$g = \frac{2c}{4 + 0.8c + c^2 + 0.2c^3}$$

As depicted in Fig. P7.28, growth goes to zero at very low concentrations due to food limitation. It also goes to zero at high concentrations due to toxicity effects. Find the value of c at which growth is a maximum.

FIGURE P7.28
The specific growth rate of a yeast that produces an antibiotic versus the food concentration.

FIGURE P7.31
A dissolved oxygen "sag" below a point discharge of sewage into a river.

7.29 A compound A will be converted into B in a stirred tank reactor. The product B and unreacted A are purified in a separation unit. Unreacted A is recycled to the reactor. A process engineer has found that the initial cost of the system is a function of the conversion x_A. Find the conversion that will result in the lowest cost system. C is a proportionality constant.

$$\text{Cost} = C \left[\left(\frac{1}{(1 - x_A)^2} \right)^{0.6} + 6 \left(\frac{1}{x_A} \right)^{0.6} \right]$$

7.30 A finite-element model of a cantilever beam subject to loading and moments (Fig. P7.30) is given by optimizing

$$f(x, y) = 5x^2 - 5xy + 2.5y^2 - x - 1.5y$$

where x = end displacement and y = end moment. Find the values of x and y that minimize $f(x, y)$.

7.31 The *Streeter-Phelps model* can be used to compute the dissolved oxygen concentration in a river below a point discharge of sewage (Fig. P7.31),

$$o = o_s - \frac{k_d L_o}{k_d + k_s - k_a} \left(e^{-k_a t} - e^{-(k_d + k_s)t} \right)$$
$$- \frac{S_b}{k_a} (1 - e^{-k_a t})$$
(P7.31)

FIGURE P7.30
A cantilever beam.

where o = dissolved oxygen concentration (mg/L), o_s = oxygen saturation concentration (mg/L), t = travel time (d), L_o = biochemical oxygen demand (BOD) concentration at the mixing point (mg/L), k_d = rate of decomposition of BOD (d^{-1}), k_s = rate of settling of BOD (d^{-1}), k_a = reaeration rate (d^{-1}), and S_b = sediment oxygen demand (mg/L/d).

As indicated in Fig. P7.31, Eq. (P7.31) produces an oxygen "sag" that reaches a critical minimum level o_c, some travel time t_c below the point discharge. This point is called "critical" because it represents the location where biota that depend on oxygen (like fish) would be the most stressed. Develop a MATLAB script that **(a)** generates a plots of the function versus travel time and **(b)** uses fminbnd to determine the critical travel time and concentration, given the following values:

o_s = 10 mg/L	k_d = 0.1 d^{-1}	k_a = 0.6 d^{-1}
k_s = 0.05 d^{-1}	L_o = 50 mg/L	S_b = 1 mg/L/d

7.32 The two-dimensional distribution of pollutant concentration in a channel can be described by

$$c(x, y) = 7.9 + 0.13x + 0.21y - 0.05x^2$$
$$- 0.016y^2 - 0.007xy$$

Determine the exact location of the peak concentration given the function and the knowledge that the peak lies within the bounds $-10 \le x \le 10$ and $0 \le y \le 20$.

7.33 A total charge Q is uniformly distributed around a ring-shaped conductor with radius a. A charge q is located at a distance x from the center of the ring (Fig. P7.33). The force exerted on the charge by the ring is given by

$$F = \frac{1}{4\pi e_0} \frac{qQx}{(x^2 + a^2)^{3/2}}$$

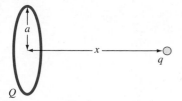

FIGURE P7.33

where $e_0 = 8.85 \times 10^{-12}$ C²/(N m²), $q = Q = 2 \times 10^{-5}$ C, and $a = 0.9$ m. Determine the distance x where the force is a maximum.

7.34 The torque transmitted to an induction motor is a function of the slip between the rotation of the stator field and the rotor speed s, where slip is defined as

$$s = \frac{n - n_R}{n}$$

where n = revolutions per second of rotating stator speed and n_R = rotor speed. Kirchhoff's laws can be used to show that the torque (expressed in dimensionless form) and slip are related by

$$T = \frac{15s(1 - s)}{(1 - s)(4s^2 - 3s + 4)}$$

Figure P7.34 shows this function. Use a numerical method to determine the slip at which the maximum torque occurs.

7.35 The total drag on an airfoil can be estimated by

$$D = 0.01\sigma V^2 + \underset{\text{Friction}}{\underbrace{}} \frac{0.95}{\sigma}\left(\frac{W}{V}\right)^2 \underset{\text{Lift}}{\underbrace{}}$$

where D = drag, σ = ratio of air density between the flight altitude and sea level, W = weight, and V = velocity. As seen in Fig. P7.35, the two factors contributing to drag are affected

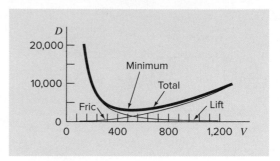

FIGURE P7.35
Plot of drag versus velocity for an airfoil.

differently as velocity increases. Whereas friction drag increases with velocity, the drag due to lift decreases. The combination of the two factors leads to a minimum drag.

(a) If $\sigma = 0.6$ and $W = 16,000$, determine the minimum drag and the velocity at which it occurs.

(b) In addition, develop a sensitivity analysis to determine how this optimum varies in response to a range of $W = 12,000$ to $20,000$ with $\sigma = 0.6$.

7.36 Roller bearings are subject to fatigue failure caused by large contact loads F (Fig. P7.36). The problem of finding the location of the maximum stress along the x axis can be shown to be equivalent to maximizing the function:

$$f(x) = \frac{0.4}{\sqrt{1 + x^2}} - \sqrt{1 + x^2}\left(1 - \frac{0.4}{1 + x^2}\right) + x$$

Find the x that maximizes $f(x)$.

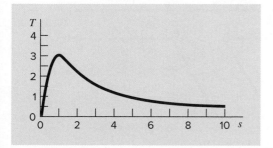

FIGURE P7.34
Torque transmitted to an inductor as a function of slip.

FIGURE P7.36
Roller bearings.

FIGURE P7.37
Two frictionless masses connected to a wall by a pair of linear elastic springs.

FIGURE P7.39
A ladder leaning against a fence and just touching a wall.

7.37 In a similar fashion to the case study described in Sec. 7.4, develop the potential energy function for the system depicted in Fig. P7.37. Develop contour and surface plots in MATLAB. Minimize the potential energy function to determine the equilibrium displacements x_1 and x_2 given the forcing function $F = 100$ N and the parameters $k_a = 20$ and $k_b = 15$ N/m.

7.38 As an agricultural engineer, you must design a trapezoidal open channel to carry irrigation water (Fig. P7.38). Determine the optimal dimensions to minimize the wetted perimeter for a cross-sectional area of 50 m². Are the relative dimensions universal?

7.39 Use the function fminsearch to determine the length of the shortest ladder that reaches from the ground over the fence to the building's wall (Fig. P7.39). Test it for the case where $h = d = 4$ m.

7.40 The length of the longest ladder that can negotiate the corner depicted in Fig. P7.40 can be determined by computing the value of θ that minimizes the following function:

$$L(\theta) = \frac{w_1}{\sin \theta} + \frac{w_2}{\sin(\pi - \alpha - \theta)}$$

For the case where $w_1 = w_2 = 2$ m, use a numerical method described in this chapter (including MATLAB's built-in capabilities) to develop a plot of L versus a range of α's from 45 to 135°.

7.41 Figure P7.41 shows a pinned-fixed beam subject to a uniform load. The equation for the resulting deflections is

$$y = -\frac{w}{48EI}(2x^4 - 3Lx^3 + L^3x)$$

FIGURE P7.38

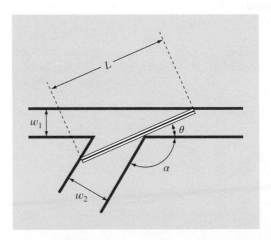

FIGURE P7.40
A ladder negotiating a corner formed by two hallways.

FIGURE P7.41

Develop a MATLAB script that uses `fminbnd` to **(a)** generate a labeled plot of deflection versus distance and **(b)** determine the location and magnitude of the maximum deflection. Employ an initial guesses of 0 and L and use `optimset` to display the iterations. Use the following parameter values in your computation (making sure that you use consistent units): $L = 400$ cm, $E = 52,000$ kN/cm^2, $I = 32,000$ cm^4, and $w = 4$ kN/cm.

7.42 For a jet in steady, level flight, thrust balances drag and lift balances weight (Fig. P7.42). Under these conditions, the optimal cruise speed occurs when the ratio of drag force to velocity is minimized. The drag, C_D, can be computed as

$$C_D = C_{D0} + \frac{C_L^2}{\pi \cdot AR}$$

where $C_{D0} =$ drag coefficient at zero lift, $C_L =$ the lift coefficient, and $AR =$ the aspect ratio. For steady level flight, the lift coefficient can be computed as

$$C_L = \frac{2W}{\rho v^2 A}$$

where $W =$ the jet's weight (N), $\rho =$ air density (kg/m^3), $v =$ velocity (m/s), and $A =$ wing planform area (m^2). The drag force can then be computed as

$$F_D = W \frac{C_D}{C_L}$$

Use these formulas to determine the optimal steady cruise velocity for a 670 kN jet flying at 10 km above sea level. Employ the following parameters in your computation: $A = 150$ m^2, $AR = 6.5$, $C_{D0} = 0.018$, and $\rho = 0.413$ kg/m^3.

7.43 Develop a MATLAB script to generate a plot of the optimal velocity of the jet from Prob. 7.42 versus elevation above sea level. Employ a mass of 68,300 kg for the jet. Note that the gravitational acceleration at 45° latitude can be computed as a function of elevation with

$$g(h) = 9.8066 \left(\frac{r_e}{r_e + h} \right)^2$$

where $g(h) =$ gravitational acceleration (m/s^2) at elevation h (m) above sea level, and $r_e =$ Earth's mean radius ($= 6.371 \times 10^6$ m). In addition, air density as a function of elevation can be calculated with

$$\rho(h) = -9.57926 \times 10^{-14} h^3 + 4.71260 \times 10^{-9} h^2 - 1.18951 \times 10^{-4} h + 1.22534$$

Employ the other parameters from Prob. 7.42 and design the plot for elevations ranging from $h = 0$ to 12 km above sea level.

7.44 As depicted in Fig. P7.44, a mobile fire hose projects a stream of water onto the roof of a building. At what angle, θ, and how far from the building, x_1, should the hose be placed in order to maximize the coverage of the roof; that is, to maximize: $x_2 - x_1$? Note that the water velocity leaving the nozzle has a constant value of 3 m/s regardless of the angle, and the other parameter values are $h_1 = 0.06$ m, $h_2 = 0.2$ m, and $L = 0.12$ m. [Hint: The coverage is maximized for the trajectory that just clears the top front corner. That is, we want to choose an x_1 and θ that just clear the top corner while maximizing $x_2 - x_1$.]

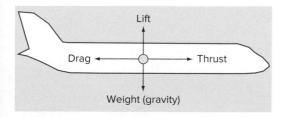

FIGURE P7.42
The four major forces on a jet in steady, level flight.

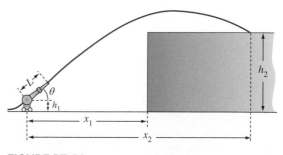

FIGURE P7.44

7.45 Since many pollutants enter lakes (and other water-bodies for that matter) at their peripheries, an important water-quality problem involves modeling the distribution of contaminants in the vicinity of a waste discharge or a river. For a vertically well-mixed, constant-depth layer, the steady-state distribution of a pollutant reacting with first-order decay is represented by

$$0 = -U_x \frac{\partial c}{\partial x} + E \left(\frac{\partial^2 c}{\partial x^2} + \frac{\partial^2 c}{\partial y^2} \right) - kc$$

where the x and y axes are defined to be parallel and perpendicular to the shoreline, respectively (Fig. P7.45). The parameters and variables are: U_x = the water velocity along the shoreline (m/d), c = concentration, E = the turbulent diffusion coefficient, and k = the first-order decay rate. For the case where a constant loading, W, enters at (0, 0), the solution for the concentration at any coordinate is given by

$$c = 2 \left\{ c(x, y) + \sum_{n=1}^{\infty} [c(x, y + 2nY) - c(x, y + 2nY)] \right\}$$

where

$$c(x, y) = \frac{W}{\pi HE} e^{\frac{U_x x}{2E}} K_0 \left(\sqrt{(x^2 + y^2) \left[\frac{k}{E} + \left(\frac{U_x}{2E} \right)^2 \right]} \right)$$

where Y = the width, H = depth, and K_0 = the modified Bessel function of the second kind. Develop a MATLAB script to generate a contour plot of concentration for a section of a lake with $Y = 4.8$ km and a length from $X = -2.4$ to 2.4 km using $\Delta x = \Delta y = 0.32$ km. Employ the following parameters for your calculation: $W = 1.2 \times 10^{12}$, $H = 20$, $E = 5 \times 10^6$, $U_x = 5 \times 10^3$, $k = 1$, and $n = 3$.

FIGURE P7.45
Plan view of a section of a lake with a point source of pollutant entering at the middle of the lower boundary.

Linear Systems

3.1 OVERVIEW

What Are Linear Algebraic Equations?

In Part Two, we determined the value x that satisfied a single equation, $f(x) = 0$. Now, we deal with the case of determining the values $x_1, x_2, \ldots, x_n$ that simultaneously satisfy a set of equations:

$$f_1(x_1, x_2, \ldots, x_n) = 0$$
$$f_2(x_1, x_2, \ldots, x_n) = 0$$
$$\vdots \qquad \vdots$$
$$f_n(x_1, x_2, \ldots, x_n) = 0$$

Such systems are either linear or nonlinear. In Part Three, we deal with *linear algebraic equations* that are of the general form

$$a_{11}x_1 + a_{12}x_2 + \cdots + a_{1n}x_n = b_1$$
$$a_{21}x_1 + a_{22}x_2 + \cdots + a_{2n}x_n = b_2$$
$$\vdots \qquad \vdots \qquad \text{(PT3.1)}$$
$$a_{n1}x_1 + a_{n2}x_2 + \cdots + a_{nn}x_n) = b_n$$

where the a's are constant coefficients, the b's are constants, the x's are unknowns, and n is the number of equations. All other algebraic equations are nonlinear.

Linear Algebraic Equations in Engineering and Science

Many of the fundamental equations of engineering and science are based on conservation laws. Some familiar quantities that conform to such laws are mass, energy, and momentum. In mathematical terms, these principles lead to balance or continuity equations that relate system behavior as represented

Feed

(a)

Feed

(b)

FIGURE PT3.1
Two types of systems that can be modeled using linear algebraic equations: (a) lumped variable system that involves coupled finite components and (b) distributed variable system that involves a continuum.

by the levels or response of the quantity being modeled to the properties or characteristics of the system and the external stimuli or forcing functions acting on the system.

As an example, the principle of mass conservation can be used to formulate a model for a series of chemical reactors (Fig. PT3.1a). For this case, the quantity being modeled is the mass of the chemical in each reactor. The system properties are the reaction characteristics of the chemical and the reactors' sizes and flow rates. The forcing functions are the feed rates of the chemical into the system.

When we studied roots of equations, you saw how single-component systems result in a single equation that can be solved using root-location techniques. Multicomponent systems result in a coupled set of mathematical equations that must be solved simultaneously. The equations are coupled because the individual parts of the system are influenced by other parts. For example, in Fig. PT3.1a, reactor 4 receives chemical inputs from reactors 2 and 3. Consequently, its response is dependent on the quantity of chemical in these other reactors.

When these dependencies are expressed mathematically, the resulting equations are often of the linear algebraic form of Eq. (PT3.1). The x's are usually measures of the magnitudes of the responses of the individual components. Using Fig. PT3.1a as an example, x_1 might quantify the amount of chemical mass in the first reactor, x_2 might quantify the amount in the second, and so forth. The a's typically represent the properties and characteristics that bear on the interactions between components. For instance, the a's for Fig. PT3.1a might be reflective of the flow rates of mass between the reactors. Finally, the b's usually represent the forcing functions acting on the system, such as the feed rate.

Multicomponent problems of these types arise from both lumped (macro-) or distributed (micro-) variable mathematical models. *Lumped variable problems* involve coupled finite components. The three interconnected bungee jumpers described at the beginning of Chap. 8 are a lumped system. Other examples include trusses, reactors, and electric circuits.

Conversely, *distributed variable problems* attempt to describe the spatial detail on a continuous or semicontinuous basis. The distribution of chemicals along the length of an elongated, rectangular reactor (Fig. PT3.1*b*) is an example of a continuous variable model. Differential equations derived from conservation laws specify the distribution of the dependent variable for such systems. These differential equations can be solved numerically by converting them to an equivalent system of simultaneous algebraic equations.

The solution of such sets of equations represents a major application area for the methods in the following chapters. These equations are coupled because the variables at one location are dependent on the variables in adjoining regions. For example, the concentration at the middle of the reactor in Fig. PT3.1*b* is a function of the concentration in adjoining regions. Similar examples could be developed for the spatial distribution of temperature, momentum, or electricity.

Aside from physical systems, simultaneous linear algebraic equations also arise in a variety of mathematical problem contexts. These result when mathematical functions are required to satisfy several conditions simultaneously. Each condition results in an equation that contains known coefficients and unknown variables. The techniques discussed in this part can be used to solve for the unknowns when the equations are linear and algebraic. Some widely used numerical techniques that employ simultaneous equations are regression analysis and spline interpolation.

3.2 PART ORGANIZATION

Due to its importance in formulating and solving linear algebraic equations, *Chap. 8* provides a brief overview of *matrix algebra*. Aside from covering the rudiments of matrix representation and manipulation, the chapter also describes how matrices are handled in MATLAB.

Chapter 9 is devoted to the most fundamental technique for solving linear algebraic systems: *Gauss elimination*. Before launching into a detailed discussion of this technique, a preliminary section deals with simple methods for solving small systems. These approaches are presented to provide you with visual insight and because one of the methods—the elimination of unknowns—represents the basis for Gauss elimination.

After this preliminary material, "naive" Gauss elimination is discussed. We start with this "stripped-down" version because it allows the fundamental technique to be elaborated on without complicating details. Then, in subsequent sections, we discuss potential problems of the naive approach and present a number of modifications to minimize and circumvent these problems. The focus of this discussion will be the process of switching rows, or *partial pivoting*. The chapter ends with a brief description of efficient methods for solving *tridiagonal matrices*.

Chapter 10 illustrates how Gauss elimination can be formulated as an *LU factorization*. Such solution techniques are valuable for cases where many right-hand-side vectors need to be evaluated. The chapter ends with a brief outline of how MATLAB solves linear systems.

Chapter 11 starts with a description of how *LU* factorization can be employed to efficiently calculate the *matrix inverse,* which has tremendous utility in analyzing stimulus-response relationships of physical systems. The remainder of the chapter is devoted to the important concept of matrix condition. The condition number is introduced as a measure of the roundoff errors that can result when solving ill-conditioned matrices.

Chapter 12 deals with iterative solution techniques, which are similar in spirit to the approximate methods for roots of equations discussed in Chap. 6. That is, they involve guessing a solution and then iterating to obtain a refined estimate. The emphasis is on the *Gauss-Seidel method,* although a description is provided of an alternative approach, the *Jacobi method.* The chapter ends with a brief description of how *nonlinear simultaneous equations* can be solved.

Finally, *Chap. 13* is devoted to *eigenvalue* problems. These have general mathematical relevance as well as many applications in engineering and science. We describe two simple methods as well as MATLAB's capabilities for determining eigenvalues and *eigenvectors.* In terms of applications, we focus on their use to study the vibrations and oscillations of mechanical systems and structures.

8

Linear Algebraic Equations and Matrices

CHAPTER OBJECTIVES

The primary objective of this chapter is to acquaint you with linear algebraic equations and their relationship to matrices and matrix algebra. Specific objectives and topics covered are

- Understanding matrix notation.
- Being able to identify the following types of matrices: identity, diagonal, symmetric, triangular, and tridiagonal.
- Knowing how to perform matrix multiplication and being able to assess when it is feasible.
- Knowing how to represent a system of linear algebraic equations in matrix form.
- Knowing how to solve linear algebraic equations with left division and matrix inversion in MATLAB.

YOU'VE GOT A PROBLEM

Suppose that three jumpers are connected by bungee cords. Figure 8.1a shows them being held in place vertically so that each cord is fully extended but unstretched. We can define three distances, x_1, x_2, and x_3, as measured downward from each of their unstretched positions. After they are released, gravity takes hold and the jumpers will eventually come to the equilibrium positions shown in Fig. 8.1b.

Suppose that you are asked to compute the displacement of each of the jumpers. If we assume that each cord behaves as a linear spring and follows Hooke's law, free-body diagrams can be developed for each jumper as depicted in Fig. 8.2.

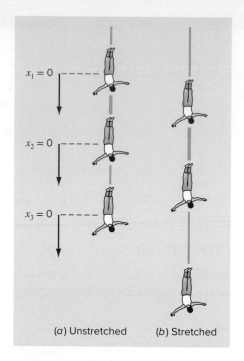

FIGURE 8.1
Three individuals connected by bungee cords.

FIGURE 8.2
Free-body diagrams.

Using Newton's second law, force balances can be written for each jumper:

$$m_1 \frac{d^2x_1}{dt^2} = m_1 g + k_2(x_2 - x_1) - k_1 x_1$$

$$m_2 \frac{d^2x_2}{dt^2} = m_2 g + k_3(x_3 - x_2) + k_2(x_1 - x_2) \qquad (8.1)$$

$$m_3 \frac{d^2x_3}{dt^2} = m_3 g + k_3(x_2 - x_3)$$

where m_i = the mass of jumper i (kg), t = time (s), k_j = the spring constant for cord j (N/m), x_i = the displacement of jumper i measured downward from the equilibrium position (m), and g = gravitational acceleration (9.81 m/s^2). Because we are interested in the steady-state solution, the second derivatives can be set to zero. Collecting terms gives

$$\begin{aligned} (k_1 + k_2)x_1 \qquad - k_2 x_2 \qquad &= m_1 g \\ -k_2 x_1 + (k_2 + k_3)x_2 - k_3 x_3 &= m_2 g \\ -k_3 x_2 + k_3 x_3 &= m_3 g \end{aligned} \qquad (8.2)$$

Thus, the problem reduces to solving a system of three simultaneous equations for the three unknown displacements. Because we have used a linear law for the cords, these equations are linear algebraic equations. Chapters 8 through 12 will introduce you to how MATLAB is used to solve such systems of equations.

8.1 MATRIX ALGEBRA OVERVIEW

Knowledge of matrices is essential for understanding the solution of linear algebraic equations. The following sections outline how matrices provide a concise way to represent and manipulate linear algebraic equations.

8.1.1 Matrix Notation

A *matrix* consists of a rectangular array of elements represented by a single symbol. As depicted in Fig. 8.3, [A] is the shorthand notation for the matrix and a_{ij} designates an individual *element* of the matrix.

A horizontal set of elements is called a *row* and a vertical set is called a *column*. The first subscript i always designates the number of the row in which the element lies. The second subscript j designates the column. For example, element a_{23} is in row 2 and column 3.

The matrix in Fig. 8.3 has m rows and n columns and is said to have a dimension of m by n (or $m \times n$). It is referred to as an m by n matrix.

Matrices with row dimension $m = 1$, such as

$$[b] = [b_1 \quad b_2 \quad \cdots \quad b_n]$$

are called *row vectors*. Note that for simplicity, the first subscript of each element is dropped. Also, it should be mentioned that there are times when it is desirable to employ a special shorthand notation to distinguish a row matrix from other types of matrices. One way to accomplish this is to employ special open-topped brackets, as in $\lfloor b \rfloor$.[1]

Matrices with column dimension $n = 1$, such as

$$[c] = \begin{bmatrix} c_1 \\ c_2 \\ \vdots \\ c_m \end{bmatrix} \qquad (8.3)$$

FIGURE 8.3
A matrix.

---

[1] In addition to special brackets, we will use case to distinguish between vectors (lowercase) and matrices (uppercase).

are referred to as *column vectors*. For simplicity, the second subscript is dropped. As with the row vector, there are occasions when it is desirable to employ a special shorthand notation to distinguish a column matrix from other types of matrices. One way to accomplish this is to employ special brackets, as in $\{c\}$.

Matrices where $m = n$ are called *square matrices*. For example, a 3×3 matrix is

$$[A] = \begin{bmatrix} a_{11} & a_{12} & a_{13} \\ a_{21} & a_{22} & a_{23} \\ a_{31} & a_{32} & a_{33} \end{bmatrix}$$

The diagonal consisting of the elements a_{11}, a_{22}, and a_{33} is termed the *principal* or *main diagonal* of the matrix.

Square matrices are particularly important when solving sets of simultaneous linear equations. For such systems, the number of equations (corresponding to rows) and the number of unknowns (corresponding to columns) must be equal for a unique solution to be possible. Consequently, square matrices of coefficients are encountered when dealing with such systems.

There are a number of special forms of square matrices that are important and should be noted:

A *symmetric matrix* is one where the rows equal the columns—that is, $a_{ij} = a_{ji}$ for all i's and j's. For example,

$$[A] = \begin{bmatrix} 5 & 1 & 2 \\ 1 & 3 & 7 \\ 2 & 7 & 8 \end{bmatrix}$$

is a 3×3 symmetric matrix.

A *diagonal matrix* is a square matrix where all elements off the main diagonal are equal to zero, as in

$$[A] = \begin{bmatrix} a_{11} & & \\ & a_{22} & \\ & & a_{33} \end{bmatrix}$$

Note that where large blocks of elements are zero, they are left blank.

An *identity matrix* is a diagonal matrix where all elements on the main diagonal are equal to 1, as in

$$[I] = \begin{bmatrix} 1 & & \\ & 1 & \\ & & 1 \end{bmatrix}$$

The identity matrix has properties similar to unity. That is,

$$[A][I] = [I][A] = [A]$$

An *upper triangular matrix* is one where all the elements below the main diagonal are zero, as in

$$[A] = \begin{bmatrix} a_{11} & a_{12} & a_{13} \\ & a_{22} & a_{23} \\ & & a_{33} \end{bmatrix}$$

A *lower triangular matrix* is one where all elements above the main diagonal are zero, as in

$$[A] = \begin{bmatrix} a_{11} & & \\ a_{21} & a_{22} & \\ a_{31} & a_{32} & a_{33} \end{bmatrix}$$

A *banded matrix* has all elements equal to zero, with the exception of a band centered on the main diagonal:

$$[A] = \begin{bmatrix} a_{11} & a_{12} & & \\ a_{21} & a_{22} & a_{23} & \\ & a_{32} & a_{33} & a_{34} \\ & & a_{43} & a_{44} \end{bmatrix}$$

The preceding matrix has a bandwidth of 3 and is given a special name—the *tridiagonal matrix*.

8.1.2 Matrix Operating Rules

Now that we have specified what we mean by a matrix, we can define some operating rules that govern its use. Two m by n matrices are equal if, and only if, every element in the first is equal to every element in the second—that is, $[A] = [B]$ if $a_{ij} = b_{ij}$ for all i and j.

Addition of two matrices, say, $[A]$ and $[B]$, is accomplished by adding corresponding terms in each matrix. The elements of the resulting matrix $[C]$ are computed as

$$c_{ij} = a_{ij} + b_{ij}$$

for $i = 1, 2, \ldots, m$ and $j = 1, 2, \ldots, n$. Similarly, the subtraction of two matrices, say, $[E]$ minus $[F]$, is obtained by subtracting corresponding terms, as in

$$d_{ij} = e_{ij} - f_{ij}$$

for $i = 1, 2, \ldots, m$ and $j = 1, 2, \ldots, n$. It follows directly from the preceding definitions that addition and subtraction can be performed only between matrices having the same dimensions.

Both addition and subtraction are commutative:

$$[A] + [B] = [B] + [A]$$

and associative:

$$([A] + [B]) + [C] = [A] + ([B] + [C])$$

The multiplication of a matrix $[A]$ by a scalar g is obtained by multiplying every element of $[A]$ by g. For example, for a 3×3 matrix:

$$[D] = g[A] = \begin{bmatrix} ga_{11} & ga_{12} & ga_{13} \\ ga_{21} & ga_{22} & ga_{23} \\ ga_{31} & ga_{32} & ga_{33} \end{bmatrix}$$

FIGURE 8.4
Visual depiction of how the rows and columns line up in matrix multiplication.

FIGURE 8.5
Matrix multiplication can be performed only if the inner dimensions are equal.

The product of two matrices is represented as $[C] = [A][B]$, where the elements of $[C]$ are defined as

$$c_{ij} = \sum_{k=1}^{n} a_{ik} b_{kj} \tag{8.4}$$

where n = the column dimension of $[A]$ and the row dimension of $[B]$. That is, the c_{ij} element is obtained by adding the product of individual elements from the ith row of the first matrix, in this case $[A]$, by the jth column of the second matrix $[B]$. Figure 8.4 depicts how the rows and columns line up in matrix multiplication.

According to this definition, matrix multiplication can be performed only if the first matrix has as many columns as the number of rows in the second matrix. Thus, if $[A]$ is an m by n matrix, $[B]$ could be an n by l matrix. For this case, the resulting $[C]$ matrix would have the dimension of m by l. However, if $[B]$ were an m by l matrix, the multiplication could not be performed. Figure 8.5 provides an easy way to check whether two matrices can be multiplied.

If the dimensions of the matrices are suitable, matrix multiplication is *associative:*

$$([A][B]) [C] = [A]([B][C])$$

and *distributive:*

$$[A]([B] + [C]) = [A][B] + [A][C]$$

or

$$([A] + [B])[C] = [A][C] + [B][C]$$

However, multiplication is not generally *commutative:*

$$[A][B] \neq [B][A]$$

That is, the order of matrix multiplication is important.

Although multiplication is possible, matrix division is not a defined operation. However, if a matrix $[A]$ is square and nonsingular, there is another matrix $[A]^{-1}$, called the *inverse* of $[A]$, for which

$$[A][A]^{-1} = [A]^{-1}[A] = [I]$$

Thus, the multiplication of a matrix by the inverse is analogous to division, in the sense that a number divided by itself is equal to 1. That is, multiplication of a matrix by its inverse leads to the identity matrix.

The inverse of a 2×2 matrix can be represented simply by

$$[A]^{-1} = \frac{1}{a_{11}a_{22} - a_{12}a_{21}} \begin{bmatrix} a_{22} & -a_{12} \\ -a_{21} & a_{11} \end{bmatrix}$$

Similar formulas for higher-dimensional matrices are much more involved. Chapter 11 will deal with techniques for using numerical methods and the computer to calculate the inverse for such systems.

The *transpose* of a matrix involves transforming its rows into columns and its columns into rows. For example, for the 3×3 matrix:

$$[A] = \begin{bmatrix} a_{11} & a_{12} & a_{13} \\ a_{21} & a_{22} & a_{23} \\ a_{31} & a_{32} & a_{33} \end{bmatrix}$$

the transpose, designated $[A]^T$, is defined as

$$[A]^T = \begin{bmatrix} a_{11} & a_{21} & a_{31} \\ a_{12} & a_{22} & a_{32} \\ a_{13} & a_{23} & a_{33} \end{bmatrix}$$

In other words, the element a_{ij} of the transpose is equal to the a_{ji} element of the original matrix.

The transpose has a variety of functions in matrix algebra. One simple advantage is that it allows a column vector to be written as a row, and vice versa. For example, if

$$\{c\} = \begin{Bmatrix} c_1 \\ c_1 \\ c_1 \end{Bmatrix}$$

then

$$\{c\}^T = \lfloor c_1 \quad c_2 \quad c_3 \rfloor$$

In addition, the transpose has numerous mathematical applications.

A *permutation matrix* (also called a *transposition matrix*) is an identity matrix with rows and columns interchanged. For example, here is a permutation matrix that is constructed by switching the first and third rows and columns of a 3×3 identity matrix:

$$[P] = \begin{bmatrix} 0 & 0 & 1 \\ 0 & 1 & 0 \\ 1 & 0 & 0 \end{bmatrix}$$

Left multiplying a matrix $[A]$ by this matrix, as in $[P][A]$, will switch the corresponding rows of $[A]$. Right multiplying, as in $[A][P]$, will switch the corresponding columns. Here is an example of left multiplication:

$$[P]\,[A] = \begin{bmatrix} 0 & 0 & 1 \\ 0 & 1 & 0 \\ 1 & 0 & 0 \end{bmatrix} \begin{bmatrix} 2 & -7 & 4 \\ 8 & 3 & -6 \\ 5 & 1 & 9 \end{bmatrix} = \begin{bmatrix} 5 & 1 & 9 \\ 8 & 3 & -6 \\ 2 & -7 & 4 \end{bmatrix}$$

The final matrix manipulation that will have utility in our discussion is *augmentation*. A matrix is augmented by the addition of a column (or columns) to the original matrix. For example, suppose we have a 3×3 matrix of coefficients. We might wish to augment this matrix $[A]$ with a 3×3 identity matrix to yield a 3×6 dimensional matrix:

$$\begin{bmatrix} a_{11} & a_{11} & a_{11} & 1 & 0 & 0 \\ a_{21} & a_{21} & a_{21} & 0 & 1 & 0 \\ a_{31} & a_{31} & a_{31} & 0 & 0 & 1 \end{bmatrix}$$

Such an expression has utility when we must perform a set of identical operations on the rows of two matrices. Thus, we can perform the operations on the single augmented matrix rather than on the two individual matrices.

EXAMPLE 8.1   MATLAB Matrix Manipulations

Problem Statement. The following example illustrates how a variety of matrix manipulations are implemented with MATLAB. It is best approached as a hands-on exercise on the computer.

Solution. Create a 3×3 matrix:

```
>> A = [1 5 6;7 4 2;-3 6 7]

A =
     1    5    6
     7    4    2
    -3    6    7
```

The transpose of $[A]$ can be obtained using the ' operator:

```
>> A'

ans =
     1    7   -3
     5    4    6
     6    2    7
```

Next we will create another 3×3 matrix on a row basis. First create three row vectors:

```
>> x = [8 6 9];
>> y = [-5 8 1];
>> z = [4 8 2];
```

Then we can combine these to form the matrix:

```
>> B = [x; y; z]

B =
     8     6     9
    -5     8     1
     4     8     2
```

We can add [A] and [B] together:

```
>> C = A+B

C =
     9    11    15
     2    12     3
     1    14     9
```

Further, we can subtract [B] from [C] to arrive back at [A]:

```
>> A = C-B

A =
     1     5     6
     7     4     2
    -3     6     7
```

Because their inner dimensions are equal, [A] and [B] can be multiplied

```
>> A*B

ans =
     7    94    26
    44    90    71
   -26    86    -7
```

Note that [A] and [B] can also be multiplied on an element-by-element basis by including a period with the multiplication operator as in

```
>> A.*B

ans =
     8    30    54
   -35    32     2
   -12    48    14
```

A 2×3 matrix can be set up

```
>> D = [1 4 3;5 8 1];
```

If [A] is multiplied times [D], an error message will occur

```
>> A*D

??? Error using ==> mtimes
Inner matrix dimensions must agree.
```

However, if we reverse the order of multiplication so that the inner dimensions match, matrix multiplication works

```
>> D*A
```

```
ans =
```

```
    20    39    35
    58    63    53
```

The matrix inverse can be computed with the `inv` function:

```
>> AI = inv(A)
```

```
AI =
```

```
     0.2462     0.0154    -0.2154
    -0.8462     0.3846     0.6154
     0.8308    -0.3231    -0.4769
```

To test that this is the correct result, the inverse can be multiplied by the original matrix to give the identity matrix:

```
>> A*AI
```

```
ans =
```

```
    1.0000   -0.0000   -0.0000
    0.0000    1.0000   -0.0000
    0.0000   -0.0000    1.0000
```

The `eye` function can be used to generate an identity matrix:

```
>> I = eye(3)
```

```
I =
```

```
    1    0    0
    0    1    0
    0    0    1
```

We can set up a permutation matrix to switch the first and third rows and columns of a 3×3 matrix as

```
>> P=[0 0 1;0 1 0;1 0 0]
```

```
P =
```

```
    0    0    1
    0    1    0
    1    0    0
```

We can then either switch the rows:

```
>> PA=P*A
```

```
PA =
```

```
    -3    6    7
     7    4    2
     1    5    6
```

or the columns:

```
>> AP=A*P

AP =

      6     5     1
      2     4     7
      7     6    -3
```

Finally, matrices can be augmented simply as in

```
>> Aug = [A I]

Aug =

      1     5     6     1     0     0
      7     4     2     0     1     0
     -3     6     7     0     0     1
```

Note that the dimensions of a matrix can be determined by the size function:

```
>> [n,m] = size(Aug)

n =

     3

m =

     6
```

8.1.3 Representing Linear Algebraic Equations in Matrix Form

It should be clear that matrices provide a concise notation for representing simultaneous linear equations. For example, a 3×3 set of linear equations,

$$
\begin{aligned}
a_{11}x_1 + a_{12}x_2 + a_{13}x_3 &= b_1 \\
a_{21}x_1 + a_{22}x_2 + a_{23}x_3 &= b_2 \\
a_{31}x_1 + a_{32}x_2 + a_{33}x_3 &= b_3
\end{aligned}
\tag{8.5}
$$

can be expressed as

$$
[A]\{x\} = \{b\}
\tag{8.6}
$$

where $[A]$ is the matrix of coefficients:

$$
[A] = \begin{bmatrix} a_{11} & a_{12} & a_{13} \\ a_{21} & a_{22} & a_{23} \\ a_{31} & a_{32} & a_{33} \end{bmatrix}
$$

$\{b\}$ is the column vector of constants:

$$\{b\}^T = \lfloor b_1 \quad b_2 \quad b_3 \rfloor$$

and $\{x\}$ is the column vector of unknowns:

$$\{x\}^T = \lfloor x_1 \quad x_2 \quad x_3 \rfloor$$

Recall the definition of matrix multiplication [Eq. (8.4)] to convince yourself that Eqs. (8.5) and (8.6) are equivalent. Also, realize that Eq. (8.6) is a valid matrix multiplication because the number of columns n of the first matrix $[A]$ is equal to the number of rows n of the second matrix $\{x\}$.

This part of the book is devoted to solving Eq. (8.6) for $\{x\}$. A formal way to obtain a solution using matrix algebra is to multiply each side of the equation by the inverse of $[A]$ to yield

$$[A]^{-1}[A]\{x\} = [A]^{-1}\{b\}$$

Because $[A]^{-1}[A]$ equals the identity matrix, the equation becomes

$$\{x\} = [A]^{-1}\{b\} \tag{8.7}$$

Therefore, the equation has been solved for $\{x\}$. This is another example of how the inverse plays a role in matrix algebra that is similar to division. It should be noted that this is not a very efficient way to solve a system of equations. Thus, other approaches are employed in numerical algorithms. However, as discussed in Sec. 11.1.2, the matrix inverse itself has great value in the engineering analyses of such systems.

It should be noted that systems with more equations (rows) than unknowns (columns), $m > n$, are said to be *overdetermined*. A typical example is least-squares regression where an equation with n coefficients is fit to m data points (x, y). Conversely, systems with less equations than unknowns, $m < n$, are said to be *underdetermined*. A typical example of underdetermined systems is numerical optimization.

8.2   SOLVING LINEAR ALGEBRAIC EQUATIONS WITH MATLAB

MATLAB provides two direct ways to solve systems of linear algebraic equations. The most efficient way is to employ the backslash, or "left-division," operator as in

```
>> x = A\b
```

The second is to use matrix inversion:

```
>> x = inv(A)*b
```

As stated at the end of Sec. 8.1.3, the matrix inverse solution is less efficient than using the backslash. Both options are illustrated in the following example.

EXAMPLE 8.2   Solving the Bungee Jumper Problem with MATLAB

Problem Statement. Use MATLAB to solve the bungee jumper problem described at the beginning of this chapter. The parameters for the problem are

Jumper	Mass (kg)	Spring Constant (N/m)	Unstretched Cord Length (m)
Top (1)	60	50	20
Middle (2)	70	100	20
Bottom (3)	80	50	20

Solution. Substituting these parameter values into Eq. (8.2) gives

$$
\begin{bmatrix} 150 & -100 & 0 \\ -100 & 150 & -50 \\ 0 & -50 & 50 \end{bmatrix} \begin{Bmatrix} x_1 \\ x_2 \\ x_3 \end{Bmatrix} = \begin{Bmatrix} 588.6 \\ 686.7 \\ 784.8 \end{Bmatrix}
$$

Start up MATLAB and enter the coefficient matrix and the right-hand-side vector:

```
>> K = [150 -100 0;-100 150 -50;0 -50 50]

K =
   150  -100     0
  -100   150   -50
     0   -50    50

>> mg = [588.6; 686.7; 784.8]

mg =
  588.6000
  686.7000
  784.8000
```

Employing left division yields

```
>> x = K\mg

x =
   41.2020
   55.9170
   71.6130
```

Alternatively, multiplying the inverse of the coefficient matrix by the right-hand-side vector gives the same result:

```
>> x = inv(K)*mg

x =
   41.2020
   55.9170
   71.6130
```

Because the jumpers were connected by 20-m cords, their initial positions relative to the platform is

```
>> xi = [20;40;60];
```

Thus, their final positions can be calculated as

```
>> xf = x+xi

xf =
   61.2020
   95.9170
  131.6130
```

The results, which are displayed in Fig. 8.6, make sense. The first cord is extended the longest because it has a lower spring constant and is subject to the most weight (all three jumpers). Notice that the second and third cords are extended about the same amount. Because it is subject to the weight of two jumpers, one might expect the second cord to be extended longer than the third. However, because it is stiffer (i.e., it has a higher spring constant), it stretches less than expected based on the weight it carries.

FIGURE 8.6
Positions of three individuals connected by bungee cords. (*a*) Unstretched and (*b*) stretched.

8.3 CASE STUDY   CURRENTS AND VOLTAGES IN CIRCUITS

Background.   Recall that in Chap. 1 (Table 1.1), we summarized some models and associated conservation laws that figure prominently in engineering. As in Fig. 8.7, each model represents a system of interacting elements. Consequently, steady-state balances derived from the conservation laws yield systems of simultaneous equations. In many cases, such systems are linear and hence can be expressed in matrix form. The present case study focuses on one such application: circuit analysis.

A common problem in electrical engineering involves determining the currents and voltages at various locations in resistor circuits. These problems are solved using *Kirchhoff's current* and *voltage rules*. The *current* (or point) *rule* states that the algebraic sum of all currents entering a node must be zero (Fig. 8.8*a*), or

$$\sum i = 0 \qquad\qquad (8.8)$$

where all current entering the node is considered positive in sign. The current rule is an application of the principle of *conservation of charge* (recall Table 1.1).

8.3 CASE STUDY   continued

FIGURE 8.7
Engineering systems which, at steady state, can be modeled with linear algebraic equations.

FIGURE 8.8
Schematic representations of (a) Kirchhoff's current rule and (b) Ohm's law.

The *voltage* (or loop) *rule* specifies that the algebraic sum of the potential differences (i.e., voltage changes) in any loop must equal zero. For a resistor circuit, this is expressed as

$$\sum \xi - \sum iR = 0 \tag{8.9}$$

where ξ is the emf (electromotive force) of the voltage sources, and R is the resistance of any resistors on the loop. Note that the second term derives from *Ohm's law* (Fig. 8.8b), which states that the voltage drop across an ideal resistor is equal to the product of the current and the resistance. Kirchhoff's voltage rule is an expression of the *conservation of energy*.

Solution. Application of these rules results in systems of simultaneous linear algebraic equations because the various loops within a circuit are interconnected. For example, consider the circuit shown in Fig. 8.9. The currents associated with this circuit are unknown both in magnitude and direction. This presents no great difficulty because one simply assumes a direction for each current. If the resultant solution from Kirchhoff's laws is negative, then the assumed direction was incorrect. For example, Fig. 8.10 shows some assumed currents.

8.3 CASE STUDY   continued

FIGURE 8.9
A resistor circuit to be solved using simultaneous
linear algebraic equations.

FIGURE 8.10
Assumed current directions.

Given these assumptions, Kirchhoff's current rule is applied at each node to yield

$$i_{12} + i_{52} + i_{32} = 0$$
$$i_{65} - i_{52} - i_{54} = 0$$
$$i_{43} - i_{32} = 0$$
$$i_{54} - i_{43} = 0$$

Application of the voltage rule to each of the two loops gives

$$-i_{54}R_{54} - i_{43}R_{43} - i_{32}R_{32} + i_{52}R_{52} = 0$$
$$-i_{65}R_{65} - i_{52}R_{52} + i_{12}R_{12} - 200 = 0$$

or, substituting the resistances from Fig. 8.9 and bringing constants to the right-hand side,

$$-15i_{54} - 5i_{43} - 10i_{32} + 10i_{52} = 0$$
$$-20i_{65} - 10i_{52} + 5i_{12} = 200$$

Therefore, the problem amounts to solving six equations with six unknown currents. These
equations can be expressed in matrix form as

$$
\begin{bmatrix}
1 & 1 & 1 & 0 & 0 & 0 \\
0 & -1 & 0 & 1 & -1 & 0 \\
0 & 0 & -1 & 0 & 0 & 1 \\
0 & 0 & 0 & 0 & 1 & -1 \\
0 & 10 & -10 & 0 & -15 & -5 \\
5 & -10 & 0 & -20 & 0 & 0
\end{bmatrix}
\begin{Bmatrix}
i_{12} \\
i_{52} \\
i_{32} \\
i_{65} \\
i_{54} \\
i_{43}
\end{Bmatrix}
=
\begin{Bmatrix}
0 \\
0 \\
0 \\
0 \\
0 \\
200
\end{Bmatrix}
$$

8.3 CASE STUDY   continued

Although impractical to solve by hand, this system is easily handled by MATLAB. The solution is

```
>> A=[1 1 1 0 0 0
0 -1 0 1 -1 0
0 0 -1 0 0 1
0 0 0 0 1 -1
0 10 -10 0 -15 -5
5 -10 0 -20 0 0];
>> b=[0 0 0 0 0 200]';
>> current=A\b

current =
     6.1538
    -4.6154
    -1.5385
    -6.1538
    -1.5385
    -1.5385
```

Thus, with proper interpretation of the signs of the result, the circuit currents and voltages are as shown in Fig. 8.11. The advantages of using MATLAB for problems of this type should be evident.

FIGURE 8.11
The solution for currents and voltages obtained using MATLAB.

$V = 153.85$ $V = 169.23$ $V = 200$
$i = 1.5385$ $i = 6.1538$
$V = 146.15$ $V = 123.08$ $V = 0$

PROBLEMS

8.1 Given a square matrix $[A]$, write a single line MATLAB command that will create a new matrix $[Aug]$ that consists of the original matrix $[A]$ augmented by an identity matrix $[I]$.

8.2 A number of matrices are defined as

$$[A] = \begin{bmatrix} 4 & 7 \\ 1 & 2 \\ 5 & 6 \end{bmatrix} \quad [B] = \begin{bmatrix} 4 & 3 & 7 \\ 1 & 2 & 7 \\ 2 & 0 & 4 \end{bmatrix}$$

$$\{C\} = \begin{Bmatrix} 3 \\ 6 \\ 1 \end{Bmatrix} \quad [D] = \begin{bmatrix} 9 & 4 & 3 & -6 \\ 2 & -1 & 7 & 5 \end{bmatrix}$$

$$[E] = \begin{bmatrix} 1 & 5 & 8 \\ 7 & 2 & 3 \\ 4 & 0 & 6 \end{bmatrix}$$

$$[F] = \begin{bmatrix} 3 & 0 & 1 \\ 1 & 7 & 3 \end{bmatrix} \quad \lfloor G \rfloor = \lfloor 7\ 6\ 4 \rfloor$$

Answer the following questions regarding these matrices:
(a) What are the dimensions of the matrices?
(b) Identify the square, column, and row matrices.
(c) What are the values of the elements: a_{12}, b_{23}, d_{32}, e_{22}, f_{12}, g_{12}?
(d) Perform the following operations:

    (1) $[E] + [B]$ (2) $[A] + [F]$ (3) $[B] - [E]$

    (4) $7 \times [B]$ (5) $\{C\}^T$ (6) $[E] \times [B]$

    (7) $[B] \times [A]$ (8) $[D]^T$ (9) $[A] \times \{C\}$

    (10) $[I] \times [B]$ (11) $[E]^T \times [E]$ (12) $\{C\}^T \times \{C\}$

8.3 Write the following set of equations in matrix form:

$$50 = 5x_3 - 7x_2$$
$$4x_2 + 7x_3 + 30 = 0$$
$$x_1 - 7x_3 = 40 - 3x_2 + 5x_1$$

Use MATLAB to solve for the unknowns. In addition, use it to compute the transpose and the inverse of the coefficient matrix.

8.4 Three matrices are defined as

$$[A] = \begin{bmatrix} 6 & -1 \\ 12 & 8 \\ -5 & 4 \end{bmatrix} \quad [B] = \begin{bmatrix} 4 & 0 \\ 0.5 & 2 \end{bmatrix} \quad [C] = \begin{bmatrix} 2 & -2 \\ 3 & 1 \end{bmatrix}$$

(a) Perform all possible multiplications that can be computed between pairs of these matrices.
(b) Justify why the remaining pairs cannot be multiplied.

(c) Use the results of (a) to illustrate why the order of multiplication is important.

8.5 Solve the following system with MATLAB:

$$\begin{bmatrix} 3 + 2i & 4 \\ -i & 1 \end{bmatrix} \begin{Bmatrix} z_1 \\ z_2 \end{Bmatrix} = \begin{Bmatrix} 2 + i \\ 3 \end{Bmatrix}$$

8.6 Develop, debug, and test your own M-file to multiply two matrices—that is, $[X] = [Y][Z]$, where $[Y]$ is m by n and $[Z]$ is n by p. Employ for...end loops to implement the multiplication and include error traps to flag bad cases. Test the program using the matrices from Prob. 8.4.

8.7 Develop, debug, and test your own M-file to generate the transpose of a matrix. Employ for...end loops to implement the transpose. Test it on the matrices from Prob. 8.4.

8.8 Develop, debug, and test your own M-file function to switch the rows of a matrix using a permutation matrix. The first lines of the function should be as follows:

```
function B = permut(A,r1,r2)
% Permut: Switch rows of matrix A
% with a permutation matrix
% B = permut(A,r1,r2)
% input:
% A = original matrix
% r1, r2 = rows to be switched
% output:
% B = matrix with rows switched
```

Include error traps for erroneous inputs (e.g., user specifies rows that exceed the dimensions of the original matrix).

8.9 Five reactors linked by pipes are shown in Fig. P8.9. The rate of mass flow through each pipe is computed as the product of flow (Q) and concentration (c). At steady state, the mass flow into and out of each reactor must be equal. For example, for the first reactor, a *mass balance* can be written as

$$Q_{01}c_{01} + Q_{31}c_3 = Q_{15}c_1 + Q_{12}c_1$$

Write mass balances for the remaining reactors in Fig. P8.9 and express the equations in matrix form. Then use MATLAB to solve for the concentrations in each reactor.

8.10 An important problem in structural engineering is that of finding the forces in a statically determinate truss (Fig. P8.10). This type of structure can be described as a system of coupled linear algebraic equations derived from force balances. The sum of the forces in both horizontal and

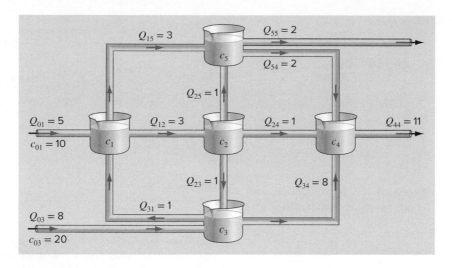

FIGURE P8.9

vertical directions must be zero at each node, because the system is at rest. Therefore, for node 1:

$$\sum F_H = 0 = -F_1 \cos 30° + F_3 \cos 60° + F_{1,h}$$
$$\sum F_V = 0 = -F_1 \sin 30° - F_3 \sin 60° + F_{1,v}$$

for node 2:

$$\sum F_H = 0 = F_2 + F_1 \cos 30° + F_{2,h} + H_2$$
$$\sum F_V = 0 = F_1 \sin 30° + F_{2,v} + V_2$$

for node 3:

$$\sum F_H = 0 = -F_2 - F_3 \cos 60° + F_{3,h}$$
$$\sum F_V = 0 = F_3 \sin 60° + F_{3,v} + V_3$$

where $F_{i,h}$ is the external horizontal force applied to node i (where a positive force is from left to right) and $F_{i,v}$ is the external vertical force applied to node i (where a positive force is upward). Thus, in this problem, the 2000-N downward force on node 1 corresponds to $F_{i,v} = -2000$. For this case, all other $F_{i,v}$'s and $F_{i,h}$'s are zero. Express this set of linear algebraic equations in matrix form and then use MATLAB to solve for the unknowns.

8.11 Consider the three mass-four spring system in Fig. P8.11. Determining the equations of motion from $\sum F_x = ma_x$ for each mass using its free-body diagram results in the

FIGURE P8.10

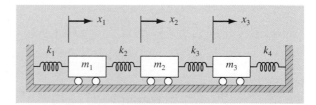

FIGURE P8.11

following differential equations:

$$\ddot{x}_1 + \left(\frac{k_1 + k_2}{m_1}\right)x_1 - \left(\frac{k_2}{m_1}\right)x_2 = 0$$

$$\ddot{x}_2 - \left(\frac{k_2}{m_2}\right)x_1 + \left(\frac{k_2 + k_3}{m_2}\right)x_2 - \left(\frac{k_3}{m_2}\right)x_3 = 0$$

$$\ddot{x}_3 - \left(\frac{k_3}{m_3}\right)x_2 + \left(\frac{k_3 + k_4}{m_3}\right)x_3 = 0$$

where $k_1 = k_4 = 10$ N/m, $k_2 = k_3 = 30$ N/m, and $m_1 = m_2 = m_3 = 1$ kg. The three equations can be written in matrix form:

$$0 = \{\text{Acceleration vector}\}$$
$$\quad + [k/m \text{ matrix}]\{\text{displacement vector } x\}$$

At a specific time where $x_1 = 0.05$ m, $x_2 = 0.04$ m, and $x_3 = 0.03$ m, this forms a tridiagonal matrix. Use MATLAB to solve for the acceleration of each mass.

8.12 Perform the same computation as in Example 8.2, but use five jumpers with the following characteristics:

Jumper	Mass (kg)	Spring Constant (N/m)	Unstretched Cord Length (m)
1	55	80	10
2	75	50	10
3	60	70	10
4	75	100	10
5	90	20	10

FIGURE P8.14

8.13 Three masses are suspended vertically by a series of identical springs where mass 1 is at the top and mass 3 is at the bottom. If $g = 9.81$ m/s^2, $m_1 = 2$ kg, $m_2 = 3$ kg, $m_3 = 2.5$ kg, and the k's = 10 kg/s^2, use MATLAB to solve for the displacements x.

8.14 Perform the same computation as in Sec. 8.3, but for the circuit in Fig. P8.14.

8.15 Perform the same computation as in Sec. 8.3, but for the circuit in Fig. P8.15.

8.16 Besides solving simultaneous equations, linear algebra has lots of other applications in engineering and science. An example from computer graphics involves rotating an object in Euclidean space. The following *rotation matrix* can be employed to rotate a group of points counter-clockwise through an angle θ about the origin of a Cartesian coordinate system,

$$R = \begin{bmatrix} \cos\theta & -\sin\theta \\ \sin\theta & \cos\theta \end{bmatrix}$$

FIGURE P8.15

To do this, each point's position must be represented by a column vector v, containing the coordinates of the point. For example, here are vectors for the x and y coordinates of the rectangle in Fig. P8.16

```
x = [1 4 4 1]; y = [1 1 4 4];
```

The rotated vector is then generated with matrix multiplication: $[R]\{v\}$. Develop a MATLAB function to perform this operation and display the initial and the rotated points as filled shapes on the same graph. Here is a script to test your function:

```
clc;clf;format compact
x = [1 4 4 1]; y = [1 1 4 4];
[xt, yt] = Rotate2D(45, x, y);
```

and here is a skeleton of the function

```
function [xr, yr] = Rotate2D(thetad, x, y)
% two dimensional rotation 2D rotate Cartesian
% [xr, yr] = rot2d(thetad, x, y)
% Rotation of a two-dimensional object the
Cartesian coordinates
% of which are contained in the vectors x and y.
% input:
% thetad = angle of rotation (degrees)
% x = vector containing objects x coordinates
% y = vector containing objects y coordinates
% output:
% xr = vector containing objects rotated x
coordinates
% yr = vector containing objects rotated y
coordinates

% convert angle to radians and set up rotation
matrix
    •
    •  .
    •
```

```
% close shape
    •
    •  .
    •
% plot original object
hold on, grid on
    •
    •  .
    •
% rotate shape
    •
    •  .
    •
% plot rotated object
    •
    •  .
    •
hold off
```

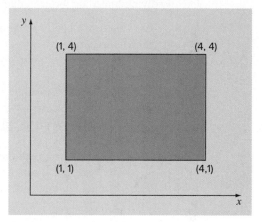

FIGURE P8.16

9

Gauss Elimination

CHAPTER OBJECTIVES

The primary objective of this chapter is to describe the Gauss elimination algorithm for solving linear algebraic equations. Specific objectives and topics covered are

- Knowing how to solve small sets of linear equations with the graphical method and Cramer's rule.
- Understanding how to implement forward elimination and back substitution as in Gauss elimination.
- Understanding how to count flops to evaluate the efficiency of an algorithm.
- Understanding the concepts of singularity and ill-condition.
- Understanding how partial pivoting is implemented and how it differs from complete pivoting.
- Knowing how to compute the determinant as part of the Gauss elimination algorithm with partial pivoting.
- Recognizing how the banded structure of a tridiagonal system can be exploited to obtain extremely efficient solutions.

At the end of Chap. 8, we stated that MATLAB provides two simple and direct methods for solving systems of linear algebraic equations: left division,

```
>> x = A\b
```

and matrix inversion,

```
>> x = inv(A)*b
```

Chapters 9 and 10 provide background on how such solutions are obtained. This material is included to provide insight into how MATLAB operates. In addition, it is intended to show how you can build your own solution algorithms in computational environments that do not have MATLAB's built-in capabilities.

The technique described in this chapter is called Gauss elimination because it involves combining equations to eliminate unknowns. Although it is one of the earliest methods for solving simultaneous equations, it remains among the most important algorithms in use today and is the basis for linear equation solving on many popular software packages including MATLAB.

9.1   SOLVING SMALL NUMBERS OF EQUATIONS

Before proceeding to Gauss elimination, we will describe several methods that are appropriate for solving small ($n \leq 3$) sets of simultaneous equations and that do not require a computer. These are the graphical method, Cramer's rule, and the elimination of unknowns.

9.1.1 The Graphical Method

A graphical solution is obtainable for two linear equations by plotting them on Cartesian coordinates with one axis corresponding to x_1 and the other to x_2. Because the equations are linear, each equation will plot as a straight line. For example, suppose that we have the following equations:

$$3x_1 + 2x_2 = 18$$
$$-x_1 + 2x_2 = 2$$

If we assume that x_1 is the abscissa, we can solve each of these equations for x_2:

$$x_2 = -\frac{3}{2}x_1 + 9$$
$$x_2 = \frac{1}{2}x_1 + 1$$

The equations are now in the form of straight lines—that is, $x_2 =$ (slope) $x_1 +$ intercept. When these equations are graphed, the values of x_1 and x_2 at the intersection of the lines represent the solution (Fig. 9.1). For this case, the solution is $x_1 = 4$ and $x_2 = 3$.

For three simultaneous equations, each equation would be represented by a plane in a three-dimensional coordinate system. The point where the three planes intersect would represent the solution. Beyond three equations, graphical methods break down and, consequently, have little practical value for solving simultaneous equations. However, they are useful in visualizing properties of the solutions.

For example, Fig. 9.2 depicts three cases that can pose problems when solving sets of linear equations. Fig. 9.2*a* shows the case where the two equations represent parallel lines. For such situations, there is no solution because the lines never cross. Figure 9.2*b* depicts the case where the two lines are coincident. For such situations there is an infinite number of solutions. Both types of systems are said to be *singular*.

In addition, systems that are very close to being singular (Fig. 9.2*c*) can also cause problems. These systems are said to be *ill-conditioned*. Graphically, this corresponds to the fact that it is difficult to identify the exact point at which the lines intersect. Ill-conditioned systems will also pose problems when they are encountered during the numerical solution of linear equations. This is because they will be extremely sensitive to roundoff error.

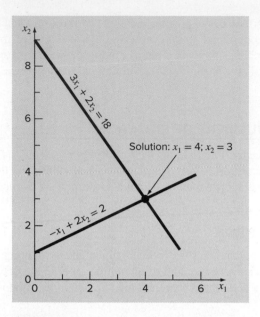

FIGURE 9.1
Graphical solution of a set of two simultaneous linear algebraic equations. The intersection of the lines represents the solution.

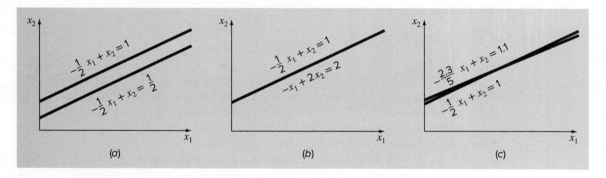

FIGURE 9.2
Graphical depiction of singular and ill-conditioned systems: (*a*) no solution, (*b*) infinite solutions, and (*c*) ill-conditioned system where the slopes are so close that the point of intersection is difficult to detect visually.

9.1.2 Determinants and Cramer's Rule

Cramer's rule is another solution technique that is best suited to small numbers of equations. Before describing this method, we will briefly review the concept of the determinant, which is used to implement Cramer's rule. In addition, the determinant has relevance to the evaluation of the ill-conditioning of a matrix.

Determinants. The determinant can be illustrated for a set of three equations:

$$[A]\{x\} = \{b\}$$

where $[A]$ is the coefficient matrix

$$[A] = \begin{bmatrix} a_{11} & a_{12} & a_{13} \\ a_{21} & a_{22} & a_{23} \\ a_{31} & a_{32} & a_{33} \end{bmatrix}$$

The *determinant* of this system is formed from the coefficients of $[A]$ and is represented as

$$D = \begin{vmatrix} a_{11} & a_{12} & a_{13} \\ a_{21} & a_{22} & a_{23} \\ a_{31} & a_{32} & a_{33} \end{vmatrix}$$

Although the determinant D and the coefficient matrix $[A]$ are composed of the same elements, they are completely different mathematical concepts. That is why they are distinguished visually by using brackets to enclose the matrix and straight lines to enclose the determinant. In contrast to a matrix, the determinant is a single number. For example, the value of the determinant for two simultaneous equations

$$D = \begin{vmatrix} a_{11} & a_{12} \\ a_{21} & a_{22} \end{vmatrix}$$

is calculated by

$$D = a_{11}a_{22} - a_{12}a_{21}$$

For the third-order case, the determinant can be computed as

$$D = a_{11} \begin{vmatrix} a_{22} & a_{23} \\ a_{32} & a_{33} \end{vmatrix} - a_{12} \begin{vmatrix} a_{21} & a_{23} \\ a_{31} & a_{33} \end{vmatrix} + a_{13} \begin{vmatrix} a_{21} & a_{22} \\ a_{31} & a_{32} \end{vmatrix} \tag{9.1}$$

where the 2×2 determinants are called *minors*.

EXAMPLE 9.1   Determinants

Problem Statement.   Compute values for the determinants of the systems represented in Figs. 9.1 and 9.2.

Solution.   For Fig. 9.1:

$$D = \begin{vmatrix} 3 & 2 \\ -1 & 2 \end{vmatrix} = 3(2) - 2(-1) = 8$$

For Fig. 9.2a:

$$D = \begin{vmatrix} -\frac{1}{2} & 1 \\ -\frac{1}{2} & 1 \end{vmatrix} = -\frac{1}{2}(1) - 1\left(\frac{-1}{2}\right) = 0$$

For Fig. 9.2b:

$$D = \begin{vmatrix} -\frac{1}{2} & 1 \\ -1 & 2 \end{vmatrix} = -\frac{1}{2}(2) - 1(-1) = 0$$

For Fig. 9.2c:

$$D = \begin{vmatrix} -\frac{1}{2} & 1 \\ -\frac{2.3}{5} & 1 \end{vmatrix} = -\frac{1}{2}(1) - 1\left(\frac{-2.3}{5}\right) = -0.04$$

In the foregoing example, the singular systems had zero determinants. Additionally, the results suggest that the system that is almost singular (Fig. 9.2c) has a determinant that is close to zero. These ideas will be pursued further in our subsequent discussion of ill-conditioning in Chap. 11.

Cramer's Rule. This rule states that each unknown in a system of linear algebraic equations may be expressed as a fraction of two determinants with denominator D and with the numerator obtained from D by replacing the column of coefficients of the unknown in question by the constants $b_1, b_2, \ldots, b_n$. For example, for three equations, x_1 would be computed as

$$x_1 = \frac{\begin{vmatrix} b_1 & a_{12} & a_{13} \\ b_2 & a_{22} & a_{23} \\ b_3 & a_{32} & a_{33} \end{vmatrix}}{D}$$

EXAMPLE 9.2   Cramer's Rule

Problem Statement. Use Cramer's rule to solve

$$0.3x_1 + 0.52x_2 + \quad x_3 = -0.01$$
$$0.5x_1 + \quad x_2 + 1.9x_3 = \quad 0.67$$
$$0.1x_1 + 0.3 \ x_2 + 0.5x_3 = -0.44$$

Solution. The determinant D can be evaluated as [Eq. (9.1)]:

$$D = 0.3 \begin{vmatrix} 1 & 1.9 \\ 0.3 & 0.5 \end{vmatrix} - 0.52 \begin{vmatrix} 0.5 & 1.9 \\ 0.1 & 0.5 \end{vmatrix} + 1 \begin{vmatrix} 0.5 & 1 \\ 0.1 & 0.3 \end{vmatrix} = -0.0022$$

The solution can be calculated as

$$x_1 = \frac{\begin{vmatrix} -0.01 & 0.52 & 1 \\ 0.67 & 1 & 1.9 \\ -0.44 & 0.3 & 0.5 \end{vmatrix}}{-0.0022} = \frac{0.03278}{-0.0022} = -14.9$$

$$x_2 = \frac{\begin{vmatrix} 0.3 & -0.01 & 1 \\ 0.5 & 0.67 & 1.9 \\ 0.1 & -0.44 & 0.5 \end{vmatrix}}{-0.0022} = \frac{0.0649}{-0.0022} = -29.5$$

$$x_3 = \frac{\begin{vmatrix} 0.3 & 0.52 & -0.01 \\ 0.5 & 1 & 0.67 \\ 0.1 & 0.3 & -0.44 \end{vmatrix}}{-0.0022} = \frac{-0.04356}{-0.0022} = 19.8$$

The det Function. The determinant can be computed directly in MATLAB with the det function. For example, using the system from the previous example:

```
>> A=[0.3 0.52 1;0.5 1 1.9;0.1 0.3 0.5];
>> D=det(A)

D =
   -0.0022
```

Cramer's rule can be applied to compute x_1 as in

```
>> A(:,1)=[-0.01;0.67;-0.44]

A =
   -0.0100    0.5200    1.0000
    0.6700    1.0000    1.9000
   -0.4400    0.3000    0.5000

>> x1=det(A)/D

x1 =
  -14.9000
```

For more than three equations, Cramer's rule becomes impractical because, as the number of equations increases, the determinants are time consuming to evaluate by hand (or by computer). Consequently, more efficient alternatives are used. Some of these alternatives are based on the last noncomputer solution technique covered in Sec. 9.1.3—the elimination of unknowns.

9.1.3 Elimination of Unknowns

The elimination of unknowns by combining equations is an algebraic approach that can be illustrated for a set of two equations:

$$a_{11}x_1 + a_{12}x_2 = b_1 \tag{9.2}$$
$$a_{21}x_1 + a_{22}x_2 = b_2 \tag{9.3}$$

The basic strategy is to multiply the equations by constants so that one of the unknowns will be eliminated when the two equations are combined. The result is a single equation that can be solved for the remaining unknown. This value can then be substituted into either of the original equations to compute the other variable.

For example, Eq. (9.2) might be multiplied by a_{21} and Eq. (9.3) by a_{11} to give

$$a_{21}a_{11}x_1 + a_{21}a_{12}x_2 = a_{21}b_1 \tag{9.4}$$
$$a_{11}a_{21}x_1 + a_{11}a_{22}x_2 = a_{11}b_2 \tag{9.5}$$

Subtracting Eq. (9.4) from Eq. (9.5) will, therefore, eliminate the x_1 term from the equations to yield

$$a_{11}a_{22}x_2 - a_{21}a_{12}x_2 = a_{11}b_2 - a_{21}b_1$$

which can be solved for

$$x_2 = \frac{a_{11}b_2 - a_{21}b_1}{a_{11}a_{22} - a_{21}a_{12}} \tag{9.6}$$

Equation (9.6) can then be substituted into Eq. (9.2), which can be solved for

$$x_1 = \frac{a_{22}b_1 - a_{12}b_2}{a_{11}a_{22} - a_{21}a_{12}} \tag{9.7}$$

Notice that Eqs. (9.6) and (9.7) follow directly from Cramer's rule:

$$x_1 = \frac{\begin{vmatrix} b_1 & a_{12} \\ b_2 & a_{22} \end{vmatrix}}{\begin{vmatrix} a_{11} & a_{12} \\ b_{21} & a_{22} \end{vmatrix}} = \frac{a_{22}b_1 - a_{12}b_2}{a_{11}a_{22} - a_{21}a_{12}}$$

$$x_2 = \frac{\begin{vmatrix} a_{11} & b_1 \\ a_{21} & b_2 \end{vmatrix}}{\begin{vmatrix} a_{11} & a_{12} \\ a_{21} & a_{22} \end{vmatrix}} = \frac{a_{11}b_2 - a_{21}b_1}{a_{11}a_{22} - a_{21}a_{12}}$$

The elimination of unknowns can be extended to systems with more than two or three equations. However, the numerous calculations that are required for larger systems make the method extremely tedious to implement by hand. However, as described in Sec. 9.2, the technique can be formalized and readily programmed for the computer.

9.2   NAIVE GAUSS ELIMINATION

In Sec. 9.1.3, the elimination of unknowns was used to solve a pair of simultaneous equations. The procedure consisted of two steps (Fig. 9.3):

1. The equations were manipulated to eliminate one of the unknowns from the equations. The result of this elimination step was that we had one equation with one unknown.
2. Consequently, this equation could be solved directly and the result back-substituted into one of the original equations to solve for the remaining unknown.

This basic approach can be extended to large sets of equations by developing a systematic scheme or algorithm to eliminate unknowns and to back-substitute. Gauss elimination is the most basic of these schemes.

This section includes the systematic techniques for forward elimination and back substitution that comprise Gauss elimination. Although these techniques are ideally suited for implementation on computers, some modifications will be required to obtain a reliable algorithm. In particular, the computer program must avoid division by zero. The following method is called "naive" Gauss elimination because it does not avoid this problem. Section 9.3 will deal with the additional features required for an effective computer program.

The approach is designed to solve a general set of n equations:

$$a_{11}x_1 + a_{12}x_2 + a_{13}x_3 + \cdots + a_{1n}x_n = b_1 \tag{9.8a}$$

$$a_{21}x_1 + a_{22}x_2 + a_{23}x_3 + \cdots + a_{2n}x_n = b_2 \tag{9.8b}$$

$$\vdots \qquad \vdots$$

$$a_{n1}x_1 + a_{n2}x_2 + a_{n3}x_3 + \cdots + a_{nn}x_n = b_n \tag{9.8c}$$

FIGURE 9.3
The two phases of Gauss elimination: (a) forward elimination and (b) back substitution.

As was the case with the solution of two equations, the technique for n equations consists of two phases: elimination of unknowns and solution through back substitution.

Forward Elimination of Unknowns. The first phase is designed to reduce the set of equations to an upper triangular system (Fig. 9.3a). The initial step will be to eliminate the first unknown x_1 from the second through the nth equations. To do this, multiply Eq. (9.8a) by a_{21}/a_{11} to give

$$a_{21}x_1 + \frac{a_{21}}{a_{11}} a_{12}x_2 + \frac{a_{21}}{a_{11}} a_{13}x_3 + \cdots + \frac{a_{21}}{a_{11}} a_{1n}x_n = \frac{a_{21}}{a_{11}} b_1 \qquad (9.9)$$

This equation can be subtracted from Eq. (9.8b) to give

$$\left(a_{22} - \frac{a_{21}}{a_{11}} a_{12}\right) x_2 + \cdots + \left(a_{2n} - \frac{a_{21}}{a_{11}} a_{1n}\right) x_n = b_2 - \frac{a_{21}}{a_{11}} b_1$$

or

$$a'_{22} x_2 + \cdots + a'_{2n} x_n = b'_2$$

where the prime indicates that the elements have been changed from their original values.

The procedure is then repeated for the remaining equations. For instance, Eq. (9.8a) can be multiplied by a_{31}/a_{11} and the result subtracted from the third equation. Repeating the procedure for the remaining equations results in the following modified system:

$$a_{11}x_1 + a_{12}x_2 + a_{13}x_3 + \cdots + a_{1n}x_n = b_1 \qquad (9.10a)$$
$$a'_{22} x_2 + a'_{23} x_3 + \cdots + a'_{2n} x_n = b'_2 \qquad (9.10b)$$

$$a'_{32} x_2 + a'_{33} x_3 + \cdots + a'_{3n} x_n = b'_3 \tag{9.10c}$$

$$\vdots \qquad \vdots$$

$$a'_{n2} x_2 + a'_{n3} x_3 + \cdots + a'_{nn} x_n = b'_n \tag{9.10d}$$

For the foregoing steps, Eq. (9.8a) is called the *pivot equation* and a_{11} is called the *pivot element*. Note that the process of multiplying the first row by a_{21}/a_{11} is equivalent to dividing it by a_{11} and multiplying it by a_{21}. Sometimes the division operation is referred to as *normalization*. We make this distinction because a zero pivot element can interfere with normalization by causing a division by zero. We will return to this important issue after we complete our description of naive Gauss elimination.

The next step is to eliminate x_2 from Eq. (9.10c) through (9.10d). To do this, multiply Eq. (9.10b) by a'_{32}/a'_{22} and subtract the result from Eq. (9.10c). Perform a similar elimination for the remaining equations to yield

$$a_{11} x_1 + a_{12} x_2 + a_{13} x_3 + \cdots + a_{1n} x_n = b_1$$

$$a'_{22} x_2 + a'_{23} x_3 + \cdots + a'_{2n} x_n = b'_2$$

$$a''_{33} x_3 + \cdots + a''_{3n} x_n = b''_3$$

$$\vdots \qquad \vdots$$

$$a''_{n3} x_3 + \cdots + a''_{nn} x_n = b''_n$$

where the double prime indicates that the elements have been modified twice.

The procedure can be continued using the remaining pivot equations. The final manipulation in the sequence is to use the $(n-1)$th equation to eliminate the x_{n-1} term from the nth equation. At this point, the system will have been transformed to an upper triangular system:

$$a_{11} x_1 + a_{12} x_2 + a_{13} x_3 + \cdots + a_{1n} x_n = b_1 \tag{9.11a}$$

$$a'_{22} x_2 + a'_{23} x_3 + \cdots + a'_{2n} x_n = b'_2 \tag{9.11b}$$

$$a''_{33} x_3 + \cdots + a''_{3n} x_n = b''_3 \tag{9.11c}$$

$$\ddots \qquad \vdots$$

$$a^{(n-1)}_{nn} x_n = b^{(n-1)}_n \tag{9.11d}$$

Back Substitution. Equation (9.11d) can now be solved for x_n:

$$x_n = \frac{b^{(n-1)}_n}{a^{(n-1)}_{nn}} \tag{9.12}$$

This result can be back-substituted into the $(n-1)$th equation to solve for x_{n-1}. The procedure, which is repeated to evaluate the remaining x's, can be represented by the following formula:

$$x_i = \frac{b_i^{(i-1)} - \displaystyle\sum_{j=i+1}^{n} a_{ij}^{(i-1)} x_j}{a_{ii}^{(i-1)}} \qquad \text{for } i = n-1, n-2, \ldots, 1 \tag{9.13}$$

EXAMPLE 9.3   Naive Gauss Elimination

Problem Statement. Use Gauss elimination to solve

$$3x_1 - 0.1x_2 - 0.2x_3 = 7.85 \qquad\qquad\qquad\qquad \text{(E9.3.1)}$$
$$0.1x_1 + 7x_2 - 0.3x_3 = -19.3 \qquad\qquad\qquad\qquad \text{(E9.3.2)}$$
$$0.3x_1 - 0.2x_2 + 10x_3 = 71.4 \qquad\qquad\qquad\qquad \text{(E9.3.3)}$$

Solution. The first part of the procedure is forward elimination. Multiply Eq. (E9.3.1) by 0.1/3 and subtract the result from Eq. (E9.3.2) to give

$$7.00333x_2 - 0.293333x_3 = -19.5617$$

Then multiply Eq. (E9.3.1) by 0.3/3 and subtract it from Eq. (E9.3.3). After these operations, the set of equations is

$$3x_1 - 0.1x_2 - 0.2x_3 = 7.85 \qquad\qquad\qquad\qquad \text{(E9.3.4)}$$
$$7.00333x_2 - 0.293333x_3 = -19.5617 \qquad\qquad\qquad\qquad \text{(E9.3.5)}$$
$$-0.190000x_2 + 10.0200x_3 = 70.6150 \qquad\qquad\qquad\qquad \text{(E9.3.6)}$$

To complete the forward elimination, x_2 must be removed from Eq. (E9.3.6). To accomplish this, multiply Eq. (E9.3.5) by $-0.190000/7.00333$ and subtract the result from Eq. (E9.3.6). This eliminates x_2 from the third equation and reduces the system to an upper triangular form, as in

$$3x_1 - 0.1x_2 - 0.2x_3 = 7.85 \qquad\qquad\qquad\qquad \text{(E9.3.7)}$$
$$7.00333x_2 - 0.293333x_3 = -19.5617 \qquad\qquad\qquad\qquad \text{(E9.3.8)}$$
$$10.0120x_3 = 70.0843 \qquad\qquad\qquad\qquad \text{(E9.3.9)}$$

We can now solve these equations by back substitution. First, Eq. (E9.3.9) can be solved for

$$x_3 = \frac{70.0843}{10.0120} = 7.00003$$

This result can be back-substituted into Eq. (E9.3.8), which can then be solved for

$$x_2 = \frac{-19.5617 + 0.293333(7.00003)}{7.00333} = -2.50000$$

Finally, $x_3 = 7.00003$ and $x_2 = -2.50000$ can be substituted back into Eq. (E9.3.7), which can be solved for

$$x_1 = \frac{7.85 + 0.1(-2.50000) + 0.2(7.00003)}{3} = 3.00000$$

Although there is a slight round off error, the results are very close to the exact solution of $x_1 = 3$, $x_2 = -2.5$, and $x_3 = 7$. This can be verified by substituting the results into the original equation set:

$$3(3) - 0.1(-2.5) - 0.2(7.00003) = 7.84999 \cong 7.85$$
$$0.1(3) + 7(-2.5) - 0.3(7.00003) = -19.30000 = -19.3$$
$$0.3(3) - 0.2(-2.5) + 10(7.00003) = 71.4003 \cong 71.4$$

```
function x = GaussNaive(A,b)
% GaussNaive: naive Gauss elimination
%   x = GaussNaive(A,b): Gauss elimination without pivoting.
% input:
%   A = coefficient matrix
%   b = right hand side vector
% output:
%   x = solution vector

[m,n] = size(A);
if m~=n, error('Matrix A must be square'); end
nb = n+1;
Aug = [A b];
% forward elimination
for k = 1:n-1
  for i = k+1:n
    factor = Aug(i,k)/Aug(k,k);
    Aug(i,k:nb) = Aug(i,k:nb)-factor*Aug(k,k:nb);
  end
end
% back substitution
x = zeros(n,1);
x(n) = Aug(n,nb)/Aug(n,n);
for i = n-1:-1:1
  x(i) = (Aug(i,nb)-Aug(i,i+1:n)*x(i+1:n))/Aug(i,i);
end
```

FIGURE 9.4
An M-file to implement naive Gauss elimination.

9.2.1 MATLAB M-file: GaussNaive

An M-file that implements naive Gauss elimination is listed in Fig. 9.4. Notice that the coefficient matrix A and the right-hand-side vector b are combined in the augmented matrix Aug. Thus, the operations are performed on Aug rather than separately on A and b.

Two nested loops provide a concise representation of the forward elimination step. An outer loop moves down the matrix from one pivot row to the next. The inner loop moves below the pivot row to each of the subsequent rows where elimination is to take place. Finally, the actual elimination is represented by a single line that takes advantage of MATLAB's ability to perform matrix operations.

The back-substitution step follows directly from Eqs. (9.12) and (9.13). Again, MATLAB's ability to perform matrix operations allows Eq. (9.13) to be programmed as a single line.

9.2.2 Operation Counting

The execution time of Gauss elimination depends on the amount of *floating-point operations* (or *flops*) involved in the algorithm. On modern computers using math coprocessors, the time consumed to perform addition/subtraction and multiplication/division is about the same.

Therefore, totaling up these operations provides insight into which parts of the algorithm are most time consuming and how computation time increases as the system gets larger.

Before analyzing naive Gauss elimination, we will first define some quantities that facilitate operation counting:

$$\sum_{i=1}^{m} cf(i) = c\sum_{i=1}^{m} f(i) \qquad \sum_{i=1}^{m} f(i) + g(i) = \sum_{i=1}^{m} f(i) + \sum_{i=1}^{m} g(i) \qquad (9.14a,b)$$

$$\sum_{i=1}^{m} 1 = 1 + 1 + 1 + \cdots + 1 = m \qquad \sum_{i=k}^{m} 1 = m - k + 1 \qquad (9.14c,d)$$

$$\sum_{i=1}^{m} i = 1 + 2 + 3 + \cdots + m = \frac{m(m+1)}{2} = \frac{m^2}{2} + O(m) \qquad (9.14e)$$

$$\sum_{i=1}^{m} i^2 = 1^2 + 2^2 + 3^2 + \cdots + m^2 = \frac{m(m+1)(2m+1)}{6} = \frac{m^3}{3} + O(m^2) \qquad (9.14f)$$

where $O(m^n)$ means "terms of order m^n and lower."

Now let us examine the naive Gauss elimination algorithm (Fig. 9.4) in detail. We will first count the flops in the elimination stage. On the first pass through the outer loop, $k = 1$. Therefore, the limits on the inner loop are from $i = 2$ to n. According to Eq. (9.14d), this means that the number of iterations of the inner loop will be

$$\sum_{i=2}^{n} 1 = n - 2 + 1 = n - 1 \qquad (9.15)$$

For every one of these iterations, there is one division to calculate the factor. The next line then performs a multiplication and a subtraction for each column element from 2 to nb. Because $nb = n + 1$, going from 2 to nb results in n multiplications and n subtractions. Together with the single division, this amounts to $n + 1$ multiplications/divisions and n addition/subtractions for every iteration of the inner loop. The total for the first pass through the outer loop is therefore $(n - 1)(n + 1)$ multiplication/divisions and $(n - 1)(n)$ addition/subtractions.

Similar reasoning can be used to estimate the flops for the subsequent iterations of the outer loop. These can be summarized as

Outer Loop k	Inner Loop i	Addition/Subtraction Flops	Multiplication/Division Flops
1	2, n	$(n-1)(n)$	$(n-1)(n+1)$
2	3, n	$(n-2)(n-1)$	$(n-2)(n)$
⋮	⋮		
k	$k+1, n$	$(n-k)(n+1-k)$	$(n-k)(n+2-k)$
⋮			
$n-1$	n, n	$(1)(2)$	$(1)(3)$

Therefore, the total addition/subtraction flops for elimination can be computed as

$$\sum_{k=1}^{n-1} (n-k)(n+1-k) = \sum_{k=1}^{n-1} [n(n+1) - k(2n+1) + k^2] \qquad (9.16)$$

or

$$n(n+1) \sum_{k=1}^{n-1} 1 - (2n+1) \sum_{k=1}^{n-1} k + \sum_{k=1}^{n-1} k^2 \qquad (9.17)$$

Applying some of the relationships from Eq. (9.14) yields

$$[n^3 + O(n)] - [n^3 + O(n^2)] + \left[\frac{1}{3}n^3 + O(n^2)\right] = \frac{n^3}{3} + O(n) \qquad (9.18)$$

A similar analysis for the multiplication/division flops yields

$$[n^3 + O(n^2)] - [n^3 + O(n)] + \left[\frac{1}{3}n^3 + O(n^2)\right] = \frac{n^3}{3} + O(n^2) \qquad (9.19)$$

Summing these results gives

$$\frac{2n^3}{3} + O(n^2) \qquad (9.20)$$

Thus, the total number of flops is equal to $2n^3/3$ plus an additional component proportional to terms of order n^2 and lower. The result is written in this way because as n gets large, the $O(n^2)$ and lower terms become negligible. We are therefore justified in concluding that for large n, the effort involved in forward elimination converges on $2n^3/3$.

Because only a single loop is used, back substitution is much simpler to evaluate. The number of addition/subtraction flops is equal to $n(n-1)/2$. Because of the extra division prior to the loop, the number of multiplication/division flops is $n(n+1)/2$. These can be added to arrive at a total of

$$n^2 + O(n) \qquad (9.21)$$

Thus, the total effort in naive Gauss elimination can be represented as

$$\underbrace{\frac{2n^3}{3} + O(n^2)}_{\substack{\text{Forward} \\ \text{elimination}}} + \underbrace{n^2 + O(n)}_{\substack{\text{Back} \\ \text{substitution}}} \xrightarrow{\text{as } n \text{ increases}} \frac{2n^3}{3} + O(n^2) \qquad (9.22)$$

Two useful general conclusions can be drawn from this analysis:

1. As the system gets larger, the computation time increases greatly. As in Table 9.1, the amount of flops increases nearly three orders of magnitude for every order of magnitude increase in the number of equations.

TABLE 9.1 Number of flops for naive Gauss elimination.

n	Elimination	Back Substitution	Total Flops	$2n^3/3$	Percent Due to Elimination
10	705	100	805	667	87.58
100	671550	10000	681550	666667	98.53
1000	6.67×10^8	1×10^6	6.68×10^8	6.67×10^8	99.85

2. Most of the effort is incurred in the elimination step. Thus, efforts to make the method more efficient should probably focus on this step.

9.3 PIVOTING

The primary reason that the foregoing technique is called "naive" is that during both the elimination and the back-substitution phases, it is possible that a division by zero can occur. For example, if we use naive Gauss elimination to solve

$$2x_2 + 3x_3 = 8$$
$$4x_1 + 6x_2 + 7x_3 = -3$$
$$2x_1 - 3x_2 + 6x_3 = 5$$

the normalization of the first row would involve division by $a_{11} = 0$. Problems may also arise when the pivot element is close, rather than exactly equal, to zero because if the magnitude of the pivot element is small compared to the other elements, then round off errors can be introduced.

Therefore, before each row is normalized, it is advantageous to determine the coefficient with the largest absolute value in the column below the pivot element. The rows can then be switched so that the largest element is the pivot element. This is called *partial pivoting*.

If columns as well as rows are searched for the largest element and then switched, the procedure is called *complete pivoting*. Complete pivoting is rarely used because most of the improvement comes from partial pivoting. In addition, switching columns changes the order of the x's and, consequently, adds significant and usually unjustified complexity to the computer program.

The following example illustrates the advantages of partial pivoting. Aside from avoiding division by zero, pivoting also minimizes round off error. As such, it also serves as a partial remedy for ill-conditioning.

EXAMPLE 9.4 Partial Pivoting

Problem Statement. Use Gauss elimination to solve

$$0.0003x_1 + 3.0000x_2 = 2.0001$$
$$1.0000x_1 + 1.0000x_2 = 1.0000$$

Note that in this form the first pivot element, $a_{11} = 0.0003$, is very close to zero. Then repeat the computation, but partial pivot by reversing the order of the equations. The exact solution is $x_1 = 1/3$ and $x_2 = 2/3$.

Solution. Multiplying the first equation by $1/(0.0003)$ yields

$$x_1 + 10,000x_2 = 6667$$

which can be used to eliminate x_1 from the second equation:

$$-9999x_2 = -6666$$

which can be solved for $x_2 = 2/3$. This result can be substituted back into the first equation to evaluate x_1:

$$x_1 = \frac{2.0001 - 3(2/3)}{0.0003} \tag{E9.4.1}$$

Due to subtractive cancellation, the result is very sensitive to the number of significant figures carried in the computation:

Significant Figures	x_2	x_1	Absolute Value of Percent Relative Error for x_1
3	0.667	−3.33	1099
4	0.6667	0.0000	100
5	0.66667	0.30000	10
6	0.666667	0.330000	1
7	0.6666667	0.3330000	0.1

Note how the solution for x_1 is highly dependent on the number of significant figures. This is because in Eq. (E9.4.1), we are subtracting two almost-equal numbers.

On the other hand, if the equations are solved in reverse order, the row with the larger pivot element is normalized. The equations are

$$1.0000x_1 + 1.0000x_2 = 1.0000$$
$$0.0003x_1 + 3.0000x_2 = 2.0001$$

Elimination and substitution again yields $x_2 = 2/3$. For different numbers of significant figures, x_1 can be computed from the first equation, as in

$$x_1 = \frac{1 - (2/3)}{1}$$

This case is much less sensitive to the number of significant figures in the computation:

Significant Figures	x_2	x_1	Absolute Value of Percent Relative Error for x_1
3	0.667	0.333	0.1
4	0.6667	0.3333	0.01
5	0.66667	0.33333	0.001
6	0.666667	0.333333	0.0001
7	0.6666667	0.3333333	0.0000

Thus, a pivot strategy is much more satisfactory.

9.3.1 MATLAB M-file: GaussPivot

An M-file that implements Gauss elimination with partial pivoting is listed in Fig. 9.5. It is identical to the M-file for naive Gauss elimination presented previously in Sec. 9.2.1 with the exception of the bold portion that implements partial pivoting.

Notice how the built-in MATLAB function max is used to determine the largest available coefficient in the column below the pivot element. The max function has the syntax

    [y,i] = max(x)

where y is the largest element in the vector x, and i is the index corresponding to that element.

```
function x = GaussPivot(A,b)
% GaussPivot: Gauss elimination pivoting
%   x = GaussPivot(A,b): Gauss elimination with pivoting.
% input:
%   A = coefficient matrix
%   b = right hand side vector
% output:
%   x = solution vector
[m,n]=size(A);
if m~=n, error('Matrix A must be square'); end
nb=n+1;
Aug=[A b];
% forward elimination
for k = 1:n-1
  % partial pivoting
  [big,i] = max(abs(Aug(k:n,k)));
  ipr = i+k-1;
  if ipr~ = k
    Aug([k,ipr],:) = Aug([ipr,k],:);
  end
    for i = k+1:n
      factor = Aug(i,k)/Aug(k,k);
      Aug(i,k:nb) = Aug(i,k:nb)-factor*Aug(k,k:nb);
    end
end
% back substitution
x=zeros(n,1);
x(n)=Aug(n,nb)/Aug(n,n);
for i = n-1:-1:1
  x(i)=(Aug(i,nb)-Aug(i,i+1:n)*x(i+1:n))/Aug(i,i);
end
```

FIGURE 9.5
An M-file to implement Gauss elimination with partial pivoting.

9.3.2 Determinant Evaluation with Gauss Elimination

At the end of Sec. 9.1.2, we suggested that determinant evaluation by expansion of minors was impractical for large sets of equations. However, because the determinant has value in assessing system condition, it would be useful to have a practical method for computing this quantity.

Fortunately, Gauss elimination provides a simple way to do this. The method is based on the fact that the determinant of a triangular matrix can be simply computed as the product of its diagonal elements:

$$D = a_{11}a_{22}a_{33} \cdots a_{nn}$$

The validity of this formulation can be illustrated for a 3×3 system:

$$D = \begin{vmatrix} a_{11} & a_{12} & a_{13} \\ 0 & a_{22} & a_{23} \\ 0 & 0 & a_{33} \end{vmatrix}$$

where the determinant can be evaluated as [recall Eq. (9.1)]:

$$D = a_{11} \begin{vmatrix} a_{22} & a_{23} \\ 0 & a_{33} \end{vmatrix} - a_{12} \begin{vmatrix} 0 & a_{23} \\ 0 & a_{33} \end{vmatrix} + a_{13} \begin{vmatrix} 0 & a_{22} \\ 0 & 0 \end{vmatrix}$$

or, by evaluating the minors:

$$D = a_{11}a_{22}a_{33} - a_{12}(0) + a_{13}(0) = a_{11}a_{22}a_{33}$$

Recall that the forward-elimination step of Gauss elimination results in an upper triangular system. Because the value of the determinant is not changed by the forward-elimination process, the determinant can be simply evaluated at the end of this step via

$$D = a_{11}a'_{22}a''_{33} \cdots a_{nn}^{(n-1)}$$

where the superscripts signify the number of times that the elements have been modified by the elimination process. Thus, we can capitalize on the effort that has already been expended in reducing the system to triangular form and, in the bargain, come up with a simple estimate of the determinant.

There is a slight modification to the above approach when the program employs partial pivoting. For such cases, the determinant changes sign every time a row is switched. One way to represent this is by modifying the determinant calculation as in

$$D = a_{11}a'_{22}a''_{33} \cdots a_{nn}^{(n-1)} (-1)^p$$

where p represents the number of times that rows are pivoted. This modification can be incorporated simply into a program by merely keeping track of the number of pivots that take place during the course of the computation.

9.4 TRIDIAGONAL SYSTEMS

Certain matrices have a particular structure that can be exploited to develop efficient solution schemes. For example, a banded matrix is a square matrix that has all elements equal to zero, with the exception of a band centered on the main diagonal.

A *tridiagonal* system has a bandwidth of 3 and can be expressed generally as

$$\begin{bmatrix} f_1 & g_1 & & & & & \\ e_2 & f_2 & g_2 & & & & \\ & e_3 & f_3 & g_3 & & & \\ & & \cdot & \cdot & \cdot & & \\ & & & \cdot & \cdot & \cdot & \\ & & & & \cdot & \cdot & \cdot \\ & & & & e_{n-1} & f_{n-1} & g_{n-1} \\ & & & & & e_n & f_n \end{bmatrix} \begin{Bmatrix} x_1 \\ x_2 \\ x_3 \\ \cdot \\ \cdot \\ \cdot \\ x_{n-1} \\ x_n \end{Bmatrix} = \begin{Bmatrix} r_1 \\ r_2 \\ r_3 \\ \cdot \\ \cdot \\ \cdot \\ r_{n-1} \\ r_n \end{Bmatrix} \qquad (9.23)$$

Notice that we have changed our notation for the coefficients from a's and b's to e's, f's, g's, and r's. This was done to avoid storing large numbers of useless zeros in the square matrix of a's. This space-saving modification is advantageous because the resulting algorithm requires less computer memory.

An algorithm to solve such systems can be directly patterned after Gauss elimination—that is, using forward elimination and back substitution. However, because most of the matrix elements are already zero, much less effort is expended than for a full matrix. This efficiency is illustrated in the following example.

EXAMPLE 9.5   Solution of a Tridiagonal System

Problem Statement. Solve the following tridiagonal system:

$$
\begin{bmatrix}
2.04 & -1 & & \\
-1 & 2.04 & -1 & \\
& -1 & 2.04 & -1 \\
& & -1 & 2.04
\end{bmatrix}
\begin{Bmatrix}
x_1 \\ x_2 \\ x_3 \\ x_4
\end{Bmatrix}
=
\begin{Bmatrix}
40.8 \\ 0.8 \\ 0.8 \\ 200.8
\end{Bmatrix}
$$

Solution. As with Gauss elimination, the first step involves transforming the matrix to upper triangular form. This is done by multiplying the first equation by the factor e_2/f_1 and subtracting the result from the second equation. This creates a zero in place of e_2 and transforms the other coefficients to new values,

$$
f_2 = f_2 - \frac{e_2}{f_1} g_1 = 2.04 - \frac{-1}{2.04}(-1) = 1.550
$$

$$
r_2 = r_2 - \frac{e_2}{f_1} r_1 = 0.8 - \frac{-1}{2.04}(40.8) = 20.8
$$

Notice that g_2 is unmodified because the element above it in the first row is zero.

After performing a similar calculation for the third and fourth rows, the system is transformed to the upper triangular form

$$
\begin{bmatrix}
2.04 & -1 & & \\
& 1.550 & -1 & \\
& & 1.395 & -1 \\
& & & 1.323
\end{bmatrix}
\begin{Bmatrix}
x_1 \\ x_2 \\ x_3 \\ x_4
\end{Bmatrix}
=
\begin{Bmatrix}
40.8 \\ 20.8 \\ 14.221 \\ 210.996
\end{Bmatrix}
$$

Now back substitution can be applied to generate the final solution:

$$
x_4 = \frac{r_4}{f_4} = \frac{210.996}{1.323} = 159.480
$$

$$
x_3 = \frac{r_3 - g_3 x_4}{f_3} = \frac{14.221 - (-1)159.480}{1.395} = 124.538
$$

$$
x_2 = \frac{r_2 - g_2 x_3}{f_2} = \frac{20.800 - (-1)124.538}{1.550} = 93.778
$$

$$
x_1 = \frac{r_1 - g_1 x_2}{f_1} = \frac{40.800 - (-1)93.778}{2.040} = 65.970
$$

```
function x = Tridiag(e,f,g,r)
% Tridiag: Tridiagonal equation solver banded system
%   x = Tridiag(e,f,g,r): Tridiagonal system solver.
% input:
%   e = subdiagonal vector
%   f = diagonal vector
%   g = superdiagonal vector
%   r = right hand side vector
% output:
%   x = solution vector
n=length(f);
% forward elimination
for k = 2:n
  factor = e(k)/f(k-1);
  f(k) = f(k) - factor*g(k-1);
  r(k) = r(k) - factor*r(k-1);
end
% back substitution
x(n) = r(n)/f(n);
for k = n-1:-1:1
  x(k) = (r(k)-g(k)*x(k+1))/f(k);
end
```

FIGURE 9.6
An M-file to solve a tridiagonal system.

9.4.1 MATLAB M-file: Tridiag

An M-file that solves a tridiagonal system of equations is listed in Fig. 9.6. Note that the algorithm does not include partial pivoting. Although pivoting is sometimes required, most tridiagonal systems routinely solved in engineering and science do not require pivoting.

Recall that the computational effort for Gauss elimination was proportional to n^3. Because of its sparseness, the effort involved in solving tridiagonal systems is proportional to n. Consequently, the algorithm in Fig. 9.6 executes much, much faster than Gauss elimination, particularly for large systems.

9.5 CASE STUDY   MODEL OF A HEATED ROD

Background. Linear algebraic equations can arise when modeling distributed systems. For example, Fig. 9.7 shows a long, thin rod positioned between two walls that are held at constant temperatures. Heat flows through the rod as well as between the rod and the surrounding air. For the steady-state case, a differential equation based on heat conservation can be written for such a system as

$$\frac{d^2T}{dx^2} + h'(T_a - T) = 0 \tag{9.24}$$

9.5 CASE STUDY   continued

FIGURE 9.7
A noninsulated uniform rod positioned between two walls of constant but different temperature. The finite-difference representation employs four interior nodes.

where T = temperature (°C), x = distance along the rod (m), h' = a heat transfer coefficient between the rod and the surrounding air (m^{-2}), and T_a = the air temperature (°C).

Given values for the parameters, forcing functions, and boundary conditions, calculus can be used to develop an analytical solution. For example, if $h' = 0.01$, $T_a = 20$, $T(0) = 40$, and $T(10) = 200$, the solution is

$$T = 73.4523e^{0.1x} - 53.4523e^{-0.1x} + 20 \tag{9.25}$$

Although it provided a solution here, calculus does not work for all such problems. In such instances, numerical methods provide a valuable alternative. In this case study, we will use finite differences to transform this differential equation into a tridiagonal system of linear algebraic equations which can be readily solved using the numerical methods described in this chapter.

Solution. Equation (9.24) can be transformed into a set of linear algebraic equations by conceptualizing the rod as consisting of a series of nodes. For example, the rod in Fig. 9.7 is divided into six equispaced nodes. Since the rod has a length of 10, the spacing between nodes is $\Delta x = 2$.

Calculus was necessary to solve Eq. (9.24) because it includes a second derivative. As we learned in Sec. 4.3.4, finite-difference approximations provide a means to transform derivatives into algebraic form. For example, the second derivative at each node can be approximated as

$$\frac{d^2T}{dx^2} = \frac{T_{i+1} - 2T_i + T_{i-1}}{\Delta x^2}$$

where T_i designates the temperature at node i. This approximation can be substituted into Eq. (9.24) to give

$$\frac{T_{i+1} - 2T_i + T_{i-1}}{\Delta x^2} + h'(T_a - T_i) = 0$$

continued

Collecting terms and substituting the parameters gives

$$-T_{i-1} + 2.04T_i - T_{i+1} = 0.8 \tag{9.26}$$

Thus, Eq. (9.24) has been transformed from a differential equation into an algebraic equation. Equation (9.26) can now be applied to each of the interior nodes:

$$
\begin{aligned}
-T_0 + 2.04T_1 - T_2 &= 0.8 \\
-T_1 + 2.04T_2 - T_3 &= 0.8 \\
-T_2 + 2.04T_3 - T_4 &= 0.8 \\
-T_3 + 2.04T_4 - T_5 &= 0.8
\end{aligned}
\tag{9.27}
$$

The values of the fixed end temperatures, $T_0 = 40$ and $T_5 = 200$, can be substituted and moved to the right-hand side. The results are four equations with four unknowns expressed in matrix form as

$$
\begin{bmatrix}
2.04 & -1 & 0 & 0 \\
-1 & 2.04 & -1 & 0 \\
0 & -1 & 2.04 & -1 \\
0 & 0 & -1 & 2.04
\end{bmatrix}
\begin{Bmatrix}
T_1 \\ T_2 \\ T_3 \\ T_4
\end{Bmatrix}
=
\begin{Bmatrix}
40.8 \\ 0.8 \\ 0.8 \\ 200.8
\end{Bmatrix}
\tag{9.28}
$$

So our original differential equation has been converted into an equivalent system of linear algebraic equations. Consequently, we can use the techniques described in this chapter to solve for the temperatures. For example, using MATLAB

```
>> A=[2.04 -1 0 0
-1 2.04 -1 0
0 -1 2.04 -1
0 0 -1 2.04];
>> b=[40.8 0.8 0.8 200.8]';
>> T=(A\b)'

T =
    65.9698   93.7785   124.5382   159.4795
```

A plot can also be developed comparing these results with the analytical solution obtained with Eq. (9.25),

```
>> T=[40 T 200];
>> x=[0:2:10];
>> xanal=[0:10];
>> TT=@(x) 73.4523*exp(0.1*x)-53.4523* ...
       exp(-0.1*x)+20;
>> Tanal=TT(xanal);
>> plot(x,T,'o',xanal,Tanal)
```

As in Fig. 9.8, the numerical results are quite close to those obtained with calculus.

9.5 CASE STUDY   continued

FIGURE 9.8
A plot of temperature versus distance along a heated rod. Both analytical (line) and numerical (points) solutions are displayed.

In addition to being a linear system, notice that Eq. (9.28) is also tridiagonal. We can use an efficient solution scheme like the M-file in Fig. 9.6 to obtain the solution:

```
>> e=[0 -1 -1 -1];
>> f=[2.04 2.04 2.04 2.04];
>> g=[-1 -1 -1 0];
>> r=[40.8 0.8 0.8 200.8];
>> Tridiag(e,f,g,r)

ans =
    65.9698    93.7785   124.5382   159.4795
```

The system is tridiagonal because each node depends only on its adjacent nodes. Because we numbered the nodes sequentially, the resulting equations are tridiagonal. Such cases often occur when solving differential equations based on conservation laws.

PROBLEMS

9.1 Determine the number of total flops as a function of the number of equations n for the tridiagonal algorithm (Fig. 9.6).

9.2 Use the graphical method to solve

$$4x_1 - 8x_2 = -24$$
$$x_1 + 6x_2 = 34$$

Check your results by substituting them back into the equations.

9.3 Given the system of equations

$$-1.1x_1 + 10x_2 = 120$$
$$-2x_1 + 17.4x_2 = 174$$

(a) Solve graphically and check your results by substituting them back into the equations.
(b) On the basis of the graphical solution, what do you expect regarding the condition of the system?
(c) Compute the determinant.

9.4 Given the system of equations

$$-3x_2 + 7x_3 = 4$$
$$x_1 + 2x_2 - x_3 = 0$$
$$5x_1 - 2x_2 = 3$$

(a) Compute the determinant.
(b) Use Cramer's rule to solve for the x's.
(c) Use Gauss elimination with partial pivoting to solve for the x's. As part of the computation, calculate the determinant in order to verify the value computed in (a).
(d) Substitute your results back into the original equations to check your solution.

9.5 Given the equations

$$0.5x_1 - x_2 = -9.5$$
$$1.02x_1 - 2x_2 = -18.8$$

(a) Solve graphically.
(b) Compute the determinant.
(c) On the basis of (a) and (b), what would you expect regarding the system's condition?
(d) Solve by the elimination of unknowns.
(e) Solve again, but with a_{11} modified slightly to 0.52. Interpret your results.

9.6 Given the equations

$$10x_1 + 2x_2 - x_3 = 27$$
$$-3x_1 - 5x_2 + 2x_3 = -61.5$$
$$x_1 + x_2 + 6x_3 = -21.5$$

(a) Solve by naive Gauss elimination. Show all steps of the computation.
(b) Substitute your results into the original equations to check your answers.

9.7 Given the equations

$$2x_1 - 6x_2 - x_3 = -38$$
$$-3x_1 - x_2 + 7x_3 = -34$$
$$-8x_1 + x_2 - 2x_3 = -20$$

(a) Solve by Gauss elimination with partial pivoting. As part of the computation, use the diagonal elements to calculate the determinant. Show all steps of the computation.
(b) Substitute your results into the original equations to check your answers.

9.8 Perform the same calculations as in Example 9.5, but for the tridiagonal system:

$$\begin{bmatrix} 0.8 & -0.4 & \\ -0.4 & 0.8 & -0.4 \\ & -0.4 & 0.8 \end{bmatrix} \begin{Bmatrix} x_1 \\ x_2 \\ x_3 \end{Bmatrix} = \begin{Bmatrix} 41 \\ 25 \\ 105 \end{Bmatrix}$$

9.9 Figure P9.9 shows three reactors linked by pipes. As indicated, the rate of transfer of chemicals through each pipe is equal to a flow rate (Q, with units of cubic meters per second) multiplied by the concentration of the reactor from which the flow originates (c, with units of milligrams per cubic meter). If the system is at a steady state, the transfer into each reactor will balance the transfer out. Develop mass-balance equations for the reactors and solve the three simultaneous linear algebraic equations for their concentrations.

9.10 A civil engineer involved in construction requires 4800, 5800, and 5700 m^3 of sand, fine gravel, and coarse gravel, respectively, for a building project. There are three

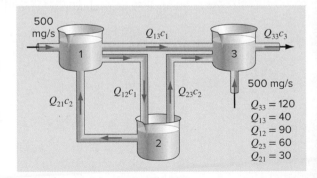

FIGURE P9.9
Three reactors linked by pipes. The rate of mass transfer through each pipe is equal to the product of flow Q and concentration c of the reactor from which the flow originates.

pits from which these materials can be obtained. The composition of these pits is

	Sand %	Fine Gravel %	Coarse Gravel %
Pit1	55	30	15
Pit2	25	45	30
Pit3	25	20	55

How many cubic meters must be hauled from each pit in order to meet the engineer's needs?

9.11 An electrical engineer supervises the production of three types of electrical components. Three kinds of material—metal, plastic, and rubber—are required for production. The amounts needed to produce each component are

Component	Metal (g/ component)	Plastic (g/ component)	Rubber (g/ component)
1	15	0.30	1.0
2	17	0.40	1.2
3	19	0.55	1.5

If totals of 3.89, 0.095, and 0.282 kg of metal, plastic, and rubber, respectively, are available each day, how many components can be produced per day?

9.12 As described in Sec. 9.4, linear algebraic equations can arise in the solution of differential equations. For example, the following differential equation results from a steady-state mass balance for a chemical in a one-dimensional canal:

$$0 = D\frac{d^2c}{dx^2} - U\frac{dc}{dx} - kc$$

where c = concentration, t = time, x = distance, D = diffusion coefficient, U = fluid velocity, and k = a first-order decay rate. Convert this differential equation to an equivalent system of simultaneous algebraic equations. Given $D = 2$, $U = 1$, $k = 0.2$, $c(0) = 80$ and $c(10) = 20$, solve these equations from $x = 0$ to 10 and develop a plot of concentration versus distance.

9.13 A stage extraction process is depicted in Fig. P9.13. In such systems, a stream containing a weight fraction y_{in} of a chemical enters from the left at a mass flow rate of F_1. Simultaneously, a solvent carrying a weight fraction x_{in} of the same chemical enters from the right at a flow rate of F_2. Thus, for stage i, a mass balance can be represented as

$$F_1 y_{i-1} + F_2 x_{i+1} = F_1 y_i + F_2 x_i \qquad (P9.13a)$$

At each stage, an equilibrium is assumed to be established between y_i and x_i as in

$$K = \frac{x_i}{y_i} \qquad (P9.13b)$$

where K is called a distribution coefficient. Equation (P9.13b) can be solved for x_i and substituted into Eq. (P9.13a) to yield

$$y_{i-1} - \left(1 + \frac{F_2}{F_1}K\right)y_i + \left(\frac{F_2}{F_1}K\right)y_{i+1} = 0 \qquad (P9.13c)$$

If $F_1 = 500$ kg/h, $y_{in} = 0.1$, $F_2 = 1000$ kg/h, $x_{in} = 0$, and $K = 4$, determine the values of y_{out} and x_{out} if a five-stage reactor is used. Note that Eq. (P9.13c) must be modified to account for the inflow weight fractions when applied to the first and last stages.

9.14 A peristaltic pump delivers a unit flow (Q_1) of a highly viscous fluid. The network is depicted in Fig. P9.14. Every pipe section has the same length and diameter. The mass and mechanical energy balance can be simplified to obtain the

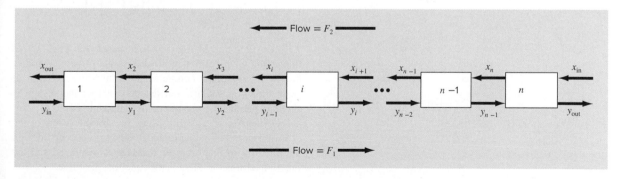

FIGURE P9.13
A stage extraction process.

FIGURE P9.14

flows in every pipe. Solve the following system of equations to obtain the flow in every stream.

$Q_3 + 2Q_4 - 2Q_2 = 0$	$Q_1 = Q_2 + Q_3$
$Q_5 + 2Q_6 - 2Q_4 = 0$	$Q_3 = Q_4 + Q_5$
$3Q_7 - 2Q_6 = 0$	$Q_5 = Q_6 + Q_7$

9.15 A truss is loaded as shown in Fig. P9.15. Using the following set of equations, solve for the 10 unknowns, AB, BC, AD, BD, CD, DE, CE, A_x, A_y, and E_y.

$A_x + AD = 0$	$-24 - CD - (4/5)CE = 0$
$A_y + AB = 0$	$-AD + DE - (3/5)BD = 0$
$74 + BC + (3/5)BD = 0$	$CD + (4/5)BD = 0$
$-AB - (4/5)BD = 0$	$-DE - (3/5)CE = 0$
$-BC + (3/5)CE = 0$	$E_y + (4/5)CE = 0$

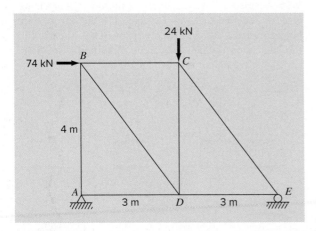

FIGURE P9.15

9.16 A *pentadiagonal* system with a bandwidth of five can be expressed generally as

$$\begin{bmatrix} f_1 & g_1 & h_1 & & & & \\ e_2 & f_2 & g_2 & h_2 & & & \\ d_3 & e_3 & f_3 & g_3 & h_3 & & \\ & & \cdot & \cdot & \cdot & & \\ & & & \cdot & \cdot & \cdot & \\ & & & & \cdot & \cdot & \cdot \\ & & & d_{n-1} & e_{n-1} & f_{n-1} & g_{n-1} \\ & & & & d_n & e_n & f_n \end{bmatrix}$$

$$\times \begin{Bmatrix} x_1 \\ x_2 \\ x_3 \\ \cdot \\ \cdot \\ \cdot \\ x_{n-1} \\ x_n \end{Bmatrix} = \begin{Bmatrix} r_1 \\ r_2 \\ r_3 \\ \cdot \\ \cdot \\ \cdot \\ r_{n-1} \\ r_n \end{Bmatrix}$$

Develop an M-file to efficiently solve such systems without pivoting in a similar fashion to the algorithm used for tridiagonal matrices in Sec. 9.4.1. Test it for the following case:

$$\begin{bmatrix} 8 & -2 & -1 & 0 & 0 \\ -2 & 9 & -4 & -1 & 0 \\ -1 & -3 & 7 & -1 & -2 \\ 0 & -4 & -2 & 12 & -5 \\ 0 & 0 & -7 & -3 & -15 \end{bmatrix} \begin{Bmatrix} x_1 \\ x_2 \\ x_3 \\ x_4 \\ x_5 \end{Bmatrix} = \begin{Bmatrix} 5 \\ 2 \\ 1 \\ 1 \\ 5 \end{Bmatrix}$$

9.17 Develop an M-file function based on Fig. 9.5 to implement Gauss elimination with partial pivoting. Modify the function so that it computes and returns the determinant (with the correct sign), and detects whether the system is singular based on a near-zero determinant. For the latter, define "near-zero" as being when the absolute value of the determinant is below a tolerance. When this occurs, design the function so that an error message is displayed and the function terminates. Here is the functions first line:

```
function [x, D] = GaussPivotNew(A, b, tol)
```

where D = the determinant and tol = the tolerance. Test your program for Prob. 9.5 with tol = 1×10^{-5}.

9.18 As described in Sec. 9.5, linear algebraic equations can arise in the solution of differential equations. The following differential equation results from a steady-state mass balance for a chemical in a one-dimensional canal,

$$0 = D \frac{d^2c}{dx^2} - U \frac{dc}{dx} - kc$$

where x = distance along the canal (m), c = concentration, t = time, x = distance, D = diffusion coefficient, U = fluid velocity, and k = a first-order decay rate.

(a) Convert this differential equation to an equivalent system of simultaneous algebraic equations using centered difference approximations for the derivatives.

(b) Develop a function to solve these equations from $x = 0$ to L and return the resulting distances and concentrations. The first line of your function should be

```
function [x,c]=YourLastName_reactor(D, U, k,
c0, cL, L, dx)
```

(c) Develop a script that invokes this function and then plots the results.

(d) Test your script for the following parameters: $L = 10$ m, $\Delta x = 0.5$ m, $D = 2$ m²/d, $U = 1$ m/d, $k = 0.2$/d, $c(0) = 80$ mg/L, and $c(10) = 20$ mg/L.

9.19 The following differential equation results from a force balance for a beam with a uniform loading,

$$0 = EI \frac{d^2y}{dx^2} - \frac{wLx}{2} + \frac{wx^2}{2}$$

where x = distance along the beam (m), y = deflection (m), L = length (m), E = modulus of elasticity (N/m²), I = moment of inertia (m⁴), and w = uniform load (N/m).

(a) Convert this differential equation to an equivalent system of simultaneous algebraic equations using a centered-difference approximation for the second derivative.

(b) Develop a function to solve these equations from $x = 0$ to L and return the resulting distances and deflections. The first line of your function should be

```
function [x, y]=YourLastName_beam(E, I, w, y0,
yL, L, dx)
```

(c) Develop a script that invokes this function and then plots the results.

(d) Test your script for the following parameters: $L = 3$ m, $\Delta x = 0.2$ m, $E = 250 \times 10^9$ N/m², $I = 3 \times 10^{-4}$ m⁴, $w = 22{,}500$ N/m, $y(0) = 0$, and $y(3) = 0$.

9.20 Heat is conducted along a metal rod positioned between two fixed temperature walls. Aside from conduction, heat is transferred between the rod and the surrounding air by convection. Based on a heat balance, the distribution of temperature along the rod is described by the following second-order differential equation

$$0 = \frac{d^2T}{dx^2} + h'(T_\infty - T)$$

where T = temperature (K), h' = a bulk heat transfer coefficient reflecting the relative importance of convection to conduction (m⁻²), x = distance along the rod (m), and T_∞ = temperature of the surrounding fluid (K).

(a) Convert this differential equation to an equivalent system of simultaneous algebraic equations using a centered difference approximation for the second derivative.

(b) Develop a function to solve these equations from $x = 0$ to L and return the resulting distances and temperatures. The first line of your function should be

```
function [x, y]=YourLastName_rod(hp, Tinf, T0,
TL, L, dx)
```

(c) Develop a script that invokes this function and then plots the results.

(d) Test your script for the following parameters: $h' = 0.0425$ m⁻², $L = 12$ m, $T_\infty = 220$ K, $T(0) = 320$ K, $T(L) = 450$ K, and $\Delta x = 0.5$ m.

10

LU Factorization

CHAPTER OBJECTIVES

The primary objective of this chapter is to acquaint you with *LU* factorization.[1]
Specific objectives and topics covered are

- Understanding that *LU* factorization involves decomposing the coefficient matrix into two triangular matrices that can then be used to efficiently evaluate different right-hand-side vectors.
- Knowing how to express Gauss elimination as an *LU* factorization.
- Given an *LU* factorization, knowing how to evaluate multiple right-hand-side vectors.
- Recognizing that Cholesky's method provides an efficient way to decompose a symmetric matrix and that the resulting triangular matrix and its transpose can be used to evaluate right-hand-side vectors efficiently.
- Understanding in general terms what happens when MATLAB's backslash operator is used to solve linear systems.

As described in Chap. 9, Gauss elimination is designed to solve systems of linear algebraic equations:

$$[A]\{x\} = \{b\} \tag{10.1}$$

Although it certainly represents a sound way to solve such systems, it becomes inefficient when solving equations with the same coefficients [A], but with different right-hand-side constants {b}.

---

[1] In the parlance of numerical methods, the terms "factorization" and "decomposition" are synonymous. To be consistent with the MATLAB documentation, we have chosen to employ the terminology *LU factorization* for the subject of this chapter. Note that *LU decomposition* is very commonly used to describe the same approach.

Recall that Gauss elimination involves two steps: forward elimination and back substitution (Fig. 9.3). As we learned in Sec. 9.2.2, the forward-elimination step comprises the bulk of the computational effort. This is particularly true for large systems of equations.

LU factorization methods separate the time-consuming elimination of the matrix [*A*] from the manipulations of the right-hand side {*b*}. Thus, once [*A*] has been "factored" or "decomposed," multiple right-hand-side vectors can be evaluated in an efficient manner.

Interestingly, Gauss elimination itself can be expressed as an *LU* factorization. Before showing how this can be done, let us first provide a mathematical overview of the factorization strategy.

10.1 OVERVIEW OF *LU* FACTORIZATION

Just as was the case with Gauss elimination, *LU* factorization requires pivoting to avoid division by zero. However, to simplify the following description, we will omit pivoting. In addition, the following explanation is limited to a set of three simultaneous equations. The results can be directly extended to n-dimensional systems.

Equation (10.1) can be rearranged to give

$$[A]\{x\} - \{b\} = 0 \qquad (10.2)$$

Suppose that Eq. (10.2) could be expressed as an upper triangular system. For example, for a 3×3 system:

$$\begin{bmatrix} u_{11} & u_{12} & u_{13} \\ 0 & u_{22} & u_{23} \\ 0 & 0 & u_{33} \end{bmatrix} \begin{Bmatrix} x_1 \\ x_2 \\ x_3 \end{Bmatrix} = \begin{Bmatrix} d_1 \\ d_2 \\ d_3 \end{Bmatrix} \qquad (10.3)$$

Recognize that this is similar to the manipulation that occurs in the first step of Gauss elimination. That is, elimination is used to reduce the system to upper triangular form. Equation (10.3) can also be expressed in matrix notation and rearranged to give

$$[U]\{x\} - \{d\} = 0 \qquad (10.4)$$

Now assume that there is a lower diagonal matrix with 1's on the diagonal,

$$[L] = \begin{bmatrix} 1 & 0 & 0 \\ l_{21} & 1 & 0 \\ l_{31} & l_{32} & 1 \end{bmatrix} \qquad (10.5)$$

that has the property that when Eq. (10.4) is premultiplied by it, Eq. (10.2) is the result. That is,

$$[L]\{[U]\{x\} - \{d\}\} = [A]\{x\} - \{b\} \qquad (10.6)$$

If this equation holds, it follows from the rules for matrix multiplication that

$$[L][U] = [A] \qquad (10.7)$$

and

$$[L]\{d\} = \{b\} \qquad (10.8)$$

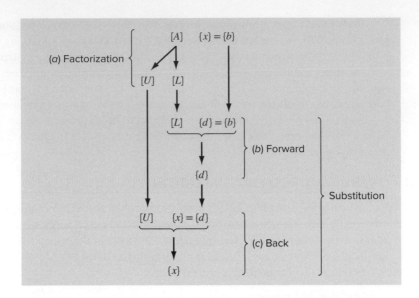

FIGURE 10.1
The steps in *LU* factorization.

A two-step strategy (see Fig. 10.1) for obtaining solutions can be based on Eqs. (10.3), (10.7), and (10.8):

1. *LU* factorization step. [A] is factored or "decomposed" into lower [L] and upper [U] triangular matrices.
2. Substitution step. [L] and [U] are used to determine a solution {x} for a right-hand side {b}. This step itself consists of two steps. First, Eq. (10.8) is used to generate an intermediate vector {d} by forward substitution. Then, the result is substituted into Eq. (10.3) which can be solved by back substitution for {x}.

Now let us show how Gauss elimination can be implemented in this way.

10.2   GAUSS ELIMINATION AS *LU* FACTORIZATION

Although it might appear at face value to be unrelated to *LU* factorization, Gauss elimination can be used to decompose [A] into [L] and [U]. This can be easily seen for [U], which is a direct product of the forward elimination. Recall that the forward-elimination step is intended to reduce the original coefficient matrix [A] to the form

$$[U] = \begin{bmatrix} a_{11} & a_{12} & a_{13} \\ 0 & a'_{22} & a'_{23} \\ 0 & 0 & a''_{33} \end{bmatrix} \tag{10.9}$$

which is in the desired upper triangular format.

Though it might not be as apparent, the matrix [L] is also produced during the step. This can be readily illustrated for a three-equation system,

$$\begin{bmatrix} a_{11} & a_{12} & a_{13} \\ a_{21} & a_{22} & a_{23} \\ a_{31} & a_{32} & a_{33} \end{bmatrix} \begin{Bmatrix} x_1 \\ x_2 \\ x_3 \end{Bmatrix} = \begin{Bmatrix} b_1 \\ b_2 \\ b_3 \end{Bmatrix}$$

The first step in Gauss elimination is to multiply row 1 by the factor [recall Eq. (9.9)]

$$f_{21} = \frac{a_{21}}{a_{11}}$$

and subtract the result from the second row to eliminate a_{21}. Similarly, row 1 is multiplied by

$$f_{31} = \frac{a_{31}}{a_{11}}$$

and the result subtracted from the third row to eliminate a_{31}. The final step is to multiply the modified second row by

$$f_{32} = \frac{a'_{32}}{a'_{22}}$$

and subtract the result from the third row to eliminate a'_{32}.

Now suppose that we merely perform all these manipulations on the matrix [A]. Clearly, if we do not want to change the equations, we also have to do the same to the right-hand side {b}. But there is absolutely no reason that we have to perform the manipulations simultaneously. Thus, we could save the f's and manipulate {b} later.

Where do we store the factors f_{21}, f_{31}, and f_{32}? Recall that the whole idea behind the elimination was to create zeros in a_{21}, a_{31}, and a_{32}. Thus, we can store f_{21} in a_{21}, f_{31} in a_{31}, and f_{32} in a_{32}. After elimination, the [A] matrix can therefore be written as

$$\begin{bmatrix} a_{11} & a_{12} & a_{13} \\ f_{21} & a'_{22} & a'_{23} \\ f_{31} & f_{32} & a''_{33} \end{bmatrix} \tag{10.10}$$

This matrix, in fact, represents an efficient storage of the *LU* factorization of [A],

$$[A] \rightarrow [L][U] \tag{10.11}$$

where

$$[U] = \begin{bmatrix} a_{11} & a_{12} & a_{13} \\ 0 & a'_{22} & a'_{22} \\ 0 & 0 & a''_{33} \end{bmatrix} \tag{10.12}$$

and

$$[L] = \begin{bmatrix} 1 & 0 & 0 \\ f_{21} & 1 & 0 \\ f_{31} & f_{32} & 1 \end{bmatrix} \tag{10.13}$$

The following example confirms that $[A] = [L][U]$.

EXAMPLE 10.1   *LU* Factorization with Gauss Elimination

Problem Statement. Derive an *LU* factorization based on the Gauss elimination performed previously in Example 9.3.

Solution. In Example 9.3, we used Gauss elimination to solve a set of linear algebraic equations that had the following coefficient matrix:

$$[A] = \begin{bmatrix} 3 & -0.1 & -0.2 \\ 0.1 & 7 & -0.3 \\ 0.3 & -0.2 & 10 \end{bmatrix}$$

After forward elimination, the following upper triangular matrix was obtained:

$$[U] = \begin{bmatrix} 3 & -0.1 & -0.2 \\ 0 & 7.00333 & -0.293333 \\ 0 & 0 & 10.0120 \end{bmatrix}$$

The factors employed to obtain the upper triangular matrix can be assembled into a lower triangular matrix. The elements a_{21} and a_{31} were eliminated by using the factors

$$f_{21} = \frac{0.1}{3} = 0.0333333 \qquad f_{31} = \frac{0.3}{3} = 0.1000000$$

and the element a_{32} was eliminated by using the factor

$$f_{32} = \frac{-0.19}{7.00333} = -0.0271300$$

Thus, the lower triangular matrix is

$$[L] = \begin{bmatrix} 1 & 0 & 0 \\ 0.0333333 & 1 & 0 \\ 0.100000 & -0.0271300 & 1 \end{bmatrix}$$

Consequently, the *LU* factorization is

$$[A] = [L][U] = \begin{bmatrix} 1 & 0 & 0 \\ 0.0333333 & 1 & 0 \\ 0.100000 & -0.0271300 & 1 \end{bmatrix} \begin{bmatrix} 3 & -0.1 & -0.2 \\ 0 & 7.00333 & -0.293333 \\ 0 & 0 & 10.0120 \end{bmatrix}$$

This result can be verified by performing the multiplication of $[L][U]$ to give

$$[L][U] = \begin{bmatrix} 3 & -0.1 & -0.2 \\ 0.0999999 & 7 & -0.3 \\ 0.3 & -0.2 & 9.99996 \end{bmatrix}$$

where the minor discrepancies are due to roundoff.

After the matrix is decomposed, a solution can be generated for a particular right-hand-side vector {*b*}. This is done in two steps. First, a forward-substitution step is executed by solving Eq. (10.8) for {*d*}. It is important to recognize that this merely amounts to performing the elimination manipulations on {*b*}. Thus, at the end of this step, the right-hand side will be in the same state that it would have been had we performed forward manipulation on [A] and {b} simultaneously.

The forward-substitution step can be represented concisely as

$$d_i = b_i - \sum_{j=1}^{i-1} l_{ij} d_j \qquad \text{for } i = 1, 2, \ldots, n$$

The second step then merely amounts to implementing back substitution to solve Eq. (10.3). Again, it is important to recognize that this is identical to the back-substitution phase of conventional Gauss elimination [compare with Eqs. (9.12) and (9.13)]:

$$x_n = d_n / u_{nn}$$

$$x_i = \frac{d_i - \sum\limits_{j=i+1}^{n} u_{ij} x_j}{u_{ii}} \qquad \text{for } i = n - 1, n - 2, \ldots, 1$$

EXAMPLE 10.2   The Substitution Steps

Problem Statement. Complete the problem initiated in Example 10.1 by generating the final solution with forward and back substitution.

Solution. As just stated, the intent of forward substitution is to impose the elimination manipulations that we had formerly applied to [A] on the right-hand-side vector {b}. Recall that the system being solved is

$$\begin{bmatrix} 3 & -0.1 & -0.2 \\ 0.1 & 7 & -0.3 \\ 0.3 & -0.2 & 10 \end{bmatrix} \begin{Bmatrix} x_1 \\ x_2 \\ x_3 \end{Bmatrix} = \begin{Bmatrix} 7.85 \\ -19.3 \\ 71.4 \end{Bmatrix}$$

and that the forward-elimination phase of conventional Gauss elimination resulted in

$$\begin{bmatrix} 3 & -0.1 & -0.2 \\ 0 & 7.00333 & -0.293333 \\ 0 & 0 & 10.0120 \end{bmatrix} \begin{Bmatrix} x_1 \\ x_2 \\ x_3 \end{Bmatrix} = \begin{Bmatrix} 7.85 \\ -19.5617 \\ 70.0843 \end{Bmatrix}$$

The forward-substitution phase is implemented by applying Eq. (10.8):

$$\begin{bmatrix} 1 & 0 & 0 \\ 0.0333333 & 1 & 0 \\ 0.100000 & -0.0271300 & 1 \end{bmatrix} \begin{Bmatrix} d_1 \\ d_2 \\ d_3 \end{Bmatrix} = \begin{Bmatrix} 7.85 \\ -19.3 \\ 71.4 \end{Bmatrix}$$

or multiplying out the left-hand side:

$$\begin{aligned} d_1 &= 7.85 \\ 0.0333333 d_1 + d_2 &= -19.3 \\ 0.100000 d_1 - 0.0271300 d_2 + d_3 &= 71.4 \end{aligned}$$

We can solve the first equation for $d_1 = 7.85$, which can be substituted into the second equation to solve for

$$d_2 = -19.3 - 0.0333333(7.85) = -19.5617$$

Both d_1 and d_2 can be substituted into the third equation to give

$$d_3 = 71.4 - 0.1(7.85) + 0.02713(-19.5617) = 70.0843$$

Thus,

$$\{d\} = \left\{ \begin{array}{c} 7.85 \\ -19.5617 \\ 70.0843 \end{array} \right\}$$

This result can then be substituted into Eq. (10.3), $[U]\{x\} = \{d\}$:

$$\begin{bmatrix} 3 & -0.1 & -0.2 \\ 0 & 7.00333 & -0.293333 \\ 0 & 0 & 10.0120 \end{bmatrix} \left\{ \begin{array}{c} x_1 \\ x_2 \\ x_3 \end{array} \right\} = \left\{ \begin{array}{c} 7.85 \\ -19.5617 \\ 70.0843 \end{array} \right\}$$

which can be solved by back substitution (see Example 9.3 for details) for the final solution:

$$\{x\} = \left\{ \begin{array}{c} 3 \\ -2.5 \\ 7.00003 \end{array} \right\}$$

10.2.1 *LU* Factorization with Pivoting

Just as for standard Gauss elimination, partial pivoting is necessary to obtain reliable solutions with *LU* factorization. One way to do this involves using a permutation matrix (recall Sec. 8.1.2). The approach consists of the following steps:

1. **Elimination.** The *LU* factorization with pivoting of a matrix [A] can be represented in matrix form as

   $$[P][A] = [L][U]$$

   The upper triangular matrix, [U], is generated by elimination with partial pivoting, while storing the multiplier factors in [L] and employing the permutation matrix, [P], to keep track of the row switches.

2. **Forward substitution.** The matrices [L] and [P] are used to perform the elimination step with pivoting on $\{b\}$ in order to generate the intermediate right-hand-side vector, $\{d\}$. This step can be represented concisely as the solution of the following matrix formulation:

   $$[L]\{d\} = [P]\{b\}$$

3. **Back substitution.** The final solution is generated in the same fashion as done previously for Gauss elimination. This step can also be represented concisely as the solution of the matrix formulation:

   $$[U]\{x\} = \{d\}$$

The approach is illustrated in the following example.

EXAMPLE 10.3   *LU* Factorization with Pivoting

Problem Statement. Compute the *LU* factorization and find the solution for the same system analyzed in Example 9.4

$$\begin{bmatrix} 0.0003 & 3.0000 \\ 1.0000 & 1.0000 \end{bmatrix} \begin{Bmatrix} x_1 \\ x_2 \end{Bmatrix} = \begin{Bmatrix} 2.0001 \\ 1.0000 \end{Bmatrix}$$

Solution. Before elimination, we set up the initial permutation matrix:

$$[P] = \begin{bmatrix} 1.0000 & 0.0000 \\ 0.0000 & 1.0000 \end{bmatrix}$$

We immediately see that pivoting is necessary, so prior to elimination we switch the rows:

$$[A] = \begin{bmatrix} 1.0000 & 1.0000 \\ 0.0003 & 3.0000 \end{bmatrix}$$

At the same time, we keep track of the pivot by switching the rows of the permutation matrix:

$$[P] = \begin{bmatrix} 0.0000 & 1.0000 \\ 1.0000 & 0.0000 \end{bmatrix}$$

We then eliminate a_{21} by subtracting the factor $l_{21} = a_{21}/a_{11} = 0.0003/1 = 0.0003$ from the second row of *A*. In so doing, we compute that the new value of $a_{22}' = 3 - 0.0003(1) = 2.9997$. Thus, the elimination step is complete with the result:

$$[U] = \begin{bmatrix} 1 & 1 \\ 0 & 2.9997 \end{bmatrix} \qquad [L] = \begin{bmatrix} 1 & 0 \\ 0.0003 & 1 \end{bmatrix}$$

Before implementing forward substitution, the permutation matrix is used to reorder the right-hand-side vector to reflect the pivots as in

$$[P]\{b\} = \begin{bmatrix} 0.0000 & 1.0000 \\ 1.0000 & 0.0000 \end{bmatrix} \begin{Bmatrix} 2.0001 \\ 1 \end{Bmatrix} = \begin{Bmatrix} 1 \\ 2.0001 \end{Bmatrix}$$

Then, forward substitution is applied as in

$$\begin{bmatrix} 1 & 0 \\ 0.0003 & 1 \end{bmatrix} \begin{Bmatrix} d_1 \\ d_2 \end{Bmatrix} = \begin{Bmatrix} 1 \\ 2.0001 \end{Bmatrix}$$

which can be solved for $d_1 = 1$ and $d_2 = 2.0001 - 0.0003(1) = 1.9998$. At this point, the system is

$$\begin{bmatrix} 1 & 1 \\ 0 & 2.9997 \end{bmatrix} \begin{Bmatrix} x_1 \\ x_2 \end{Bmatrix} = \begin{Bmatrix} 1 \\ 1.9998 \end{Bmatrix}$$

Applying back substitution gives the final result:

$$x_2 = \frac{1.9998}{2.9997} = 0.66667$$

$$x_1 = \frac{1 - 1(0.66667)}{1} = 0.33333$$

The *LU* factorization algorithm requires the same total flops as for Gauss elimination. The only difference is that a little less effort is expended in the factorization phase since the operations are not applied to the right-hand side. Conversely, the substitution phase takes a little more effort.

10.2.2 MATLAB Function: lu

MATLAB has a built-in function lu that generates the *LU* factorization. It has the general syntax:

    [L,U] = lu(X)

where *L* and *U* are the lower triangular and upper triangular matrices, respectively, derived from the *LU* factorization of the matrix *X*. Note that this function uses partial pivoting to avoid division by zero. The following example shows how it can be employed to generate both the factorization and a solution for the same problem that was solved in Examples 10.1 and 10.2.

EXAMPLE 10.4   *LU* Factorization with MATLAB

Problem Statement.   Use MATLAB to compute the *LU* factorization and find the solution for the same linear system analyzed in Examples 10.1 and 10.2:

$$\begin{bmatrix} 3 & -0.1 & -0.2 \\ 0.1 & 7 & -0.3 \\ 0.3 & -0.2 & 10 \end{bmatrix} \begin{Bmatrix} x_1 \\ x_2 \\ x_3 \end{Bmatrix} = \begin{Bmatrix} 7.85 \\ -19.3 \\ 71.4 \end{Bmatrix}$$

Solution.   The coefficient matrix and the right-hand-side vector can be entered in standard fashion as

```
>> A = [3 -.1 -.2;.1 7 -.3;.3 -.2 10];
>> b = [7.85; -19.3; 71.4];
```

Next, the *LU* factorization can be computed with

```
>> [L,U] = lu(A)

L =
    1.0000         0         0
    0.0333    1.0000         0
    0.1000   -0.0271    1.0000


U =
    3.0000   -0.1000   -0.2000
         0    7.0033   -0.2933
         0         0   10.0120
```

This is the same result that we obtained by hand in Example 10.1. We can test that it is correct by computing the original matrix as

```
>> L*U
```

```
ans =

    3.0000   -0.1000   -0.2000
    0.1000    7.0000   -0.3000
    0.3000   -0.2000   10.0000
```

To generate the solution, we first compute

```
>> d = L\b

d =

    7.8500
  -19.5617
   70.0843
```

And then use this result to compute the solution

```
>> x = U\d

x =

    3.0000
   -2.5000
    7.0000
```

These results conform to those obtained by hand in Example 10.2.

10.3 CHOLESKY FACTORIZATION

Recall from Chap. 8 that a symmetric matrix is one where $a_{ij} = a_{ji}$ for all i and j. In other words, $[A] = [A]^T$. Such systems occur commonly in both mathematical and engineering/science problem contexts.

Special solution techniques are available for such systems. They offer computational advantages because only half the storage is needed and only half the computation time is required for their solution.

One of the most popular approaches involves *Cholesky factorization* (also called Cholesky decomposition). This algorithm is based on the fact that a symmetric matrix can be decomposed, as in

$$[A] = [U]^T [U] \tag{10.14}$$

That is, the resulting triangular factors are the transpose of each other.

The terms of Eq. (10.14) can be multiplied out and set equal to each other. The factorization can be generated efficiently by recurrence relations. For the ith row:

$$u_{ii} = \sqrt{a_{ii} - \sum_{k=1}^{i-1} u_{ki}^2} \tag{10.15}$$

$$u_{ij} = \frac{a_{ij} - \sum_{k=1}^{i-1} u_{ki} u_{kj}}{u_{ii}} \qquad \text{for } j = i+1, \ldots, n \tag{10.16}$$

EXAMPLE 10.5   Cholesky Factorization

Problem Statement. Compute the Cholesky factorization for the symmetric matrix

$$[A] = \begin{bmatrix} 6 & 15 & 55 \\ 15 & 55 & 225 \\ 55 & 225 & 979 \end{bmatrix}$$

Solution. For the first row ($i = 1$), Eq. (10.15) is employed to compute

$$u_{11} = \sqrt{a_{11}} = \sqrt{6} = 2.44949$$

Then, Eq. (10.16) can be used to determine

$$u_{12} = \frac{a_{12}}{u_{11}} = \frac{15}{2.44949} = 6.123724$$

$$u_{13} = \frac{a_{13}}{u_{11}} = \frac{55}{2.44949} = 22.45366$$

For the second row ($i = 2$):

$$u_{22} = \sqrt{a_{22} - u_{12}^2} = \sqrt{55 - (6.123724)^2} = 4.1833$$

$$u_{23} = \frac{a_{23} - u_{12}u_{13}}{u_{22}} = \frac{225 - 6.123724(22.45366)}{4.1833} = 20.9165$$

For the third row ($i = 3$):

$$u_{33} = \sqrt{a_{33} - u_{13}^2 - u_{23}^2} = \sqrt{979 - (22.45366)^2 - (20.9165)^2} = 6.110101$$

Thus, the Cholesky factorization yields

$$[U] = \begin{bmatrix} 2.44949 & 6.123724 & 22.45366 \\ & 4.1833 & 20.9165 \\ & & 6.110101 \end{bmatrix}$$

The validity of this factorization can be verified by substituting it and its transpose into Eq. (10.14) to see if their product yields the original matrix $[A]$. This is left for an exercise.

After obtaining the factorization, it can be used to determine a solution for a right-hand-side vector $\{b\}$ in a manner similar to *LU* factorization. First, an intermediate vector $\{d\}$ is created by solving

$$[U]^T \{d\} = \{b\} \tag{10.17}$$

Then, the final solution can be obtained by solving

$$[U]\{x\} = \{d\} \tag{10.18}$$

10.3.1 MATLAB Function: chol

MATLAB has a built-in function chol that generates the Cholesky factorization. It has the general syntax,

U = chol(X)

where U is an upper triangular matrix so that $U'*U = X$. The following example shows how it can be employed to generate both the factorization and a solution for the same matrix that we looked at in the previous example.

EXAMPLE 10.6   Cholesky Factorization with MATLAB

Problem Statement. Use MATLAB to compute the Cholesky factorization for the same matrix we analyzed in Example 10.5.

$$[A] = \begin{bmatrix} 6 & 15 & 55 \\ 15 & 55 & 225 \\ 55 & 225 & 979 \end{bmatrix}$$

Also obtain a solution for a right-hand-side vector that is the sum of the rows of $[A]$. Note that for this case, the answer will be a vector of ones.

Solution. The matrix is entered in standard fashion as

>> A = [6 15 55; 15 55 225; 55 225 979];

A right-hand-side vector that is the sum of the rows of $[A]$ can be generated as

>> b = [sum(A(1,:)); sum(A(2,:)); sum(A(3,:))]

b =
            76
           295
          1259

Next, the Cholesky factorization can be computed with

>> U = chol(A)

U =
     2.4495   6.1237   22.4537
          0   4.1833   20.9165
          0   0   6.1101

We can test that this is correct by computing the original matrix as

>> U'*U

ans =
     6.0000   15.0000   55.0000
    15.0000   55.0000   225.0000
    55.0000   225.0000   979.0000

To generate the solution, we first compute

```
>> d = U'\b

d =
   31.0269
   25.0998
    6.1101
```

And then use this result to compute the solution

```
>> x = U\d

x =
    1.0000
    1.0000
    1.0000
```

10.4   MATLAB LEFT DIVISION

We previously introduced left division without any explanation of how it works. Now that we have some background on matrix solution techniques, we can provide a simplified description of its operation.

When we implement left division with the backslash operator, MATLAB invokes a highly sophisticated algorithm to obtain a solution. In essence, MATLAB examines the structure of the coefficient matrix and then implements an optimal method to obtain the solution. Although the details of the algorithm are beyond our scope, a simplified overview can be outlined.

First, MATLAB checks to see whether [A] is in a format where a solution can be obtained without full Gauss elimination. These include systems that are (a) sparse and banded, (b) triangular (or easily transformed into triangular form), or (c) symmetric. If any of these cases are detected, the solution is obtained with the efficient techniques that are available for such systems. Some of the techniques include banded solvers, back and forward substitution, and Cholesky factorization.

If none of these simplified solutions are possible and the matrix is square,[2] a general triangular factorization is computed by Gauss elimination with partial pivoting and the solution obtained with substitution.

---

[2] It should be noted that in the event that [A] is not square, a least-squares solution is obtained with an approach called *QR factorization*.

PROBLEMS

10.1 Determine the total flops as a function of the number of equations n for the **(a)** factorization, **(b)** forward substitution, and **(c)** back-substitution phases of the LU factorization version of Gauss elimination.

10.2 Use the rules of matrix multiplication to prove that Eqs. (10.7) and (10.8) follow from Eq. (10.6).

10.3 Use naive Gauss elimination to factor the following system according to the description in Sec. 10.2:

$$10x_1 + 2x_2 - x_3 = 27$$
$$-3x_1 - 6x_2 + 2x_3 = -61.5$$
$$x_1 + x_2 + 5x_3 = -21.5$$

Then, multiply the resulting $[L]$ and $[U]$ matrices to determine that $[A]$ is produced.

10.4 (a) Use LU factorization to solve the system of equations in Prob. 10.3. Show all the steps in the computation. **(b)** Also solve the system for an alternative right-hand-side vector

$$\{b\}^T = \lfloor 12 \quad 18 \quad -6 \rfloor$$

10.5 Solve the following system of equations using LU factorization with partial pivoting:

$$2x_1 - 6x_2 - x_3 = -38$$
$$-3x_1 - x_2 + 7x_3 = -34$$
$$-8x_1 + x_2 - 2x_3 = -40$$

10.6 Develop your own M-file to determine the LU factorization of a square matrix without partial pivoting. That is, develop a function that is passed the square matrix and returns the triangular matrices $[L]$ and $[U]$. Test your function by using it to solve the system in Prob. 10.3. Confirm that your function is working properly by verifying that $[L][U] = [A]$ and by using the built-in function lu.

10.7 Confirm the validity of the Cholesky factorization of Example 10.5 by substituting the results into Eq. (10.14) to verify that the product of $[U]^T$ and $[U]$ yields $[A]$.

10.8 (a) Perform a Cholesky factorization of the following symmetric system by hand:

$$\begin{bmatrix} 8 & 20 & 15 \\ 20 & 80 & 50 \\ 15 & 50 & 60 \end{bmatrix} \begin{Bmatrix} x_1 \\ x_2 \\ x_3 \end{Bmatrix} = \begin{Bmatrix} 50 \\ 250 \\ 100 \end{Bmatrix}$$

(b) Verify your hand calculation with the built-in chol function. **(c)** Employ the results of the factorization $[U]$ to determine the solution for the right-hand-side vector.

10.9 Develop your own M-file to determine the Cholesky factorization of a symmetric matrix without pivoting. That

is, develop a function that is passed the symmetric matrix and returns the matrix $[U]$. Test your function by using it to solve the system in Prob. 10.8 and use the built-in function chol to confirm that your function is working properly.

10.10 Solve the following set of equations with LU factorization with pivoting:

$$3x_1 - 2x_2 + x_3 = -10$$
$$2x_1 + 6x_2 - 4x_3 = 44$$
$$-x_1 - 2x_2 + 5x_3 = -26$$

10.11 (a) Determine the LU factorization without pivoting by hand for the following matrix and check your results by validating that $[L][U] = [A]$.

$$\begin{bmatrix} 8 & 2 & 1 \\ 3 & 7 & 2 \\ 2 & 3 & 9 \end{bmatrix}$$

(b) Employ the result of **(a)** to compute the determinant. **(c)** Repeat **(a)** and **(b)** using MATLAB.

10.12 Use the following LU factorization to **(a)** compute the determinant and **(b)** solve $[A]\{x\} = \{b\}$ with $\{b\}^T = \lfloor -10 \quad 44 \quad -26 \rfloor$

$$[A] = [L][U] = \begin{bmatrix} 1 & & \\ 0.6667 & 1 & \\ -0.3333 & -0.3636 & 1 \end{bmatrix}$$

$$\times \begin{bmatrix} 3 & -2 & 1 \\ & 7.3333 & -4.6667 \\ & & 3.6364 \end{bmatrix}$$

10.13 Use Cholesky factorization to determine $[U]$ so that

$$[A] = [U]^T[U] = \begin{bmatrix} 2 & -1 & 0 \\ -1 & 2 & -1 \\ 0 & -1 & 2 \end{bmatrix}$$

10.14 Compute the Cholesky factorization of

$$[A] = \begin{bmatrix} 9 & 0 & 0 \\ 0 & 25 & 0 \\ 0 & 0 & 4 \end{bmatrix}$$

Do your results make sense in terms of Eqs. (10.15) and (10.16)?

11

Matrix Inverse and Condition

CHAPTER OBJECTIVES

The primary objective of this chapter is to show how to compute the matrix inverse
and to illustrate how it can be used to analyze complex linear systems that occur in
engineering and science. In addition, a method to assess a matrix solution's sensitivity
to roundoff error is described. Specific objectives and topics covered are

- Knowing how to determine the matrix inverse in an efficient manner based on *LU*
  factorization.
- Understanding how the matrix inverse can be used to assess stimulus-response
  characteristics of engineering systems.
- Understanding the meaning of matrix and vector norms and how they are computed.
- Knowing how to use norms to compute the matrix condition number.
- Understanding how the magnitude of the condition number can be used to esti-
  mate the precision of solutions of linear algebraic equations.

11.1 THE MATRIX INVERSE

In our discussion of matrix operations (Sec. 8.1.2), we introduced the notion that if a matrix
$[A]$ is square, there is another matrix $[A]^{-1}$, called the inverse of $[A]$, for which

$$[A][A]^{-1} = [A]^{-1}[A] = [I] \tag{11.1}$$

Now we will focus on how the inverse can be computed numerically. Then we will explore
how it can be used for engineering analysis.

11.1.1 Calculating the Inverse

The inverse can be computed in a column-by-column fashion by generating solutions with
unit vectors as the right-hand-side constants. For example, if the right-hand-side constant

has a 1 in the first position and zeros elsewhere,

$$\{b\} = \begin{Bmatrix} 1 \\ 0 \\ 0 \end{Bmatrix} \tag{11.2}$$

the resulting solution will be the first column of the matrix inverse. Similarly, if a unit vector with a 1 at the second row is used

$$\{b\} = \begin{Bmatrix} 0 \\ 1 \\ 0 \end{Bmatrix} \tag{11.3}$$

the result will be the second column of the matrix inverse.

The best way to implement such a calculation is with LU factorization. Recall that one of the great strengths of LU factorization is that it provides a very efficient means to evaluate multiple right-hand-side vectors. Thus, it is ideal for evaluating the multiple unit vectors needed to compute the inverse.

EXAMPLE 11.1   Matrix Inversion

Problem Statement. Employ LU factorization to determine the matrix inverse for the system from Example 10.1:

$$[A] = \begin{bmatrix} 3 & -0.1 & -0.2 \\ 0.1 & 7 & -0.3 \\ 0.3 & -0.2 & 10 \end{bmatrix}$$

Recall that the factorization resulted in the following lower and upper triangular matrices:

$$[U] = \begin{bmatrix} 3 & -0.1 & -0.2 \\ 0 & 7.00333 & -0.293333 \\ 0 & 0 & 10.0120 \end{bmatrix} \qquad [L] = \begin{bmatrix} 1 & 0 & 0 \\ 0.0333333 & 1 & 0 \\ 0.100000 & -0.0271300 & 1 \end{bmatrix}$$

Solution. The first column of the matrix inverse can be determined by performing the forward-substitution solution procedure with a unit vector (with 1 in the first row) as the right-hand-side vector. Thus, the lower triangular system can be set up as [recall Eq. (10.8)]

$$\begin{bmatrix} 1 & 0 & 0 \\ 0.0333333 & 1 & 0 \\ 0.100000 & -0.0271300 & 1 \end{bmatrix} \begin{Bmatrix} d_1 \\ d_2 \\ d_3 \end{Bmatrix} = \begin{Bmatrix} 1 \\ 0 \\ 0 \end{Bmatrix}$$

and solved with forward substitution for $\{d\}^T = \lfloor 1 \ -0.03333 \ -0.1009 \rfloor$. This vector can then be used as the right-hand side of the upper triangular system [recall Eq. (10.3)]

$$\begin{bmatrix} 3 & -0.1 & -0.2 \\ 0 & 7.00333 & -0.293333 \\ 0 & 0 & 10.0120 \end{bmatrix} \begin{Bmatrix} x_1 \\ x_2 \\ x_3 \end{Bmatrix} = \begin{Bmatrix} 1 \\ -0.03333 \\ -0.1009 \end{Bmatrix}$$

which can be solved by back substitution for $\{x\}^T = \lfloor 0.33249 \ -0.00518 \ -0.01008 \rfloor$, which is the first column of the matrix inverse:

$$[A]^{-1} = \begin{bmatrix} 0.33249 & 0 & 0 \\ -0.00518 & 0 & 0 \\ -0.01008 & 0 & 0 \end{bmatrix}$$

To determine the second column, Eq. (10.8) is formulated as

$$
\begin{bmatrix}
1 & 0 & 0 \\
0.0333333 & 1 & 0 \\
0.100000 & -0.0271300 & 1
\end{bmatrix}
\begin{Bmatrix} d_1 \\ d_2 \\ d_3 \end{Bmatrix}
= \begin{Bmatrix} 0 \\ 1 \\ 0 \end{Bmatrix}
$$

This can be solved for $\{d\}$, and the results are used with Eq. (10.3) to determine $\{x\}^T = \lfloor 0.004944 \quad 0.142903 \quad 0.00271 \rfloor$, which is the second column of the matrix inverse:

$$
[A]^{-1} =
\begin{bmatrix}
0.33249 & 0.004944 & 0 \\
-0.00518 & 0.142903 & 0 \\
-0.01008 & 0.002710 & 0
\end{bmatrix}
$$

Finally, the same procedures can be implemented with $\{b\}^T = \lfloor 0\ 0\ 1 \rfloor$ to solve for $\{x\}^T = \lfloor 0.006798 \quad 0.004183 \quad 0.09988 \rfloor$, which is the final column of the matrix inverse:

$$
[A]^{-1} =
\begin{bmatrix}
0.33249 & 0.004944 & 0.006798 \\
-0.00518 & 0.142903 & 0.004183 \\
-0.01008 & 0.002710 & 0.099880
\end{bmatrix}
$$

The validity of this result can be checked by verifying that $[A][A]^{-1} = [I]$.

11.1.2 Stimulus-Response Computations

As discussed in PT 3.1, many of the linear systems of equations arising in engineering and science are derived from conservation laws. The mathematical expression of these laws is some form of balance equation to ensure that a particular property—mass, force, heat, momentum, electrostatic potential—is conserved. For a force balance on a structure, the properties might be horizontal or vertical components of the forces acting on each node of the structure. For a mass balance, the properties might be the mass in each reactor of a chemical process. Other fields of engineering and science would yield similar examples.

A single balance equation can be written for each part of the system, resulting in a set of equations defining the behavior of the property for the entire system. These equations are interrelated, or coupled, in that each equation may include one or more of the variables from the other equations. For many cases, these systems are linear and, therefore, of the exact form dealt with in this chapter:

$$
[A]\{x\} = \{b\} \tag{11.4}
$$

Now, for balance equations, the terms of Eq. (11.4) have a definite physical interpretation. For example, the elements of $\{x\}$ are the levels of the property being balanced for each part of the system. In a force balance of a structure, they represent the horizontal and vertical forces in each member. For the mass balance, they are the mass of chemical in each reactor. In either case, they represent the system's *state* or *response,* which we are trying to determine.

The right-hand-side vector $\{b\}$ contains those elements of the balance that are independent of behavior of the system—that is, they are constants. In many problems, they represent the *forcing functions* or *external stimuli* that drive the system.

Finally, the matrix of coefficients [A] usually contains the *parameters* that express how the parts of the system *interact* or are coupled. Consequently, Eq. (11.4) might be reexpressed as

[Interactions]{response} = {stimuli}

As we know from previous chapters, there are a variety of ways to solve Eq. (11.4). However, using the matrix inverse yields a particularly interesting result. The formal solution can be expressed as

$$\{x\} = [A]^{-1}\{b\}$$

or (recalling our definition of matrix multiplication from Sec. 8.1.2)

$$x_1 = a_{11}^{-1} b_1 + a_{12}^{-1} b_2 + a_{13}^{-1} b_3$$

$$x_2 = a_{21}^{-1} b_1 + a_{22}^{-1} b_2 + a_{23}^{-1} b_3$$

$$x_3 = a_{31}^{-1} b_1 + a_{32}^{-1} b_2 + a_{33}^{-1} b_3$$

Thus, we find that the inverted matrix itself, aside from providing a solution, has extremely useful properties. That is, each of its elements represents the response of a single part of the system to a unit stimulus of any other part of the system.

Notice that these formulations are linear and, therefore, superposition and proportionality hold. *Superposition* means that if a system is subject to several different stimuli (the b's), the responses can be computed individually and the results summed to obtain a total response. *Proportionality* means that multiplying the stimuli by a quantity results in the response to those stimuli being multiplied by the same quantity. Thus, the coefficient a_{11}^{-1} is a proportionality constant that gives the value of x_1 due to a unit level of b_1. This result is independent of the effects of b_2 and b_3 on x_1, which are reflected in the coefficients a_{12}^{-1} and a_{13}^{-1}, respectively. Therefore, we can draw the general conclusion that the element a_{ij}^{-1} of the inverted matrix represents the value of x_i due to a unit quantity of b_j.

Using the example of the structure, element a_{ij}^{-1} of the matrix inverse would represent the force in member i due to a unit external force at node j. Even for small systems, such behavior of individual stimulus-response interactions would not be intuitively obvious. As such, the matrix inverse provides a powerful technique for understanding the interrelationships of component parts of complicated systems.

EXAMPLE 11.2   Analyzing the Bungee Jumper Problem

Problem Statement. At the beginning of Chap. 8, we set up a problem involving three individuals suspended vertically connected by bungee cords. We derived a system of linear algebraic equations based on force balances for each jumper,

$$\begin{bmatrix} 150 & -100 & 0 \\ -100 & 150 & -50 \\ 0 & -50 & 50 \end{bmatrix} \begin{Bmatrix} x_1 \\ x_2 \\ x_3 \end{Bmatrix} = \begin{Bmatrix} 588.6 \\ 686.7 \\ 784.8 \end{Bmatrix}$$

In Example 8.2, we used MATLAB to solve this system for the vertical positions of the jumpers (the x's). In the present example, use MATLAB to compute the matrix inverse and interpret what it means.

Solution.   Start up MATLAB and enter the coefficient matrix:

```
>> K = [150 -100  0;-100  150 -50;0  -50 50];
```

The inverse can then be computed as

```
>> KI = inv(K)

KI =
    0.0200    0.0200    0.0200
    0.0200    0.0300    0.0300
    0.0200    0.0300    0.0500
```

Each element of the inverse, k_{ij}^{-1} of the inverted matrix represents the vertical change in position (in meters) of jumper i due to a unit change in force (in Newtons) applied to jumper j.

First, observe that the numbers in the first column ($j = 1$) indicate that the position of all three jumpers would increase by 0.02 m if the force on the first jumper was increased by 1 N. This makes sense, because the additional force would only elongate the first cord by that amount.

In contrast, the numbers in the second column ($j = 2$) indicate that applying a force of 1 N to the second jumper would move the first jumper down by 0.02 m, but the second and third by 0.03 m. The 0.02-m elongation of the first jumper makes sense because the first cord is subject to an extra 1 N regardless of whether the force is applied to the first or second jumper. However, for the second jumper the elongation is now 0.03 m because along with the first cord, the second cord also elongates due to the additional force. And of course, the third jumper shows the identical translation as the second jumper as there is no additional force on the third cord that connects them.

As expected, the third column ($j = 3$) indicates that applying a force of 1 N to the third jumper results in the first and second jumpers moving the same distances as occurred when the force was applied to the second jumper. However, now because of the additional elongation of the third cord, the third jumper is moved farther downward.

Superposition and proportionality can be demonstrated by using the inverse to determine how much farther the third jumper would move downward if additional forces of 10, 50, and 20 N were applied to the first, second, and third jumpers, respectively. This can be done simply by using the appropriate elements of the third row of the inverse to compute,

$$\Delta x_3 = k_{31}^{-1} \Delta F_1 + k_{32}^{-1} \Delta F_2 + k_{33}^{-1} \Delta F_3 = 0.02(10) + 0.03(50) + 0.05(20) = 2.7 \text{ m}$$

11.2   ERROR ANALYSIS AND SYSTEM CONDITION

Aside from its engineering and scientific applications, the inverse also provides a means to discern whether systems are ill-conditioned. Three direct methods can be devised for this purpose:

1.   Scale the matrix of coefficients $[A]$ so that the largest element in each row is 1. Invert the scaled matrix and if there are elements of $[A]^{-1}$ that are several orders of magnitude greater than one, it is likely that the system is ill-conditioned.

2.   Multiply the inverse by the original coefficient matrix and assess whether the result is close to the identity matrix. If not, it indicates ill-conditioning.
3.   Invert the inverted matrix and assess whether the result is sufficiently close to the original coefficient matrix. If not, it again indicates that the system is ill-conditioned.

Although these methods can indicate ill-conditioning, it would be preferable to obtain a single number that could serve as an indicator of the problem. Attempts to formulate such a matrix condition number are based on the mathematical concept of the norm.

11.2.1 Vector and Matrix Norms

A *norm* is a real-valued function that provides a measure of the size or "length" of multicomponent mathematical entities such as vectors and matrices.

   A simple example is a vector in three-dimensional Euclidean space (Fig. 11.1) that can be represented as

$$\lfloor F \rfloor = \lfloor a \quad b \quad c \rfloor$$

where a, b, and c are the distances along the x, y, and z axes, respectively. The length of this vector—that is, the distance from the coordinate $(0, 0, 0)$ to (a, b, c)—can be simply computed as

$$\|F\|_e = \sqrt{a^2 + b^2 + c^2}$$

where the nomenclature $\|F\|_e$ indicates that this length is referred to as the *Euclidean norm* of $[F]$.

   Similarly, for an n-dimensional vector $\lfloor X \rfloor = \lfloor x_1 \quad x_2 \quad \cdots \quad x_n \rfloor$, a Euclidean norm would be computed as

$$\|X\|_e = \sqrt{\sum_{i=1}^{n} x_i^2}$$

FIGURE 11.1
Graphical depiction of a vector in Euclidean space.

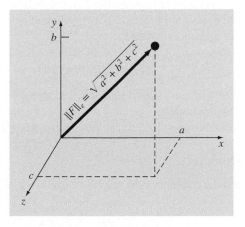

The concept can be extended further to a matrix $[A]$, as in

$$\|A\|_f = \sqrt{\sum_{i=1}^{n} \sum_{j=1}^{n} a_{i,j}^2} \tag{11.5}$$

which is given a special name—the *Frobenius norm*. As with the other vector norms, it provides a single value to quantify the "size" of $[A]$.

It should be noted that there are alternatives to the Euclidean and Frobenius norms. For vectors, there are alternatives called p norms that can be represented generally by

$$\|X\|_p = \left(\sum_{i=1}^{n} |x_i|^p \right)^{1/p}$$

We can see that the Euclidean norm and the 2 norm, $\|X\|_2$, are identical for vectors.

Other important examples are $(p = 1)$

$$\|X\|_1 = \sum_{i=1}^{n} |x_i|$$

which represents the norm as the sum of the absolute values of the elements. Another is the maximum-magnitude or uniform-vector norm $(p = \infty)$,

$$\|X\|_\infty = \max_{1 \leq i \leq n} |x_i|$$

which defines the norm as the element with the largest absolute value.

Using a similar approach, norms can be developed for matrices. For example,

$$\|A\|_1 = \max_{1 \leq j \leq n} \sum_{i=1}^{n} |a_{ij}|$$

That is, a summation of the absolute values of the coefficients is performed for each column, and the largest of these summations is taken as the norm. This is called the *column-sum norm*.

A similar determination can be made for the rows, resulting in a uniform-matrix or *row-sum norm:*

$$\|A\|_\infty = \max_{1 \leq i \leq n} \sum_{j=1}^{n} |a_{ij}|$$

It should be noted that, in contrast to vectors, the 2 norm and the Frobenius norm for a matrix are not the same. Whereas the Frobenius norm $\|A\|_f$ can be easily determined by Eq. (11.5), the matrix 2 norm $\|A\|_2$ is calculated as

$$\|A\|_2 = (\mu_{max})^{1/2}$$

where μ_{max} is the largest eigenvalue of $[A]^T[A]$. In Chap. 13, we will learn more about eigenvalues. For the time being, the important point is that the $\|A\|_2$, or *spectral norm*, is the minimum norm and, therefore, provides the tightest measure of size (Ortega, 1972).

11.2.2 Matrix Condition Number

Now that we have introduced the concept of the norm, we can use it to define

$$\text{Cond}[A] = \|A\| \cdot \|A^{-1}\|$$

where Cond[A] is called the *matrix condition number*. Note that for a matrix [A], this number will be greater than or equal to 1. It can be shown (Ralston and Rabinowitz, 1978; Gerald and Wheatley, 1989) that

$$\frac{\|\Delta X\|}{\|X\|} \le \text{Cond}\,[A]\,\frac{\|\Delta A\|}{\|A\|}$$

That is, the relative error of the norm of the computed solution can be as large as the relative error of the norm of the coefficients of [A] multiplied by the condition number. For example, if the coefficients of [A] are known to t-digit precision (i.e., rounding errors are on the order of 10^{-t}) and Cond[A] = 10^c, the solution [X] may be valid to only $t - c$ digits (rounding errors $\approx 10^{c-t}$).

EXAMPLE 11.3   Matrix Condition Evaluation

Problem Statement. The Hilbert matrix, which is notoriously ill-conditioned, can be represented generally as

$$\begin{bmatrix} 1 & \frac{1}{2} & \frac{1}{3} & \cdots & \frac{1}{n} \\ \frac{1}{2} & \frac{1}{3} & \frac{1}{4} & \cdots & \frac{1}{n+1} \\ \vdots & \vdots & \vdots & & \vdots \\ \frac{1}{n} & \frac{1}{n+1} & \frac{1}{n+2} & \cdots & \frac{1}{2n-1} \end{bmatrix}$$

Use the row-sum norm to estimate the matrix condition number for the 3×3 Hilbert matrix:

$$[A] = \begin{bmatrix} 1 & \frac{1}{2} & \frac{1}{3} \\ \frac{1}{2} & \frac{1}{3} & \frac{1}{4} \\ \frac{1}{3} & \frac{1}{4} & \frac{1}{5} \end{bmatrix}$$

Solution. First, the matrix can be normalized so that the maximum element in each row is 1:

$$[A] = \begin{bmatrix} 1 & \frac{1}{2} & \frac{1}{3} \\ 1 & \frac{2}{3} & \frac{1}{2} \\ 1 & \frac{3}{4} & \frac{3}{5} \end{bmatrix}$$

Summing each of the rows gives 1.833, 2.1667, and 2.35. Thus, the third row has the largest sum and the row-sum norm is

$$\|A\|_\infty = 1 + \frac{3}{4} + \frac{3}{5} = 2.35$$

The inverse of the scaled matrix can be computed as

$$[A]^{-1} = \begin{bmatrix} 9 & -18 & 10 \\ -36 & 96 & -60 \\ 30 & -90 & 60 \end{bmatrix}$$

Note that the elements of this matrix are larger than the original matrix. This is also reflected in its row-sum norm, which is computed as

$$\|A^{-1}\|_\infty = |-36| + |96| + |-60| = 192$$

Thus, the condition number can be calculated as

$$\text{Cond}[A] = 2.35(192) = 451.2$$

The fact that the condition number is much greater than unity suggests that the system is ill-conditioned. The extent of the ill-conditioning can be quantified by calculating $c = \log 451.2 = 2.65$. Hence, the last three significant digits of the solution could exhibit rounding errors. Note that such estimates almost always overpredict the actual error. However, they are useful in alerting you to the possibility that roundoff errors may be significant.

11.2.3 Norms and Condition Number in MATLAB

MATLAB has built-in functions to compute both norms and condition numbers:

```
>> norm(X,p)
```

and

```
>> cond(X,p)
```

where X is the vector or matrix and p designates the type of norm or condition number (1, 2, inf, or 'fro'). Note that the cond function is equivalent to

```
>> norm(X,p) * norm(inv(X),p)
```

Also, note that if p is omitted, it is automatically set to 2.

EXAMPLE 11.4   Matrix Condition Evaluation with MATLAB

Problem Statement. Use MATLAB to evaluate both the norms and condition numbers for the scaled Hilbert matrix previously analyzed in Example 11.3:

$$[A] = \begin{bmatrix} 1 & \frac{1}{2} & \frac{1}{3} \\ 1 & \frac{2}{3} & \frac{1}{2} \\ 1 & \frac{3}{4} & \frac{3}{5} \end{bmatrix}$$

(a) As in Example 11.3, first compute the row-sum versions ($p = $ inf). (b) Also compute the Frobenius ($p = $ 'fro') and the spectral ($p = 2$) condition numbers.

Solution: (a) First, enter the matrix:

```
>> A = [1 1/2 1/3;1 2/3 1/2;1 3/4 3/5];
```

Then, the row-sum norm and condition number can be computed as

```
>> norm(A,inf)

ans =
    2.3500

>> cond(A,inf)

ans =
  451.2000
```

These results correspond to those that were calculated by hand in Example 11.3.

(*b*) The condition numbers based on the Frobenius and spectral norms are

```
>> cond(A,'fro')

ans =
   368.0866

>> cond(A)

ans =
   366.3503
```

11.3 CASE STUDY   INDOOR AIR POLLUTION

Background. As the name implies, indoor air pollution deals with air contamination in enclosed spaces such as homes, offices, and work areas. Suppose that you are studying the ventilation system for Bubba's Gas 'N Guzzle, a truck-stop restaurant located adjacent to an eight-lane freeway.

As depicted in Fig. 11.2, the restaurant serving area consists of two rooms for smokers and kids and one elongated room. Room 1 and section 3 have sources of carbon monoxide from smokers and a faulty grill, respectively. In addition, rooms 1 and 2 gain carbon monoxide from air intakes that unfortunately are positioned alongside the freeway.

FIGURE 11.2
Overhead view of rooms in a restaurant. The one-way arrows represent volumetric airflows, whereas the two-way arrows represent diffusive mixing. The smoker and grill loads add carbon monoxide mass to the system but negligible airflow.

Write steady-state mass balances for each room and solve the resulting linear algebraic equations for the concentration of carbon monoxide in each room. In addition, generate the matrix inverse and use it to analyze how the various sources affect the kids' room. For example, determine what percent of the carbon monoxide in the kids' section is due to (1) the smokers, (2) the grill, and (3) the intake vents. In addition, compute the improvement in the kids' section concentration if the carbon monoxide load is decreased by banning smoking and fixing the grill. Finally, analyze how the concentration in the kids' area would change if a screen is constructed so that the mixing between areas 2 and 4 is decreased to 5 m^3/hr.

Solution. Steady-state mass balances can be written for each room. For example, the balance for the smoking section (room 1) is

$$0 = W_{smoker} + \quad Q_a c_a \quad - \quad Q_a c_1 \quad + E_{13}(c_3 - c_1)$$
$$(\text{Load}) + (\text{Inflow}) - (\text{Outflow}) + (\text{Mixing})$$

Similar balances can be written for the other rooms:

$$0 = Q_b c_b + (Q_a - Q_d)c_4 - Q_c c_2 + E_{24}(c_4 - c_2)$$
$$0 = W_{grill} + Q_a c_1 + E_{13}(c_1 - c_3) + E_{34}(c_4 - c_3) - Q_a c_3$$
$$0 = Q_a c_3 + E_{34}(c_3 - c_4) + E_{24}(c_2 - c_4) - Q_a c_4$$

Substituting the parameters yields the final system of equation:

$$\begin{bmatrix} 225 & 0 & -25 & 0 \\ 0 & 175 & 0 & -125 \\ -225 & 0 & 275 & -50 \\ 0 & -25 & -250 & 275 \end{bmatrix} \begin{Bmatrix} c_1 \\ c_2 \\ c_3 \\ c_4 \end{Bmatrix} = \begin{Bmatrix} 1400 \\ 100 \\ 2000 \\ 0 \end{Bmatrix}$$

MATLAB can be used to generate the solution. First, we can compute the inverse. Note that we use the "short g" format in order to obtain five significant digits of precision:

```
>> format short g
>> A=[225 0 -25 0
0 175 0 -125
-225 0 275 -50
0 -25 -250 275];
>> AI=inv(A)

AI =
       0.0049962     1.5326e-005    0.00055172     0.00010728
       0.0034483     0.0062069      0.0034483      0.0034483
       0.0049655     0.00013793     0.0049655      0.00096552
       0.0048276     0.00068966     0.0048276      0.0048276
```

11.3 CASE STUDY continued

The solution can then be generated as

```
>> b=[1400 100 2000 0]';
>> c=AI*b

c =
        8.0996
       12.345
       16.897
       16.483
```

Thus, we get the surprising result that the smoking section has the lowest carbon monoxide levels! The highest concentrations occur in rooms 3 and 4 with section 2 having an intermediate level. These results take place because (a) carbon monoxide is conservative and (b) the only air exhausts are out of sections 2 and 4 (Q_c and Q_d). Room 3 is so bad because not only does it get the load from the faulty grill, but it also receives the effluent from room 1.

Although the foregoing is interesting, the real power of linear systems comes from using the elements of the matrix inverse to understand how the parts of the system interact. For example, the elements of the matrix inverse can be used to determine the percent of the carbon monoxide in the kids' section due to each source:

The smokers:

$$c_{2,\text{smokers}} = a_{21}^{-1} W_{\text{smokers}} = 0.0034483(1000) = 3.4483$$

$$\%_{\text{smokers}} = \frac{3.4483}{12.345} \times 100\% = 27.93\%$$

The grill:

$$c_{2,\text{grill}} = a_{23}^{-1} W_{\text{grill}} = 0.0034483(2000) = 6.897$$

$$\%_{\text{grill}} = \frac{6.897}{12.345} \times 100\% = 55.87\%$$

The intakes:

$$c_{2,\text{intakes}} = a_{21}^{-1} Q_a c_a + a_{22}^{-1} Q_b c_b = 0.0034483(200)2 + 0.0062069(50)2$$

$$= 1.37931 + 0.62069 = 2$$

$$\%_{\text{grill}} = \frac{2}{12.345} \times 100\% = 16.20\%$$

The faulty grill is clearly the most significant source.

The inverse can also be employed to determine the impact of proposed remedies such as banning smoking and fixing the grill. Because the model is linear, superposition holds and the results can be determined individually and summed:

$$\Delta c_2 = a_{21}^{-1} \Delta W_{\text{smokers}} + a_{23}^{-1} \Delta W_{\text{grill}} = 0.0034483(-1000) + 0.0034483(-2000)$$

$$= -3.4483 - 6.8966 = -10.345$$

Note that the same computation would be made in MATLAB as

```
>> AI(2,1)*(-1000)+AI(2,3)*(-2000)

ans =
      -10.345
```

Implementing both remedies would reduce the concentration by 10.345 mg/m^3. The result would bring the kids' room concentration to $12.345 - 10.345 = 2$ mg/m^3. This makes sense, because in the absence of the smoker and grill loads, the only sources are the air intakes which are at 2 mg/m^3.

Because all the foregoing calculations involved changing the forcing functions, it was not necessary to recompute the solution. However, if the mixing between the kids' area and zone 4 is decreased, the matrix is changed

$$\begin{bmatrix} 225 & 0 & -25 & 0 \\ 0 & 155 & 0 & -105 \\ -225 & 0 & 275 & -50 \\ 0 & -5 & -250 & 255 \end{bmatrix} \begin{Bmatrix} c_1 \\ c_2 \\ c_3 \\ c_4 \end{Bmatrix} = \begin{Bmatrix} 1400 \\ 100 \\ 2000 \\ 0 \end{Bmatrix}$$

The results for this case involve a new solution. Using MATLAB, the result is

$$\begin{Bmatrix} c_1 \\ c_2 \\ c_3 \\ c_4 \end{Bmatrix} = \begin{Bmatrix} 8.1084 \\ 12.0800 \\ 16.9760 \\ 16.8800 \end{Bmatrix}$$

Therefore, this remedy would only improve the kids' area concentration by a paltry 0.265 mg/m^3.

PROBLEMS

11.1 Determine the matrix inverse for the following system:

$$10x_1 + 2x_2 - x_3 = 27$$
$$-3x_1 - 6x_2 + 2x_3 = -61.5$$
$$x_1 + x_2 + 5x_3 = -21.5$$

Check your results by verifying that $[A][A]^{-1} = [I]$. Do not use a pivoting strategy.

11.2 Determine the matrix inverse for the following system:

$$-8x_1 + x_2 - 2x_3 = -20$$
$$2x_1 - 6x_2 - x_3 = -38$$
$$-3x_1 - x_2 + 7x_3 = -34$$

11.3 The following system of equations is designed to determine concentrations (the c's in g/m^3) in a series of coupled reactors as a function of the amount of mass input to each reactor (the right-hand sides in g/day):

$$15c_1 - 3c_2 - c_3 = 4000$$
$$-3c_1 + 18c_2 - 6c_3 = 1200$$
$$-4c_1 - c_2 + 12c_3 = 2350$$

(a) Determine the matrix inverse.
(b) Use the inverse to determine the solution.
(c) Determine how much the rate of mass input to reactor 3 must be increased to induce a 10 g/m³ rise in the concentration of reactor 1.
(d) How much will the concentration in reactor 3 be reduced if the rate of mass input to reactors 1 and 2 is reduced by 500 and 250 g/day, respectively?

11.4 Determine the matrix inverse for the system described in Prob. 8.9. Use the matrix inverse to determine the concentration in reactor 5 if the inflow concentrations are changed to $c_{01} = 20$ and $c_{03} = 50$.

11.5 Determine the matrix inverse for the system described in Prob. 8.10. Use the matrix inverse to determine the force in the three members (F_1, F_2, and F_3) if the vertical load at node 1 is doubled to $F_{1,v} = -2000$ N and a horizontal load of $F_{3,h} = -500$ N is applied to node 3.

11.6 Determine $\|A\|_f$, $\|A\|_1$, and $\|A\|_\infty$ for

$$[A] = \begin{bmatrix} 8 & 2 & -10 \\ -9 & 1 & 3 \\ 15 & -1 & 6 \end{bmatrix}$$

Before determining the norms, scale the matrix by making the maximum element in each row equal to one.

11.7 Determine the Frobenius and row-sum norms for the systems in Probs. 11.2 and 11.3.

11.8 Use MATLAB to determine the spectral condition number for the following system. Do not normalize the system:

$$\begin{bmatrix} 1 & 4 & 9 & 16 & 25 \\ 4 & 9 & 16 & 25 & 36 \\ 9 & 16 & 25 & 36 & 49 \\ 16 & 25 & 36 & 49 & 64 \\ 25 & 36 & 49 & 64 & 81 \end{bmatrix}$$

Compute the condition number based on the row-sum norm.

11.9 Besides the Hilbert matrix, there are other matrices that are inherently ill-conditioned. One such case is the *Vandermonde matrix*, which has the following form:

$$\begin{bmatrix} x_1^2 & x_1 & 1 \\ x_2^2 & x_2 & 1 \\ x_3^2 & x_3 & 1 \end{bmatrix}$$

(a) Determine the condition number based on the row-sum norm for the case where $x_1 = 4$, $x_2 = 2$, and $x_3 = 7$.
(b) Use MATLAB to compute the spectral and Frobenius condition numbers.

11.10 Use MATLAB to determine the spectral condition number for a 10-dimensional Hilbert matrix. How many digits of precision are expected to be lost due to ill-conditioning? Determine the solution for this system for the case where each element of the right-hand-side vector $\{b\}$ consists of the summation of the coefficients in its row. In other words, solve for the case where all the unknowns should be exactly one. Compare the resulting errors with those expected based on the condition number.

11.11 Repeat Prob. 11.10, but for the case of a six-dimensional Vandermonde matrix (see Prob. 11.9) where $x_1 = 4$, $x_2 = 2$, $x_3 = 7$, $x_4 = 10$, $x_5 = 3$, and $x_6 = 5$.

11.12 The Lower Colorado River consists of a series of four reservoirs as shown in Fig. P11.12.

Mass balances can be written for each reservoir, and the following set of simultaneous linear algebraic equations results:

$$\begin{bmatrix} 13.422 & 0 & 0 & 0 \\ -13.422 & 12.252 & 0 & 0 \\ 0 & -12.252 & 12.377 & 0 \\ 0 & 0 & -12.377 & 11.797 \end{bmatrix}$$

$$\times \begin{Bmatrix} c_1 \\ c_2 \\ c_3 \\ c_4 \end{Bmatrix} = \begin{Bmatrix} 750.5 \\ 300 \\ 102 \\ 30 \end{Bmatrix}$$

where the right-hand-side vector consists of the loadings of chloride to each of the four lakes and c_1, c_2, c_3, and c_4 = the resulting chloride concentrations for Lakes Powell, Mead, Mohave, and Havasu, respectively.

(a) Use the matrix inverse to solve for the concentrations in each of the four lakes.
(b) How much must the loading to Lake Powell be reduced for the chloride concentration of Lake Havasu to be 75?
(c) Using the column-sum norm, compute the condition number and how many suspect digits would be generated by solving this system.

11.13 (a) Determine the matrix inverse and condition number for the following matrix:

$$\begin{bmatrix} 1 & 2 & 3 \\ 4 & 5 & 6 \\ 7 & 8 & 9 \end{bmatrix}$$

(b) Repeat (a) but change a_{33} slightly to 9.1.

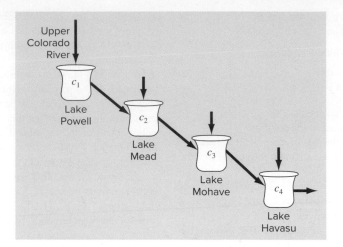

FIGURE P11.12
The Lower Colorado River.

11.14 Polynomial interpolation consists of determining the unique $(n-1)$th-order polynomial that fits n data points. Such polynomials have the general form,

$$f(x) = p_1 x^{n-1} + p_2 x^{n-2} + \cdots + p_{n-1} x + p_n \qquad \text{(P11.14)}$$

where the p's are constant coefficients. A straightforward way for computing the coefficients is to generate n linear algebraic equations that we can solve simultaneously for the coefficients. Suppose that we want to determine the coefficients of the fourth-order polynomial $f(x) = p_1 x^4 + p_2 x^3 + p_3 x^2 + p_4 x + p_5$ that passes through the following five points: $(200, 0.746)$, $(250, 0.675)$, $(300, 0.616)$, $(400, 0.525)$, and $(500, 0.457)$. Each of these pairs can be substituted into Eq. (P11.14) to yield a system of five equations with five unknowns (the p's). Use this approach to solve for the coefficients. In addition, determine and interpret the condition number.

11.15 A chemical constituent flows between three reactors as depicted in Fig. P11.15. Steady-state mass balances can be written for a substance that reacts with first-order kinetics. For example, the mass balance for reactor 1 is

$$Q_{1,in}c_{1,in} - Q_{1,2}c_1 - Q_{1,3}c_1 + Q_{2,1}c_2 - kV_1c_1 = 0 \qquad \text{(P11.15)}$$

where $Q_{1,in}$ = the volumetric inflow to reactor 1 (m³/min), $c_{1,in}$ = the inflow concentration to reactor 1 (g/m³), $Q_{i,j}$ = the flow from reactor i to reactor j (m³/min), c_i = the concentration of reactor i (g/m³), k = a first-order decay rate (/min), and V_i = the volume of reactor i (m³).

(a) Write the mass balances for reactors 2 and 3.

(b) If $k = 0.1$/min, write the mass balances for all three reactors as a system of linear algebraic equations.

(c) Compute the LU decomposition for this system.

(d) Use the LU decomposition to compute the matrix inverse.

(e) Use the matrix inverse to answer the following questions: (*i*) What are the steady-state concentrations for the three reactors? (*ii*) If the inflow concentration to the second reactor is set to zero, what is the resulting reduction in concentration of reactor 1? (*iii*) If the inflow concentration to reactor 1 is doubled, and the inflow concentration to reactor 2 is halved, what is the concentration of reactor 3?

FIGURE P11.15

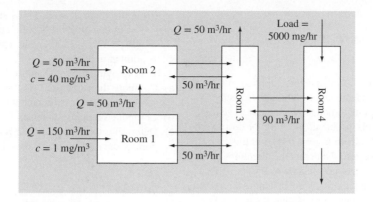

FIGURE P11.18

11.16 As described in Examples 8.2 and 11.2, use the matrix inverse to answer the following:
(a) Determine the change in position of the first jumper, if the mass of the third jumper is increased to 100 kg.
(b) What force must be applied to the third jumper so that the final position of the third jumper is 140 m?

11.17 Determine the matrix inverse for the electric circuit formulated in Sec. 8.3. Use the inverse to determine the new current between nodes 2 and 5 (i_{52}), if a voltage of 200 V is applied at node 6 and the voltage at node 1 is halved.

11.18 **(a)** Using the same approach as described in Sec. 11.3, develop steady-state mass balances for the room configuration depicted in Fig. P11.18.
(b) Determine the matrix inverse and use it to calculate the resulting concentrations in the rooms.
(c) Use the matrix inverse to determine how much the room 4 load must be reduced to maintain a concentration of 20 mg/m³ in room 2.

11.19 Write your own well-structured MATLAB Function procedure named Fnorm to calculate the Frobenius norm of an $m \times n$ matrix with for...end loops,

$$\|A\|_f = \sqrt{\sum_{i=1}^{m} \sum_{j=1}^{n} a_{i,j}^2}$$

Have the function scale the matrix before computing the norm. Test your function with the following script:

```
A = [5 7 -9; 1 8 4; 7 6 2];
Fn = Fnorm(A)
```

Here is the first line of your function

```
function Norm = Fnorm(x)
```

11.20 Figure P11.20 shows a statically determinate truss.

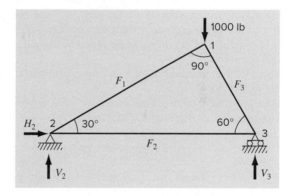

FIGURE P11.20
Forces on a statically determinate truss.

This type of structure can be described as a system of coupled linear algebraic equations by developing free-body force diagrams for the forces at each node in Fig. P11.20. The sum of the forces in both horizontal and vertical directions must be zero at each node, because the system is at rest. Therefore, for node 1,

$$F_H = 0 = -F_1 \cos 30° + F_3 \cos 60° + F_{1,h}$$

$$F_V = 0 = -F_1 \sin 30° - F_3 \sin 60° + F_{1,v}$$

for node 2,

$$F_H = 0 = F_2 + F_1 \cos 30° + F_{2,h} + H_2$$

$$F_V = 0 = F_1 \sin 30° + F_{2,v} + V_2$$

for node 3,

$$F_H = 0 = -F_2 - F_3 \cos 60° + F_{3,h}$$

$$F_V = 0 = F_3 \sin 60° + F_{3,v} + V_3$$

where $F_{i,h}$ is the external horizontal force applied to node i (where a positive force is from left to right) and $F_{1,v}$ is the external vertical force applied to node i (where a positive force is upward). Thus, in this problem, the 1000-N downward force on node 1 corresponds to $F_{1,v} = -1000$. For this case all other $F_{i,v}$'s and $F_{i,h}$'s are zero. Note that the directions of the internal forces and reactions are unknown. Proper application of Newton's laws requires only consistent assumptions regarding direction. Solutions are negative if the directions are assumed incorrectly. Also note that in this problem, the forces in all members are assumed to be in tension and act to pull adjoining nodes together. A negative solution therefore corresponds to compression. When the external forces are substituted and the trigonometric functions evaluated, this problem reduces to a set of six linear algebraic equations with six unknowns.

(a) Solve for the forces and reactions for the case displayed in Fig. P11.20.

(b) Determine the system's matrix inverse. What is your interpretation of the zeros in the second row of the inverse?

(c) Use the elements matrix inverse to answer the following questions:

(i) If the force at node 1 was reversed (i.e., directed upward), compute the impact on H_2 and V_2.

(ii) If the force at node 1 was set to zero and horizontal forces of 1500 N were applied at nodes 1 and 2 ($F_{i,h} = F_{2,h} = 1500$), what would be the vertical reaction at node 3 (V_3).

11.21 Employing the same approach as in Prob. 11.20,

(a) Compute the forces and reactions for the members and supports for the truss depicted in Fig. P11.21.

(b) Compute the matrix inverse.

(c) Determine the change in the reactions at the two supports if the force at the peak is directed upward.

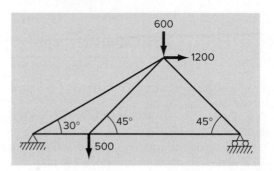

FIGURE P11.21

12

Iterative Methods

CHAPTER OBJECTIVES

The primary objective of this chapter is to acquaint you with iterative methods for solving simultaneous equations. Specific objectives and topics covered are

- Understanding the difference between the Gauss-Seidel and Jacobi methods.
- Knowing how to assess diagonal dominance and knowing what it means.
- Recognizing how relaxation can be used to improve the convergence of iterative methods.
- Understanding how to solve systems of nonlinear equations with successive substitution, Newton-Raphson, and the MATLAB fsolve function.

terative or approximate methods provide an alternative to the elimination methods described to this point. Such approaches are similar to the techniques we developed to obtain the roots of a single equation in Chaps. 5 and 6. Those approaches consisted of guessing a value and then using a systematic method to obtain a refined estimate of the root. Because the present part of the book deals with a similar problem—obtaining the values that simultaneously satisfy a set of equations—we might suspect that such approximate methods could be useful in this context. In this chapter, we will present approaches for solving both linear and nonlinear simultaneous equations.

12.1 LINEAR SYSTEMS: GAUSS-SEIDEL

The *Gauss-Seidel method* is the most commonly used iterative method for solving linear algebraic equations. Assume that we are given a set of n equations:

$$[A]\{x\} = \{b\}$$

Suppose that for conciseness we limit ourselves to a 3×3 set of equations. If the diagonal elements are all nonzero, the first equation can be solved for x_1, the second for x_2, and the third for x_3 to yield

$$x_1^j = \frac{b_1 - a_{12}x_2^{j-1} - a_{13}x_3^{j-1}}{a_{11}} \tag{12.1a}$$

$$x_2^j = \frac{b_2 - a_{21}x_1^j - a_{23}x_3^{j-1}}{a_{22}} \tag{12.1b}$$

$$x_3^j = \frac{b_3 - a_{31}x_1^j - a_{32}x_2^j}{a_{33}} \tag{12.1c}$$

where j and $j-1$ are the present and previous iterations.

To start the solution process, initial guesses must be made for the x's. A simple approach is to assume that they are all zero. These zeros can be substituted into Eq. (12.1a), which can be used to calculate a new value for $x_1 = b_1/a_{11}$. Then we substitute this new value of x_1 along with the previous guess of zero for x_3 into Eq. (12.1b) to compute a new value for x_2. The process is repeated for Eq. (12.1c) to calculate a new estimate for x_3. Then we return to the first equation and repeat the entire procedure until our solution converges closely enough to the true values. Convergence can be checked using the criterion that for all i,

$$\varepsilon_{a,i} = \left| \frac{x_i^j - x_i^{j-1}}{x_i^j} \right| \times 100\% \le \varepsilon_s \tag{12.2}$$

EXAMPLE 12.1   Gauss-Seidel Method

Problem Statement. Use the Gauss-Seidel method to obtain the solution for

$$3x_1 - 0.1x_2 - 0.2x_3 = 7.85$$
$$0.1x_1 + 7x_2 - 0.3x_3 = -19.3$$
$$0.3x_1 - 0.2x_2 + 10x_3 = 71.4$$

Note that the solution is $x_1 = 3$, $x_2 = -2.5$, and $x_3 = 7$.

Solution. First, solve each of the equations for its unknown on the diagonal:

$$x_1 = \frac{7.85 + 0.1x_2 + 0.2x_3}{3} \tag{E12.1.1}$$

$$x_2 = \frac{-19.3 - 0.1x_1 + 0.3x_3}{7} \tag{E12.1.2}$$

$$x_3 = \frac{71.4 - 0.3x_1 + 0.2x_2}{10} \tag{E12.1.3}$$

By assuming that x_2 and x_3 are zero, Eq. (E12.1.1) can be used to compute

$$x_1 = \frac{7.85 + 0.1(0) + 0.2(0)}{3} = 2.616667$$

This value, along with the assumed value of $x_3 = 0$, can be substituted into Eq. (E12.1.2) to calculate

$$x_2 = \frac{-19.3 - 0.1(2.616667) + 0.3(0)}{7} = -2.794524$$

The first iteration is completed by substituting the calculated values for x_1 and x_2 into Eq. (E12.1.3) to yield

$$x_3 = \frac{71.4 - 0.3(2.616667) + 0.2(-2.794524)}{10} = 7.005610$$

For the second iteration, the same process is repeated to compute

$$x_1 = \frac{7.85 + 0.1(-2.794524) + 0.2(7.005610)}{3} = 2.990557$$

$$x_2 = \frac{-19.3 - 0.1(2.990557) + 0.3(7.005610)}{7} = -2.499625$$

$$x_3 = \frac{71.4 - 0.3(2.990557) + 0.2(-2.499625)}{10} = 7.000291$$

The method is, therefore, converging on the true solution. Additional iterations could be applied to improve the answers. However, in an actual problem, we would not know the true answer *a priori*. Consequently, Eq. (12.2) provides a means to estimate the error. For example, for x_1:

$$\varepsilon_{a,1} = \left| \frac{2.990557 - 2.616667}{2.990557} \right| \times 100\% = 12.5\%$$

For x_2 and x_3, the error estimates are $\varepsilon_{a,2} = 11.8\%$ and $\varepsilon_{a,3} = 0.076\%$. Note that, as was the case when determining roots of a single equation, formulations such as Eq. (12.2) usually provide a conservative appraisal of convergence. Thus, when they are met, they ensure that the result is known to at least the tolerance specified by ε_s.

As each new x value is computed for the Gauss-Seidel method, it is immediately used in the next equation to determine another x value. Thus, if the solution is converging, the best available estimates will be employed. An alternative approach, called *Jacobi itera-tion,* utilizes a somewhat different tactic. Rather than using the latest available x's, this technique uses Eq. (12.1) to compute a set of new x's on the basis of a set of old x's. Thus, as new values are generated, they are not immediately used but rather are retained for the next iteration.

The difference between the Gauss-Seidel method and Jacobi iteration is depicted in Fig. 12.1. Although there are certain cases where the Jacobi method is useful, Gauss-Seidel's utilization of the best available estimates usually makes it the method of preference.

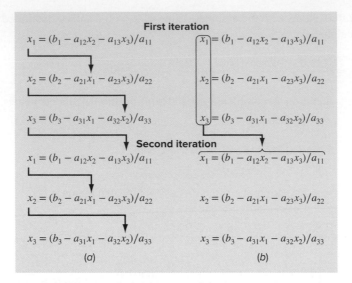

FIGURE 12.1
Graphical depiction of the difference between (a) the Gauss-Seidel and (b) the Jacobi iterative methods for solving simultaneous linear algebraic equations.

12.1.1 Convergence and Diagonal Dominance

Note that the Gauss-Seidel method is similar in spirit to the technique of simple fixed-point iteration that was used in Sec. 6.1 to solve for the roots of a single equation. Recall that simple fixed-point iteration was sometimes nonconvergent. That is, as the iterations progressed, the answer moved farther and farther from the correct result.

Although the Gauss-Seidel method can also diverge, because it is designed for linear systems, its ability to converge is much more predictable than for fixed-point iteration of nonlinear equations. It can be shown that if the following condition holds, Gauss-Seidel will converge:

$$|a_{ii}| > \sum_{\substack{j=1 \\ j \neq i}}^{n} |a_{ij}| \tag{12.3}$$

That is, the absolute value of the diagonal coefficient in each of the equations must be larger than the sum of the absolute values of the other coefficients in the equation. Such systems are said to be *diagonally dominant*. This criterion is sufficient but not necessary for convergence. That is, although the method may sometimes work if Eq. (12.3) is not met, convergence is guaranteed if the condition is satisfied. Fortunately, many engineering and scientific problems of practical importance fulfill this requirement. Therefore, Gauss-Seidel represents a feasible approach to solve many problems in engineering and science.

12.1.2 MATLAB M-file: GaussSeidel

Before developing an algorithm, let us first recast Gauss-Seidel in a form that is compatible with MATLAB's ability to perform matrix operations. This is done by expressing Eq. (12.1) as

$$x_1^{\text{new}} = \frac{b_1}{a_{11}} \qquad\qquad\qquad - \frac{a_{12}}{a_{11}}x_2^{\text{old}} - \frac{a_{13}}{a_{11}}x_3^{\text{old}}$$

$$x_2^{\text{new}} = \frac{b_2}{a_{22}} - \frac{a_{21}}{a_{22}}x_1^{\text{new}} \qquad\qquad - \frac{a_{23}}{a_{22}}x_3^{\text{old}}$$

$$x_3^{\text{new}} = \frac{b_3}{a_{33}} - \frac{a_{31}}{a_{33}}x_1^{\text{new}} \qquad - \frac{a_{32}}{a_{33}}x_2^{\text{new}}$$

Notice that the solution can be expressed concisely in matrix form as

$$\{x\} = \{d\} - [C]\{x\} \tag{12.4}$$

where

$$\{d\} = \begin{Bmatrix} b_1/a_{11} \\ b_2/a_{22} \\ b_3/a_{33} \end{Bmatrix}$$

and

$$[C] = \begin{bmatrix} 0 & a_{12}/a_{11} & a_{13}/a_{11} \\ a_{21}/a_{22} & 0 & a_{23}/a_{22} \\ a_{31}/a_{33} & a_{32}/a_{33} & 0 \end{bmatrix}$$

An M-file to implement Eq. (12.4) is listed in Fig. 12.2.

12.1.3 Relaxation

Relaxation represents a slight modification of the Gauss-Seidel method that is designed to enhance convergence. After each new value of x is computed using Eq. (12.1), that value is modified by a weighted average of the results of the previous and the present iterations:

$$x_i^{\text{new}} = \lambda x_i^{\text{new}} + (1 - \lambda)x_i^{\text{old}} \tag{12.5}$$

where λ is a weighting factor that is assigned a value between 0 and 2.

If $\lambda = 1$, $(1 - \lambda)$ is equal to 0 and the result is unmodified. However, if λ is set at a value between 0 and 1, the result is a weighted average of the present and the previous results. This type of modification is called *underrelaxation*. It is typically employed to make a nonconvergent system converge or to hasten convergence by dampening out oscillations.

For values of λ from 1 to 2, extra weight is placed on the present value. In this instance, there is an implicit assumption that the new value is moving in the correct direction toward the true solution but at too slow a rate. Thus, the added weight of λ is intended to improve the estimate by pushing it closer to the truth. Hence, this type of modification, which is called *overrelaxation*, is designed to accelerate the convergence of an already convergent system. The approach is also called *successive overrelaxation*, or SOR.

```
function x = GaussSeidel(A,b,es,maxit)
% GaussSeidel: Gauss Seidel method
%   x = GaussSeidel(A,b): Gauss Seidel without relaxation
% input:
%   A = coefficient matrix
%   b = right hand side vector
%   es = stop criterion (default = 0.00001%)
%   maxit = max iterations (default = 50)
% output:
%   x = solution vector

if nargin<2,error('at least 2 input arguments required'),end
if nargin<4|isempty(maxit),maxit=50;end
if nargin<3|isempty(es),es=0.00001;end
[m,n] = size(A);
if m~=n, error('Matrix A must be square'); end
C = A;
for i = 1:n
  C(i,i) = 0;
  x(i) = 0;
end
x = x';
for i = 1:n
  C(i,1:n) = C(i,1:n)/A(i,i);
end
for i = 1:n
  d(i) = b(i)/A(i,i);
end
iter = 0;
while (1)
  xold = x;
  for i = 1:n
    x(i) = d(i)-C(i,:)*x;
    if x(i) ~= 0
      ea(i) = abs((x(i) - xold(i))/x(i)) * 100;
    end
  end
  iter = iter+1;
  if max(ea)<=es | iter >= maxit, break, end
end
```

FIGURE 12.2
MATLAB M-file to implement Gauss-Seidel.

The choice of a proper value for λ is highly problem-specific and is often determined empirically. For a single solution of a set of equations it is often unnecessary. However, if the system under study is to be solved repeatedly, the efficiency introduced by a wise choice of λ can be extremely important. Good examples are the very large systems of linear algebraic equations that can occur when solving partial differential equations in a variety of engineering and scientific problem contexts.

EXAMPLE 12.2   Gauss-Seidel Method with Relaxation

Problem Statement. Solve the following system with Gauss-Seidel using overrelaxation $(\lambda = 1.2)$ and a stopping criterion of $\varepsilon_s = 10\%$:

$$-3x_1 + 12x_2 = 9$$
$$10x_1 - 2x_2 = 8$$

Solution. First rearrange the equations so that they are diagonally dominant and solve the first equation for x_1 and the second for x_2:

$$x_1 = \frac{8 + 2x_2}{10} = 0.8 + 0.2x_2$$

$$x_2 = \frac{9 + 3x_1}{12} = 0.75 + 0.25x_1$$

First iteration: Using initial guesses of $x_1 = x_2 = 0$, we can solve for x_1:

$$x_1 = 0.8 + 0.2(0) = 0.8$$

Before solving for x_2, we first apply relaxation to our result for x_1:

$$x_{1,r} = 1.2(0.8) - 0.2(0) = 0.96$$

We use the subscript r to indicate that this is the "relaxed" value. This result is then used to compute x_2:

$$x_2 = 0.75 + 0.25(0.96) = 0.99$$

We then apply relaxation to this result to give

$$x_{2,r} = 1.2(0.99) - 0.2(0) = 1.188$$

At this point, we could compute estimated errors with Eq. (12.2). However, since we started with assumed values of zero, the errors for both variables will be 100%.

Second iteration: Using the same procedure as for the first iteration, the second iteration yields

$$x_1 = 0.8 + 0.2(1.188) = 1.0376$$
$$x_{1,r} = 1.2(1.0376) - 0.2(0.96) = 1.05312$$
$$\varepsilon_{a,1} = \left| \frac{1.05312 - 0.96}{1.05312} \right| \times 100\% = 8.84\%$$
$$x_2 = 0.75 + 0.25(1.05312) = 1.01328$$
$$x_{2,r} = 1.2(1.01328) - 0.2(1.188) = 0.978336$$
$$\varepsilon_{a,2} = \left| \frac{0.978336 - 1.188}{0.978336} \right| \times 100\% = 21.43\%$$

Because we have now have nonzero values from the first iteration, we can compute approximate error estimates as each new value is computed. At this point, although the error estimate for the first unknown has fallen below the 10% stopping criterion, the second has not. Hence, we must implement another iteration.

Third iteration:

$$x_1 = 0.8 + 0.2(0.978336) = 0.995667$$

$$x_{1,r} = 1.2(0.995667) - 0.2(1.05312) = 0.984177$$

$$\varepsilon_{a,1} = \left| \frac{0.984177 - 1.05312}{0.984177} \right| \times 100\% = 7.01\%$$

$$x_2 = 0.75 + 0.25(0.984177) = 0.996044$$

$$x_{2,r} = 1.2(0.996044) - 0.2(0.978336) = 0.999586$$

$$\varepsilon_{a,2} = \left| \frac{0.999586 - 0.978336}{0.999586} \right| \times 100\% = 2.13\%$$

At this point, we can terminate the computation because both error estimates have fallen below the 10% stopping criterion. The results at this juncture, $x_1 = 0.984177$ and $x_2 = 0.999586$, are converging on the exact solution of $x_1 = x_2 = 1$.

12.2   NONLINEAR SYSTEMS

The following is a set of two simultaneous nonlinear equations with two unknowns:

$$x_1^2 + x_1 x_2 = 10 \tag{12.6a}$$

$$x_2 + 3x_1 x_2^2 = 57 \tag{12.6b}$$

In contrast to linear systems which plot as straight lines (recall Fig. 9.1), these equations plot as curves on an x_2 versus x_1 graph. As in Fig. 12.3, the solution is the intersection of the curves.

FIGURE 12.3
Graphical depiction of the solution of two simultaneous nonlinear equations.

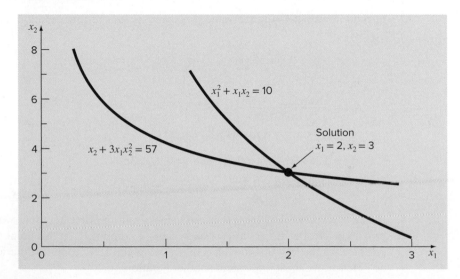

Just as we did when we determined roots for single nonlinear equations, such systems of equations can be expressed generally as

$$f_1(x_1, x_2, \ldots, x_n) = 0$$
$$f_2(x_1, x_2, \ldots, x_n) = 0$$
$$\vdots \tag{12.7}$$
$$f_n(x_1, x_2, \ldots, x_n) = 0$$

Therefore, the solution are the values of the x's that make the equations equal to zero.

12.2.1 Successive Substitution

A simple approach for solving Eq. (12.7) is to use the same strategy that was employed for fixed-point iteration and the Gauss-Seidel method. That is, each one of the nonlinear equations can be solved for one of the unknowns. These equations can then be implemented iteratively to compute new values which (hopefully) will converge on the solutions. This approach, which is called *successive substitution,* is illustrated in the following example.

EXAMPLE 12.3   Successive Substitution for a Nonlinear System

Problem Statement. Use successive substitution to determine the roots of Eq. (12.6). Note that a correct pair of roots is $x_1 = 2$ and $x_2 = 3$. Initiate the computation with guesses of $x_1 = 1.5$ and $x_2 = 3.5$.

Solution. Equation (12.6a) can be solved for

$$x_1 = \frac{10 - x_1^2}{x_2} \tag{E12.3.1}$$

and Eq. (12.6b) can be solved for

$$x_2 = 57 - 3x_1 x_2^2 \tag{E12.3.2}$$

On the basis of the initial guesses, Eq. (E12.3.1) can be used to determine a new value of x_1:

$$x_1 = \frac{10 - (1.5)^2}{3.5} = 2.21429$$

This result and the initial value of $x_2 = 3.5$ can be substituted into Eq. (E12.3.2) to determine a new value of x_2:

$$x_2 = 57 - 3(2.21429)(3.5)^2 = -24.37516$$

Thus, the approach seems to be diverging. This behavior is even more pronounced on the second iteration:

$$x_1 = \frac{10 - (2.21429)^2}{-24.37516} = -0.20910$$

$$x_2 = 57 - 3(-0.20910)(-24.37516)^2 = 429.709$$

Obviously, the approach is deteriorating.

Now we will repeat the computation but with the original equations set up in a different format. For example, an alternative solution of Eq. (12.6a) is

$$x_1 = \sqrt{10 - x_1 x_2}$$

and of Eq. (12.6b) is

$$x_2 = \sqrt{\frac{57 - x_2}{3x_1}}$$

Now the results are more satisfactory:

$$x_1 = \sqrt{10 - 1.5(3.5)} = 2.17945$$

$$x_2 = \sqrt{\frac{57 - 3.5}{3(2.17945)}} = 2.86051$$

$$x_1 = \sqrt{10 - 2.17945(2.86051)} = 1.94053$$

$$x_2 = \sqrt{\frac{57 - 2.86051}{3(1.94053)}} = 3.04955$$

Thus, the approach is converging on the true values of $x_1 = 2$ and $x_2 = 3$.

The previous example illustrates the most serious shortcoming of successive substitution—that is, convergence often depends on the manner in which the equations are formulated. Additionally, even in those instances where convergence is possible, divergence can occur if the initial guesses are insufficiently close to the true solution. These criteria are so restrictive that fixed-point iteration has limited utility for solving nonlinear systems.

12.2.2 Newton-Raphson

Just as fixed-point iteration can be used to solve systems of nonlinear equations, other open root location methods such as the Newton-Raphson method can be used for the same purpose. Recall that the Newton-Raphson method was predicated on employing the derivative (i.e., the slope) of a function to estimate its intercept with the axis of the independent variable—that is, the root. In Chap. 6, we used a graphical derivation to compute this estimate. An alternative is to derive it from a first-order Taylor series expansion:

$$f(x_{i+1}) = f(x_i) + (x_{i+1} - x_i) f'(x_i) \tag{12.8}$$

where x_i is the initial guess at the root and x_{i+1} is the point at which the slope intercepts the x axis. At this intercept, $f(x_{i+1})$ by definition equals zero and Eq. (12.8) can be rearranged to yield

$$x_{i+1} = x_i - \frac{f(x_i)}{f'(x_i)} \tag{12.9}$$

which is the single-equation form of the Newton-Raphson method.

The multiequation form is derived in an identical fashion. However, a multivariable Taylor series must be used to account for the fact that more than one independent variable contributes to the determination of the root. For the two-variable case, a first-order Taylor series can be written for each nonlinear equation as

$$f_{1,i+1} = f_{1,i} + (x_{1,i+1} - x_{1,i})\frac{\partial f_{1,i}}{\partial x_1} + (x_{2,i+1} - x_{2,i})\frac{\partial f_{1,i}}{\partial x_2} \tag{12.10a}$$

$$f_{2,i+1} = f_{2,i} + (x_{1,i+1} - x_{1,i})\frac{\partial f_{2,i}}{\partial x_1} + (x_{2,i+1} - x_{2,i})\frac{\partial f_{2,i}}{\partial x_2} \tag{12.10b}$$

Just as for the single-equation version, the root estimate corresponds to the values of x_1 and x_2, where $f_{1,i+1}$ and $f_{2,i+1}$ equal zero. For this situation, Eq. (12.10) can be rearranged to give

$$\frac{\partial f_{1,i}}{\partial x_1}x_{1,i+1} + \frac{\partial f_{1,i}}{\partial x_2}x_{2,i+1} = -f_{1,i} + x_{1,i}\frac{\partial f_{1,i}}{\partial x_1} + x_{2,i}\frac{\partial f_{1,i}}{\partial x_2} \tag{12.11a}$$

$$\frac{\partial f_{2,i}}{\partial x_1}x_{1,i+1} + \frac{\partial f_{2,i}}{\partial x_2}x_{2,i+1} = -f_{2,i} + x_{1,i}\frac{\partial f_{2,i}}{\partial x_1} + x_{2,i}\frac{\partial f_{2,i}}{\partial x_2} \tag{12.11b}$$

Because all values subscripted with i's are known (they correspond to the latest guess or approximation), the only unknowns are $x_{1,i+1}$ and $x_{2,i+1}$. Thus, Eq. (12.11) is a set of two linear equations with two unknowns. Consequently, algebraic manipulations (e.g., Cramer's rule) can be employed to solve for

$$x_{1,i+1} = x_{1,i} - \frac{f_{1,i}\dfrac{\partial f_{2,i}}{\partial x_2} - f_{2,i}\dfrac{\partial f_{1,i}}{\partial x_2}}{\dfrac{\partial f_{1,i}}{\partial x_1}\dfrac{\partial f_{2,i}}{\partial x_2} - \dfrac{\partial f_{1,i}}{\partial x_2}\dfrac{\partial f_{2,i}}{\partial x_1}} \tag{12.12a}$$

$$x_{2,i+1} = x_{2,i} - \frac{f_{2,i}\dfrac{\partial f_{1,i}}{\partial x_1} - f_{1,i}\dfrac{\partial f_{2,i}}{\partial x_1}}{\dfrac{\partial f_{1,i}}{\partial x_1}\dfrac{\partial f_{2,i}}{\partial x_2} - \dfrac{\partial f_{1,i}}{\partial x_2}\dfrac{\partial f_{2,i}}{\partial x_1}} \tag{12.12b}$$

The denominator of each of these equations is formally referred to as the determinant of the *Jacobian* of the system.

Equation (12.12) is the two-equation version of the Newton-Raphson method. As in the following example, it can be employed iteratively to home in on the roots of two simultaneous equations.

EXAMPLE 12.4 Newton-Raphson for a Nonlinear System

Problem Statement. Use the multiple-equation Newton-Raphson method to determine roots of Eq. (12.6). Initiate the computation with guesses of $x_1 = 1.5$ and $x_2 = 3.5$.

Solution. First compute the partial derivatives and evaluate them at the initial guesses of x and y:

$$\frac{\partial f_{1,0}}{\partial x_1} = 2x_1 + x_2 = 2(1.5) + 3.5 = 6.5 \qquad \frac{\partial f_{1,0}}{\partial x_2} = x_1 = 1.5$$

$$\frac{\partial f_{2,0}}{\partial x_1} = 3x_2^2 = 3(3.5)^2 = 36.75 \qquad \frac{\partial f_{2,0}}{\partial x_2} = 1 + 6x_1x_2 = 1 + 6(1.5)(3.5) = 32.5$$

Thus, the determinant of the Jacobian for the first iteration is

$$6.5(32.5) - 1.5(36.75) = 156.125$$

The values of the functions can be evaluated at the initial guesses as

$$f_{1,0} = (1.5)^2 + 1.5(3.5) - 10 = -2.5$$

$$f_{2,0} = 3.5 + 3(1.5)(3.5)^2 - 57 = 1.625$$

These values can be substituted into Eq. (12.12) to give

$$x_1 = 1.5 - \frac{-2.5(32.5) - 1.625(1.5)}{156.125} = 2.03603$$

$$x_2 = 3.5 - \frac{1.625(6.5) - (-2.5)(36.75)}{156.125} = 2.84388$$

Thus, the results are converging to the true values of $x_1 = 2$ and $x_2 = 3$. The computation can be repeated until an acceptable accuracy is obtained.

When the multiequation Newton-Raphson works, it exhibits the same speedy quadratic convergence as the single-equation version. However, just as with successive substitution, it can diverge if the initial guesses are not sufficiently close to the true roots. Whereas graphical methods could be employed to derive good guesses for the single-equation case, no such simple procedure is available for the multiequation version. Although there are some advanced approaches for obtaining acceptable first estimates, often the initial guesses must be obtained on the basis of trial and error and knowledge of the physical system being modeled.

The two-equation Newton-Raphson approach can be generalized to solve n simultaneous equations. To do this, Eq. (12.11) can be written for the kth equation as

$$\frac{\partial f_{k,i}}{\partial x_1} x_{1,i+1} + \frac{\partial f_{k,i}}{\partial x_2} x_{2,i+1} + \cdots + \frac{\partial f_{k,i}}{\partial x_n} x_{n,i+1} = -f_{k,i} + x_{1,i} \frac{\partial f_{k,i}}{\partial x_1} + x_{2,i} \frac{\partial f_{k,i}}{\partial x_2}$$

$$+ \cdots + x_{n,i} \frac{\partial f_{k,i}}{\partial x_n} \qquad (12.13)$$

where the first subscript k represents the equation or unknown and the second subscript denotes whether the value or function in question is at the present value (i) or at the next value ($i + 1$). Notice that the only unknowns in Eq. (12.13) are the $x_{k,i+1}$ terms on the left-hand side. All other quantities are located at the present value (i) and, thus, are known at any iteration. Consequently, the set of equations generally represented by Eq. (12.13) (i.e., with $k = 1, 2, \ldots, n$) constitutes a set of linear simultaneous equations that can be solved numerically by the elimination methods elaborated in previous chapters.

Matrix notation can be employed to express Eq. (12.13) concisely as

$$[J]\{x_{i+1}\} = -\{f\} + [J]\{x_i\} \qquad (12.14)$$

where the partial derivatives evaluated at i are written as the *Jacobian matrix* consisting of the partial derivatives:

$$[J] = \begin{bmatrix} \dfrac{\partial f_{1,i}}{\partial x_1} & \dfrac{\partial f_{1,i}}{\partial x_2} & \cdots & \dfrac{\partial f_{1,i}}{\partial x_n} \\[2mm] \dfrac{\partial f_{2,i}}{\partial x_1} & \dfrac{\partial f_{2,i}}{\partial x_2} & \cdots & \dfrac{\partial f_{2,i}}{\partial x_n} \\[2mm] \vdots & \vdots & \vdots & \vdots \\[2mm] \dfrac{\partial f_{n,i}}{\partial x_1} & \dfrac{\partial f_{n,i}}{\partial x_2} & \cdots & \dfrac{\partial f_{n,i}}{\partial x_n} \end{bmatrix} \tag{12.15}$$

The initial and final values are expressed in vector form as

$$\{x_i\}^T = \lfloor x_{1,i} \quad x_{2,i} \quad \cdots \quad x_{n,i} \rfloor$$

and

$$\{x_{i+1}\}^T = \lfloor x_{1,i+1} \quad x_{2,i+1} \quad \cdots \quad x_{n,i+1} \rfloor$$

Finally, the function values at i can be expressed as

$$\{f\}^T = \lfloor f_{1,i} \quad f_{2,i} \quad \cdots \quad f_{n,i} \rfloor$$

Equation (12.14) can be solved using a technique such as Gauss elimination. This process can be repeated iteratively to obtain refined estimates in a fashion similar to the two-equation case in Example 12.4.

Insight into the solution can be obtained by solving Eq. (12.14) with matrix inversion. Recall that the single-equation version of the Newton-Raphson method is

$$x_{i+1} = x_i - \frac{f(x_i)}{f'(x_i)} \tag{12.16}$$

If Eq. (12.14) is solved by multiplying it by the inverse of the Jacobian, the result is

$$\{x_{i+1}\} = \{x_i\} - [J]^{-1}\{f\} \tag{12.17}$$

Comparison of Eqs. (12.16) and (12.17) clearly illustrates the parallels between the two equations. In essence, the Jacobian is analogous to the derivative of a multivariate function.

Such matrix calculations can be implemented very efficiently in MATLAB. We can illustrate this by using MATLAB to duplicate the calculations from Example 12.4. After defining the initial guesses, we can compute the Jacobian and the function values as

```
>> x=[1.5;3.5];
>> J=[2*x(1)+x(2) x(1);3*x(2)^2 1+6*x(1)*x(2)]

J =
     6.5000    1.5000
    36.7500   32.5000

>> f=[x(1)^2+x(1)*x(2)-10;x(2)+3*x(1)*x(2)^2-57]
```

```
f =
    -2.5000
     1.6250
```

Then, we can implement Eq. (12.17) to yield the improved estimates

```
>> x=x-J\f

x =
     2.0360
     2.8439
```

Although we could continue the iterations in the command mode, a nicer alternative is to express the algorithm as an M-file. As in Fig. 12.4, this routine is passed an M-file that computes the function values and the Jacobian at a given value of x. It then calls this function and implements Eq. (12.17) in an iterative fashion. The routine iterates until an upper limit of iterations (maxit) or a specified percent relative error (es) is reached.

FIGURE 12.4
MATLAB M-file to implement Newton-Raphson method for nonlinear systems of equations.

```
function [x,f,ea,iter]=newtmult(func,x0,es,maxit,varargin)
% newtmult: Newton-Raphson root zeroes nonlinear systems
%   [x,f,ea,iter]=newtmult(func,x0,es,maxit,p1,p2,...):
%     uses the Newton-Raphson method to find the roots of
%     a system of nonlinear equations
% input:
%   func = name of function that returns f and J
%   x0 = initial guess
%   es = desired percent relative error (default = 0.0001%)
%   maxit = maximum allowable iterations (default = 50)
%   p1,p2,... = additional parameters used by function
% output:
%   x = vector of roots
%   f = vector of functions evaluated at roots
%   ea = approximate percent relative error (%)
%   iter = number of iterations

if nargin<2,error('at least 2 input arguments required'),end
if nargin<3|isempty(es),es=0.0001;end
if nargin<4|isempty(maxit),maxit=50;end
iter = 0;
x=x0;
while (1)
  [J,f]=func(x,varargin{:});
  dx=J\f;
  x=x-dx;
  iter = iter + 1;
  ea=100*max(abs(dx./x));
  if iter>=maxit|ea<=es, break, end
end
```

We should note that there are two shortcomings to the foregoing approach. First, Eq. (12.15) is sometimes inconvenient to evaluate. Therefore, variations of the Newton-Raphson approach have been developed to circumvent this dilemma. As might be expected, most are based on using finite-difference approximations for the partial derivatives that comprise $[J]$. The second shortcoming of the multiequation Newton-Raphson method is that excellent initial guesses are usually required to ensure convergence. Because these are sometimes difficult or inconvenient to obtain, alternative approaches that are slower than Newton-Raphson but which have better convergence behavior have been developed. One approach is to reformulate the nonlinear system as a single function:

$$F(x) = \sum_{i=1}^{n} [f_i(x_1, x_2, \ldots, x_n)]^2$$

where $f_i(x_1, x_2, \ldots, x_n)$ is the ith member of the original system of Eq. (12.7). The values of x that minimize this function also represent the solution of the nonlinear system. Therefore, nonlinear optimization techniques can be employed to obtain solutions.

12.2.3 MATLAB Function: fsolve

The fsolve function solves systems of nonlinear equations with several variables. A general representation of its syntax is

```
[x, fx] = fsolve(function, x0, options)
```

where $[x, fx]$ = a vector containing the roots x and a vector containing the values of the functions evaluated at the roots, *function* = the name of the function containing a vector holding the equations being solved, *x0* is a vector holding the initial guesses for the unknowns, and *options* is a data structure created by the optimset function. Note that if you desire to pass function parameters but not use the options, pass an empty vector [] in its place.

The optimset function has the syntax

```
options = optimset('par₁',val₁,'par₂',val₂,...)
```

where the parameter *par*$_i$ has the value *val*$_i$. A complete listing of all the possible parameters can be obtained by merely entering *optimset* at the command prompt. The parameters commonly used with the *fsolve* function are

display: When set to 'iter' displays a detailed record of all the iterations.

tolx: A positive scalar that sets a termination tolerance on x.

tolfun: A positive scalar that sets a termination tolerance on fx.

As an example, we can solve the system from Eq. (12.6)

$$f(x_1, x_2) = 2x_1 + x_1 x_2 - 10$$
$$f(x_1, x_2) = x_2 + 3x_1 x_2^2 - 57$$

First, set up a function to hold the equations

```
function f = fun(x)
f = [x(1)^2+x(1)*x(2)-10;x(2)+3*x(1)*x(2)^2-57];
```

A script can then be used to generate the solution,

```
clc, format compact
[x,fx] = fsolve(@fun,[1.5;3.5])
```

with the result

```
x =
    2.0000
    3.0000
fx =
    1.0e-13 *
        0
    0.1421
```

12.3 CASE STUDY   CHEMICAL REACTIONS

Background. Nonlinear systems of equations occur frequently in the characterization of chemical reactions. For example, the following chemical reactions take place in a closed system:

$$2A + B \underset{\leftarrow}{\rightarrow} C \tag{12.18}$$

$$A + D \underset{\leftarrow}{\rightarrow} C \tag{12.19}$$

At equilibrium, they can be characterized by

$$K_1 = \frac{c_c}{c_a^2 c_b} \tag{12.20}$$

$$K_2 = \frac{c_c}{c_a c_d} \tag{12.21}$$

where the nomenclature c_i represents the concentration of constituent i. If x_1 and x_2 are the number of moles of C that are produced due to the first and second reactions, respectively, formulate the equilibrium relationships as a pair of two simultaneous nonlinear equations. If $K_1 = 4 \times 10^{-4}$, $K_2 = 3.7 \times 10^{-2}$, $c_{a,0} = 50$, $c_{b,0} = 20$, $c_{c,0} = 5$, and $c_{d,0} = 10$, employ the Newton-Raphson method to solve these equations.

Solution. Using the stoichiometry of Eqs. (12.18) and (12.19), the concentrations of each constituent can be represented in terms of x_1 and x_2 as

$$c_a = c_{a,0} - 2x_1 - x_2 \tag{12.22}$$

$$c_b = c_{b,0} - x_1 \tag{12.23}$$

$$c_c = c_{c,0} + x_1 + x_2 \tag{12.24}$$

$$c_d = c_{d,0} - x_2 \tag{12.25}$$

12.3 CASE STUDY continued

where the subscript 0 designates the initial concentration of each constituent. These values can be substituted into Eqs. (12.20) and (12.21) to give

$$K_1 = \frac{(c_{c,0} + x_1 + x_2)}{(c_{a,0} - 2x_1 - x_2)^2 (c_{b,0} - x_1)}$$

$$K_2 = \frac{(c_{c,0} + x_1 + x_2)}{(c_{a,0} - 2x_1 - x_2)(c_{d,0} - x_2)}$$

Given the parameter values, these are two nonlinear equations with two unknowns. Thus, the solution to this problem involves determining the roots of

$$f_1(x_1, x_2) = \frac{5 + x_1 + x_2}{(50 - 2x_1 - x_2)^2 (20 - x_1)} - 4 \times 10^{-4} \tag{12.26}$$

$$f_2(x_1, x_2) = \frac{5 + x_1 + x_2}{(50 - 2x_1 - x_2)(10 - x_2)} - 3.7 \times 10^{-2} \tag{12.27}$$

In order to use Newton-Raphson, we must determine the Jacobian by taking the partial derivatives of Eqs. (12.26) and (12.27). Although this is certainly possible, evaluating the derivatives is time consuming. An alternative is to represent them by finite differences in a fashion similar to the approach used for the modified secant method in Sec. 6.3. For example, the partial derivatives comprising the Jacobian can be evaluated as

$$\frac{\partial f_1}{\partial x_1} = \frac{f_1(x_1 + \delta x_1, x_2) - f_1(x_1, x_2)}{\delta x_1} \qquad \frac{\partial f_1}{\partial x_2} = \frac{f_1(x_1, x_2 + \delta x_2) - f_1(x_1, x_2)}{\delta x_2}$$

$$\frac{\partial f_2}{\partial x_1} = \frac{f_2(x_1 + \delta x_1, x_2) - f_2(x_1, x_2)}{\delta x_1} \qquad \frac{\partial f_2}{\partial x_2} = \frac{f_2(x_1, x_2 + \delta x_2) - f_2(x_1, x_2)}{\delta x_2}$$

These relationships can then be expressed as an M-file to compute both the function values and the Jacobian as

```
function [J,f]=jfreact(x,varargin)
del=0.000001;
df1dx1 = (u(x(1)+del*x(1),x(2))-u(x(1),x(2)))/(del*x(1));
df1dx2 = (u(x(1),x(2)+del*x(2))-u(x(1),x(2)))/(del*x(2));
df2dx1 = (v(x(1)+del*x(1),x(2))-v(x(1),x(2)))/(del*x(1));
df2dx2 = (v(x(1),x(2)+del*x(2))-v(x(1),x(2)))/(del*x(2));
J=[df1dx1 df1dx2;df2dx1 df2dx2];
f1=u(x(1),x(2));
f2=v(x(1),x(2));
f=[f1;f2];

function f=u(x,y)
f = (5 + x + y) / (50 - 2 * x - y) ^ 2 / (20 - x) - 0.0004;

function f=v(x,y)
f = (5 + x + y) / (50 - 2 * x - y) / (10 - y) - 0.037;
```

12.3 CASE STUDY   continued

The function newtmult (Fig. 12.4) can then be employed to determine the roots given initial guesses of $x_1 = x_2 = 3$:

```
>>> format short e, x0 =[3; 3];
>> [x,f,ea,iter]=newtmult(@jfreact,x0)

x =
  3.3366e+000
  2.6772e+000

f =
 -7.1286e-017
  8.5973e-014

ea =
  5.2237e-010

iter =
     4
```

After four iterations, a solution of $x_1 = 3.3366$ and $x_2 = 2.6772$ is obtained. These values can then be substituted into Eq. (12.22) through (12.25) to compute the equilibrium concentrations of the four constituents:

$$c_a = 50 - 2(3.3366) - 2.6772 = 40.6496$$

$$c_b = 20 - 3.3366 = 16.6634$$

$$c_c = 5 + 3.3366 + 2.6772 = 11.0138$$

$$c_d = 10 - 2.6772 = 7.3228$$

Finally, the fsolve function can also be used to obtain the solution by first writing a MATLAB file function to hold the system of nonlinear equations as a vector

```
function F=myfun(x)
F=[(5+x(1)+x(2))/(50-2*x(1)-x(2))^2/(20-x(1))-0.0004;...
    (5+x(1)+x(2))/(50-2*x(1)-x(2))/(10-x(2))-0.037];
```

The solution can then be generated by

```
[x,fx] = fsolve(@myfun,[3;3])
```

with the result

```
x =
    3.3372
    2.6834
fx =
   1.0e-04 *
    0.0041
    0.6087
```

PROBLEMS

12.1 Solve the following system using three iterations with Gauss-Seidel using overrelaxation ($\lambda = 1.25$). If necessary, rearrange the equations and show all the steps in your solution including your error estimates. At the end of the computation, compute the true error of your final results.

$$3x_1 + 8x_2 = 11$$
$$7x_1 - x_2 = 5$$

12.2 **(a)** Use the Gauss-Seidel method to solve the following system until the percent relative error falls below $\varepsilon_s = 5\%$:

$$\begin{bmatrix} 0.8 & -0.4 & \\ -0.4 & 0.8 & -0.4 \\ & -0.4 & 0.8 \end{bmatrix} \begin{Bmatrix} x_1 \\ x_2 \\ x_3 \end{Bmatrix} = \begin{Bmatrix} 41 \\ 25 \\ 105 \end{Bmatrix}$$

(b) Repeat **(a)** but use overrelaxation with $\lambda = 1.2$.

12.3 Use the Gauss-Seidel method to solve the following system until the percent relative error falls below $\varepsilon_s = 5\%$:

$$10x_1 + 2x_2 - x_3 = 27$$
$$-3x_1 - 6x_2 + 2x_3 = -61.5$$
$$x_1 + x_2 + 5x_3 = -21.5$$

12.4 Repeat Prob. 12.3 but use Jacobi iteration.

12.5 The following system of equations is designed to determine concentrations (the c's in g/m^3) in a series of coupled reactors as a function of the amount of mass input to each reactor (the right-hand sides in g/day):

$$15c_1 - 3c_2 - c_3 = 3800$$
$$-3c_1 + 18c_2 - 6c_3 = 1200$$
$$-4c_1 - c_2 + 12c_3 = 2350$$

Solve this problem with the Gauss-Seidel method to $\varepsilon_s = 5\%$.

12.6 Use the Gauss-Seidel method **(a)** without relaxation and **(b)** with relaxation ($\lambda = 1.2$) to solve the following system to a tolerance of $\varepsilon_s = 5\%$. If necessary, rearrange the equations to achieve convergence.

$$2x_1 - 6x_2 - x_3 = -38$$
$$-3x_1 - x_2 + 7x_3 = -34$$
$$-8x_1 + x_2 - 2x_3 = -20$$

12.7 Of the following three sets of linear equations, identify the set(s) that you could not solve using an iterative method such as Gauss-Seidel. Show using any number of iterations that is necessary that your solution does not converge. Clearly state your convergence criteria (how you know it is not converging).

Set One	Set Two	Set Three
$8x + 3y + z = 13$	$x + y + 6z = 8$	$-3x + 4y + 5z = 6$
$-6x + 8z = 2$	$x + 5y - z = 5$	$-2x + 2y - 3z = -3$
$2x + 5y - z = 6$	$4x + 2y - 2z = 4$	$2y - z = 1$

12.8 Determine the solution of the simultaneous nonlinear equations

$$y = -x^2 + x + 0.75$$
$$y + 5xy = x^2$$

Use the Newton-Raphson method and employ initial guesses of $x = y = 1.2$.

12.9 Determine the solution of the simultaneous nonlinear equations:

$$x^2 = 5 - y^2$$
$$y + 1 = x^2$$

(a) Graphically.
(b) Successive substitution using initial guesses of $x = y = 1.5$.
(c) Newton-Raphson using initial guesses of $x = y = 1.5$.

12.10 Figure P12.10 depicts a chemical exchange process consisting of a series of reactors in which a gas flowing from left to right is passed over a liquid flowing from right to left. The transfer of a chemical from the gas into the liquid occurs at a rate that is proportional to the difference between the gas and liquid concentrations in each reactor. At steady state, a mass balance for the first reactor can be written for the gas as

$$Q_G c_{G0} - Q_G c_{G1} + D(c_{L1} - c_{G1}) = 0$$

and for the liquid as

$$Q_L c_{L2} - Q_L c_{L1} + D(c_{G1} - c_{L1}) = 0$$

where Q_G and Q_L are the gas and liquid flow rates, respectively, and D = the gas-liquid exchange rate. Similar balances can be written for the other reactors. Use Gauss-Seidel without relaxation to solve for the concentrations given the following values: $Q_G = 2$, $Q_L = 1$, $D = 0.8$, $c_{G0} = 100$, $c_{L6} = 10$.

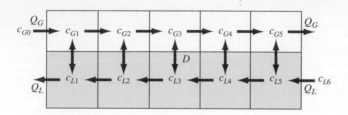

FIGURE P12.10

12.11 The steady-state distribution of temperature on a heated plate can be modeled by the *Laplace equation:*

$$0 = \frac{\partial^2 T}{\partial x^2} + \frac{\partial^2 T}{\partial y^2}$$

If the plate is represented by a series of nodes (Fig. P12.11), centered finite differences can be substituted for the second derivatives, which result in a system of linear algebraic equations. Use the Gauss-Seidel method to solve for the temperatures of the nodes in Fig. P12.11.

12.12 Develop your own M-file function for the Gauss-Seidel method without relaxation based on Fig. 12.2, but

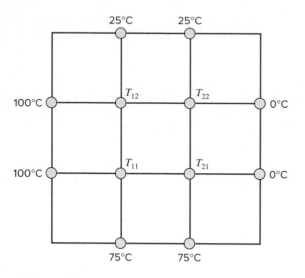

FIGURE P12.11

change the first line so that it returns the approximate error and the number of iterations:

```
function [x,ea,iter] = ...
    GaussSeidel(A,b,es, maxit)
```

Test it by duplicating Example 12.1 and then use it to solve Prob. 12.2*a*.

12.13 Develop your own M-file function for Gauss-Seidel with relaxation. Here is the function's first line:

```
function [x,ea,iter] = ...
    GaussSeidelR(A,b,lambda,es,maxit)
```

In the event that the user does not enter a value for λ, set the default value as $\lambda = 1$. Test it by duplicating Example 12.2 and then use it to solve Prob. 12.2*b*.

12.14 Develop your own M-file function for the Newton-Raphson method for nonlinear systems of equations based on Fig. 12.4. Test it by solving Example 12.4 and then use it to solve Prob. 12.8.

12.15 Determine the roots of the following simultaneous nonlinear equations using (**a**) fixed-point iteration, (**b**) the Newton-Raphson method, and (**c**) the fsolve function:

$$y = -x^2 + x + 0.75 \qquad\qquad y + 5xy = x^2$$

Employ initial guesses of $x = y = 1.2$ and discuss the results.

12.16 Determine the roots of the simultaneous nonlinear equations

$$(x - 4)^2 + (y - 4)^2 = 5 \qquad\qquad x^2 + y^2 = 16$$

Use a graphical approach to obtain your initial guesses. Determine refined estimates with (**a**) the two-equation Newton-Raphson method and (**b**) the fsolve function.

12.17 Repeat Prob. 12.16 except determine the positive root of

$$y = x^2 + 1 \qquad\qquad y = 2\cos x$$

12.18 The following chemical reactions take place in a closed system

$$2A + B \rightleftharpoons C$$
$$A + D \rightleftharpoons C$$

At equilibrium, they can be characterized by

$$K_1 = \frac{c_c}{c_a^2 c_b} \qquad\qquad K_2 = \frac{c_c}{c_a c_d}$$

where the nomenclature c_i represents the concentration of constituent i. If x_1 and x_2 are the number of moles of C that are produced due to the first and second reactions, respectively, use an approach to reformulate the equilibrium relationships in terms of the initial concentrations of the constituents. Then, solve the pair of simultaneous nonlinear equations for x_1 and x_2 if $K_1 = 4 \times 10^{-4}$, $K_2 = 3.7 \times 10^{-2}$, $c_{a,0} = 50$, $c_{b,0} = 20$, $c_{c,0} = 5$, and $c_{d,0} = 10$.

(a) Use a graphical approach to develop your initial guesses. Then use these guesses as the starting point to determine refined estimates with

(b) the Newton-Raphson method, and

(c) the fsolve function.

12.19 As previously described in Sec. 5.6, the following system of five nonlinear equations govern the chemistry of rainwater,

$$K_1 = 10^6 \frac{[H^+][HCO_3^-]}{K_H p_{co_2}} \quad K_2 = \frac{[H^+][CO_3^{-2}]}{[HCO_3^-]} \quad K_w = [H^+][OH^-]$$

$$c_T = \frac{K_H p_{co_2}}{10^6} + [HCO_3^-] + [CO_3^{-2}]$$

$$0 = [HCO_3^-] + 2\,[CO_3^{-2}] + [OH^-] + [H^+]$$

where K_H = Henry's constant, K_1, K_2, and K_w = equilibrium coefficients, c_T = total inorganic carbon, $[HCO_3^-]$ = bicarbonate, $[CO_3^{-2}]$ = carbonate, $[H^+]$ = hydrogen ion, and $[OH^-]$ = hydroxyl ion. Notice how the partial pressure of CO_2 shows up in the equations indicating the impact of this greenhouse gas on the acidity of rain. Use these equations and the fsolve function to compute the pH of rainwater given that $K_H = 10^{-1.46}$, $K_1 = 10^{-6.3}$, $K_2 = 10^{-10.3}$, and $K_w = 10^{-14}$. Compare the results in 1958 when the p_{CO_2} was 315 and in 2015 when it was about 400 ppm. Note that this is a difficult problem to solve because the concentrations tend to be very small and vary over many orders of magnitude. Therefore, it is useful to use the trick based on expressing the unknowns in a negative log scale, $pK = -\log_{10}(K)$. That is, the five unknowns, c_T = total inorganic carbon, = bicarbonate, carbonate, $[H^+]$, $[OH^-]$, $[HCO_3^-]$, $[CO_3^{-2}]$ and c_T can be reexpressed as the unknowns pH, pOH, pHCO$_3$, pCO$_3$, and pc$_T$ as in

$$[H^+] = 10^{-pH} \qquad [OH^-] = 10^{-pOH}$$

$$[HCO_3^-] = 10^{-pHCO_3}$$

$$[CO_3^{-2}] = 10^{-pCO_3} \qquad c_T = 10^{-pcT}$$

In addition, it is helpful to use optimset to set a stringent criterion for the function tolerance as in the following script which you can use to generate your solutions

```
clc, format compact
xguess = [7;7;3;7;3];
options = optimset('tolfun',1e-12)
[x1,fx1] = fsolve(@funpH,xguess,options,315);
[x2,fx2] = fsolve(@funpH,xguess,options,400);
x1',fx1',x2',fx2'
```

13

Eigenvalues

CHAPTER OBJECTIVES

The primary objective of this chapter is to introduce you to eigenvalues. Specific objectives and topics covered are

- Understanding the mathematical definition of eigenvalues and eigenvectors.
- Understanding the physical interpretation of eigenvalues and eigenvectors within the context of engineering systems that vibrate or oscillate.
- Knowing how to implement the polynomial method.
- Knowing how to implement the power method to evaluate the largest and smallest eigenvalues and their respective eigenvectors.
- Knowing how to use and interpret MATLAB's eig function.

YOU'VE GOT A PROBLEM

At the beginning of Chap. 8, we used Newton's second law and force balances to predict the equilibrium positions of three bungee jumpers connected by cords. Because we assumed that the cords behaved like ideal springs (i.e., followed Hooke's law), the steady-state solution reduced to solving a system of linear algebraic equations [recall Eq. (8.1) and Example 8.2]. In mechanics, this is referred to as a *statics* problem.

Now let's look at a *dynamics* problem involving the same system. That is, we'll study the jumpers' motion as a function of time. To do this, their initial conditions (i.e., their initial positions and velocities) must be prescribed. For example, we can set the jumpers' initial positions at the equilibrium values computed in Example 8.2. If we then set their initial velocities to zero, nothing would happen because the system would be at equilibrium.

Because we are now interested in examining the system's dynamics, we must set the initial conditions to values that induce motion. Although we set the jumpers' initial positions to the equilibrium values and the middle jumper's initial velocity to zero, we set the upper and bottom jumper's initial velocities to some admittedly extreme values. That is, we

FIGURE 13.1
The (a) positions and (b) velocities versus time for the system of three interconnected bungee jumpers from Example 8.2.

impose a downward velocity of 200 m/s on jumper 1 and an upward velocity of 100 m/s on jumper 3. (Safety tip: Don't try this at home!) We then used MATLAB to solve the differential equations [Eq. (8.1)] to generate the resulting positions and velocities as a function of time.[1]

As displayed in Fig. 13.1, the outcome is that the jumpers oscillate wildly. Because there are no friction forces (e.g., no air drag or spring dampening), they lurch up and down around their equilibrium positions in a persistent manner that at least visually borders on the chaotic. Closer inspection of the individual trajectories suggests that there may be some pattern to the oscillations. For example, the distances between peaks and troughs might be constant. But when viewed as a time series, it is difficult to perceive whether there is anything systematic and predictable going on.

In this chapter, we deal with one approach for extracting something fundamental out of such seemingly chaotic behavior. This entails determining the *eigenvalues*, or *characteristic values*, for such systems. As we will see, this involves formulating and solving systems of linear algebraic equations in a fashion that differs from what we've done to this point. To do this, let's first describe exactly what is meant by eigenvalues from a mathematical standpoint.

[1] We will show how this is done when we cover ordinary differential equations in Part Six.

13.1 MATHEMATICAL BACKGROUND

Chapters 8 through 12 have dealt with methods for solving sets of linear algebraic equations of the general form

$$[A]\{x\} = \{b\} \tag{13.1}$$

Such systems are called *nonhomogeneous* because of the presence of the vector $\{b\}$ on the right-hand side of the equality. If the equations comprising such a system are linearly independent (i.e., have a nonzero determinant), they will have a unique solution. In other words, there is one set of x values that will make the equations balance. As we've already seen in Sec. 9.1.1, for two equations with two unknowns, the solution can be visualized as the intersection of two straight lines represented by the equations (recall Fig. 9.1).

In contrast, a *homogeneous* linear algebraic system has a right-hand side equal to zero:

$$[A]\{x\} = 0 \tag{13.2}$$

At face value, this equation suggests that the only possible solution would be the trivial case for which all x's $= 0$. Graphically this would correspond to two straight lines that intersected at zero.

Although this is certainly true, eigenvalue problems associated with engineering are typically of the general form

$$[[A] - \lambda[I]]\{x\} = 0 \tag{13.3}$$

where the parameter λ is the *eigenvalue*. Thus, rather than setting the x's to zero, we can determine the value of λ that drives the left-hand side to zero! One way to accomplish this is based on the fact that, for nontrivial solutions to be possible, the determinant of the matrix must equal zero:

$$|[A] - \lambda[I]| = 0 \tag{13.4}$$

Expanding the determinant yields a polynomial in λ, which is called the *characteristic polynomial*. The roots of this polynomial are the solutions for the eigenvalues.

In order to better understand these concepts, it is useful to examine the two-equation case,

$$\begin{aligned}(a_{11} - \lambda)x_1 + \quad a_{12}x_2 &= 0 \\ a_{21}x_1 + (a_{22} - \lambda)x_2 &= 0\end{aligned} \tag{13.5}$$

Expanding the determinant of the coefficient matrix gives

$$\begin{vmatrix} a_{11} - \lambda & a_{12} \\ a_{21} & a_{22} - \lambda \end{vmatrix} = \lambda^2 - (a_{11} + a_{22})\lambda - a_{12}a_{21} \tag{13.6}$$

which is the *characteristic polynomial*. The quadratic formula can then be used to solve for the two eigenvalues:

$$\begin{aligned}\lambda_1 \\ \lambda_2\end{aligned} = \frac{(a_{11} - a_{22}) \pm \sqrt{(a_{11} - a_{22})^2 - 4a_{12}a_{21}}}{2} \tag{13.7}$$

These are the values that solve Eq. (13.5). Before proceeding, let's convince ourselves that this approach (which, by the way, is called the *polynomial method*) is correct.

EXAMPLE 13.1   The Polynomial Method

Problem Statement. Use the polynomial method to solve for the eigenvalues of the following homogeneous system:

$$(10 - \lambda)x_1 \quad\quad - 5x_2 = 0$$
$$-5x_1 + (10 - \lambda)x_2 = 0$$

Solution. Before determining the correct solution, let's first investigate the case where we have an incorrect eigenvalue. For example, if $\lambda = 3$, the equations become

$$7x_1 - 5x_2 = 0$$
$$-5x_1 + 7x_2 = 0$$

Plotting these equations yields two straight lines that intersect at the origin (Fig. 13.2a). Thus, the only solution is the trivial case where $x_1 = x_2 = 0$.

To determine the correct eigenvalues, we can expand the determinant to give the characteristic polynomial:

$$\begin{vmatrix} 10 - \lambda & -5 \\ -5 & 10 - \lambda \end{vmatrix} = \lambda^2 - 20\lambda + 75$$

which can be solved for

$$\frac{\lambda_1}{\lambda_2} = \frac{20 \pm \sqrt{20^2 - 4(1)75}}{2} = 15, 5$$

Therefore, the eigenvalues for this system are 15 and 5.

We can now substitute either of these values back into the system and examine the result. For $\lambda_1 = 15$, we obtain

$$-5x_1 - 5x_2 = 0$$
$$-5x_1 - 5x_2 = 0$$

Thus, a correct eigenvalue makes the two equations identical (Fig. 13.2b). In essence as we move toward a correct eigenvalue the two lines rotate until they lie on top of each other. Mathematically, this means that there are an infinite number of solutions. But solving either of the equations yields the interesting result that all the solutions have the property that $x_1 = -x_2$. Although at first glance this might appear trivial, it's actually quite interesting as it tells us that the ratio of the unknowns is a constant. This result can be expressed in vector form as

$$\{x\} = \begin{Bmatrix} -1 \\ 1 \end{Bmatrix}$$

which is referred to as the *eigenvector* corresponding to the eigenvalue $\lambda = 15$.

In a similar fashion, substituting the second eigenvalue, $\lambda_2 = 5$, gives

$$5x_1 - 5x_2 = 0$$
$$-5x_1 + 5x_2 = 0$$

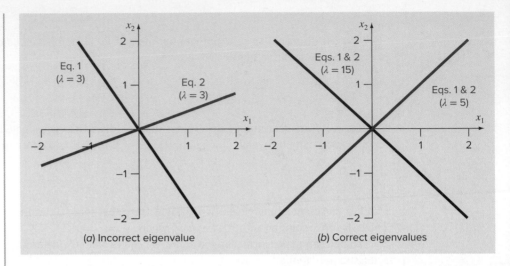

FIGURE 13.2
Plots of a system of two homogeneous linear equations from Example 13.1. (*a*) An incorrect eigenvalue ($\lambda = 3$) means that the two equations, which are labeled as Eqs. 1 and 2 in the figure, plot as separate lines and the only solution is the trivial case ($x_1 = x_2 = 0$). (*b*) In contrast, the cases with correct eigenvalues ($\lambda = 5$ and 15), the equations fall on top of each other.

Again, the eigenvalue makes the two equations identical (Fig. 13.2*b*) and we can see that the solution for this case corresponds to $x_1 = x_2$, and the eigenvector is

$$\{x\} = \begin{Bmatrix} 1 \\ 1 \end{Bmatrix}$$

We should recognize that MATLAB has built-in functions to facilitate the polynomial method. For Example 13.1, the `poly` function can be used to generate the characteristic polynomial as in

```
>> A = [10 -5;-5 10];
>> p = poly(A)

p =
     1    -20    75
```

Then, the `roots` function can be employed to compute the eigenvalues:

```
>> d = roots(p)

d =
     15
      5
```

The previous example yields the useful mathematical insight that the solution of n homogeneous equations of the form of Eq. (13.3) consists of a set of n eigenvalues and their associated eigenvectors. Further, it showed that the eigenvectors provide the ratios of the unknowns representing the solution.

In the next section, we will show how such information has utility in engineering and science by turning back to our physical problem setting of oscillating objects. However, before doing so, we'd like to make two more mathematical points.

First, inspection of Fig. 13.2b indicates that the straight lines representing each eigenvalue solution are at right angles to each other. That is, they are *orthogonal*. This property is true for symmetric matrices with distinct eigenvalues.

Second, multiplying out Eq. (13.3) and separating terms gives

$$[A]\{x\} = \lambda\{x\}$$

When viewed in this way, we can see that solving for the eigenvalues and eigenvectors amounts to translating the information content of a matrix $[A]$ into a scalar λ. This might not seem significant for the 2×2 system we have been examining, but it is pretty remarkable when we consider that the size of $[A]$ can potentially be much larger.

13.2 PHYSICAL BACKGROUND

The mass-spring system in Fig. 13.3a is a simple context to illustrate how eigenvalues occur in physical problem settings. It will also help to demonstrate some of the mathematical concepts introduced in the previous section.

To simplify the analysis, assume that each mass has no external or damping forces acting on it. In addition, assume that each spring has the same natural length l and the same spring constant k. Finally, assume that the displacement of each spring is measured relative to its own local coordinate system with an origin at the spring's equilibrium position (Fig. 13.3a). Under these assumptions, Newton's second law can be employed to develop a force balance for each mass:

$$m_1 \frac{d^2x_1}{dt^2} = -kx_1 + k(x_2 - x_1) \tag{13.8a}$$

FIGURE 13.3
A two mass–three spring system with frictionless rollers vibrating between two fixed walls. The position of the masses can be referenced to local coordinates with origins at their respective equilibrium positions (a). As in (b), positioning the masses away from equilibrium creates forces in the springs that on release lead to oscillations of the masses.

$$m_2 \frac{d^2x_2}{dt^2} = -k(x_2 - x_1) - kx_2 \tag{13.8b}$$

where x_i is the displacement of mass i away from its equilibrium position (Fig. 13.3b).

From vibration theory, it is known that solutions to Eq. (13.8) can take the form

$$x_i = X_i \sin(\omega t) \tag{13.9}$$

where X_i = the *amplitude* of the oscillation of mass i (m) and ω = the *angular frequency* of the oscillation (radians/time), which is equal to

$$\omega = \frac{2\pi}{T_p} \tag{13.10}$$

where T_p = the *period* (time/cycle). Note that the inverse of the period is called the *ordinary frequency* f (cycles/time). If time is measured in seconds, the unit for f is the cycles/s, which is referred to as a *Hertz* (Hz).

Equation (13.9) can be differentiated twice and substituted into Eq. (13.8). After collection of terms, the result is

$$\left(\frac{2k}{m_1} - \omega^2\right) X_1 - \frac{k}{m_1} X_2 = 0 \tag{13.11a}$$

$$-\frac{k}{m_2} X_1 + \left(\frac{2k}{m_2} - \omega^2\right) X_2 = 0 \tag{13.11b}$$

Comparison of Eq. (13.11) with Eq. (13.3) indicates that at this point, the solution has been reduced to an eigenvalue problem—where, for this case, the eigenvalue is the square of the frequency. For a two-degree-of-freedom system such as Fig. 13.3, there will be two such values along with their eigenvectors. As shown in the following example, the latter establish the unique relationship between the unknowns.

EXAMPLE 13.2   Physical Interpretation of Eigenvalues and Eigenvectors

Problem Statement.   If $m_1 = m_2 = 40$ kg and $k = 200$ N/m, Eq. (13.11) is

$$(10 - \lambda)x_1 - 5x_2 = 0$$
$$-5x_1 + (10 - \lambda)x_2 = 0$$

Mathematically, this is the same system we already solved with the polynomial methods in Example 13.2. Thus, the two eigenvalues are $\omega^2 = 15$ and 5 s^{-2} and the corresponding eigenvectors are $X_1 = X_2$ and $X_1 = -X_2$. Interpret these results as they relate to the mass-spring system of Fig. 13.3.

Solution.   This example provides valuable information regarding the behavior of the system in Fig. 13.3. First, it tells us that the system has two primary modes of oscillation with angular frequencies of $\omega = 3.873$ and 2.36 radians s^{-1}, respectively. These values can be also expressed as periods (1.62 and 2.81 s, respectively) or ordinary frequencies (0.6164 and 0.3559 Hz, respectively).

As stated in Sec. 13.1, a unique set of values cannot be obtained for the unknown amplitudes X. However, their ratios are specified by the eigenvectors. Thus, if the system

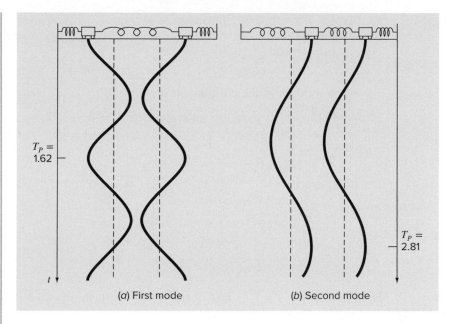

FIGURE 13.4
The principal modes of vibration of two equal masses connected by three identical springs between fixed walls.

is vibrating in the first mode, the first eigenvector tells us that the amplitude of the second mass will be equal but of opposite sign to the amplitude of the first. As in Fig. 13.4a, the masses vibrate apart and then together indefinitely (like two hands clapping every 1.62 s).

In the second mode, the eigenvector specifies that the two masses have equal amplitudes at all times. Thus, as in Fig. 13.4b, they vibrate back and forth in unison every 2.81 s. We should note that the configuration of the amplitudes provides guidance on how to set their initial values to attain pure motion in either of the two modes. Any other configuration will lead to superposition of the modes. It is this superposition that leads to the apparently chaotic behavior of systems like the bungee jumpers in Fig. 13.1. But as this example should make clear, there is an underlying systematic behavior that is embodied by the eigenvalues.

13.3   THE POWER METHOD

The power method is an iterative approach that can be employed to determine the largest or dominant eigenvalue. With slight modification, it can also be employed to determine the smallest value. It has the additional benefit that the corresponding eigenvector is obtained as a by-product of the method. To implement the power method, the system being analyzed is expressed in the form

$$[A]\{x\} = \lambda\{x\} \tag{13.12}$$

As illustrated by the following example, Eq. (13.12) forms the basis for an iterative solution technique that eventually yields the highest eigenvalue and its associated eigenvector.

EXAMPLE 13.3   Power Method for Highest Eigenvalue

Problem Statement. Using the same approach as in Sec. 13.2, we can derive the following homogeneous set of equations for a three mass–four spring system between two fixed walls:

$$\left(\frac{2k}{m_1} - \omega^2\right) X_1 \qquad -\frac{k}{m_1} X_2 \qquad\qquad = 0$$

$$-\frac{k}{m_2} X_1 + \left(\frac{2k}{m_2} - \omega^2\right) X_2 \qquad -\frac{k}{m_2} X_3 = 0$$

$$-\frac{k}{m_3} X_2 + \left(\frac{k}{m_3} - \omega^2\right) X_3 = 0$$

If all the masses $m = 1$ kg and all the spring constants $k = 20$ N/m, the system can be expressed in the matrix format of Eq. (13.4) as

$$\begin{bmatrix} 40 & -20 & 0 \\ -20 & 40 & -20 \\ 0 & -20 & 40 \end{bmatrix} - \lambda[I] = 0$$

where the eigenvalue λ is the square of the angular frequency ω^2. Employ the power method to determine the highest eigenvalue and its associated eigenvector.

Solution. The system is first written in the form of Eq. (13.12):

$$\begin{aligned} 40X_1 - 20X_2 \qquad &= \lambda X_1 \\ -20X_1 + 40X_2 - 20X_3 &= \lambda X_2 \\ -20X_2 + 40X_3 &= \lambda X_3 \end{aligned}$$

At this point, we can specify initial values of the X's and use the left-hand side to compute an eigenvalue and eigenvector. A good first choice is to assume that all the X's on the left-hand side of the equation are equal to one:

$$\begin{aligned} 40(1) - 20(1) \qquad &= 20 \\ -20(1) + 40(1) - 20(1) &= 0 \\ -20(1) + 40(1) &= 20 \end{aligned}$$

Next, the right-hand side is normalized by 20 to make the largest element equal to one:

$$\begin{Bmatrix} 20 \\ 0 \\ 20 \end{Bmatrix} = 20 \begin{Bmatrix} 1 \\ 0 \\ 1 \end{Bmatrix}$$

Thus, the normalization factor is our first estimate of the eigenvalue (20) and the corresponding eigenvector is $\{1\ 0\ 1\}^T$. This iteration can be expressed concisely in matrix form as

$$\begin{bmatrix} 40 & -20 & 0 \\ -20 & 40 & -20 \\ 0 & -20 & 40 \end{bmatrix} \begin{Bmatrix} 1 \\ 1 \\ 1 \end{Bmatrix} = \begin{Bmatrix} 20 \\ 0 \\ 20 \end{Bmatrix} = 20 \begin{Bmatrix} 1 \\ 0 \\ 1 \end{Bmatrix}$$

The next iteration consists of multiplying the matrix by the eigenvector from the last iteration, $\{1\ 0\ 1\}^T$ to give

$$\begin{bmatrix} 40 & -20 & 0 \\ -20 & 40 & -20 \\ 0 & -20 & 40 \end{bmatrix} \begin{Bmatrix} 1 \\ 0 \\ 1 \end{Bmatrix} = \begin{Bmatrix} 40 \\ -40 \\ 40 \end{Bmatrix} = 40 \begin{Bmatrix} 1 \\ -1 \\ 1 \end{Bmatrix}$$

Therefore, the eigenvalue estimate for the second iteration is 40, which can be employed to determine an error estimate:

$$|\varepsilon_a| = \left| \frac{40 - 20}{40} \right| \times 100\% = 50\%$$

The process can then be repeated.

Third iteration:

$$\begin{bmatrix} 40 & -20 & 0 \\ -20 & 40 & -20 \\ 0 & -20 & 40 \end{bmatrix} \begin{Bmatrix} 1 \\ -1 \\ 1 \end{Bmatrix} = \begin{Bmatrix} 60 \\ -80 \\ 60 \end{Bmatrix} = -80 \begin{Bmatrix} -0.75 \\ 1 \\ -0.75 \end{Bmatrix}$$

where $|\varepsilon_a| = 150\%$ (which is high because of the sign change).

Fourth iteration:

$$\begin{bmatrix} 40 & -20 & 0 \\ -20 & 40 & -20 \\ 0 & -20 & 40 \end{bmatrix} \begin{Bmatrix} -0.75 \\ 1 \\ -0.75 \end{Bmatrix} = \begin{Bmatrix} -50 \\ 70 \\ -50 \end{Bmatrix} = 70 \begin{Bmatrix} -0.71429 \\ 1 \\ -0.71429 \end{Bmatrix}$$

where $|\varepsilon_a| = 214\%$ (another sign change).

Fifth iteration:

$$\begin{bmatrix} 40 & -20 & 0 \\ -20 & 40 & -20 \\ 0 & -20 & 40 \end{bmatrix} \begin{Bmatrix} -0.71429 \\ 1 \\ -0.71429 \end{Bmatrix} = \begin{Bmatrix} -48.51714 \\ 68.51714 \\ -48.51714 \end{Bmatrix} = 68.51714 \begin{Bmatrix} -0.70833 \\ 1 \\ -0.70833 \end{Bmatrix}$$

where $|\varepsilon_a| = 2.08\%$.

Thus, the eigenvalue is converging. After several more iterations, it stabilizes on a value of 68.28427 with a corresponding eigenvector of $\{-0.707107\ 1\ -0.707107\}^T$.

Note that there are some instances where the power method will converge to the second-largest eigenvalue instead of to the largest. James, Smith, and Wolford (1985) provide an illustration of such a case. Other special cases are discussed in Fadeev and Fadeeva (1963).

In addition, there are sometimes cases where we are interested in determining the smallest eigenvalue. This can be done by applying the power method to the matrix inverse of $[A]$. For this case, the power method will converge on the largest value of $1/\lambda$—in other words, the smallest value of λ. An application to find the smallest eigenvalue will be left as a problem exercise.

Finally, after finding the largest eigenvalue, it is possible to determine the next highest by replacing the original matrix by one that includes only the remaining eigenvalues. The process of removing the largest known eigenvalue is called *deflation*.

We should mention that although the power method can be used to locate intermediate values, better methods are available for cases where we need to determine all the eigenvalues as described in the next section. Thus, the power method is primarily used when we want to locate the largest or the smallest eigenvalue.

13.4   MATLAB FUNCTION: eig

As might be expected, MATLAB has powerful and robust capabilities for evaluating eigenvalues and eigenvectors. The function eig, which is used for this purpose, can be employed to generate a vector of the eigenvalues as in

```
>> e = eig(A)
```

where e is a vector containing the eigenvalues of a square matrix A. Alternatively, it can be invoked as

```
>> [V,D] = eig(A)
```

where D is a diagonal matrix of the eigenvalues and V is a full matrix whose columns are the corresponding eigenvectors.

It should be noted that MATLAB scales the eigenvectors by dividing them by their Euclidean distance. Thus, as shown in the following example, although their magnitude may be different from values computed with say the polynomial method, the ratio of their elements will be identical.

EXAMPLE 13.4   Eigenvalues and Eigenvectors with MATLAB

Problem Statement. Use MATLAB to determine all the eigenvalues and eigenvectors for the system described in Example 13.3.

Solution. Recall that the matrix to be analyzed is

$$\begin{bmatrix} 40 & -20 & 0 \\ -20 & 40 & -20 \\ 0 & -20 & 40 \end{bmatrix}$$

The matrix can be entered as

```
>> A = [40 -20 0;-20 40 -20;0 -20 40];
```

If we just desire the eigenvalues, we can enter

```
>> e = eig(A)

e =
   11.7157
   40.0000
   68.2843
```

Notice that the highest eigenvalue (68.2843) is consistent with the value previously determined with the power method in Example 13.3.

If we want both the eigenvalues and eigenvectors, we can enter

```
>> [v,d] = eig(A)

v =
    0.5000   -0.7071   -0.5000
    0.7071   -0.0000    0.7071
    0.5000    0.7071   -0.5000

d =
   11.7157         0         0
         0   40.0000         0
         0         0    8.2843
```

Although the results are scaled differently, the eigenvector corresponding to the highest eigenvalue $\{-0.5\ 0.7071\ -0.5\}^T$ is consistent with the value previously determined with the power method in Example 13.3: $\{-0.707107\ 1\ -0.707107\}^T$. The can be demonstrated by dividing the eigenvector from the power method by its Euclidean norm:

```
>> vpower = [-0.7071 1 -0.7071]';
>> vMATLAB = vpower/norm(vpower)

vMATLAB =
   -0.5000
    0.7071
   -0.5000
```

Thus, although the magnitudes of the elements differ, their ratios are identical.

13.5 CASE STUDY   EIGENVALUES AND EARTHQUAKES

Background. Engineers and scientists use mass-spring models to gain insight into the dynamics of structures under the influence of disturbances such as earthquakes. Figure 13.5 shows such a model for a three-story building. Each floor mass is represented by m_i, and each floor stiffness is represented by k_i for $i = 1$ to 3.

FIGURE 13.5
A three-story building modeled as a mass-spring system.

$m_3 = 8000$ kg

$k_3 = 1800$ kN/m

$m_2 = 10000$ kg

$k_2 = 2400$ kN/m

$m_1 = 12000$ kg

$k_1 = 3000$ kN/m

13.5 CASE STUDY continued

For this case, the analysis is limited to horizontal motion of the structure as it is subjected to horizontal base motion due to earthquakes. Using the same approach as developed in Sec. 13.2, dynamic force balances can be developed for this system as

$$\left(\frac{k_1 + k_2}{m_1} - \omega_n^2\right) X_1 \qquad -\frac{k_2}{m_1} X_2 \qquad\qquad = 0$$

$$-\frac{k_2}{m_2} X_1 + \left(\frac{k_2 + k_3}{m_2} - \omega_n^2\right) X_2 \qquad -\frac{k_3}{m_2} X_3 = 0$$

$$-\frac{k_3}{m_3} X_2 + \left(\frac{k_3}{m_3} - \omega_n^2\right) X_3 = 0$$

where X_i represent horizontal floor translations (m), and ω_n is the *natural*, or *resonant, frequency* (radians/s). The resonant frequency can be expressed in Hertz (cycles/s) by dividing it by 2π radians/cycle.

Use MATLAB to determine the eigenvalues and eigenvectors for this system. Graphically represent the modes of vibration for the structure by displaying the amplitudes versus height for each of the eigenvectors. Normalize the amplitudes so that the translation of the third floor is one.

Solution. The parameters can be substituted into the force balances to give

$$(450 - \omega_n^2) X_1 \qquad - 200 X_2 \qquad\qquad = 0$$
$$-240 X_1 + (420 - \omega_n^2) X_2 \qquad - 180 X_3 = 0$$
$$-225 X_2 + (225 - \omega_n^2) X_3 = 0$$

A MATLAB session can be conducted to evaluate the eigenvalues and eigenvectors as

```
>> A=[450 -200 0;-240 420 -180;0 -225 225];
>> [v,d]=eig(A)

v =
    -0.5879      -0.6344       0.2913
     0.7307      -0.3506       0.5725
    -0.3471       0.6890       0.7664

d =
  698.5982            0            0
        0     339.4779            0
        0            0      56.9239
```

Therefore, the eigenvalues are 698.6, 339.5, and 56.92 and the resonant frequencies in Hz are

```
>> wn=sqrt(diag(d))'/2/pi
wn =
    4.2066      2.9324       1.2008
```

13.5 CASE STUDY   continued

FIGURE 13.6
The three primary modes of oscillation of the three-story building.

The corresponding eigenvectors are (normalizing so that the amplitude for the third floor is one)

$$\left\{ \begin{array}{c} 1.6934 \\ -2.1049 \\ 1 \end{array} \right\} \left\{ \begin{array}{c} -0.9207 \\ -0.5088 \\ 1 \end{array} \right\} \left\{ \begin{array}{c} 0.3801 \\ 0.7470 \\ 1 \end{array} \right\}$$

A graph can be made showing the three modes (Fig. 13.6). Note that we have ordered them from the lowest to the highest natural frequency as is customary in structural engineering.

Natural frequencies and mode shapes are characteristics of structures in terms of their tendencies to resonate at these frequencies. The frequency content of an earthquake typically has the most energy between 0 and 20 Hz and is influenced by the earthquake magnitude, the epicentral distance, and other factors. Rather than a single frequency, they contain a spectrum of all frequencies with varying amplitudes. Buildings are more receptive to vibration at their lower modes of vibrations due to their simpler deformed shapes and requiring less strain energy to deform in the lower modes. When these amplitudes coincide with the natural frequencies of buildings, large dynamic responses are induced, creating large stresses and strains in the structure's beams, columns, and foundations. Based on analyses like the one in this case study, structural engineers can more wisely design buildings to withstand earthquakes with a good factor of safety.

PROBLEMS

13.1 Repeat Example 13.1 but for three masses with the m's $= 40$ kg and the k's $= 240$ N/m. Produce a plot like Fig. 13.4 to identify the principle modes of vibration.

13.2 Use the power method to determine the highest eigenvalue and corresponding eigenvector for

$$\begin{bmatrix} 2-\lambda & 8 & 10 \\ 8 & 4-\lambda & 5 \\ 10 & 5 & 7-\lambda \end{bmatrix}$$

13.3 Use the power method to determine the lowest eigenvalue and corresponding eigenvector for the system from Prob. 13.2.

13.4 Derive the set of differential equations for a three mass–four spring system (Fig. P13.4) that describes their time motion. Write the three differential equations in matrix form

{Acceleration vector} + [k/m matrix]
    {displacement vector x} = 0

Note each equation has been divided by the mass. Solve for the eigenvalues and natural frequencies for the following values of mass and spring constants: $k_1 = k_4 = 15$ N/m, $k_2 = k_3 = 35$ N/m, and $m_1 = m_2 = m_3 = 1.5$ kg.

13.5 Consider the mass-spring system in Fig. P13.5. The frequencies for the mass vibrations can be determined by solving for the eigenvalues and by applying $M\ddot{x} + kx = 0$, which yields

$$\begin{bmatrix} m_1 & 0 & 0 \\ 0 & m_2 & 0 \\ 0 & 0 & m_3 \end{bmatrix} \begin{Bmatrix} \ddot{x}_1 \\ \ddot{x}_2 \\ \ddot{x}_3 \end{Bmatrix} + \begin{bmatrix} 2k & -k & -k \\ -k & 2k & -k \\ -k & -k & 2k \end{bmatrix} \begin{Bmatrix} x_1 \\ x_2 \\ x_3 \end{Bmatrix} = \begin{Bmatrix} 0 \\ 0 \\ 0 \end{Bmatrix}$$

Applying the guess $x = x_0 e^{i\omega t}$ as a solution, we get the following matrix:

$$\begin{bmatrix} 2k - m_1\omega^2 & -k & -k \\ -k & 2k - m_2\omega^2 & -k \\ -k & -k & 2k - m_3\omega^2 \end{bmatrix} \begin{Bmatrix} x_{01} \\ x_{02} \\ x_{03} \end{Bmatrix} e^{i\omega t} = \begin{Bmatrix} 0 \\ 0 \\ 0 \end{Bmatrix}$$

FIGURE P13.4

FIGURE P13.5

FIGURE P13.6

Use MATLAB's eig command to solve for the eigenvalues of the $k - m\omega^2$ matrix above. Then use these eigenvalues to solve for the frequencies (ω). Let $m_1 = m_2 = m_3 = 1$ kg, and $k = 2$ N/m.

13.6 As displayed in Fig. P13.6, an LC circuit can be modeled by the following system of differential equations:

$$L_1 \frac{d^2 i_1}{dt^2} + \frac{1}{C_1}(i_1 - i_2) = 0$$

$$L_2 \frac{d^2 i_2}{dt^2} + \frac{1}{C_2}(i_2 - i_3) - \frac{1}{C_1}(i_1 - i_2) = 0$$

$$L_3 \frac{d^2 i_3}{dt^2} + \frac{1}{C_3} i_3 - \frac{1}{C_2}(i_2 - i_3) = 0$$

where L = inductance (H), t = time (s), i = current (A), and C = capacitance (F). Assuming that a solution is of the form $i_j = I_j \sin(\omega t)$, determine the eigenvalues and eigenvectors for this system with $L = 1$ H and $C = 0.25$C. Draw the network, illustrating how the currents oscillate in their primary modes.

13.7 Repeat Prob. 13.6 but with only two loops. That is, omit the i_3 loop. Draw the network, illustrating how the currents oscillate in their primary modes.

13.8 Repeat the problem in Sec. 13.5 but leave off the third floor.

13.9 Repeat the problem in Sec. 13.5 but add a fourth floor with $m_4 = 6000$ and $k_4 = 1200$ kN/m.

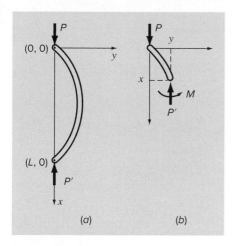

FIGURE P13.10

(a) A slender rod. (b) A freebody diagram of a rod.

13.10 The curvature of a slender column subject to an axial load P (Fig. P13.10) can be modeled by

$$\frac{d^2y}{dx^2} + p^2y = 0$$

where

$$p^2 = \frac{P}{EI}$$

where E = the modulus of elasticity, and I = the moment of inertia of the cross section about its neutral axis.

This model can be converted into an eigenvalue problem by substituting a centered finite-difference approximation for the second derivative to give

$$\frac{y_{i+1} - 2y_i + y_{i-1}}{\Delta x^2} + p^2y_i = 0$$

where i = a node located at a position along the rod's interior, and Δx = the spacing between nodes. This equation can be expressed as

$$y_{i-1} - (2 - \Delta x^2 p^2)y_i + y_{i+1} = 0$$

Writing this equation for a series of interior nodes along the axis of the column yields a homogeneous system of equations. For example, if the column is divided into five segments (i.e., four interior nodes), the result is

$$\begin{bmatrix} (2 - \Delta x^2 p^2) & -1 & 0 & 0 \\ -1 & (2 - \Delta x^2 p^2) & -1 & 0 \\ 0 & -1 & (2 - \Delta x^2 p^2) & -1 \\ 0 & 0 & -1 & (2 - \Delta x^2 p^2) \end{bmatrix} \begin{Bmatrix} y_1 \\ y_2 \\ y_3 \\ y_4 \end{Bmatrix} = 0$$

An axially loaded wooden column has the following characteristics: $E = 10 \times 10^9$ Pa, $I = 1.25 \times 10^{-5}$ m⁴, and $L = 3$ m. For the five-segment, four-node representation:

(a) Implement the polynomial method with MATLAB to determine the eigenvalues for this system.

(b) Use the MATLAB eig function to determine the eigenvalues and eigenvectors.

(c) Use the power method to determine the largest eigenvalue and its corresponding eigenvector.

13.11 A system of two homogeneous linear ordinary differential equations with constant coefficients can be written as

$$\frac{dy_1}{dt} = -5y_1 + 3y_2, \qquad y_1(0) = 50$$

$$\frac{dy_2}{dt} = 100y_1 - 301y_2, \qquad y_2(0) = 100$$

If you have taken a course in differential equations, you know that the solutions for such equations have the form

$$y_i = c_i e^{\lambda t}$$

where c and λ are constants to be determined. Substituting this solution and its derivative into the original equations converts the system into an eigenvalue problem. The resulting eigenvalues and eigenvectors can then be used to derive the general solution to the differential equations. For example, for the two-equation case, the general solution can be written in terms of vectors as

$$\{y\} = c_1\{v_1\}e^{\lambda_1 t} + c_2\{v_2\}e^{\lambda_2 t}$$

where $\{v_i\}$ = the eigenvector corresponding to the ith eigenvalue (λ_i) and the c's are unknown coefficients that can be determined with the initial conditions.

(a) Convert the system into an eigenvalue problem.

(b) Use MATLAB to solve for the eigenvalues and eigenvectors.

(c) Employ the results of (b) and the initial conditions to determine the general solution.

(d) Develop a MATLAB plot of the solution for $t = 0$ to 1.

13.12 Water flows between the North American Great Lakes as depicted in Fig. P13.12. Based on mass balances, the following differential equations can be written for the concentrations in each of the lakes for a pollutant that decays with first-order kinetics:

FIGURE P13.12
The North American Great Lakes. The arrows indicate how water flows between the lakes.

$$\frac{dc_1}{dt} = -(0.0056 + k)c_1$$

$$\frac{dc_2}{dt} = -(0.01 + k)c_2$$

$$\frac{dc_3}{dt} = 0.01902c_1 + 0.01387c_2 - (0.047 + k)c_3$$

$$\frac{dc_4}{dt} = 0.33597c_3 - (0.376 + k)c_4$$

$$\frac{dc_5}{dt} = 0.11364c_4 - (0.133 + k)c_5$$

where k = the first-order decay rate (/yr), which is equal to 0.69315/(half-life). Note that the constants in each of the equations account for the flow between the lakes. Due to the testing of nuclear weapons in the atmosphere, the concentrations of strontium-90 (^{90}Sr) in the five lakes in 1963 were approximately $\{c\} = \{17.7 \ 30.5 \ 43.9 \ 136.3 \ 30.1\}^T$ in units of Bq/m^3. Assuming that no additional ^{90}Sr entered the system thereafter, use MATLAB and the approach outlined in Prob. 13.11 to compute the concentrations in each of the lakes from 1963 through 2010. Note that ^{90}Sr has a half-life of 28.8 years.

13.13 Develop an M-file function to determine the largest eigenvalue and its associated eigenvector with the power method. Test the program by duplicating Example 13.3 and then use it to solve Prob. 13.2.

13.14 Repeat the computations in Sec. 13.5 but remove the third floor.

13.15 Repeat the computations in Sec. 13.5 but add a fourth floor with a mass of $m_4 = 6000$ kg connected with the third floor by a spring with $k_4 = 1200$ kN/m.

13.16 Recall that at the start of Chap. 8, we suspended three bungee jumpers with zero drag from frictionless cords that followed Hooke's law. Determine the resulting eigenvalues and eigenvectors that would eventually characterize the jumpers' oscillating motions and relative positions if they were instantaneously released from their starting positions as depicted in Fig. 8.1a (i.e., with each cord fully extended, but not stretched). Although bungee cords do not actually behave like true springs, assume that they stretch and compress in linear proportion to the applied force. Use the parameters from Example 8.2.

PART FOUR

Curve Fitting

4.1 OVERVIEW

What Is Curve Fitting?

Data are often given for discrete values along a continuum. However, you may require estimates at points between the discrete values. Chapters 14 through 18 describe techniques to fit curves to such data to obtain intermediate estimates. In addition, you may require a simplified version of a complicated function. One way to do this is to compute values of the function at a number of discrete values along the range of interest. Then, a simpler function may be derived to fit these values. Both of these applications are known as *curve fitting*.

There are two general approaches for curve fitting that are distinguished from each other on the basis of the amount of error associated with the data. First, where the data exhibit a significant degree of error or "scatter," the strategy is to derive a single curve that represents the general trend of the data. Because any individual data point may be incorrect, we make no effort to intersect every point. Rather, the curve is designed to follow the pattern of the points taken as a group. One approach of this nature is called *least-squares regression* (Fig. PT4.1*a*).

Second, where the data are known to be very precise, the basic approach is to fit a curve or a series of curves that pass directly through each of the points. Such data usually originate from

tables. Examples are values for the density of water or for the heat capacity of gases as a function of temperature. The estimation of values between well-known discrete points is called *interpolation* (Fig. PT4.1*b* and *c*).

Curve Fitting and Engineering and Science. Your first exposure to curve fitting may have been to determine intermediate values from tabulated data—for instance, from interest tables for engineering economics or from steam tables for thermodynamics. Throughout the remainder of your career, you will have frequent occasion to estimate intermediate values from such tables.

Although many of the widely used engineering and scientific properties have been

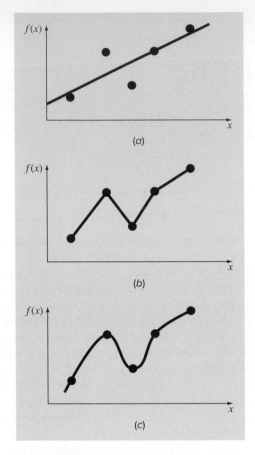

FIGURE PT4.1
Three attempts to fit a "best" curve through five data points: (a) least-squares regression, (b) linear interpolation, and (c) curvilinear interpolation.

tabulated, there are a great many more that are not available in this convenient form. Special cases and new problem contexts often require that you measure your own data and develop your own predictive relationships. Two types of applications are generally encountered when fitting experimental data: trend analysis and hypothesis testing.

Trend analysis represents the process of using the pattern of the data to make predictions. For cases where the data are measured with high precision, you might utilize interpolating polynomials. Imprecise data are often analyzed with least-squares regression.

Trend analysis may be used to predict or forecast values of the dependent variable. This can involve extrapolation beyond the limits of the observed data or interpolation within the range of the data. All fields of engineering and science involve problems of this type.

A second application of experimental curve fitting is *hypothesis testing*. Here, an existing mathematical model is compared with measured data. If the model coefficients are unknown, it may be necessary to determine values that best fit the observed data. On the other hand, if estimates of the model coefficients are already available, it may be appropriate to compare predicted values of the model with observed values to test the adequacy of the model. Often, alternative models are compared and the "best" one is selected on the basis of empirical observations.

In addition to the foregoing engineering and scientific applications, curve fitting is important in other numerical methods such as integration and the approximate solution of differential equations. Finally, curve-fitting techniques can be used to derive simple functions to approximate complicated functions.

4.2 PART ORGANIZATION

After a brief review of statistics, *Chap. 14* focuses on *linear regression;* that is, how to determine the "best" straight line through a set of uncertain data points. Besides discussing how to calculate the slope and intercept of this straight line, we also present quantitative and visual methods for evaluating the validity of the results. In addition, we describe *random number generation* as well as several approaches for the linearization of nonlinear equations.

Chapter 15 begins with brief discussions of polynomial and multiple linear regression. *Polynomial regression* deals with developing a best fit of parabolas, cubics, or higher-order polynomials. This is followed by a description of *multiple linear regression,* which is designed for the case where the dependent variable y is a linear function of two or more independent variables $x_1, x_2, \ldots, x_m$. This approach has special utility for evaluating experimental data where the variable of interest is dependent on a number of different factors.

After multiple regression, we illustrate how polynomial and multiple regression are both subsets of a *general linear least-squares model*. Among other things, this will allow us to introduce a concise matrix representation of regression and discuss its general statistical properties. Finally, the last sections of Chap. 15 are devoted to *nonlinear regression*. This approach is designed to compute a least-squares fit of a nonlinear equation to data.

Chapter 16 deals with *Fourier analysis* which involves fitting periodic functions to data. Our emphasis will be on the *fast Fourier transform* or *FFT*. This method, which is readily implemented with MATLAB, has many engineering applications, ranging from vibration analysis of structures to signal processing.

In *Chap. 17,* the alternative curve-fitting technique called *interpolation* is described. As discussed previously, interpolation is used for estimating intermediate values between precise data points. In Chap. 17, polynomials are derived for this purpose. We introduce the basic concept of polynomial interpolation by using straight lines and parabolas to connect points. Then, we develop a generalized procedure for fitting an nth-order polynomial. Two formats are presented for expressing these polynomials in equation form. The first, called *Newton's interpolating polynomial,* is preferable when the appropriate order of the polynomial is unknown. The second, called the *Lagrange interpolating polynomial,* has advantages when the proper order is known beforehand.

Finally, *Chap. 18* presents an alternative technique for fitting precise data points. This technique, called *spline interpolation,* fits polynomials to data but in a piecewise fashion. As such, it is particularly well suited for fitting data that are generally smooth but exhibit abrupt local changes. The chapter ends with an overview of how piecewise interpolation is implemented in MATLAB.

14

Linear Regression

CHAPTER OBJECTIVES

The primary objective of this chapter is to introduce you to how least-squares regression can be used to fit a straight line to measured data. Specific objectives and topics covered are

- Familiarizing yourself with some basic descriptive statistics and the normal distribution.
- Knowing how to compute the slope and intercept of a best-fit straight line with linear regression.
- Knowing how to generate random numbers with MATLAB and how they can be employed for Monte Carlo simulations.
- Knowing how to compute and understand the meaning of the coefficient of determination and the standard error of the estimate.
- Understanding how to use transformations to linearize nonlinear equations so that they can be fit with linear regression.
- Knowing how to implement linear regression with MATLAB.

YOU'VE GOT A PROBLEM

In Chap. 1, we noted that a free-falling object such as a bungee jumper is subject to the upward force of air resistance. As a first approximation, we assumed that this force was proportional to the square of velocity as in

$$F_U = c_d v^2 \tag{14.1}$$

where F_U = the upward force of air resistance [N = kg m/s^2], c_d = a drag coefficient (kg/m), and v = velocity [m/s].

Expressions such as Eq. (14.1) come from the field of fluid mechanics. Although such relationships derive in part from theory, experiments play a critical role in their formulation. One such experiment is depicted in Fig. 14.1. An individual is suspended in a wind

FIGURE 14.1
Wind tunnel experiment to measure how the force of air resistance depends on velocity.

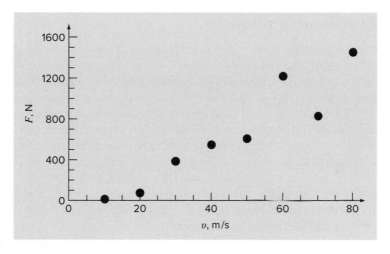

FIGURE 14.2
Plot of force versus wind velocity for an object suspended in a wind tunnel.

TABLE 14.1 Experimental data for force (N) and velocity (m/s) from a wind tunnel experiment.

v, m/s	10	20	30	40	50	60	70	80
F, N	25	70	380	550	610	1220	830	1450

tunnel (any volunteers?) and the force measured for various levels of wind velocity. The result might be as listed in Table 14.1.

The relationship can be visualized by plotting force versus velocity. As in Fig. 14.2, several features of the relationship bear mention. First, the points indicate that the force increases as velocity increases. Second, the points do not increase smoothly, but exhibit rather significant scatter, particularly at the higher velocities. Finally, although it may not be obvious, the relationship between force and velocity may not be linear. This conclusion becomes more apparent if we assume that force is zero for zero velocity.

In Chaps. 14 and 15, we will explore how to fit a "best" line or curve to such data. In so doing, we will illustrate how relationships like Eq. (14.1) arise from experimental data.

14.1 STATISTICS REVIEW

Before describing least-squares regression, we will first review some basic concepts from the field of statistics. These include the mean, standard deviation, residual sum of the squares, and the normal distribution. In addition, we describe how simple descriptive statistics and distributions can be generated in MATLAB. If you are familiar with these subjects, feel free to skip the following pages and proceed directly to Sec. 14.2. If you are unfamiliar with these concepts or are in need of a review, the following material is designed as a brief introduction.

14.1.1 Descriptive Statistics

Suppose that in the course of an engineering study, several measurements were made of a particular quantity. For example, Table 14.2 contains 24 readings of the coefficient of thermal expansion of a structural steel. Taken at face value, the data provide a limited amount of information—that is, that the values range from a minimum of 6.395 to a maximum of 6.775. Additional insight can be gained by summarizing the data in one or more well-chosen statistics that convey as much information as possible about specific characteristics of the data set. These descriptive statistics are most often selected to represent (1) the location of the center of the distribution of the data and (2) the degree of spread of the data set.

Measure of Location. The most common measure of central tendency is the arithmetic mean. The *arithmetic mean* ($\bar{y}$) of a sample is defined as the sum of the individual data points (y_i) divided by the number of points (n), or

$$\bar{y} = \frac{\sum y_i}{n} \tag{14.2}$$

where the summation (and all the succeeding summations in this section) is from $i = 1$ through n.

There are several alternatives to the arithmetic mean. The *median* is the midpoint of a group of data. It is calculated by first putting the data in ascending order. If the number of measurements is odd, the median is the middle value. If the number is even, it is the arithmetic mean of the two middle values. The median is sometimes called the *50th percentile*.

The *mode* is the value that occurs most frequently. The concept usually has direct utility only when dealing with discrete or coarsely rounded data. For continuous variables such as the data in Table 14.2, the concept is not very practical. For example, there are actually

TABLE 14.2 Measurements of the coefficient of thermal expansion of structural steel.

6.495	6.595	6.615	6.635	6.485	6.555
6.665	6.505	6.435	6.625	6.715	6.655
6.755	6.625	6.715	6.575	6.655	6.605
6.565	6.515	6.555	6.395	6.775	6.685

four modes for these data: 6.555, 6.625, 6.655, and 6.715, which all occur twice. If the numbers had not been rounded to 3 decimal digits, it would be unlikely that any of the values would even have repeated twice. However, if continuous data are grouped into equispaced intervals, it can be an informative statistic. We will return to the mode when we describe histograms later in this section.

Measures of Spread. The simplest measure of spread is the *range,* the difference between the largest and the smallest value. Although it is certainly easy to determine, it is not considered a very reliable measure because it is highly sensitive to the sample size and is very sensitive to extreme values.

The most common measure of spread for a sample is the *standard deviation* (s_y) about the mean:

$$s_y = \sqrt{\frac{S_t}{n-1}} \tag{14.3}$$

where S_t is the total sum of the squares of the residuals between the data points and the mean, or

$$S_t = \sum(y_i - \bar{y})^2 \tag{14.4}$$

Thus, if the individual measurements are spread out widely around the mean, S_t (and, consequently, s_y) will be large. If they are grouped tightly, the standard deviation will be small. The spread can also be represented by the square of the standard deviation, which is called the *variance:*

$$s_y^2 = \frac{\sum(y_i - \bar{y})^2}{n-1} \tag{14.5}$$

Note that the denominator in both Eqs. (14.3) and (14.5) is $n-1$. The quantity $n-1$ is referred to as the *degrees of freedom.* Hence S_t and s_y are said to be based on $n-1$ degrees of freedom. This nomenclature derives from the fact that the sum of the quantities upon which S_t is based (i.e., $\bar{y} - y_1, \bar{y} - y_2, \ldots, \bar{y} - y_n$) is zero. Consequently, if $\bar{y}$ is known and $n-1$ of the values are specified, the remaining value is fixed. Thus, only $n-1$ of the values are said to be freely determined. Another justification for dividing by $n-1$ is the fact that there is no such thing as the spread of a single data point. For the case where $n = 1$, Eqs. (14.3) and (14.5) yield a meaningless result of infinity.

We should note that an alternative, more convenient formula is available to compute the variance:

$$s_y^2 = \frac{\sum y_i^2 - (\sum y_i)^2/n}{n-1} \tag{14.6}$$

This version does not require precomputation of $\bar{y}$ and yields an identical result as Eq. (14.5).

A final statistic that has utility in quantifying the spread of data is the coefficient of variation (c.v.). This statistic is the ratio of the standard deviation to the mean. As such, it provides a normalized measure of the spread. It is often multiplied by 100 so that it can be expressed in the form of a percent:

$$\text{c.v.} = \frac{s_y}{\bar{y}} \times 100\% \tag{14.7}$$

EXAMPLE 14.1   Simple Statistics of a Sample

Problem Statement. Compute the mean, median, variance, standard deviation, and coefficient of variation for the data in Table 14.2.

Solution. The data can be assembled in tabular form and the necessary sums computed as in Table 14.3.
    The mean can be computed as [Eq. (14.2)],

$$\bar{y} = \frac{158.4}{24} = 6.6$$

Because there are an even number of values, the median is computed as the arithmetic mean of the middle two values: $(6.605 + 6.615)/2 = 6.61$.
    As in Table 14.3, the sum of the squares of the residuals is 0.217000, which can be used to compute the standard deviation [Eq. (14.3)]:

$$s_y = \sqrt{\frac{0.217000}{24 - 1}} = 0.097133$$

TABLE 14.3   Data and summations for computing simple descriptive statistics for the coefficients of thermal expansion from Table 14.2.

i	y_i	$(y_i - \bar{y})^2$	y_i^2
1	6.395	0.04203	40.896
2	6.435	0.02723	41.409
3	6.485	0.01323	42.055
4	6.495	0.01103	42.185
5	6.505	0.00903	42.315
6	6.515	0.00723	42.445
7	6.555	0.00203	42.968
8	6.555	0.00203	42.968
9	6.565	0.00123	43.099
10	6.575	0.00063	43.231
11	6.595	0.00003	43.494
12	6.605	0.00002	43.626
13	6.615	0.00022	43.758
14	6.625	0.00062	43.891
15	6.625	0.00062	43.891
16	6.635	0.00122	44.023
17	6.655	0.00302	44.289
18	6.655	0.00302	44.289
19	6.665	0.00422	44.422
20	6.685	0.00722	44.689
21	6.715	0.01322	45.091
22	6.715	0.01322	45.091
23	6.755	0.02402	45.630
24	6.775	0.03062	45.901
Σ	158.400	0.21700	1045.657

the variance [Eq. (14.5)]:

$$s_y^2 = (0.097133)^2 = 0.009435$$

and the coefficient of variation [Eq. (14.7)]:

$$\text{c.v.} = \frac{0.097133}{6.6} \times 100\% = 1.47\%$$

The validity of Eq. (14.6) can also be verified by computing

$$s_y^2 = \frac{1045.657 - (158.400)^2/24}{24 - 1} = 0.009435$$

14.1.2 The Normal Distribution

Another characteristic that bears on the present discussion is the data distribution—that is, the shape with which the data are spread around the mean. A histogram provides a simple visual representation of the distribution. A *histogram* is constructed by sorting the measurements into intervals, or *bins*. The units of measurement are plotted on the abscissa and the frequency of occurrence of each interval is plotted on the ordinate.

As an example, a histogram can be created for the data from Table 14.2. The result (Fig. 14.3) suggests that most of the data are grouped close to the mean value of 6.6. Notice also, that now that we have grouped the data, we can see that the bin with the most values is from 6.6 to 6.64. Although we could say that the mode is the midpoint of this bin, 6.62, it is more common to report the most frequent range as the *modal class interval*.

If we have a very large set of data, the histogram often can be approximated by a smooth curve. The symmetric, bell-shaped curve superimposed on Fig. 14.3 is one such characteristic shape—the *normal distribution*. Given enough additional measurements, the histogram for this particular case could eventually approach the normal distribution.

FIGURE 14.3
A histogram used to depict the distribution of data. As the number of data points increases, the histogram often approaches the smooth, bell-shaped curve called the normal distribution.

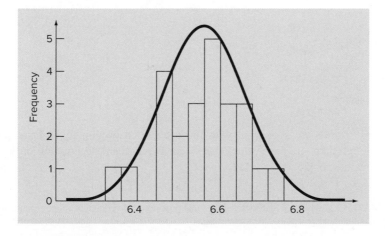

The concepts of the mean, standard deviation, residual sum of the squares, and normal distribution all have great relevance to engineering and science. A very simple example is their use to quantify the confidence that can be ascribed to a particular measurement. If a quantity is normally distributed, the range defined by $\bar{y} - s_y$ to $\bar{y} + s_y$ will encompass approximately 68% of the total measurements. Similarly, the range defined by $\bar{y} - 2s_y$ to $\bar{y} + 2s_y$ will encompass approximately 95%.

For example, for the data in Table 14.2, we calculated in Example 14.1 that $\bar{y} = 6.6$ and $s_y = 0.097133$. Based on our analysis, we can tentatively make the statement that approximately 95% of the readings should fall between 6.405734 and 6.794266. Because it is so far outside these bounds, if someone told us that they had measured a value of 7.35, we would suspect that the measurement might be erroneous.

14.1.3 Descriptive Statistics in MATLAB

Standard MATLAB has several functions to compute descriptive statistics.[1] For example, the arithmetic mean is computed as mean(*x*). If *x* is a vector, the function returns the mean of the vector's values. If it is a matrix, it returns a row vector containing the arithmetic mean of each column of *x*. The following is the result of using mean and the other statistical functions to analyze a column vector s that holds the data from Table 14.2:

```
>> format short g
>> mean(s),median(s),mode(s)

ans =
          6.6
ans =
         6.61
ans =
        6.555

>> min(s),max(s)

ans =
        6.395
ans =
        6.775

>> range=max(s) -min(s)

range =
         0.38

>> var(s),std(s)

ans =
    0.0094348

ans =
    0.097133
```

These results are consistent with those obtained previously in Example 14.1. Note that although there are four values that occur twice, the mode function only returns the first of the values: 6.555.

---

[1] MATLAB also offers a Statistics Toolbox that provides a wide range of common statistical tasks, from random number generation, to curve fitting, to design of experiments and statistical process control.

FIGURE 14.4
Histogram generated with the MATLAB hist function.

MATLAB can also be used to generate a histogram based on the hist function. The hist function has the syntax

```
[n, x] = hist(y, x)
```

where n = the number of elements in each bin, x = a vector specifying the midpoint of each bin, and y is the vector being analyzed. For the data from Table 14.2, the result is

```
>> [n,x] =hist(s)

n =
     1    1    3    1    4    3    5    2    2    2
x =
   6.414 6.452 6.49 6.528 6.566 6.604 6.642 6.68 6.718 6.756
```

The resulting histogram depicted in Fig. 14.4 is similar to the one we generated by hand in Fig. 14.3. Note that all the arguments and outputs with the exception of y are optional. For example, hist(y) without output arguments just produces a histogram bar plot with 10 bins determined automatically based on the range of values in y.

14.2   RANDOM NUMBERS AND SIMULATION

In this section, we will describe two MATLAB functions that can be used to produce a sequence of random numbers. The first (rand) generates numbers that are uniformly distributed, and the second (randn) generates numbers that have a normal distribution.

14.2.1 MATLAB Function: rand

This function generates a sequence of numbers that are uniformly distributed between 0 and 1. A simple representation of its syntax is

```
r = rand(m, n)
```

where r = an *m*-by-*n* matrix of random numbers. The following formula can then be used to generate a uniform distribution on another interval:

```
runiform = low + (up - low) * rand(m, n)
```

where low = the lower bound and up = the upper bound.

EXAMPLE 14.2   Generating Uniform Random Values of Drag

Problem Statement. If the initial velocity is zero, the downward velocity of the free-falling bungee jumper can be predicted with the following analytical solution [Eq. (1.9)]:

$$v = \sqrt{\frac{gm}{c_d}} \tanh\left(\sqrt{\frac{gc_d}{m}}\, t\right)$$

Suppose that $g = 9.81\,\mathrm{m/s^2}$, and $m = 68.1$ kg, but c_d is not known precisely. For example, you might know that it varies uniformly between 0.225 and 0.275 (i.e., $\pm$ 10% around a mean value of 0.25 kg/m). Use the rand function to generate 1000 random uniformly distributed values of c_d and then employ these values along with the analytical solution to compute the resulting distribution of velocities at $t = 4$ s.

Solution.   Before generating the random numbers, we can first compute the mean velocity:

$$v_{\mathrm{mean}} = \sqrt{\frac{9.81(68.1)}{0.25}} \tanh\left(\sqrt{\frac{9.81(0.25)}{68.1}}\, 4\right) = 33.1118\,\frac{\mathrm{m}}{\mathrm{s}}$$

We can also generate the range:

$$v_{\mathrm{low}} = \sqrt{\frac{9.81(68.1)}{0.275}} \tanh\left(\sqrt{\frac{9.81(0.275)}{68.1}}\, 4\right) = 32.6223\,\frac{\mathrm{m}}{\mathrm{s}}$$

$$v_{\mathrm{high}} = \sqrt{\frac{9.81(68.1)}{0.225}} \tanh\left(\sqrt{\frac{9.81(0.225)}{68.1}}\, 4\right) = 33.6198\,\frac{\mathrm{m}}{\mathrm{s}}$$

Thus, we can see that the velocity varies by

$$\Delta v = \frac{33.6198 - 32.6223}{2(33.1118)} \times 100\% = 1.5063\%$$

The following script generates the random values for c_d, along with their mean, standard deviation, percent variation, and a histogram:

```
clc,format short g
n=1000;t=4;m=68.1;g=9.81;
cd=0.25;cdmin=cd-0.025,cdmax=cd+0.025
r=rand(n,1);
```

```
cdrand=cdmin+(cdmax-cdmin)*r;
meancd=mean(cdrand),stdcd=std(cdrand)
Deltacd=(max(cdrand)-min(cdrand))/meancd/2*100.
subplot(2,1,1)
hist(cdrand),title('(a) Distribution of drag')
xlabel('cd (kg/m)')
```

The results are

```
meancd =
      0.25018
stdcd =
      0.014528
Deltacd =
        9.9762
```

These results, as well as the histogram (Fig. 14.5a) indicate that rand has yielded 1000 uniformly distributed values with the desired mean value and range. The values can then be employed along with the analytical solution to compute the resulting distribution of velocities at $t = 4$ s.

```
vrand=sqrt(g*m./cdrand).*tanh(sqrt(g*cdrand/m)*t);
meanv=mean(vrand)
Deltav=(max(vrand)-min(vrand))/meanv/2*100.
subplot(2,1,2)
hist(vrand),title('(b) Distribution of velocity')
xlabel('v (m/s)')
```

FIGURE 14.5
Histograms of (a) uniformly distributed drag coefficients and (b) the resulting distribution of velocity.

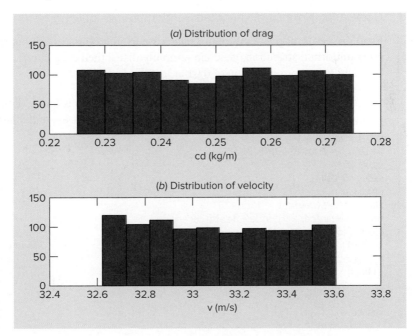

The results are

```
meanv =
    33.1151
Deltav =
    1.5048
```

These results, as well as the histogram (Fig. 14.5b), closely conform to our hand calculations.

The foregoing example is formally referred to as a *Monte Carlo simulation*. The term, which is a reference to Monaco's Monte Carlo casino, was first used by physicists working on nuclear weapons projects in the 1940s. Although it yields intuitive results for this simple example, there are instances where such computer simulations yield surprising outcomes and provide insights that would otherwise be impossible to determine. The approach is feasible only because of the computer's ability to implement tedious, repetitive computations in an efficient manner.

14.2.2 MATLAB Function: randn

This function generates a sequence of numbers that are normally distributed with a mean of 0 and a standard deviation of 1. A simple representation of its syntax is

```
r = randn(m, n)
```

where r = an *m*-by-*n* matrix of random numbers. The following formula can then be used to generate a normal distribution with a different mean (mn) and standard deviation (s),

```
rnormal = mn + s * randn(m, n)
```

EXAMPLE 14.3   Generating Normally Distributed Random Values of Drag

Problem Statement. Analyze the same case as in Example 14.2, but rather than employing a uniform distribution, generate normally distributed drag coefficients with a mean of 0.25 and a standard deviation of 0.01443.

Solution. The following script generates the random values for c_d, along with their mean, standard deviation, coefficient of variation (expressed as a %), and a histogram:

```
clc,format short g
n=1000;t=4;m=68.1;g=9.81;
cd=0.25;
stdev=0.01443;
r=randn(n,1);
cdrand=cd+stdev*r;
meancd=mean(cdrand),stdevcd=std(cdrand)
cvcd=stdevcd/meancd*100.
subplot(2,1,1)
hist(cdrand),title('(a) Distribution of drag')
xlabel('cd (kg/m)')
```

The results are

```
meancd =
     0.24988
```

```
stdevcd =
      0.014465
cvcd =
      5.7887
```

These results, as well as the histogram (Fig. 14.6a) indicate that randn has yielded 1000 uniformly distributed values with the desired mean, standard deviation, and coefficient of variation. The values can then be employed along with the analytical solution to compute the resulting distribution of velocities at $t = 4$ s.

```
vrand=sqrt(g*m./cdrand).*tanh(sqrt(g*cdrand/m)*t);
meanv=mean(vrand),stdevv=std(vrand)
cvv=stdevv/meanv*100.
subplot(2,1,2)
hist(vrand),title('(b) Distribution of velocity')
xlabel('v (m/s)')
```

The results are

```
meanv =
      33.117
stdevv =
      0.28839
cvv =
      0.8708
```

These results, as well as the histogram (Fig. 14.6b), indicate that the velocities are also normally distributed with a mean that is close to the value that would be computed using the mean and the analytical solution. In addition, we compute the associated standard deviation which corresponds to a coefficient of variation of ±0.8708%.

FIGURE 14.6
Histograms of (a) normally distributed drag coefficients and (b) the resulting distribution of velocity.

Although simple, the foregoing examples illustrate how random numbers can be easily generated within MATLAB. We will explore additional applications in the end-of-chapter problems.

14.3   LINEAR LEAST-SQUARES REGRESSION

Where substantial error is associated with data, the best curve-fitting strategy is to derive an approximating function that fits the shape or general trend of the data without necessarily matching the individual points. One approach to do this is to visually inspect the plotted data and then sketch a "best" line through the points. Although such "eyeball" approaches have commonsense appeal and are valid for "back-of-the-envelope" calculations, they are deficient because they are arbitrary. That is, unless the points define a perfect straight line (in which case, interpolation would be appropriate), different analysts would draw different lines.

To remove this subjectivity, some criterion must be devised to establish a basis for the fit. One way to do this is to derive a curve that minimizes the discrepancy between the data points and the curve. To do this, we must first quantify the discrepancy. The simplest example is fitting a straight line to a set of paired observations: $(x_1, y_1), (x_2, y_2), \ldots, (x_n, y_n)$. The mathematical expression for the straight line is

$$y = a_0 + a_1 x + e \tag{14.8}$$

where a_0 and a_1 are coefficients representing the intercept and the slope, respectively, and e is the error, or *residual,* between the model and the observations, which can be represented by rearranging Eq. (14.8) as

$$e = y - a_0 - a_1 x \tag{14.9}$$

Thus, the residual is the discrepancy between the true value of y and the approximate value, $a_0 + a_1 x$, predicted by the linear equation.

14.3.1 Criteria for a "Best" Fit

One strategy for fitting a "best" line through the data would be to minimize the sum of the residual errors for all the available data, as in

$$\sum_{i=1}^{n} e_i = \sum_{i=1}^{n} (y_i - a_0 - a_1 x_i) \tag{14.10}$$

where n = total number of points. However, this is an inadequate criterion, as illustrated by Fig. 14.7a, which depicts the fit of a straight line to two points. Obviously, the best fit is the line connecting the points. However, any straight line passing through the midpoint of the connecting line (except a perfectly vertical line) results in a minimum value of Eq. (14.10) equal to zero because positive and negative errors cancel.

One way to remove the effect of the signs might be to minimize the sum of the absolute values of the discrepancies, as in

$$\sum_{i=1}^{n} |e_i| = \sum_{i=1}^{n} |y_i - a_0 - a_1 x_i| \tag{14.11}$$

Figure 14.7b demonstrates why this criterion is also inadequate. For the four points shown, any straight line falling within the dashed lines will minimize the sum of the absolute values of the residuals. Thus, this criterion also does not yield a unique best fit.

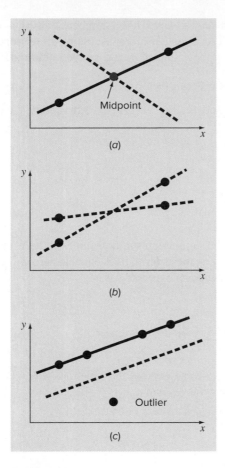

FIGURE 14.7
Examples of some criteria for "best fit" that are inadequate for regression: (*a*) minimizes
the sum of the residuals, (*b*) minimizes the sum of the absolute values of the residuals, and
(*c*) minimizes the maximum error of any individual point.

A third strategy for fitting a best line is the *minimax* criterion. In this technique, the
line is chosen that minimizes the maximum distance that an individual point falls from
the line. As depicted in Fig. 14.7*c*, this strategy is ill-suited for regression because it gives
undue influence to an outlier—that is, a single point with a large error. It should be noted
that the minimax principle is sometimes well-suited for fitting a simple function to a complicated function (Carnahan, Luther, and Wilkes, 1969).

A strategy that overcomes the shortcomings of the aforementioned approaches is to
minimize the sum of the squares of the residuals:

$$S_r = \sum_{i=1}^{n} e_i^2 = \sum_{i=1}^{n} (y_i - a_0 - a_1 x_i)^2 \tag{14.12}$$

This criterion, which is called *least squares,* has a number of advantages, including that it yields a unique line for a given set of data. Before discussing these properties, we will present a technique for determining the values of a_0 and a_1 that minimize Eq. (14.12).

14.3.2 Least-Squares Fit of a Straight Line

To determine values for a_0 and a_1, Eq. (14.12) is differentiated with respect to each unknown coefficient:

$$\frac{\partial S_r}{\partial a_0} = -2 \sum (y_i - a_0 - a_1 x_i)$$

$$\frac{\partial S_r}{\partial a_1} = -2 \sum [(y_i - a_0 - a_1 x_i) x_i]$$

Note that we have simplified the summation symbols; unless otherwise indicated, all summations are from $i = 1$ to n. Setting these derivatives equal to zero will result in a minimum S_r. If this is done, the equations can be expressed as

$$0 = \sum y_i - \sum a_0 - \sum a_1 x_i$$

$$0 = \sum x_i y_i - \sum a_0 x_i - \sum a_1 x_i^2$$

Now, realizing that $\sum a_0 = na_0$ we can express the equations as a set of two simultaneous linear equations with two unknowns (a_0 and a_1):

$$n \quad a_0 + \left(\sum x_i\right) a_1 = \sum y_i \tag{14.13}$$

$$\left(\sum x_i\right) a_0 + \left(\sum x_i^2\right) a_1 = \sum x_i y_i \tag{14.14}$$

These are called the *normal equations.* They can be solved simultaneously for

$$a_1 = \frac{n \sum x_i y_i - \sum x_i \sum y_i}{n \sum x_i^2 - \left(\sum x_i\right)^2} \tag{14.15}$$

This result can then be used in conjunction with Eq. (14.13) to solve for

$$a_0 = \bar{y} - a_1 \bar{x} \tag{14.16}$$

where $\bar{y}$ and $\bar{x}$ are the means of y and x, respectively.

EXAMPLE 14.4　Linear Regression

Problem Statement.　Fit a straight line to the values in Table 14.1.

Solution.　In this application, force is the dependent variable (y) and velocity is the independent variable (x). The data can be set up in tabular form and the necessary sums computed as in Table 14.4.

TABLE 14.4 Data and summations needed to compute the best-fit line for the data from Table 14.1.

i	x_i	y_i	x_i^2	$x_i y_i$
1	10	25	100	250
2	20	70	400	1,400
3	30	380	900	11,400
4	40	550	1,600	22,000
5	50	610	2,500	30,500
6	60	1,220	3,600	73,200
7	70	830	4,900	58,100
8	80	1,450	6,400	116,000
Σ	360	5,135	20,400	312,850

The means can be computed as

$$\bar{x} = \frac{360}{8} = 45 \qquad \bar{y} = \frac{5,135}{8} = 641.875$$

The slope and the intercept can then be calculated with Eqs. (14.15) and (14.16) as

$$a_1 = \frac{8(312,850) - 360(5,135)}{8(20,400) - (360)^2} = 19.47024$$

$$a_0 = 641.875 - 19.47024(45) = -234.2857$$

Using force and velocity in place of y and x, the least-squares fit is

$$F = -234.2857 + 19.47024v$$

The line, along with the data, is shown in Fig. 14.8.

FIGURE 14.8
Least-squares fit of a straight line to the data from Table 14.1

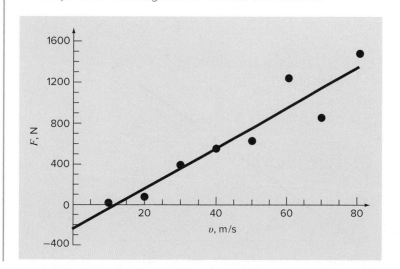

Notice that although the line fits the data well, the zero intercept means that the equation predicts physically unrealistic negative forces at low velocities. In Sec. 14.4, we will show how transformations can be employed to derive an alternative best-fit line that is more physically realistic.

14.3.3 Quantification of Error of Linear Regression

Any line other than the one computed in Example 14.4 results in a larger sum of the squares of the residuals. Thus, the line is unique and in terms of our chosen criterion is a "best" line through the points. A number of additional properties of this fit can be elucidated by examining more closely the way in which residuals were computed. Recall that the sum of the squares is defined as [Eq. (14.12)]

$$S_r = \sum_{i=1}^{n} (y_i - a_0 - a_1 x_i)^2 \tag{14.17}$$

Notice the similarity between this equation and Eq. (14.4)

$$S_t = \sum (y_i - \bar{y})^2 \tag{14.18}$$

In Eq. (14.18), the square of the residual represented the square of the discrepancy between the data and a single estimate of the measure of central tendency—the mean. In Eq. (14.17), the square of the residual represents the square of the vertical distance between the data and another measure of central tendency—the straight line (Fig. 14.9).

The analogy can be extended further for cases where (1) the spread of the points around the line is of similar magnitude along the entire range of the data and (2) the distribution of these points about the line is normal. It can be demonstrated that if these criteria

---

FIGURE 14.9
The residual in linear regression represents the vertical distance between a data point and the straight line.

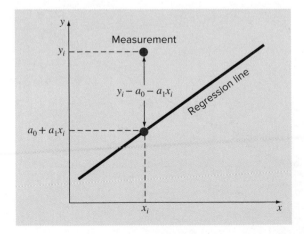

are met, least-squares regression will provide the best (i.e., the most likely) estimates of a_0 and a_1 (Draper and Smith, 1981). This is called the *maximum likelihood principle* in statistics. In addition, if these criteria are met, a "standard deviation" for the regression line can be determined as [compare with Eq. (14.3)]

$$s_{y/x} = \sqrt{\frac{S_r}{n-2}}$$

(14.19)

where $s_{y/x}$ is called the *standard error of the estimate*. The subscript notation "y/x" designates that the error is for a predicted value of y corresponding to a particular value of x. Also, notice that we now divide by $n-2$ because two data-derived estimates—a_0 and a_1—were used to compute S_r; thus, we have lost two degrees of freedom. As with our discussion of the standard deviation, another justification for dividing by $n-2$ is that there is no such thing as the "spread of data" around a straight line connecting two points. Thus, for the case where $n=2$, Eq. (14.19) yields a meaningless result of infinity.

Just as was the case with the standard deviation, the standard error of the estimate quantifies the spread of the data. However, $s_{y/x}$ quantifies the spread *around the regression line* as shown in Fig. 14.10b in contrast to the standard deviation s_y that quantified the spread *around the mean* (Fig. 14.10a).

These concepts can be used to quantify the "goodness" of our fit. This is particularly useful for comparison of several regressions (Fig. 14.11). To do this, we return to the original data and determine the total sum of the squares around the mean for the dependent variable (in our case, y). As was the case for Eq. (14.18), this quantity is designated S_t. This is the magnitude of the residual error associated with the dependent variable prior to regression. After performing the regression, we can compute S_r, the sum of the squares of the residuals around the regression line with Eq. (14.17). This characterizes the residual

---

FIGURE 14.10
Regression data showing (*a*) the spread of the data around the mean of the dependent variable and (*b*) the spread of the data around the best-fit line. The reduction in the spread in going from (*a*) to (*b*), as indicated by the bell-shaped curves at the right, represents the improvement due to linear regression.

(*a*)   (*b*)

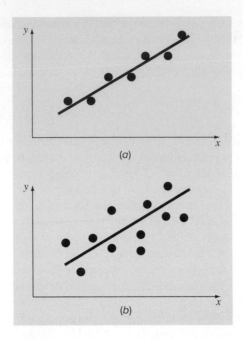

FIGURE 14.11
Examples of linear regression with (a) small and (b) large residual errors.

error that remains after the regression. It is, therefore, sometimes called the unexplained sum of the squares. The difference between the two quantities, $S_t - S_r$, quantifies the improvement or error reduction due to describing the data in terms of a straight line rather than as an average value. Because the magnitude of this quantity is scale-dependent, the difference is normalized to S_t to yield

$$r^2 = \frac{S_t - S_r}{S_t} \tag{14.20}$$

where r^2 is called the *coefficient of determination* and r is the *correlation coefficient* ($= \sqrt{r^2}$). For a perfect fit, $S_r = 0$ and $r^2 = 1$, signifying that the line explains 100% of the variability of the data. For $r^2 = 0$ $S_r = S_t$ and the fit represents no improvement. An alternative formulation for r that is more convenient for computer implementation is

$$r = \frac{n \sum (x_i y_i) - (\sum x_i)(\sum y_i)}{\sqrt{n \sum x_i^2 - (\sum x_i)^2}\sqrt{n \sum y_i^2 - (\sum y_i)^2}} \tag{14.21}$$

EXAMPLE 14.5   Estimation of Errors for the Linear Least-Squares Fit

Problem Statement. Compute the total standard deviation, the standard error of the estimate, and the correlation coefficient for the fit in Example 14.4.

Solution. The data can be set up in tabular form and the necessary sums computed as in Table 14.5.

TABLE 14.5 Data and summations needed to compute the goodness-of-fit statistics for the data from Table 14.1.

i	x_i	y_i	$a_0 + a_1 x_i$	$(y_i - \bar{y})^2$	$(y_i - a_0 - a_1 x_i)^2$
1	10	25	−39.58	380,535	4,171
2	20	70	155.12	327,041	7,245
3	30	380	349.82	68,579	911
4	40	550	544.52	8,441	30
5	50	610	739.23	1,016	16,699
6	60	1,220	933.93	334,229	81,837
7	70	830	1,128.63	35,391	89,180
8	80	1,450	1,323.33	653,066	16,044
Σ	360	5,135		1,808,297	216,118

The standard deviation is [Eq. (14.3)]

$$s_y = \sqrt{\frac{1,808,297}{8-1}} = 508.26$$

and the standard error of the estimate is [Eq. (14.19)]

$$s_{y/x} = \sqrt{\frac{216,118}{8-2}} = 189.79$$

Thus, because $s_{y/x} < s_y$, the linear regression model has merit. The extent of the improvement is quantified by [Eq. (14.20)]

$$r^2 = \frac{1,808,297 - 216,118}{1,808,297} = 0.8805$$

or $r = \sqrt{0.8805} = 0.9383$ These results indicate that 88.05% of the original uncertainty has been explained by the linear model.

Before proceeding, a word of caution is in order. Although the coefficient of determination provides a handy measure of goodness-of-fit, you should be careful not to ascribe more meaning to it than is warranted. Just because r^2 is "close" to 1 does not mean that the fit is necessarily "good." For example, it is possible to obtain a relatively high value of r^2 when the underlying relationship between y and x is not even linear. Draper and Smith (1981) provide guidance and additional material regarding assessment of results for linear regression. In addition, at the minimum, you should always inspect a plot of the data along with your regression curve.

A nice example was developed by Anscombe (1973). As in Fig. 14.12, he came up with four data sets consisting of 11 data points each. Although their graphs are very different, all have the same best-fit equation, $y = 3 + 0.5x$, and the same coefficient of determination, $r^2 = 0.67$! This example dramatically illustrates why developing plots is so valuable.

FIGURE 14.12
Anscombe's four data sets along with the best-fit line, $y = 3 + 0.5x$.

14.4 LINEARIZATION OF NONLINEAR RELATIONSHIPS

Linear regression provides a powerful technique for fitting a best line to data. However, it is predicated on the fact that the relationship between the dependent and independent variables is linear. This is not always the case, and the first step in any regression analysis should be to plot and visually inspect the data to ascertain whether a linear model applies. In some cases, techniques such as polynomial regression, which is described in Chap. 15, are appropriate. For others, transformations can be used to express the data in a form that is compatible with linear regression.

One example is the *exponential model:*

$$y = \alpha_1 e^{\beta_1 x} \tag{14.22}$$

where α_1 and β_1 are constants. This model is used in many fields of engineering and science to characterize quantities that increase (positive β_1) or decrease (negative β_1) at a rate that is directly proportional to their own magnitude. For example, population growth or radioactive decay can exhibit such behavior. As depicted in Fig. 14.13a, the equation represents a nonlinear relationship (for $\beta_1 \neq 0$) between y and x.

Another example of a nonlinear model is the simple *power equation:*

$$y = \alpha_2 x^{\beta_2} \tag{14.23}$$

where α_2 and β_2 are constant coefficients. This model has wide applicability in all fields of engineering and science. It is very frequently used to fit experimental data when the underlying model is not known. As depicted in Fig. 14.13b, the equation (for $\beta_2 \neq 0$) is nonlinear.

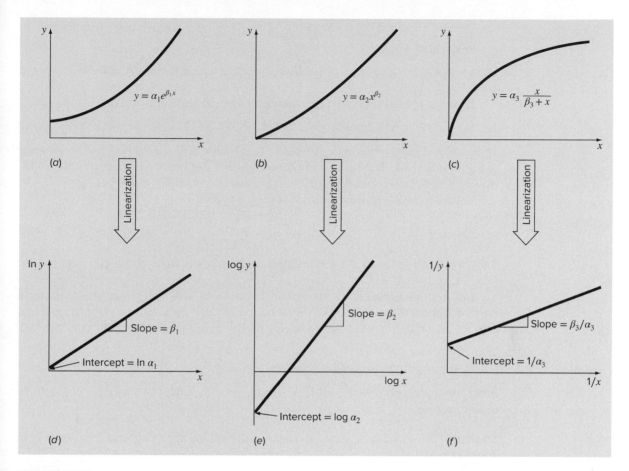

FIGURE 14.13
(a) The exponential equation, (b) the power equation, and (c) the saturation-growth-rate equation. Parts (d), (e), and (f) are linearized versions of these equations that result from simple transformations.

A third example of a nonlinear model is the *saturation-growth-rate equation:*

$$y = \alpha_3 \frac{x}{\beta_3 + x} \tag{14.24}$$

where α_3 and β_3 are constant coefficients. This model, which is particularly well-suited for characterizing population growth rate under limiting conditions, also represents a nonlinear relationship between y and x (Fig. 14.13c) that levels off, or "saturates," as x increases. It has many applications, particularly in biologically related areas of both engineering and science.

Nonlinear regression techniques are available to fit these equations to experimental data directly. However, a simpler alternative is to use mathematical manipulations to transform the equations into a linear form. Then linear regression can be employed to fit the equations to data.

For example, Eq. (14.22) can be linearized by taking its natural logarithm to yield

$$\ln y = \ln \alpha_1 + \beta_1 x \tag{14.25}$$

Thus, a plot of $\ln y$ versus x will yield a straight line with a slope of β_1 and an intercept of $\ln \alpha_1$ (Fig. 14.13d).

Equation (14.23) is linearized by taking its base-10 logarithm to give

$$\log y = \log \alpha_2 + \beta_2 \log x \tag{14.26}$$

Thus, a plot of $\log y$ versus $\log x$ will yield a straight line with a slope of β_2 and an intercept of $\log \alpha_2$ (Fig. 14.13e). Note that any base logarithm can be used to linearize this model. However, as done here, the base-10 logarithm is most commonly employed.

Equation (14.24) is linearized by inverting it to give

$$\frac{1}{y} = \frac{1}{\alpha_3} + \frac{\beta_3}{\alpha_3}\frac{1}{x} \tag{14.27}$$

Thus, a plot of $1/y$ versus $1/x$ will be linear, with a slope of β_3/α_3 and an intercept of $1/\alpha_3$ (Fig. 14.13f).

In their transformed forms, these models can be fit with linear regression to evaluate the constant coefficients. They can then be transformed back to their original state and used for predictive purposes. The following illustrates this procedure for the power model.

EXAMPLE 14.6   Fitting Data with the Power Equation

Problem Statement. Fit Eq. (14.23) to the data in Table 14.1 using a logarithmic transformation.

Solution. The data can be set up in tabular form and the necessary sums computed as in Table 14.6.

The means can be computed as

$$\bar{x} = \frac{12.606}{8} = 1.5757 \qquad \bar{y} = \frac{20.515}{8} = 2.5644$$

TABLE 14.6   Data and summations needed to fit the power model to the data from Table 14.1

i	x_i	y_i	$\log x_i$	$\log y_i$	$(\log x_i)^2$	$\log x_i \log y_i$
1	10	25	1.000	1.398	1.000	1.398
2	20	70	1.301	1.845	1.693	2.401
3	30	380	1.477	2.580	2.182	3.811
4	40	550	1.602	2.740	2.567	4.390
5	50	610	1.699	2.785	2.886	4.732
6	60	1220	1.778	3.086	3.162	5.488
7	70	830	1.845	2.919	3.404	5.386
8	80	1450	1.903	3.161	3.622	6.016
Σ			12.606	20.515	20.516	33.622

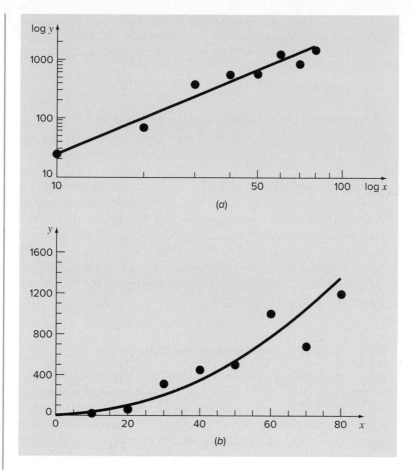

FIGURE 14.14
Least-squares fit of a power model to the data from Table 14.1. (*a*) The fit of the transformed data. (*b*) The power equation fit along with the data.

The slope and the intercept can then be calculated with Eqs. (14.15) and (14.16) as

$$a_1 = \frac{8(33.622) - 12.606(20.515)}{8(20.516) - (12.606)^2} = 1.9842$$

$$a_0 = 2.5644 - 1.9842(1.5757) = -0.5620$$

The least-squares fit is

$$\log y = -0.5620 + 1.9842 \log x$$

The fit, along with the data, is shown in Fig. 14.14*a*.

We can also display the fit using the untransformed coordinates. To do this, the coefficients of the power model are determined as $\alpha_2 = 10^{-0.5620} = 0.2741$ and $\beta_2 = 1.9842$. Using force and velocity in place of y and x, the least-squares fit is

$$F = 0.2741v^{1.9842}$$

This equation, along with the data, is shown in Fig. 14.14b.

The fits in Example 14.6 (Fig. 14.14) should be compared with the one obtained previously in Example 14.4 (Fig. 14.8) using linear regression on the untransformed data. Although both results would appear to be acceptable, the transformed result has the advantage that it does not yield negative force predictions at low velocities. Further, it is known from the discipline of fluid mechanics that the drag force on an object moving through a fluid is often well described by a model with velocity squared. Thus, knowledge from the field you are studying often has a large bearing on the choice of the appropriate model equation you use for curve fitting.

14.4.1 General Comments on Linear Regression

Before proceeding to curvilinear and multiple linear regression, we must emphasize the introductory nature of the foregoing material on linear regression. We have focused on the simple derivation and practical use of equations to fit data. You should be cognizant of the fact that there are theoretical aspects of regression that are of practical importance but are beyond the scope of this book. For example, some statistical assumptions that are inherent in the linear least-squares procedures are

1. Each x has a fixed value; it is not random and is known without error.
2. The y values are independent random variables and all have the same variance.
3. The y values for a given x must be normally distributed.

Such assumptions are relevant to the proper derivation and use of regression. For example, the first assumption means that (1) the x values must be error-free and (2) the regression of y versus x is not the same as x versus y. You are urged to consult other references such as Draper and Smith (1981) to appreciate aspects and nuances of regression that are beyond the scope of this book.

14.5   COMPUTER APPLICATIONS

Linear regression is so commonplace that it can be implemented on most pocket calculators. In this section, we will show how a simple M-file can be developed to determine the slope and intercept as well as to create a plot of the data and the best-fit line. We will also show how linear regression can be implemented with the built-in polyfit function.

14.5.1 MATLAB M-file: linregr

An algorithm for linear regression can be easily developed (Fig. 14.15). The required summations are readily computed with MATLAB's sum function. These are then used to compute the slope and the intercept with Eqs. (14.15) and (14.16). The routine displays the intercept and slope, the coefficient of determination, and a plot of the best-fit line along with the measurements.

A simple example of the use of this M-file would be to fit the force-velocity data analyzed in Example 14.4:

```
>> x = [10 20 30 40 50 60 70 80];
>> y = [25 70 380 550 610 1220 830 1450];
>> [a, r2] = linregr(x,y)

a =
    19.4702    -234.2857


r2 =
     0.8805
```

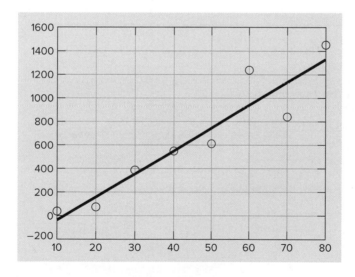

It can just as easily be used to fit the power model (Example 14.6) by applying the log10 function to the data as in

```
>> [a, r2] = linregr(log10(x),log10(y))

a =
     1.9842    -0.5620

r2 =
     0.9481
```

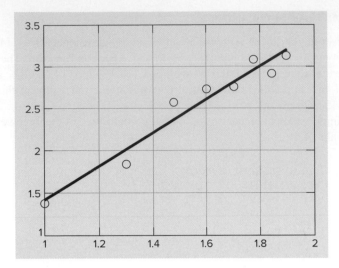

FIGURE 14.15
An M-file to implement linear regression.

```
function [a, r2] = linregr(x,y)
% linregr: linear regression curve fitting
%    [a, r2] = linregr(x,y): Least squares fit of straight
%               line to data by solving the normal equations
%
% input:
%    x = independent variable
%    y = dependent variable
% output:
%    a = vector of slope, a(1), and intercept, a(2)
%    r2 = coefficient of determination

n = length(x);
if length(y)~=n, error('x and y must be same length'); end
x = x(:); y = y(:);      % convert to column vectors
sx = sum(x); sy = sum(y);
sx2 = sum(x.*x); sxy = sum(x.*y); sy2 = sum(y.*y);
a(1) = (n*sxy-sx*sy)/(n*sx2-sx^2);
a(2) = sy/n-a(1)*sx/n;
r2 = ((n*sxy-sx*sy)/sqrt(n*sx2-sx^2)/sqrt(n*sy2-sy^2))^2;
% create plot of data and best fit line
xp = linspace(min(x),max(x),2);
yp = a(1)*xp+a(2);
plot(x,y,'o',xp,yp)
grid on
```

14.5.2 MATLAB Functions: polyfit and polyval

MATLAB has a built-in function polyfit that fits a least-squares nth-order polynomial to data. It can be applied as in

```
>> p = polyfit(x, y, n)
```

where x and y are the vectors of the independent and the dependent variables, respectively, and $n = $ the order of the polynomial. The function returns a vector p containing the polynomial's coefficients. We should note that it represents the polynomial using decreasing powers of x as in the following representation:

$$f(x) = p_1 x^n + p_2 x^{n-1} + \cdots + p_n x + p_{n+1}$$

Because a straight line is a first-order polynomial, polyfit(x,y,1) will return the slope and the intercept of the best-fit straight line.

```
>> x = [10 20 30 40 50 60 70 80];
>> y = [25 70 380 550 610 1220 830 1450];
>> a = polyfit(x,y,1)

a =
   19.4702 -234.2857
```

Thus, the slope is 19.4702 and the intercept is −234.2857.

Another function, polyval, can then be used to compute a value using the coefficients. It has the general format:

```
>> y = polyval(p, x)
```

where $p = $ the polynomial coefficients, and $y = $ the best-fit value at x. For example,

```
>> y = polyval(a,45)

y =
   641.8750
```

14.6 CASE STUDY   ENZYME KINETICS

Background. *Enzymes* act as catalysts to speed up the rate of chemical reactions in living cells. In most cases, they convert one chemical, the *substrate,* into another, the *product.* The *Michaelis-Menten* equation is commonly used to describe such reactions:

$$v = \frac{v_m[S]}{k_s + [S]} \tag{14.28}$$

where $v = $ the initial reaction velocity, $v_m = $ the maximum initial reaction velocity, $[S] = $ substrate concentration, and $k_s = $ a half-saturation constant. As in Fig. 14.16, the equation describes a saturating relationship which levels off with increasing $[S]$. The graph also illustrates that the *half-saturation constant* corresponds to the substrate concentration at which the velocity is half the maximum.

14.6 CASE STUDY continued

FIGURE 14.16
Two versions of the Michaelis-Menten model of enzyme kinetics.

Although the Michaelis-Menten model provides a nice starting point, it has been refined and extended to incorporate additional features of enzyme kinetics. One simple extension involves so-called *allosteric enzymes,* where the binding of a substrate molecule at one site leads to enhanced binding of subsequent molecules at other sites. For cases with two interacting bonding sites, the following second-order version often results in a better fit:

$$v = \frac{v_m [S]^2}{k_s^2 + [S]^2} \tag{14.29}$$

This model also describes a saturating curve but, as depicted in Fig. 14.16, the squared concentrations tend to make the shape more *sigmoid,* or S-shaped.

Suppose that you are provided with the following data:

$[S]$	1.3	1.8	3	4.5	6	8	9
v	0.07	0.13	0.22	0.275	0.335	0.35	0.36

Employ linear regression to fit these data with linearized versions of Eqs. (14.28) and (14.29). Aside from estimating the model parameters, assess the validity of the fits with both statistical measures and graphs.

Solution. Equation (14.28), which is in the format of the saturation-growth-rate model (Eq. 14.24), can be linearized by inverting it to give [recall Eq. (14.27)]

$$\frac{1}{v} = \frac{1}{v_m} + \frac{k_s}{v_m} \frac{1}{[S]}$$

14.6 CASE STUDY continued

The linregr function from Fig. 14.15 can then be used to determine the least-squares fit:

```
>> S=[1.3 1.8 3 4.5 6 8 9];
>> v=[0.07 0.13 0.22 0.275 0.335 0.35 0.36];
>> [a,r2]=linregr(1./S,1./v)

a =
    16.4022    0.1902
r2 =
    0.9344
```

The model coefficients can then be calculated as

```
>> vm=1/a(2)

vm =
    5.2570

>> ks=vm*a(1)

ks =
    86.2260
```

Thus, the best-fit model is

$$v = \frac{5.2570[S]}{86.2260 + [S]}$$

Although the high value of r^2 might lead you to believe that this result is acceptable, inspection of the coefficients might raise doubts. For example, the maximum velocity (5.2570) is much greater than the highest observed velocity (0.36). In addition, the half-saturation rate (86.2260) is much bigger than the maximum substrate concentration (9).

The problem is underscored when the fit is plotted along with the data. Figure 14.17a shows the transformed version. Although the straight line follows the upward trend, the data clearly appear to be curved. When the original equation is plotted along with the data in the untransformed version (Fig. 14.17b), the fit is obviously unacceptable. The data are clearly leveling off at about 0.36 or 0.37. If this is correct, an eyeball estimate would suggest that v_m should be about 0.36, and k_s should be in the range of 2 to 3.

Beyond the visual evidence, the poorness of the fit is also reflected by statistics like the coefficient of determination. For the untransformed case, a much less acceptable result of $r^2 = 0.6406$ is obtained.

The foregoing analysis can be repeated for the second-order model. Equation (14.28) can also be linearized by inverting it to give

$$\frac{1}{v} = \frac{1}{v_m} + \frac{k_s^2}{v_m}\frac{1}{[S]^2}$$

The linregr function from Fig. 14.15 can again be used to determine the least-squares fit:

```
>> [a,r2]=linregr(1./S.^2,1./v)

a =
    19.3760    2.4492
r2 =
    0.9929
```

14.6 CASE STUDY continued

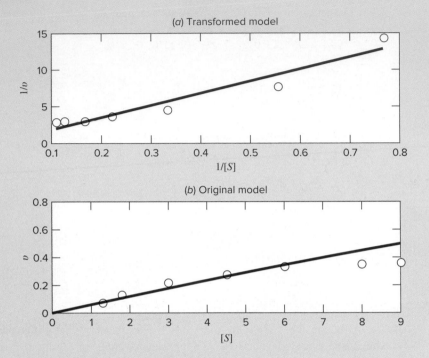

FIGURE 14.17
Plots of least-squares fit (line) of the Michaelis-Menten model along with data (points). The plot in (a) shows the transformed fit, and (b) shows how the fit looks when viewed in the untransformed, original form.

The model coefficients can then be calculated as

```
>> vm = 1/a(2)

vm =
    0.4083
>> ks = sqrt(vm*a(1))

ks =
    2.8127
```

Substituting these values into Eq. (14.29) gives

$$v = \frac{0.4083[S]^2}{7.911 + [S]^2}$$

Although we know that a high r^2 does not guarantee of a good fit, the fact that it is very high (0.9929) is promising. In addition, the parameters values also seem consistent with the trends in the data; that is, the k_m is slightly greater than the highest observed velocity and the half-saturation rate is lower than the maximum substrate concentration (9).

14.6 CASE STUDY continued

(a) Transformed model

(b) Original model

FIGURE 14.18
Plots of least-squares fit (line) of the second-order Michaelis-Menten model along with data (points). The plot in (a) shows the transformed fit, and (b) shows the untransformed, original form.

The adequacy of the fit can be assessed graphically. As in Fig. 14.18a, the transformed results appear linear. When the original equation is plotted along with the data in the untransformed version (Fig. 14.18b), the fit nicely follows the trend in the measurements. Beyond the graphs, the goodness of the fit is also reflected by the fact that the coefficient of determination for the untransformed case can be computed as $r^2 = 0.9896$.

Based on our analysis, we can conclude that the second-order model provides a good fit of this data set. This might suggest that we are dealing with an allosteric enzyme.

Beyond this specific result, there are a few other general conclusions that can be drawn from this case study. First, we should never solely rely on statistics such as r^2 as the sole basis of assessing goodness of fit. Second, regression equations should always be assessed graphically. And for cases where transformations are employed, a graph of the untransformed model and data should always be inspected.

Finally, although transformations may yield a decent fit of the transformed data, this does not always translate into an acceptable fit in the original format. The reason that this might occur is that minimizing squared residuals of transformed data is not the same as for the untransformed data. Linear regression assumes that the scatter of points around the

14.6 CASE STUDY continued

best-fit line follows a Gaussian distribution, and that the standard deviation is the same at every value of the dependent variable. These assumptions are rarely true after transforming data.

As a consequence of the last conclusion, some analysts suggest that rather than using linear transformations, nonlinear regression should be employed to fit curvilinear data. In this approach, a best-fit curve is developed that directly minimizes the untransformed residuals. We will describe how this is done in Chap. 15.

PROBLEMS

14.1 Given the data

0.90	1.42	1.30	1.55	1.63
1.32	1.35	1.47	1.95	1.66
1.96	1.47	1.92	1.35	1.05
1.85	1.74	1.65	1.78	1.71
2.29	1.82	2.06	2.14	1.27

Determine (a) the mean, (b) median, (c) mode, (d) range, (e) standard deviation, (f) variance, and (g) coefficient of variation.

14.2 Construct a histogram from the data from Prob. 14.1. Use a range from 0.8 to 2.4 with intervals of 0.2.

14.3 Given the data

29.65	28.55	28.65	30.15	29.35	29.75	29.25
30.65	28.15	29.85	29.05	30.25	30.85	28.75
29.65	30.45	29.15	30.45	33.65	29.35	29.75
31.25	29.45	30.15	29.65	30.55	29.65	29.25

Determine (a) the mean, (b) median, (c) mode, (d) range, (e) standard deviation, (f) variance, and (g) coefficient of variation.

(h) Construct a histogram. Use a range from 28 to 34 with increments of 0.4.

(i) Assuming that the distribution is normal, and that your estimate of the standard deviation is valid, compute the range (i.e., the lower and the upper values) that encompasses 68% of the readings. Determine whether this is a valid estimate for the data in this problem.

14.4 Using the same approach as was employed to derive Eqs. (14.15) and (14.16), derive the least-squares fit of the following model:

$$y = a_1 x + e$$

That is, determine the slope that results in the least-squares fit for a straight line with a zero intercept. Fit the following data with this model and display the result graphically.

| x | 2 | 4 | 6 | 7 | 10 | 11 | 14 | 17 | 20 |
| y | 4 | 5 | 6 | 5 | 8 | 8 | 6 | 9 | 12 |

14.5 Use least-squares regression to fit a straight line to

| x | 0 | 2 | 4 | 6 | 9 | 11 | 12 | 15 | 17 | 19 |
| y | 5 | 6 | 7 | 6 | 9 | 8 | 8 | 10 | 12 | 12 |

Along with the slope and intercept, compute the standard error of the estimate and the correlation coefficient. Plot the data and the regression line. Then repeat the problem, but regress x versus y—that is, switch the variables. Interpret your results.

14.6 Fit a power model to the data from Table 14.1, but use natural logarithms to perform the transformations.

14.7 The following data were gathered to determine the relationship between pressure and temperature of a fixed volume of 1 kg of nitrogen. The volume is 10 m³.

| T, °C | −40 | 0 | 40 | 80 | 120 | 160 |
| p, N/m² | 6900 | 8100 | 9350 | 10,500 | 11,700 | 12,800 |

Employ the ideal gas law $pV = nRT$ to determine R on the basis of these data. Note that for the law, T must be expressed in kelvins.

14.8 Beyond the examples in Fig. 14.13, there are other models that can be linearized using transformations. For example,

$$y = \alpha_4 x e^{\beta_4 x}$$

Linearize this model and use it to estimate α_4 and β_4 based on the following data. Develop a plot of your fit along with the data.

x	0.1	0.2	0.4	0.6	0.9	1.3	1.5	1.7	1.8
y	0.75	1.25	1.45	1.25	0.85	0.55	0.35	0.28	0.18

14.9 The concentration of E. coli bacteria in a swimming area is monitored after a storm:

t (hr)	4	8	12	16	20	24
c (CFU/100 mL)	1600	1320	1000	890	650	560

The time is measured in hours following the end of the storm and the unit CFU is a "colony forming unit." Use this data to estimate (**a**) the concentration at the end of the storm ($t = 0$) and (**b**) the time at which the concentration will reach 200 CFU/100 mL. Note that your choice of model should be consistent with the fact that negative concentrations are impossible and that the bacteria concentration always decreases with time.

14.10 Rather than using the base-e exponential model [Eq. (14.22)], a common alternative is to employ a base-10 model:

$$y = \alpha_5 10^{\beta_5 x}$$

When used for curve fitting, this equation yields identical results to the base-e version, but the value of the exponent parameter (β_5) will differ from that estimated with Eq. (14.22) (β_1). Use the base-10 version to solve Prob. 14.9. In addition, develop a formulation to relate β_1 to β_5.

14.11 Determine an equation to predict metabolism rate as a function of mass based on the following data. Use it to predict the metabolism rate of a 200-kg tiger.

Animal	Mass (kg)	Metabolism (watts)
Cow	400	270
Human	70	82
Sheep	45	50
Hen	2	4.8
Rat	0.3	1.45
Dove	0.16	0.97

14.12 On average, the surface area A of human beings is related to weight W and height H. Measurements on a number of individuals of height 180 cm and different weights (kg) give values of A (m^2) in the following table:

W (kg)	70	75	77	80	82	84	87	90
A (m²)	2.10	2.12	2.15	2.20	2.22	2.23	2.26	2.30

Show that a power law $A = aW^b$ fits these data reasonably well. Evaluate the constants a and b, and predict what the surface area is for a 95-kg person.

14.13 Fit an exponential model to

x	0.4	0.8	1.2	1.6	2	2.3
y	800	985	1490	1950	2850	3600

Plot the data and the equation on both standard and semilogarithmic graphs with the MATLAB subplot function.

14.14 An investigator has reported the data tabulated below for an experiment to determine the growth rate of bacteria k (per d) as a function of oxygen concentration c (mg/L). It is known that such data can be modeled by the following equation:

$$k = \frac{k_{max}c^2}{c_s + c^2}$$

where c_s and k_{max} are parameters. Use a transformation to linearize this equation. Then use linear regression to estimate c_s and k_{max} and predict the growth rate at $c = 2$ mg/L.

c	0.5	0.8	1.5	2.5	4
k	1.1	2.5	5.3	7.6	8.9

14.15 Develop an M-file function to compute descriptive statistics for a vector of values. Have the function determine and display number of values, mean, median, mode, range, standard deviation, variance, and coefficient of variation. In addition, have it generate a histogram. Test it with the data from Prob. 14.3.

14.16 Modify the linregr function in Fig. 14.15 so that it (**a**) computes and returns the standard error of the estimate and (**b**) uses the subplot function to also display a plot of the residuals (the predicted minus the measured y) versus x. Test it for the data from Examples 14.2 and 14.3.

14.17 Develop an M-file function to fit a power model. Have the function return the best-fit coefficient α_2 and power β_2 along with the r^2 for the untransformed model. In addition, use the subplot function to display graphs of both the transformed and untransformed equations along with the data. Test it with the data from Prob. 14.11.

14.18 The following data show the relationship between the viscosity of SAE 70 oil and temperature. After taking the log of the data, use linear regression to find the equation of the line that best fits the data and the r^2 value.

Temperature, °C	26.67	93.33	148.89	315.56
Viscosity, μ, N·s/m²	1.35	0.085	0.012	0.00075

14.19 You perform experiments and determine the following values of heat capacity c at various temperatures T for a gas:

T	−50	−30	0	60	90	110
c	1250	1280	1350	1480	1580	1700

Use regression to determine a model to predict c as a function of T.

14.20 It is known that the tensile strength of a plastic increases as a function of the time it is heat treated. The following data are collected:

Time	10	15	20	25	40	50	55	60	75
Tensile Strength	5	20	18	40	33	54	70	60	78

(a) Fit a straight line to these data and use the equation to determine the tensile strength at a time of 32 min.
(b) Repeat the analysis but for a straight line with a zero intercept.

14.21 The following data were taken from a stirred tank reactor for the reaction $A \rightarrow B$. Use the data to determine the best possible estimates for k_{01} and E_1 for the following kinetic model:

$$-\frac{dA}{dt} = k_{01} e^{-E_1/RT} A$$

where R is the gas constant and equals 0.00198 kcal/mol/K.

−dA/dt (moles/L/s)	460	960	2485	1600	1245
A (moles/L)	200	150	50	20	10
T (K)	280	320	450	500	550

14.22 Concentration data were collected at 15 time points for the polymerization reaction:

$$xA + yB \rightarrow A_x B_y$$

We assume the reaction occurs via a complex mechanism consisting of many steps. Several models have been hypothesized, and the sum of the squares of the residuals had been calculated for the fits of the models of the data. The results are shown below. Which model best describes the data (statistically)? Explain your choice.

	Model A	Model B	Model C
S_r	135	105	100
Number of Model Parameters Fit	2	3	5

14.23 Below are data taken from a batch reactor of bacterial growth (after lag phase was over). The bacteria are allowed to grow as fast as possible for the first 2.5 hours, and then they are induced to produce a recombinant protein, the production of which slows the bacterial growth significantly. The theoretical growth of bacteria can be described by

$$\frac{dX}{dt} = \mu X$$

where X is the number of bacteria, and μ is the specific growth rate of the bacteria during exponential growth. Based on the data, estimate the specific growth rate of the bacteria during the first 2 hours of growth and during the next 4 hours of growth.

Time, h	0	1	2	3	4	5	6
[Cells], g/L	0.100	0.335	1.102	1.655	2.453	3.702	5.460

14.24 A transportation engineering study was conducted to determine the proper design of bike lanes. Data were gathered on bike-lane widths and average distance between bikes and passing cars. The data from 9 streets are

Distance, m	2.4	1.5	2.4	1.8	1.8	2.9	1.2	3	1.2
Lane Width, m	2.9	2.1	2.3	2.1	1.8	2.7	1.5	2.9	1.5

(a) Plot the data.
(b) Fit a straight line to the data with linear regression. Add this line to the plot.
(c) If the minimum safe average distance between bikes and passing cars is considered to be 1.8 m, determine the corresponding minimum lane width.

14.25 In water-resources engineering, the sizing of reservoirs depends on accurate estimates of water flow in the river that is being impounded. For some rivers, long-term historical records of such flow data are difficult to obtain. In contrast, meteorological data on precipitation are often available for many years past. Therefore, it is often useful to determine a relationship between flow and precipitation. This relationship can then be used to estimate flows for years when only precipitation measurements were made. The following data are available for a river that is to be dammed:

Precip., cm/yr	88.9	108.5	104.1	139.7	127	94	116.8	99.1
Flow, m³/s	14.6	16.7	15.3	23.2	19.5	16.1	18.1	16.6

(a) Plot the data.
(b) Fit a straight line to the data with linear regression. Superimpose this line on your plot.
(c) Use the best-fit line to predict the annual water flow if the precipitation is 120 cm.
(d) If the drainage area is 1100 km², estimate what fraction of the precipitation is lost via processes such as evaporation, deep groundwater infiltration, and consumptive use.

14.26 The mast of a sailboat has a cross-sectional area of 10.65 cm² and is constructed of an experimental aluminum alloy. Tests were performed to define the relationship between stress and strain. The test results are

Strain, cm/cm	0.0032	0.0045	0.0055	0.0016	0.0085	0.0005
Stress, N/cm²	4970	5170	5500	3590	6900	1240

The stress caused by wind can be computed as F/A_c where $F =$ force in the mast and $A_c =$ mast's cross-sectional area. This value can then be substituted into Hooke's law to determine the mast's deflection, ΔL strain $\times L$, where $L =$ the mast's length. If the wind force is 25,000 N, use the data to estimate the deflection of a 9-m mast.

14.27 The following data were taken from an experiment that measured the current in a wire for various imposed voltages:

V, V	2	3	4	5	7	10
i, A	5.2	7.8	10.7	13	19.3	27.5

(a) On the basis of a linear regression of this data, determine current for a voltage of 3.5 V. Plot the line and the data and evaluate the fit.
(b) Redo the regression and force the intercept to be zero.

14.28 An experiment is performed to determine the % elongation of electrical conducting material as a function of temperature. The resulting data are listed below. Predict the % elongation for a temperature of 400 °C.

Temperature, °C	200	250	300	375	425	475	600
% Elongation		7.5	8.6	8.7	10	11.3	12.7	15.3

14.29 The population p of a small community on the outskirts of a city grows rapidly over a 20-year period:

t	0	5	10	15	20
p	100	200	450	950	2000

As an engineer working for a utility company, you must forecast the population 5 years into the future in order to anticipate the demand for power. Employ an exponential model and linear regression to make this prediction.

14.30 The velocity u of air flowing past a flat surface is measured at several distances y away from the surface. Fit a curve to this data assuming that the velocity is zero at the surface ($y = 0$). Use your result to determine the shear stress ($\mu \, du/dy$) at the surface where $\mu = 1.8 \times 10^{-5}$ N·s/m².

y, m	0.002	0.006	0.012	0.018	0.024
u, m/s	0.287	0.899	1.915	3.048	4.299

14.31 *Andrade's equation* has been proposed as a model of the effect of temperature on viscosity:

$$\mu = De^{B/T_a}$$

where $\mu =$ dynamic viscosity of water (10^{-3} N·s/m²), $T_a =$ absolute temperature (K), and D and B are parameters. Fit this model to the following data for water:

T	0	5	10	20	30	40
μ	1.787	1.519	1.307	1.002	0.7975	0.6529

14.32 Perform the same computation as in Example 14.2, but in addition to the drag coefficient, also vary the mass uniformly by ±10%.

14.33 Perform the same computation as in Example 14.3, but in addition to the drag coefficient, also vary the mass normally around its mean value with a coefficient of variation of 5.7887%.

14.34 Manning's formula for a rectangular channel can be written as

$$Q = \frac{1}{n_m} \frac{(BH)^{5/3}}{(B + 2H)^{2/3}} \sqrt{S}$$

where Q = flow (m³/s), n_m = a roughness coefficient, B = width (m), H = depth (m), and S = slope. You are applying this formula to a stream where you know that the width = 20 m and the depth = 0.3 m. Unfortunately, you know the roughness and the slope to only a ±10% precision. That is, you know that the roughness is about 0.03 with a range from 0.027 to 0.033 and the slope is 0.0003 with a range from 0.00027 to 0.00033. Assuming uniform distributions, use a Monte Carlo analysis with n = 10,000 to estimate the distribution of flow.

14.35 A Monte Carlo analysis can be used for optimization. For example, the trajectory of a ball can be computed with

$$y = (\tan\theta_0)x - \frac{g}{2v_0^2 \cos^2\theta_0} x^2 + y_0 \qquad \text{(P14.35)}$$

where y = the height (m), θ_0 = the initial angle (radians), v_0 = the initial velocity (m/s), g = the gravitational constant = 9.81 m/s², and y_0 = the initial height (m). Given y_0 = 1 m, v_0 = 25 m/s, and θ_0 = 50°, determine the maximum height and the corresponding x distance (a) analytically with calculus and (b) numerically with Monte Carlo simulation. For the latter, develop a script that generates a vector of 10,000 uniformly distributed values of x between 0 and 60 m. Use this vector and Eq. (P14.35) to generate a vector of heights. Then, employ the max function to determine the maximum height and the associated x distance.

14.36 *Stokes Settling Law* provides a means to compute the settling velocity of spherical particles under laminar conditions

$$v_s = \frac{g}{18} \frac{\rho_s - \rho}{\mu} d^2$$

where v_s = the terminal settling velocity (m/s), g = gravitational acceleration (= 9.81 m/s²), ρ = the fluid density (kg/m³), ρ_s = the particle density (kg/m³), μ = the dynamic viscosity of the fluid (N s/m²), and d = the particle diameter (m). Suppose that you conduct an experiment in which you measure the terminal settling velocities of a number of 10-μm spheres having different densities,

ρ_s, kg/m³	1500	1600	1700	1800	1900	2000	2100	2200	2300
v_s, 10⁻³ m/s	1.03	1.12	1.59	1.76	2.42	2.51	3.06	3	3.5

(a) Generate a labeled plot of the data. (b) Fit a straight line to the data with linear regression (polyfit) and superimpose this line on your plot. (c) Use the model to predict the settling velocity of a 2500 kg/m³ density sphere. (d) Use the slope and the intercept to estimate the fluid's viscosity and density.

14.37 Beyond the examples in Fig. 14.13, there are other models that can be linearized using transformations. For example, the following model applies to third-order chemical reactions in batch reactors

$$c = c_0 \frac{1}{\sqrt{1 + 2kc_0^2 t}}$$

where c = concentration (mg/L), c_0 = initial concentration (mg/L), k = reaction rate (L²/(mg² d)), and t = time (d). Linearize this model and use it to estimate k and c_0 based on the following data. Develop plots of your fit along with the data for both the transformed, linearized form and the untransformed form.

t	0	0.5	1	1.5	2	3	4	5
c	3.26	2.09	1.62	1.48	1.17	1.06	0.9	0.85

14.38 In Chap. 7 we presented optimization techniques to find the optimal values of one- and multi-dimensional functions. Random numbers provide an alternative means to solve the same sorts of problems (recall Prob. 14.35). This is done by repeatedly evaluating the function at randomly selected values of the independent variable and keeping track of the one that yields the best value of the function being optimized. If a sufficient number of samples are conducted, the optimum will eventually be located. In their simplest manifestations, such approaches are not very efficient. However, they do have the advantage that they can detect global optimums for functions with lots of local optima. Develop a function that uses random numbers to locate the maximum of the humps function

$$f(x) = \frac{1}{(x - 0.3)^2 + 0.01} + \frac{1}{(x - 0.9)^2 + 0.04} - 6$$

in the domain bounded by x = 0 to 2. Here is a script that you can use to test your function

```
clear,clc,clf,format compact
xmin=0;xmax=2;n=1000
xp=linspace(xmin,xmax,200); yp=f(xp);
plot(xp,yp)
[xopt,fopt]=RandOpt(@f,n,xmin,xmax)
```

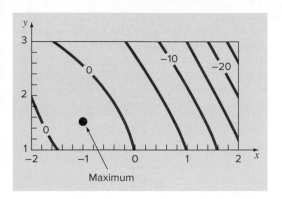

FIGURE P14.39
A two-dimensional function with a maximum of 1.25 at $x = -1$ and $y = 1.5$.

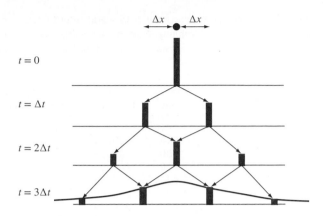

FIGURE P14.40
The one-dimensional random or "drunkard's" walk.

14.39 Using the same approach as described in Prob. 14.38, develop a function that uses random numbers to determine the maximum and corresponding x and y values of the following two-dimensional function

$$f(x, y) = y - x - 2x^2 - 2xy - y^2$$

in the domain bounded by $x = -2$ to 2 and $y = 1$ to 3. The domain is depicted in Fig. P14.39. Notice that a single maximum of 1.25 occurs at $x = -1$ and $y = 1.5$. Here is a script that you can use to test your function

```
clear,clc,format compact
xint=[-2;2];yint=[1;3];n=10000;
[xopt,yopt,fopt]=RandOpt2D(@fxy,n,xint,yint)
```

14.40 Suppose that a population of particles is confined to motion along a one-dimensional line (Fig. P14.40). Assume that each particle has an equal likelihood of moving a distance, Δx, to either the left or right over a time step, Δt. At $t = 0$ all particles are grouped at $x = 0$ and are allowed to take one step in either direction. After Δt, approximately 50% will step to the right and 50% to the left. After $2\Delta t$, 25% would be two steps to the left, 25% would be two steps to the right, and 50% would have stepped back to the origin. With additional time, the particles would spread out with the population greater near the origin and diminishing at the ends. The net result is that the distribution of the particles approaches a spreading bell-shaped distribution. This process, formally called a *random walk* (or *drunkard's walk*), describes many phenomena in engineering and science with a common example being *Brownian motion*. Develop a

MATLAB function that is given a stepsize (Δx), and a total number of particles (n) and steps (m). At each step, determine the location along the x axis of each particle and use these results to generate an animated histogram that displays how the distribution's shape evolves as the computation progresses.

14.41 Repeat Prob. 14.40, but for a two-dimensional random walk. As depicted in Fig. P14.41, have each particle take a random step of length Δ at a random angle θ ranging from 0 to 2π. Generate an animated two panel stacked plot with the location of all the particles displayed on the top plot (subplot (2,1,1)), and the histogram of the particles' x coordinates on the bottom (subplot (2,1,2)).

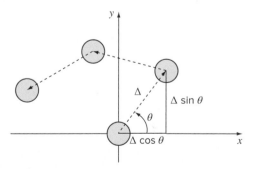

FIGURE P14.41
Depiction of the steps of a two-dimensional random or random walk.

14.42 The table below shows the 2015 world record times and holders for outdoor running. Note that all but the 100 m and the marathon (42,195 m) are run on oval tracks. Fit a power model for each gender and use it to predict the record time for a half marathon (21,097.5 m). Note that the actual records for the half marathon are 3503 s (Tadese) and 3909 s (Kiplagat) for men and women, respectively.

Event (m)	Time (s)	Men Holder	Time (s)	Women Holder
100	9.58	Bolt	10.49	Griffith-Joyner
200	19.19	Bolt	21.34	Griffith-Joyner
400	43.18	Johnson	47.60	Koch
800	100.90	Rudisha	113.28	Kratochvilova
1000	131.96	Ngeny	148.98	Masterkova
1500	206.00	El Guerrouj	230.07	Dibaba
2000	284.79	El Guerrouj	325.35	O'Sullivan
5000	757.40	Bekele	851.15	Dibaba
10,000	1577.53	Bekele	1771.78	Wang
20,000	3386.00	Gebrselassie	3926.60	Loroupe
42,195	7377.00	Kimetto	8125.00	Radcliffe

15

General Linear Least-Squares and Nonlinear Regression

CHAPTER OBJECTIVES

This chapter takes the concept of fitting a straight line and extends it to (*a*) fitting a polynomial and (*b*) fitting a variable that is a linear function of two or more independent variables. We will then show how such applications can be generalized and applied to a broader group of problems. Finally, we will illustrate how optimization techniques can be used to implement nonlinear regression. Specific objectives and topics covered are

- Knowing how to implement polynomial regression.
- Knowing how to implement multiple linear regression.
- Understanding the formulation of the general linear least-squares model.
- Understanding how the general linear least-squares model can be solved with MATLAB using either the normal equations or left division.
- Understanding how to implement nonlinear regression with optimization techniques.

15.1 POLYNOMIAL REGRESSION

In Chap.14, a procedure was developed to derive the equation of a straight line using the least-squares criterion. Some data, although exhibiting a marked pattern such as seen in Fig. 15.1, are poorly represented by a straight line. For these cases, a curve would be better suited to fit the data. As discussed in Chap. 14, one method to accomplish this objective is to use transformations. Another alternative is to fit polynomials to the data using *polynomial regression.*

The least-squares procedure can be readily extended to fit the data to a higher-order polynomial. For example, suppose that we fit a second-order polynomial or quadratic:

$$y = a_0 + a_1x + a_2x^2 + e \tag{15.1}$$

385

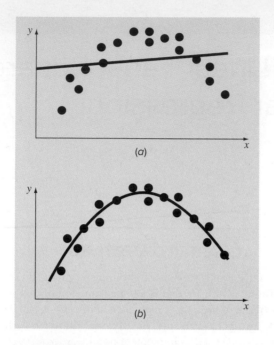

FIGURE 15.1
(a) Data that are ill-suited for linear least-squares regression. (b) Indication that a parabola is preferable.

For this case the sum of the squares of the residuals is

$$S_r = \sum_{i=1}^{n} \left(y_i - a_0 - a_1 x_i - a_2 x_i^2\right)^2 \tag{15.2}$$

To generate the least-squares fit, we take the derivative of Eq. (15.2) with respect to each of the unknown coefficients of the polynomial, as in

$$\frac{\partial S_r}{\partial a_0} = -2 \sum \left(y_i - a_0 - a_1 x_i - a_2 x_i^2\right)$$

$$\frac{\partial S_r}{\partial a_1} = -2 \sum x_i \left(y_i - a_0 - a_1 x_i - a_2 x_i^2\right)$$

$$\frac{\partial S_r}{\partial a_2} = -2 \sum x_i^2 \left(y_i - a_0 - a_1 x_i - a_2 x_i^2\right)$$

These equations can be set equal to zero and rearranged to develop the following set of normal equations:

$$(n)a_0 + \left(\sum x_i\right)a_1 + \left(\sum x_i^2\right)a_2 = \sum y_i$$

$$\left(\sum x_i\right)a_0 + \left(\sum x_i^2\right)a_1 + \left(\sum x_i^3\right)a_2 = \sum x_i y_i$$

$$\left(\sum x_i^2\right)a_0 + \left(\sum x_i^3\right)a_1 + \left(\sum x_i^4\right)a_2 = \sum x_i^2 y_i$$

where all summations are from $i = 1$ through n. Note that the preceding three equations are linear and have three unknowns: a_0, a_1, and a_2. The coefficients of the unknowns can be calculated directly from the observed data.

For this case, we see that the problem of determining a least-squares second-order polynomial is equivalent to solving a system of three simultaneous linear equations. The two-dimensional case can be easily extended to an mth-order polynomial as in

$$y = a_0 + a_1 x + a_2 x^2 + \cdots + a_m x^m + e$$

The foregoing analysis can be easily extended to this more general case. Thus, we can recognize that determining the coefficients of an mth-order polynomial is equivalent to solving a system of $m + 1$ simultaneous linear equations. For this case, the standard error is formulated as

$$s_{y/x} = \sqrt{\frac{S_r}{n - (m + 1)}} \tag{15.3}$$

This quantity is divided by $n - (m + 1)$ because $(m + 1)$ data-derived coefficients—$a_0, a_1, \ldots, a_m$—were used to compute S_r; thus, we have lost $m + 1$ degrees of freedom. In addition to the standard error, a coefficient of determination can also be computed for polynomial regression with Eq. (14.20).

EXAMPLE 15.1 Polynomial Regression

Problem Statement. Fit a second-order polynomial to the data in the first two columns of Table 15.1.

TABLE 15.1 Computations for an error analysis of the quadratic least-squares fit.

x_i	y_i	$(y_i - \bar{y})^2$	$(y_i - a_0 - a_1 x_i - a_2 x_i^2)^2$
0	2.1	544.44	0.14332
1	7.7	314.47	1.00286
2	13.6	140.03	1.08160
3	27.2	3.12	0.80487
4	40.9	239.22	0.61959
5	61.1	1272.11	0.09434
Σ	152.6	2513.39	3.74657

Solution. The following can be computed from the data:

$$m = 2 \qquad \sum x_i = 15 \qquad \sum x_i^4 = 979$$

$$n = 6 \qquad \sum y_i = 152.6 \qquad \sum x_i y_i = 585.6$$

$$\bar{x} = 2.5 \qquad \sum x_i^2 = 55 \qquad \sum x_i^2 y_i = 2488.8$$

$$\bar{y} = 25.433 \qquad \sum x_i^3 = 225$$

Therefore, the simultaneous linear equations are

$$\begin{bmatrix} 6 & 15 & 55 \\ 15 & 55 & 225 \\ 55 & 225 & 979 \end{bmatrix} \begin{Bmatrix} a_0 \\ a_1 \\ a_2 \end{Bmatrix} = \begin{Bmatrix} 152.6 \\ 585.6 \\ 2488.8 \end{Bmatrix}$$

These equations can be solved to evaluate the coefficients. For example, using MATLAB:

```
>> N = [6 15 55;15 55 225;55 225 979];
>> r = [152.6 585.6 2488.8];
>> a = N\r

a =
    2.4786
    2.3593
    1.8607
```

Therefore, the least-squares quadratic equation for this case is

$$y = 2.4786 + 2.3593x + 1.8607x^2$$

The standard error of the estimate based on the regression polynomial is [Eq. (15.3)]

$$s_{y/x} = \sqrt{\frac{3.74657}{6 - (2 + 1)}} = 1.1175$$

The coefficient of determination is

$$r^2 = \frac{2513.39 - 3.74657}{2513.39} = 0.99851$$

and the correlation coefficient is $r = 0.99925$

FIGURE 15.2
Fit of a second-order polynomial.

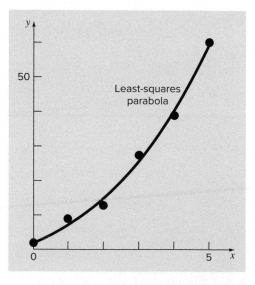

These results indicate that 99.851 percent of the original uncertainty has been explained by the model. This result supports the conclusion that the quadratic equation represents an excellent fit, as is also evident from Fig. 15.2.

15.2 MULTIPLE LINEAR REGRESSION

Another useful extension of linear regression is the case where y is a linear function of two or more independent variables. For example, y might be a linear function of x_1 and x_2, as in

$$y = a_0 + a_1 x_1 + a_2 x_2 + e$$

Such an equation is particularly useful when fitting experimental data where the variable being studied is often a function of two other variables. For this two-dimensional case, the regression "line" becomes a "plane" (Fig. 15.3).

As with the previous cases, the "best" values of the coefficients are determined by formulating the sum of the squares of the residuals:

$$S_r = \sum_{i=1}^{n} (y_i - a_0 - a_1 x_{1,i} - a_2 x_{2,i})^2 \tag{15.4}$$

and differentiating with respect to each of the unknown coefficients:

$$\frac{\partial S_r}{\partial a_0} = -2 \sum (y_i - a_0 - a_1 x_{1,i} - a_2 x_{2,i})$$

$$\frac{\partial S_r}{\partial a_1} = -2 \sum x_{1,i} (y_i - a_0 - a_1 x_{1,i} - a_2 x_{2,i})$$

$$\frac{\partial S_r}{\partial a_2} = -2 \sum x_{2,i} (y_i - a_0 - a_1 x_{1,i} - a_2 x_{2,i})$$

FIGURE 15.3
Graphical depiction of multiple linear regression where y is a linear function of x_1 and x_2.

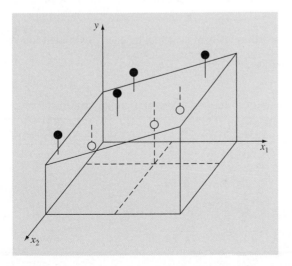

The coefficients yielding the minimum sum of the squares of the residuals are obtained by setting the partial derivatives equal to zero and expressing the result in matrix form as

$$
\begin{bmatrix}
n & \sum x_{1,i} & \sum x_{2,i} \\
\sum x_{1,i} & \sum x_{1,i}^2 & \sum x_{1,i} x_{2,i} \\
\sum x_{2,i} & \sum x_{1,i} x_{2,i} & \sum x_{2,i}^2
\end{bmatrix}
\begin{Bmatrix} a_0 \\ a_1 \\ a_2 \end{Bmatrix}
=
\begin{Bmatrix} \sum y_i \\ \sum x_{1,i} y_i \\ \sum x_{2,i} y_i \end{Bmatrix}
\tag{15.5}
$$

EXAMPLE 15.2 Multiple Linear Regression

Problem Statement. The following data were created from the equation $y = 5 + 4x_1 - 3x_2$:

x_1	x_2	y
0	0	5
2	1	10
2.5	2	9
1	3	0
4	6	3
7	2	27

Use multiple linear regression to fit this data.

Solution. The summations required to develop Eq. (15.5) are computed in Table 15.2. Substituting them into Eq. (15.5) gives

$$
\begin{bmatrix}
6 & 16.5 & 14 \\
16.5 & 76.25 & 48 \\
14 & 48 & 54
\end{bmatrix}
\begin{Bmatrix} a_0 \\ a_1 \\ a_2 \end{Bmatrix}
=
\begin{Bmatrix} 54 \\ 243.5 \\ 100 \end{Bmatrix}
\tag{15.6}
$$

which can be solved for

$$a_0 = 5 \qquad a_1 = 4 \qquad a_2 = -3$$

which is consistent with the original equation from which the data were derived.

The foregoing two-dimensional case can be easily extended to m dimensions, as in

$$y = a_0 + a_1 x_1 + a_2 x_2 + \cdots + a_m x_m + e$$

TABLE 15.2 Computations required to develop the normal equations for Example 15.2.

y	x_1	x_2	x_1^2	x_2^2	$x_1 x_2$	$x_1 y$	$x_2 y$
5	0	0	0	0	0	0	0
10	2	1	4	1	2	20	10
9	2.5	2	6.25	4	5	22.5	18
0	1	3	1	9	3	0	0
3	4	6	16	36	24	12	18
27	7	2	49	4	14	189	54
54	16.5	14	76.25	54	48	243.5	100

where the standard error is formulated as

$$s_{y/x} = \sqrt{\frac{S_r}{n - (m + 1)}}$$

and the coefficient of determination is computed with Eq. (14.20).

Although there may be certain cases where a variable is linearly related to two or more other variables, multiple linear regression has additional utility in the derivation of power equations of the general form

$$y = a_0 x_1^{a_1} x_2^{a_2} \cdots x_m^{a_m}$$

Such equations are extremely useful when fitting experimental data. To use multiple linear regression, the equation is transformed by taking its logarithm to yield

$$\log y = \log a_0 + a_1 \log x_1 + a_2 \log x_2 + \cdots + a_m \log x_m$$

15.3 GENERAL LINEAR LEAST SQUARES

In the preceding pages, we have introduced three types of regression: simple linear, polynomial, and multiple linear. In fact, all three belong to the following general linear least-squares model:

$$y = a_0 z_0 + a_1 z_1 + a_2 z_2 + \cdots + a_m z_m + e \tag{15.7}$$

where $z_0, z_1, \ldots, z_m$ are $m + 1$ basis functions. It can easily be seen how simple linear and multiple linear regression fall within this model—that is, $z_0 = 1, z_1 = x_1, z_2 = x_2, \ldots,$ $z_m = x_m$. Further, polynomial regression is also included if the basis functions are simple monomials as in $z_0 = 1, z_1 = x, z_2 = x^2, \ldots, z_m = x^m$.

Note that the terminology "linear" refers only to the model's dependence on its parameters—that is, the a's. As in the case of polynomial regression, the functions themselves can be highly nonlinear. For example, the z's can be sinusoids, as in

$$y = a_0 + a_1 \cos(\omega x) + a_2 \sin(\omega x)$$

Such a format is the basis of *Fourier analysis*.

On the other hand, a simple-looking model such as

$$y = a_0(1 - e^{-a_1 x})$$

is truly nonlinear because it cannot be manipulated into the format of Eq. (15.7).

Equation (15.7) can be expressed in matrix notation as

$$\{y\} = [Z]\{a\} + \{e\} \tag{15.8}$$

where $[Z]$ is a matrix of the calculated values of the basis functions at the measured values of the independent variables:

$$[Z] = \begin{bmatrix} z_{01} & z_{11} & \cdots & z_{m1} \\ z_{02} & z_{12} & \cdots & z_{m2} \\ \vdots & \vdots & & \vdots \\ z_{0n} & z_{1n} & \cdots & z_{mn} \end{bmatrix}$$

where m is the number of variables in the model and n is the number of data points. Because $n \geq m + 1$, you should recognize that most of the time, $[Z]$ is not a square matrix.

The column vector $\{y\}$ contains the observed values of the dependent variable:

$$\{y\}^T = \lfloor y_1 \quad y_2 \quad \cdots \quad y_n \rfloor$$

The column vector $\{a\}$ contains the unknown coefficients:

$$\{a\}^T = \lfloor a_0 \quad a_1 \quad \cdots \quad a_m \rfloor$$

and the column vector $\{e\}$ contains the residuals:

$$\{e\}^T = \lfloor e_1 \quad e_2 \quad \cdots \quad e_n \rfloor$$

The sum of the squares of the residuals for this model can be defined as

$$S_r = \sum_{i=1}^{n} \left(y_i - \sum_{j=0}^{n} a_j z_{ji} \right)^2 \tag{15.9}$$

This quantity can be minimized by taking its partial derivative with respect to each of the coefficients and setting the resulting equation equal to zero. The outcome of this process is the normal equations that can be expressed concisely in matrix form as

$$[[Z]^T [Z]]\{a\} = \{[Z]^T \{y\}\} \tag{15.10}$$

It can be shown that Eq. (15.10) is, in fact, equivalent to the normal equations developed previously for simple linear, polynomial, and multiple linear regression.

The coefficient of determination and the standard error can also be formulated in terms of matrix algebra. Recall that r^2 is defined as

$$r^2 = \frac{S_t - S_r}{S_t} = 1 - \frac{S_r}{S_t}$$

Substituting the definitions of S_r and S_t gives

$$r^2 = 1 - \frac{\sum(y_i - \hat{y}_i)^2}{\sum(y_i - \bar{y}_i)^2}$$

where $\hat{y}$ = the prediction of the least-squares fit. The residuals between the best-fit curve and the data, $y_i - \hat{y}$, can be expressed in vector form as

$$\{y\} - [Z]\{a\}$$

Matrix algebra can then be used to manipulate this vector to compute both the coefficient of determination and the standard error of the estimate as illustrated in the following example.

EXAMPLE 15.3   Polynomial Regression with MATLAB

Problem Statement. Repeat Example 15.1, but use matrix operations as described in this section.

Solution. First, enter the data to be fit

```
>> x = [0 1 2 3 4 5]';
>> y = [2.1 7.7 13.6 27.2 40.9 61.1]';
```

Next, create the $[Z]$ matrix:

```
>> Z = [ones(size(x)) x x.^2]
Z =
     1     0     0
     1     1     1
     1     2     4
     1     3     9
     1     4    16
     1     5    25
```

We can verify that $[Z]^T[Z]$ results in the coefficient matrix for the normal equations:

```
>> Z'*Z
ans =
      6    15    55
     15    55   225
     55   225   979
```

This is the same result we obtained with summations in Example 15.1. We can solve for the coefficients of the least-squares quadratic by implementing Eq. (15.10):

```
>> a = (Z'*Z)\(Z'*y)
ans =
    2.4786
    2.3593
    1.8607
```

In order to compute r^2 and $s_{y/x}$, first compute the sum of the squares of the residuals:

```
>> Sr = sum((y-Z*a).^2)
Sr =
    3.7466
```

Then r^2 can be computed as

```
>> r2 = 1-Sr/sum((y-mean(y)).^2)
r2 =
    0.9985
```

and $s_{y/x}$ can be computed as

```
>> syx = sqrt(Sr/(length(x)-length(a)))
syx =
    1.1175
```

Our primary motivation for the foregoing has been to illustrate the unity among the three approaches and to show how they can all be expressed simply in the same matrix notation. It also sets the stage for the next section where we will gain some insights into the preferred strategies for solving Eq. (15.10). The matrix notation will also have relevance when we turn to nonlinear regression in Sec. 15.5.

15.4 QR FACTORIZATION AND THE BACKSLASH OPERATOR

Generating a best fit by solving the normal equations is widely used and certainly adequate for many curve-fitting applications in engineering and science. It must be mentioned, however, that the normal equations can be ill-conditioned and hence sensitive to roundoff errors.

Two more advanced methods, *QR factorization* and *singular value decomposition,* are more robust in this regard. Although the description of these methods is beyond the scope of this text, we mention them here because they can be implemented with MATLAB.

Further, QR factorization is automatically used in two simple ways within MATLAB. First, for cases where you want to fit a polynomial, the built-in `polyfit` function automatically uses QR factorization to obtain its results.

Second, the general linear least-squares problem can be directly solved with the backslash operator. Recall that the general model is formulated as Eq. (15.8)

$$\{y\} = [Z]\{a\} \tag{15.11}$$

In Sec. 10.4, we used left division with the backslash operator to solve systems of linear algebraic equations where the number of equations equals the number of unknowns ($n = m$). For Eq. (15.8) as derived from general least squares, the number of equations is greater than the number of unknowns ($n > m$). Such systems are said to be *overdetermined.* When MATLAB senses that you want to solve such systems with left division, it automatically uses QR factorization to obtain the solution. The following example illustrates how this is done.

EXAMPLE 15.4   Implementing Polynomial Regression with `polyfit` and Left Division

Problem Statement. Repeat Example 15.3, but use the built-in `polyfit` function and left division to calculate the coefficients.

Solution. As in Example 15.3, the data can be entered and used to create the [Z] matrix as in

```
>> x = [0 1 2 3 4 5]';
>> y = [2.1 7.7 13.6 27.2 40.9 61.1]';
>> Z = [ones(size(x)) x x.^2];
```

The `polyfit` function can be used to compute the coefficients:

```
>> a = polyfit(x,y,2)

a =
    1.8607    2.3593    2.4786
```

The same result can also be calculated using the backslash:

```
>> a = Z\y

a =
    2.4786
    2.3593
    1.8607
```

As just stated, both these results are obtained automatically with QR factorization.

15.5 NONLINEAR REGRESSION

There are many cases in engineering and science where nonlinear models must be fit to data. In the present context, these models are defined as those that have a nonlinear dependence on their parameters. For example,

$$y = a_0(1 - e^{-a_1 x}) + e \tag{15.12}$$

This equation cannot be manipulated so that it conforms to the general form of Eq. (15.7).

As with linear least squares, nonlinear regression is based on determining the values of the parameters that minimize the sum of the squares of the residuals. However, for the nonlinear case, the solution must proceed in an iterative fashion.

There are techniques expressly designed for nonlinear regression. For example, the Gauss-Newton method uses a Taylor series expansion to express the original nonlinear equation in an approximate, linear form. Then least-squares theory can be used to obtain new estimates of the parameters that move in the direction of minimizing the residual. Details on this approach are provided elsewhere (Chapra and Canale, 2010).

An alternative is to use optimization techniques to directly determine the least-squares fit. For example, Eq. (15.12) can be expressed as an objective function to compute the sum of the squares:

$$f(a_0, a_1) = \sum_{i=1}^{n} [y_i - a_0(1 - e^{-a_1 x_i})]^2 \tag{15.13}$$

An optimization routine can then be used to determine the values of a_0 and a_1 that minimize the function.

As described previously in Sec. 7.3.1, MATLAB's fminsearch function can be used for this purpose. It has the general syntax

```
[x, fval] = fminsearch(fun,x0,options,p1,p2,...)
```

where x = a vector of the values of the parameters that minimize the function fun, fval = the value of the function at the minimum, x0 = a vector of the initial guesses for the parameters, options = a structure containing values of the optimization parameters as created with the optimset function (recall Sec. 6.5), and p1, p2, etc. = additional arguments that are passed to the objective function. Note that if options is omitted, MATLAB uses default values that are reasonable for most problems. If you would like to pass additional arguments (p1, p2, . . .), but do not want to set the options, use empty brackets [] as a place holder.

EXAMPLE 15.5   Nonlinear Regression with MATLAB

Problem Statement. Recall that in Example 14.6, we fit the power model to data from Table 14.1 by linearization using logarithms. This yielded the model:

$$F = 0.2741v^{1.9842}$$

Repeat this exercise, but use nonlinear regression. Employ initial guesses of 1 for the coefficients.

Solution. First, an M-file function must be created to compute the sum of the squares. The following file, called fSSR.m, is set up for the power equation:

```
function f = fSSR(a,xm,ym)
yp = a(1)*xm.^a(2);
f = sum((ym-yp).^2);
```

In command mode, the data can be entered as

```
>> x = [10 20 30 40 50 60 70 80];
>> y = [25 70 380 550 610 1220 830 1450];
```

The minimization of the function is then implemented by

```
>> fminsearch(@fSSR, [1, 1], [], x, y)

ans =
    2.5384    1.4359
```

The best-fit model is therefore

$$F = 2.5384v^{1.4359}$$

Both the original transformed fit and the present version are displayed in Fig. 15.4. Note that although the model coefficients are very different, it is difficult to judge which fit is superior based on inspection of the plot.

This example illustrates how different best-fit equations result when fitting the same model using nonlinear regression versus linear regression employing transformations. This is because the former minimizes the residuals of the original data whereas the latter minimizes the residuals of the transformed data.

FIGURE 15.4
Comparison of transformed and untransformed model fits for force versus velocity data from Table 14.1.

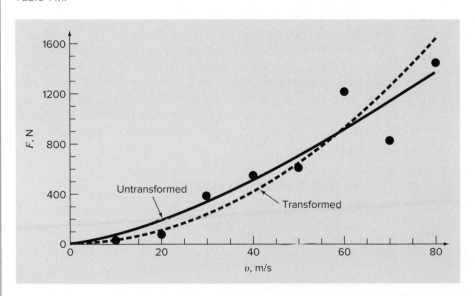

15.6 CASE STUDY FITTING EXPERIMENTAL DATA

Background. As mentioned at the end of Sec. 15.2, although there are many cases where a variable is linearly related to two or more other variables, multiple linear regression has additional utility in the derivation of multivariable power equations of the general form

$$y = a_0 x_1^{a_1} x_2^{a_2} \cdots x_m^{a_m} \tag{15.14}$$

Such equations are extremely useful when fitting experimental data. To do this, the equation is transformed by taking its logarithm to yield

$$\log y = \log a_0 + a_1 \log x_1 + a_2 \log x_2 \cdots + a_m \log x_m \tag{15.15}$$

Thus, the logarithm of the dependent variable is linearly dependent on the logarithms of the independent variables.

A simple example relates to gas transfer in natural waters such as rivers, lakes, and estuaries. In particular, it has been found that the mass-transfer coefficient of dissolved oxygen K_L (m/d) is related to a river's mean water velocity U (m/s) and depth H (m) by

$$K_L = a_0 U^{a_1} H^{a_2} \tag{15.16}$$

Taking the common logarithm yields

$$\log K_L = \log a_0 + a_1 \log U + a_2 \log H \tag{15.17}$$

The following data were collected in a laboratory flume at a constant temperature of 20 °C:

U	0.5	2	10	0.5	2	10	0.5	2	10
H	0.15	0.15	0.15	0.3	0.3	0.3	0.5	0.5	0.5
K_L	0.48	3.9	57	0.85	5	77	0.8	9	92

Use these data and general linear least squares to evaluate the constants in Eq. (15.16).

Solution. In a similar fashion to Example 15.3, we can develop a script to assign the data, create the [Z] matrix, and compute the coefficients for the least-squares fit:

```
% Compute best fit of transformed values
clc; format short g
U=[0.5 2 10 0.5 2 10 0.5 2 10]';
H=[0.15 0.15 0.15 0.3 0.3 0.3 0.5 0.5 0.5]';
KL=[0.48 3.9 57 0.85 5 77 0.8 9 92]';
logU=log10(U);logH=log10(H);logKL=log10(KL);
Z=[ones(size(logKL)) logU logH];
a=(Z'*Z)\(Z'*logKL)
```

15.6 CASE STUDY   continued

with the result:

```
a =
    0.57627
    1.562
    0.50742
```

Therefore, the best-fit model is

$$\log K_L = 0.57627 + 1.562 \log U + 0.50742 \log H$$

or in the untransformed form (note, $a_0 = 10^{0.57627} = 3.7694$),

$$K_L = 3.7694 U^{1.5620} H^{0.5074}$$

The statistics can also be determined by adding the following lines to the script:

```
% Compute fit statistics
Sr =sum((logKL-Z*a).^2)
r2 =1-Sr/sum((logKL-mean(logKL)).^2)
syx=sqrt(Sr/(length(logKL)-length(a)))

Sr =
    0.024171
r2 =
    0.99619
syx =
     0.063471
```

Finally, plots of the fit can be developed. The following statements display the model predictions versus the measured values for K_L. Subplots are employed to do this for both the transformed and untransformed versions.

```
%Generate plots
clf
KLpred=10^a(1)*U.^a(2).*H.^a(3);
KLmin=min(KL);KLmax=max(KL);
dKL=(KLmax-KLmin)/100;
KLmod=[KLmin:dKL:KLmax];
subplot(1,2,1)
loglog(KLpred,KL,'ko',KLmod,KLmod,'k-')
axis square,title('(a) log-log plot')
legend('model prediction','1:1
line','Location','NorthWest')
xlabel('log(K_L) measured'),ylabel('log(K_L) predicted')
subplot(1,2,2)
plot(KLpred,KL,'ko',KLmod,KLmod,'k-')
axis square,title('(b) untransformed plot')
legend('model prediction','1:1
line','Location','NorthWest')
xlabel('K_L measured'),ylabel('K_L predicted')
```

15.6 CASE STUDY   continued

The result is shown in Fig. 15.5.

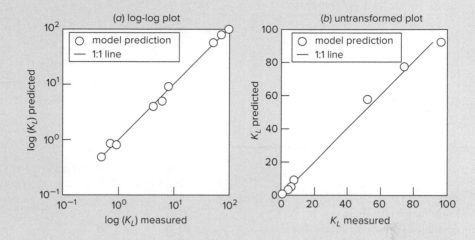

FIGURE 15.5
Plots of predicted versus measured values of the oxygen mass-transfer coefficient as computed with multiple regression. Results are shown for (a) log transformed and (b) untransformed cases. The 1:1 line, which indicates a perfect correlation, is superimposed on both plots.

PROBLEMS

15.1 Fit a parabola to the data from Table 14.1. Determine the r^2 for the fit and comment on the efficacy of the result.

15.2 Using the same approach as was employed to derive Eqs. (14.15) and (14.16), derive the least-squares fit of the following model:

$$y = a_1 x + a_2 x^2 + e$$

That is, determine the coefficients that result in the least-squares fit for a second-order polynomial with a zero intercept. Test the approach by using it to fit the data from Table 14.1.

15.3 Fit a cubic polynomial to the following data:

x	3	4	5	7	8	9	11	12
y	1.6	3.6	4.4	3.4	2.2	2.8	3.8	4.6

Along with the coefficients, determine r^2 and $s_{y/x}$.

15.4 Develop an M-file to implement polynomial regression. Pass the M-file two vectors holding the x and y values along with the desired order m. Test it by solving Prob. 15.3.

15.5 For the data from Table P15.5, use polynomial regression to derive a predictive equation for dissolved oxygen concentration as a function of temperature for the case where the chloride concentration is equal to zero. Employ a polynomial that is of sufficiently high order that the predictions match the number of significant digits displayed in the table.

15.6 Use multiple linear regression to derive a predictive equation for dissolved oxygen concentration as a function of temperature and chloride based on the data from Table P15.5. Use the equation to estimate the concentration of dissolved oxygen for a chloride concentration of 15 g/L at $T = 12$ °C. Note that the true value is 9.09 mg/L. Compute

TABLE P15.5 Dissolved oxygen concentration in water as a function of temperature (°C) and chloride concentration (g/L).

T, °C	Dissolved Oxygen (mg/L) for Temperature (°C) and Concentration of Chloride (g/L)		
	c = 0 g/L	c = 10 g/L	c = 20 g/L
0	14.6	12.9	11.4
5	12.8	11.3	10.3
10	11.3	10.1	8.96
15	10.1	9.03	8.08
20	9.09	8.17	7.35
25	8.26	7.46	6.73
30	7.56	6.85	6.20

the percent relative error for your prediction. Explain possible causes for the discrepancy.

15.7 As compared with the models from Probs. 15.5 and 15.6, a somewhat more sophisticated model that accounts for the effect of both temperature and chloride on dissolved oxygen saturation can be hypothesized as being of the form

$$o = f_3(T) + f_1(c)$$

That is, a third-order polynomial in temperature and a linear relationship in chloride is assumed to yield superior results. Use the general linear least-squares approach to fit this model to the data in Table P15.5. Use the resulting equation to estimate the dissolved oxygen concentration for a chloride concentration of 15 g/L at $T = 12$ °C. Note that the true value is 9.09 mg/L. Compute the percent relative error for your prediction.

15.8 Use multiple linear regression to fit

x_1	0	1	1	2	2	3	3	4	4
x_2	0	1	2	1	2	1	2	1	2
y	15.1	17.9	12.7	25.6	20.5	35.1	29.7	45.4	40.2

Compute the coefficients, the standard error of the estimate, and the correlation coefficient.

15.9 The following data were collected for the steady flow of water in a concrete circular pipe:

Experiment	Diameter, m	Slope, m/m	Flow, m³/s
1	0.3	0.001	0.04
2	0.6	0.001	0.24
3	0.9	0.001	0.69
4	0.3	0.01	0.13
5	0.6	0.01	0.82
6	0.9	0.01	2.38
7	0.3	0.05	0.31
8	0.6	0.05	1.95
9	0.9	0.05	5.66

Use multiple linear regression to fit the following model to this data:

$$Q = \alpha_0 D^{\alpha_1} S^{\alpha_2}$$

where Q = flow, D = diameter, and S = slope.

15.10 Three disease-carrying organisms decay exponentially in seawater according to the following model:

$$p(t) = Ae^{-1.5t} + Be^{-0.3t} + Ce^{-0.05t}$$

Use general linear least-squares to estimate the initial concentration of each organism (A, B, and C) given the following measurements:

t	0.5	1	2	3	4	5	6	7	9
$p(t)$	6	4.4	3.2	2.7	2	1.9	1.7	1.4	1.1

15.11 The following model is used to represent the effect of solar radiation on the photosynthesis rate of aquatic plants:

$$P = P_m \frac{I}{I_{sat}} e^{-\frac{I}{I_{sat}} + 1}$$

where P = the photosynthesis rate (mg m^{-3}d^{-1}), P_m = the maximum photosynthesis rate (mg m^{-3}d^{-1}), I = solar radiation (μE m^{-2}s^{-1}), and I_{sat} = optimal solar radiation (μE m^{-2}s^{-1}). Use nonlinear regression to evaluate P_m and I_{sat} based on the following data:

I	50	80	130	200	250	350	450	550	700
P	99	177	202	248	229	219	173	142	72

15.12 The following data are provided

x	1	2	3	4	5
y	2.2	2.8	3.6	4.5	5.5

Fit the following model to this data using MATLAB and the general linear least-squares model

$$y = a + bx + \frac{c}{x}$$

15.13 In Prob. 14.8 we used transformations to linearize and fit the following model:

$$y = \alpha_4 x e^{\beta_4 x}$$

Use nonlinear regression to estimate α_4 and β_4 based on the following data. Develop a plot of your fit along with the data.

x	0.1	0.2	0.4	0.6	0.9	1.3	1.5	1.7	1.8
y	0.75	1.25	1.45	1.25	0.85	0.55	0.35	0.28	0.18

15.14 Enzymatic reactions are used extensively to characterize biologically mediated reactions. The following is an example of a model that is used to fit such reactions:

$$v_0 = \frac{k_m[S]^3}{K + [S]^3}$$

where v_0 = the initial rate of the reaction (M/s), $[S]$ = the substrate concentration (M), and k_m and K are parameters. The following data can be fit with this model:

[S], M	v_0, M/s
0.01	6.078×10^{-11}
0.05	7.595×10^{-9}
0.1	6.063×10^{-8}
0.5	5.788×10^{-6}
1	1.737×10^{-5}
5	2.423×10^{-5}
10	2.430×10^{-5}
50	2.431×10^{-5}
100	2.431×10^{-5}

(a) Use a transformation to linearize the model and evaluate the parameters. Display the data and the model fit on a graph.
(b) Perform the same evaluation as in (a) but use nonlinear regression.

15.15 Given the data

x	5	10	15	20	25	30	35	40	45	50
y	17	24	31	33	37	37	40	40	42	41

use least-squares regression to fit (a) a straight line, (b) a power equation, (c) a saturation-growth-rate equation, and (d) a parabola. For (b) and (c), employ transformations to linearize the data. Plot the data along with all the curves. Is any one of the curves superior? If so, justify.

15.16 The following data represent the bacterial growth in a liquid culture over of number of days:

Day	0	4	8	12	16	20
Amount × 10^6		67.38	74.67	82.74	91.69	101.60	112.58

Find a best-fit equation to the data trend. Try several possibilities—linear, quadratic, and exponential. Determine the best equation to predict the amount of bacteria after 35 days.

15.17 Dynamic viscosity of water $\mu(10^{-3}\ N \cdot s/m^2)$ is related to temperature $T(°C)$ in the following manner:

T	0	5	10	20	30	40
μ	1.787	1.519	1.307	1.002	0.7975	0.6529

(a) Plot this data.
(b) Use linear interpolation to predict μ at $T = 7.5\ °C$.
(c) Use polynomial regression to fit a parabola to the data in order to make the same prediction.

15.18 Use general linear least squares to find the best possible virial constants (A_1 and A_2) for the following equation of state. $R = 82.05$ mL atm/gmol K, and $T = 303$ K.

$$\frac{PV}{RT} = 1 + \frac{A_1}{V} + \frac{A_2}{V^2}$$

P (atm)	0.985	1.108	1.363	1.631
V (mL)	25,000	22,200	18,000	15,000

15.19 Environmental scientists and engineers dealing with the impacts of acid rain must determine the value of the ion product of water K_w as a function of temperature. Scientists have suggested the following equation to model this relationship:

$$-\log_{10} K_w = \frac{a}{T_a} + b \log_{10} T_a + cT_a + d$$

where T_a = absolute temperature (K), and a, b, c, and d are parameters. Employ the following data and regression to estimate the parameters with MATLAB. Also, generate a plot of predicted K_w versus the data.

T (°C)	K_w
0	1.164×10^{-15}
10	2.950×10^{-15}
20	6.846×10^{-15}
30	1.467×10^{-14}
40	2.929×10^{-14}

15.20 The distance required to stop an automobile consists of both thinking and braking components, each of which is a function of its speed. The following experimental data were collected to quantify this relationship. Develop best-fit equations for both the thinking and braking components. Use these equations to estimate the total stopping distance for a car traveling at 110 km/hr.

Speed, km/hr	30	45	60	75	90	120
Thinking, m	5.6	8.5	11.1	14.5	16.7	22.4
Braking, m	5.0	12.3	21.0	32.9	47.6	84.7

15.21 An investigator has reported the data tabulated below. It is known that such data can be modeled by the following equation

$$x = e^{(y-b)/a}$$

where a and b are parameters. Use nonlinear regression to determine a and b. Based on your analysis predict y at $x = 2.6$.

x	1	2	3	4	5
y	0.5	2	2.9	3.5	4

15.22 It is known that the data tabulated below can be modeled by the following equation

$$y = \left(\frac{a + \sqrt{x}}{b\sqrt{x}}\right)^2$$

Use nonlinear regression to determine the parameters a and b. Based on your analysis predict y at $x = 1.6$.

x	0.5	1	2	3	4
y	10.4	5.8	3.3	2.4	2

15.23 An investigator has reported the data tabulated below for an experiment to determine the growth rate of bacteria k (per d), as a function of oxygen concentration c (mg/L). It is known that such data can be modeled by the following equation:

$$k = \frac{k_{max}c^2}{c_s + c^2}$$

Use nonlinear regression to estimate c_s and k_{max} and predict the growth rate at $c = 2$ mg/L.

c	0.5	0.8	1.5	2.5	4
k	1.1	2.4	5.3	7.6	8.9

15.24 A material is tested for cyclic fatigue failure whereby a stress, in MPa, is applied to the material and the number of cycles needed to cause failure is measured. The results are in the table below. Use nonlinear regression to fit a power model to this data.

N, cycles	1	10	100	1000	10,000	100,000	1,000,000
Stress, MPa	1100	1000	925	800	625	550	420

15.25 The following data shows the relationship between the viscosity of SAE 70 oil and temperature. Use nonlinear regression to fit a power equation to this data.

Temperature, T, °C	26.67	93.33	148.89	315.56
Viscosity, μ, N·s/m^2	1.35	0.085	0.012	0.00075

15.26 The concentration of E. coli bacteria in a swimming area is monitored after a storm:

t (hr)	4	8	12	16	20	24
c(CFU/100 mL)	1590	1320	1000	900	650	560

The time is measured in hours following the end of the storm and the unit CFU is a "colony forming unit." Employ nonlinear regression to fit an exponential model [Eq. (14.22)] to this data. Use the model to estimate **(a)** the concentration at the end of the storm ($t = 0$) and **(b)** the time at which the concentration will reach 200 CFU/100 mL.

15.27 Employ nonlinear regression and the following set of pressure-volume data to find the best possible virial constants (A_1 and A_2) for the equation of state shown below. $R = 82.05$ mL atm/gmol K and $T = 303$ K.

$$\frac{PV}{RT} = 1 + \frac{A_1}{V} + \frac{A_2}{V^2}$$

P (atm)	0.985	1.108	1.363	1.631
V (mL)	25,000	22,200	18,000	15,000

15.28 Three disease-carrying organisms decay exponentially in lake water according to the following model:

$$p(t) = Ae^{-1.5t} + Be^{-0.3t} + Ce^{-0.05t}$$

Use nonlinear regression to estimate the initial population of each organism (A, B, and C) given the following measurements:

t, hr	0.5	1	2	3	4	5	6	7	9
p(t)	6.0	4.4	3.2	2.7	2.2	1.9	1.7	1.4	1.1

15.29 The *Antoine equation* describes the relation between vapor pressure and temperature for pure components as

$$\ln(p) = A - \frac{B}{C + T}$$

where p is the vapor pressure, T is temperature (K), and A, B, and C are component-specific constants. Use MATLAB to determine the best values for the constants for carbon monoxide based on the following measurements

T (K)	50	60	70	80	90	100	110	120	130
p (Pa)	82	2300	18,500	80,500	2.3×10^5	5×10^5	9.6×10^5	1.5×10^6	2.4×10^6

In addition to the constants determine the r^2 and $s_{y/x}$ for your fit.

15.30 The following model, based on a simplification of the *Arrhenius equation*, is frequently used in environmental engineering to parameterize the effect of temperature, $T(°C)$, on pollutant decay rates, k (per day),

$$k = k_{20}\, \theta^{T-20}$$

where the parameters k_{20} = the decay rate at 20 °C, and θ = the dimensionless temperature dependence coefficient. The following data are collected in the laboratory

T (°C)	6	12	18	24	30
k (per d)	0.15	0.20	0.32	0.45	0.70

(a) Use a transformation to linearize this equation and then employ linear regression to estimate k_{20} and θ. (b) Employ nonlinear regression to estimate the same parameters. For both (a) and (b) employ the equation to predict the reaction rate at $T = 17$ °C.

16

Fourier Analysis

CHAPTER OBJECTIVES

The primary objective of this chapter is to introduce you to Fourier analysis. The subject, which is named after Joseph Fourier, involves identifying cycles or patterns within a time series of data. Specific objectives and topics covered in this chapter are

- Understanding sinusoids and how they can be used for curve fitting.
- Knowing how to use least-squares regression to fit a sinusoid to data.
- Knowing how to fit a Fourier series to a periodic function.
- Understanding the relationship between sinusoids and complex exponentials based on Euler's formula.
- Recognizing the benefits of analyzing mathematical function or signals in the frequency domain (i.e., as a function of frequency).
- Understanding how the Fourier integral and transform extend Fourier analysis to aperiodic functions.
- Understanding how the discrete Fourier transform (DFT) extends Fourier analysis to discrete signals.
- Recognizing how discrete sampling affects the ability of the DFT to distinguish frequencies. In particular, know how to compute and interpret the Nyquist frequency.
- Recognizing how the fast Fourier transform (FFT) provides a highly efficient means to compute the DFT for cases where the data record length is a power of 2.
- Knowing how to use the MATLAB function `fft` to compute a DFT and understand how to interpret the results.
- Knowing how to compute and interpret a power spectrum.

YOU'VE GOT A PROBLEM

At the beginning of Chap. 8, we used Newton's second law and force balances to predict the equilibrium positions of three bungee jumpers connected by cords. Then, in Chap. 13, we determined the same system's eigenvalues and eigenvectors in order to identify its resonant frequencies and principal modes of vibration. Although this analysis certainly provided useful results, it required detailed system information including knowledge of the underlying model and parameters (i.e., the jumpers' masses and the cords' spring constants).

So suppose that you have measurements of the jumpers' positions or velocities at discrete, equally spaced times (recall Fig. 13.1). Such information is referred to as a *time series*. However, suppose further that you do not know the underlying model or the parameters needed to compute the eigenvalues. For such cases, is there any way to use the time series to learn something fundamental about the system's dynamics?

In this chapter, we describe such an approach, *Fourier analysis*, which provides a way to accomplish this objective. The approach is based on the premise that more complicated functions (e.g., a time series) can be represented by the sum of simpler trigonometric functions. As a prelude to outlining how this is done, it is useful to explore how data can be fit with sinusoidal functions.

16.1 CURVE FITTING WITH SINUSOIDAL FUNCTIONS

A periodic function $f(t)$ is one for which

$$f(t) = f(t + T) \tag{16.1}$$

where T is a constant called the *period* that is the smallest value of time for which Eq. (16.1) holds. Common examples include both artificial and natural signals (Fig. 16.1a).

The most fundamental are sinusoidal functions. In this discussion, we will use the term *sinusoid* to represent any waveform that can be described as a sine or cosine. There is no

---

FIGURE 16.1
Aside from trigonometric functions such as sines and cosines, periodic functions include idealized waveforms like the square wave depicted in (a). Beyond such artificial forms, periodic signals in nature can be contaminated by noise like the air temperatures shown in (b).

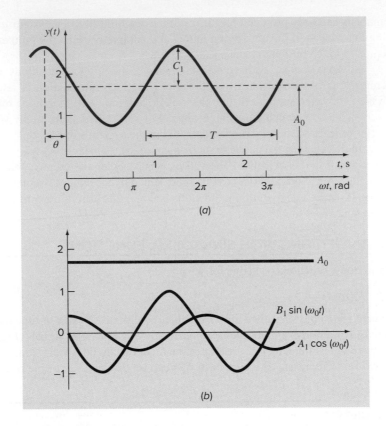

FIGURE 16.2
(a) A plot of the sinusoidal function $y(t) = A_0 + C_1 \cos(\omega_0 t + \theta)$. For this case, $A_0 = 1.7$, $C_1 = 1$, $\omega_0 = 2\pi/T = 2\pi/(1.5 \text{ s})$, and $\theta = \pi/3$ radians $= 1.0472$ ($= 0.25$ s). Other parameters used to describe the curve are the frequency $f = \omega_0/(2\pi)$, which for this case is 1 cycle/(1.5 s) $= 0.6667$ Hz and the period $T = 1.5$ s. (b) An alternative expression of the same curve is $y(t) = A_0 + A_1 \cos(\omega_0 t) + B_1 \sin(\omega_0 t)$. The three components of this function are depicted in (b), where $A_1 = 0.5$ and $B_1 = -0.866$. The summation of the three curves in (b) yields the single curve in (a).

clear-cut convention for choosing either function, and in any case, the results will be identical because the two functions are simply offset in time by $\pi/2$ radians. For this chapter, we will use the cosine, which can be expressed generally as

$$f(t) = A_0 + C_1 \cos(\omega_0 t + \theta) \tag{16.2}$$

Inspection of Eq. (16.2) indicates that four parameters serve to uniquely characterize the sinusoid (Fig. 16.2a):

- The *mean value* A_0 sets the average height above the abscissa.
- The *amplitude* C_1 specifies the height of the oscillation.
- The *angular frequency* ω_0 characterizes how often the cycles occur.
- The *phase angle* (or *phase shift*) θ parameterizes the extent to which the sinusoid is shifted horizontally.

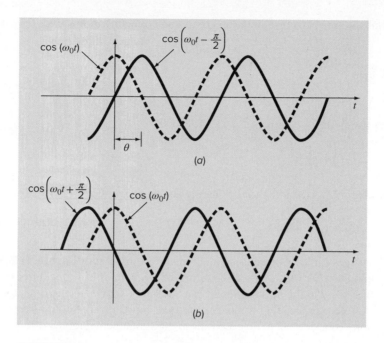

FIGURE 16.3
Graphical depictions of (*a*) a lagging phase angle and (*b*) a leading phase angle. Note that the lagging curve in (*a*) can be alternatively described as $\cos(\omega_0 t + 3\pi/2)$. In other words, if a curve lags by an angle of α, it can also be represented as leading by $2\pi - \alpha$.

Note that the *angular frequency* (in radians/time) is related to the *ordinary frequency* f (in cycles/time)[1] by

$$\omega_0 = 2\pi f \tag{16.3}$$

and the ordinary frequency in turn is related to the period T by

$$f = \frac{1}{T} \tag{16.4}$$

In addition, the *phase angle* represents the distance in radians from $t = 0$ to the point at which the cosine function begins a new cycle. As depicted in Fig. 16.3*a*, a negative value is referred to as a *lagging phase angle* because the curve $\cos(\omega_0 t - \theta)$ begins a new cycle θ radians after $\cos(\omega_0 t)$. Thus, $\cos(\omega_0 t - \theta)$ is said to lag $\cos(\omega_0 t)$. Conversely, as in Fig. 16.3*b*, a positive value is referred to as a *leading phase angle*.

Although Eq. (16.2) is an adequate mathematical characterization of a sinusoid, it is awkward to work with from the standpoint of curve fitting because the phase shift is included in the argument of the cosine function. This deficiency can be overcome by invoking the trigonometric identity:

$$C_1 \cos(\omega_0 t + \theta) = C_1[\cos(\omega_0 t)\cos(\theta) - \sin(\omega_0 t)\sin(\theta)] \tag{16.5}$$

[1] When the time unit is seconds, the unit for the ordinary frequency is a cycle/s or *Hertz* (Hz).

Substituting Eq. (16.5) into Eq. (16.2) and collecting terms gives (Fig. 16.2b)

$$f(t) = A_0 + A_1\cos(\omega_0 t) + B_1\sin(\omega_0 t) \tag{16.6}$$

where

$$A_1 = C_1\cos(\theta) \qquad\qquad B_1 = -C_1\sin(\theta) \tag{16.7}$$

Dividing the two parts of Eq. (16.7) gives

$$\theta = \arctan\left(-\frac{B_1}{A_1}\right) \tag{16.8}$$

where, if $A_1 < 0$, add π to θ. Squaring and summing Eq. (16.7) leads to

$$C_1 = \sqrt{A_1^2 + B_1^2} \tag{16.9}$$

Thus, Eq. (16.6) represents an alternative formulation of Eq. (16.2) that still requires four parameters but that is cast in the format of a general linear model [recall Eq. (15.7)]. As we will discuss in the next section, it can be simply applied as the basis for a least-squares fit.

Before proceeding to the next section, however, we should stress that we could have employed a sine rather than a cosine as our fundamental model of Eq. (16.2). For example,

$$f(t) = A_0 + C_1\sin(\omega_0 t + \delta)$$

could have been used. Simple relationships can be applied to convert between the two forms:

$$\sin(\omega_0 t + \delta) = \cos\left(\omega_0 t + \delta - \frac{\pi}{2}\right)$$

and

$$\cos(\omega_0 t + \delta) = \sin\left(\omega_0 t + \delta + \frac{\pi}{2}\right) \tag{16.10}$$

In other words, $\theta = \delta - \pi/2$. The only important consideration is that one or the other format should be used consistently. Thus, we will use the cosine version throughout our discussion.

16.1.1 Least-Squares Fit of a Sinusoid

Equation (16.6) can be thought of as a linear least-squares model:

$$y = A_0 + A_1\cos(\omega_0 t) + B_1\sin(\omega_0 t) + e \tag{16.11}$$

which is just another example of the general model [recall Eq. (15.7)]

$$y = a_0 z_0 + a_1 z_1 + a_2 z_2 + \cdots + a_m z_m + e$$

where $z_0 = 1$, $z_1 = \cos(\omega_0 t)$, $z_2 = \sin(\omega_0 t)$, and all other z's $= 0$. Thus, our goal is to determine coefficient values that minimize

$$S_r = \sum_{i=1}^{N} \{y_i - [A_0 + A_1\cos(\omega_0 t) + B_1\sin(\omega_0 t)]\}^2$$

The normal equations to accomplish this minimization can be expressed in matrix form as [recall Eq. (15.10)]

$$
\begin{bmatrix}
N & \sum \cos(\omega_0 t) & \sum \sin(\omega_0 t) \\
\sum \cos(\omega_0 t) & \sum \cos^2(\omega_0 t) & \sum \cos(\omega_0 t)\sin(\omega_0 t) \\
\sum \sin(\omega_0 t) & \sum \cos(\omega_0 t)\sin(\omega_0 t) & \sum \sin^2(\omega_0 t)
\end{bmatrix}
\begin{Bmatrix} A_0 \\ B_1 \\ B_1 \end{Bmatrix} =
\begin{Bmatrix} \sum y \\ \sum y\cos(\omega_0 t) \\ \sum y\sin(\omega_0 t) \end{Bmatrix}
$$

(16.12)

These equations can be employed to solve for the unknown coefficients. However, rather than do this, we can examine the special case where there are N observations equispaced at intervals of Δt and with a total record length of $T = (N - 1)\Delta t$. For this situation, the following average values can be determined (see Prob. 16.5):

$$
\frac{\sum \sin(\omega_0 t)}{N} = 0 \qquad \frac{\sum \cos(\omega_0 t)}{N} = 0
$$

$$
\frac{\sum \sin^2(\omega_0 t)}{N} = \frac{1}{2} \qquad \frac{\sum \cos^2(\omega_0 t)}{N} = \frac{1}{2}
$$

(16.13)

$$
\frac{\sum \cos(\omega_0 t)\sin(\omega_0 t)}{N} = 0
$$

Thus, for equispaced points the normal equations become

$$
\begin{bmatrix}
N & 0 & 0 \\
0 & N/2 & 0 \\
0 & 0 & N/2
\end{bmatrix}
\begin{Bmatrix} A_0 \\ B_1 \\ B_2 \end{Bmatrix} =
\begin{Bmatrix} \sum y \\ \sum y\cos(\omega_0 t) \\ \sum y\sin(\omega_0 t) \end{Bmatrix}
$$

The inverse of a diagonal matrix is merely another diagonal matrix whose elements are the reciprocals of the original. Thus, the coefficients can be determined as

$$
\begin{Bmatrix} A_0 \\ B_1 \\ B_2 \end{Bmatrix} =
\begin{bmatrix}
1/N & 0 & 0 \\
0 & 2/N & 0 \\
0 & 0 & 2/N
\end{bmatrix}
\begin{Bmatrix} \sum y \\ \sum y\cos(\omega_0 t) \\ \sum y\sin(\omega_0 t) \end{Bmatrix}
$$

or

$$
A_0 = \frac{\sum y}{N}
$$

(16.14)

$$
A_1 = \frac{2}{N} \sum y\cos(\omega_0 t)
$$

(16.15)

$$
B_1 = \frac{2}{N} \sum y\sin(\omega_0 t)
$$

(16.16)

Notice that the first coefficient represents the function's average value.

EXAMPLE 16.1   Least-Squares Fit of a Sinusoid

Problem Statement. The curve in Fig. 16.2a is described by $y = 1.7 + \cos(4.189t + 1.0472)$. Generate 10 discrete values for this curve at intervals of $\Delta t = 0.15$ for the range $t = 0$ to 1.35. Use this information to evaluate the coefficients of Eq. (16.11) by a least-squares fit.

Solution. The data required to evaluate the coefficients with $\omega = 4.189$ are

t	y	$y \cos(\omega_0 t)$	$y \sin(\omega_0 t)$
0	2.200	2.200	0.000
0.15	1.595	1.291	0.938
0.30	1.031	0.319	0.980
0.45	0.722	−0.223	0.687
0.60	0.786	−0.636	0.462
0.75	1.200	−1.200	0.000
0.90	1.805	−1.460	−1.061
1.05	2.369	−0.732	−2.253
1.20	2.678	0.829	−2.547
1.35	2.614	2.114	−1.536
$\Sigma =$	17.000	2.502	−4.330

These results can be used to determine [Eqs. (16.14) through (16.16)]

$$A_0 = \frac{17.000}{10} = 1.7 \quad A_1 = \frac{2}{10} 2.502 = 0.500 \quad B_1 = \frac{2}{10}(-4.330) = -0.866$$

Thus, the least-squares fit is

$$y = 1.7 + 0.500 \cos(\omega_0 t) - 0.866 \sin(\omega_0 t)$$

The model can also be expressed in the format of Eq. (16.2) by calculating [Eq. (16.8)]

$$\theta = \arctan\left(\frac{-0.866}{0.500}\right) = 1.0472$$

and [Eq. (16.9)]

$$C_1 = \sqrt{0.5^2 + (-0.866)^2} = 1.00$$

to give

$$y = 1.7 + \cos(\omega_0 t + 1.0472)$$

or alternatively, as a sine by using [Eq. (16.10)]

$$y = 1.7 + \sin(\omega_0 t + 2.618)$$

The foregoing analysis can be extended to the general model

$$f(t) = A_0 + A_1\cos(\omega_0 t) + B_1\sin(\omega_0 t) + A_2\cos(2\omega_0 t) + B_2\sin(2\omega_0 t)$$
$$+ \cdots + A_m\cos(m\omega_0 t) + B_m\sin(m\omega_0 t)$$

where, for equally spaced data, the coefficients can be evaluated by

$$A_0 = \frac{\Sigma y}{N}$$

$$\left.\begin{array}{l} A_j = \dfrac{2}{N} \Sigma y \cos(j\omega_0)t \\[2mm] B_j = \dfrac{2}{N} \Sigma y \sin(j\omega_0 t) \end{array}\right\} \quad j = 1, 2, \ldots, m$$

Although these relationships can be used to fit data in the regression sense (i.e., $N > 2m + 1$), an alternative application is to employ them for interpolation or collocation—that is, to use them for the case where the number of unknowns $2m + 1$ is equal to the number of data points N. This is the approach used in the continuous Fourier series, as described next.

16.2   CONTINUOUS FOURIER SERIES

In the course of studying heat-flow problems, Fourier showed that an arbitrary periodic function can be represented by an infinite series of sinusoids of harmonically related frequencies. For a function with period T, a continuous Fourier series can be written

$$f(t) = a_0 + a_1\cos(\omega_0 t) + b_1\sin(\omega_0 t) + a_2\cos(2\omega_0 t) + b_2\sin(2\omega_0 t) + \cdots$$

or more concisely,

$$f(t) = a_0 + \sum_{k=1}^{\infty}[a_k\cos(k\omega_0 t) + b_k\sin(k\omega_0 t)] \tag{16.17}$$

where the angular frequency of the first mode ($\omega_0 = 2\pi/T$) is called the *fundamental frequency* and its constant multiples $2\omega_0$, $3\omega_0$, etc., are called *harmonics*. Thus, Eq. (16.17) expresses $f(t)$ as a linear combination of the basis functions: 1, $\cos(\omega_0 t)$, $\sin(\omega_0 t)$, $\cos(2\omega_0 t)$, $\sin(2\omega_0 t)$, ...

The coefficients of Eq. (16.17) can be computed via

$$a_k = \frac{2}{T}\int_0^T f(t)\cos(k\omega_0 t)\, dt \tag{16.18}$$

and

$$b_k = \frac{2}{T}\int_0^T f(t)\sin(k\omega_0 t)\, dt \tag{16.19}$$

for $k = 1, 2, \ldots$ and

$$a_0 = \frac{1}{T}\int_0^T f(t)\, dt \tag{16.20}$$

EXAMPLE 16.2   Continuous Fourier Series Approximation

Problem Statement. Use the continuous Fourier series to approximate the square or rectangular wave function (Fig. 16.1a) with a height of 2 and a period $T = 2\pi/\omega_0$:

$$f(t) = \begin{cases} -1 & -T/2 < t < -T/4 \\ 1 & -T/4 < t < T/4 \\ -1 & T/4 < t < T/2 \end{cases}$$

Solution.   Because the average height of the wave is zero, a value of $a_0 = 0$ can be obtained directly. The remaining coefficients can be evaluated as [Eq. (16.18)]

$$a_k = \frac{2}{T} \int_{-T/2}^{T/2} f(t) \cos(k\omega_0 t)\, dt$$

$$= \frac{2}{T} \left[-\int_{-T/2}^{-T/4} \cos(k\omega_0 t)\, dt + \int_{-T/4}^{T/4} \cos(k\omega_0 t)\, dt - \int_{T/4}^{T/2} \cos(k\omega_0 t)\, dt \right]$$

The integrals can be evaluated to give

$$a_k = \begin{cases} 4/(k\pi) & \text{for } k = 1, 5, 9, \ldots \\ -4/(k\pi) & \text{for } k = 3, 7, 11, \ldots \\ 0 & \text{for } k = \text{even integers} \end{cases}$$

FIGURE 16.4
The Fourier series approximation of a square wave. The series of plots shows the summation up to and including the (a) first, (b) second, and (c) third terms. The individual terms that were added or subtracted at each stage are also shown.

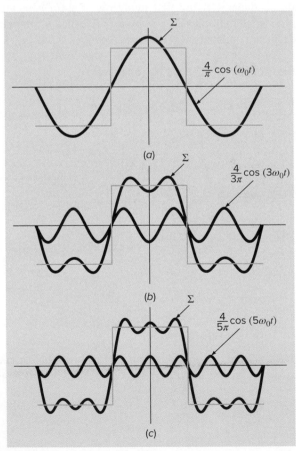

Similarly, it can be determined that all the b's $= 0$. Therefore, the Fourier series approximation is

$$f(t) = \frac{4}{\pi} \cos(\omega_0 t) - \frac{4}{3\pi} \cos(3\omega_0 t) + \frac{4}{5\pi} \cos(5\omega_0 t) - \frac{4}{7\pi} \cos(7\omega_0 t) + \cdots$$

The results up to the first three terms are shown in Fig. 16.4.

Before proceeding, the Fourier series can also be expressed in a more compact form using complex notation. This is based on *Euler's formula* (Fig. 16.5):

$$e^{\pm ix} = \cos x \pm i \sin x \tag{16.21}$$

where $i = \sqrt{-1}$, and x is in radians. Equation (16.21) can be used to express the Fourier series concisely as (Chapra and Canale, 2010)

$$f(t) = \sum_{k=-\infty}^{\infty} \tilde{c}_k e^{ik\omega_0 t} \tag{16.22}$$

where the coefficients are

$$\tilde{c}_k = \frac{1}{T} \int_{-T/2}^{T/2} f(t) e^{-ik\omega_0 t} \, dt \tag{16.23}$$

Note that the tildes ~ are included to stress that the coefficients are complex numbers. Because it is more concise, we will primarily use the complex form in the rest of the chapter. Just remember, that it is identical to the sinusoidal representation.

FIGURE 16.5
Graphical depiction of Euler's formula. The rotating vector is called a phasor.

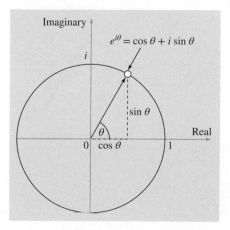

16.3   FREQUENCY AND TIME DOMAINS

To this point, our discussion of Fourier analysis has been limited to the *time domain*. We have done this because most of us are fairly comfortable conceptualizing a function's behavior in this dimension. Although it is not as familiar, the *frequency domain* provides an alternative perspective for characterizing the behavior of oscillating functions.

Just as amplitude can be plotted versus time, it can also be plotted versus frequency. Both types of expression are depicted in Fig. 16.6*a*, where we have drawn a three-dimensional graph of a sinusoidal function:

$$f(t) = C_1 \cos\left(t + \frac{\pi}{2}\right)$$

FIGURE 16.6
(*a*) A depiction of how a sinusoid can be portrayed in the time and the frequency domains. The time projection is reproduced in (*b*), whereas the amplitude-frequency projection is reproduced in (*c*). The phase-frequency projection is shown in (*d*).

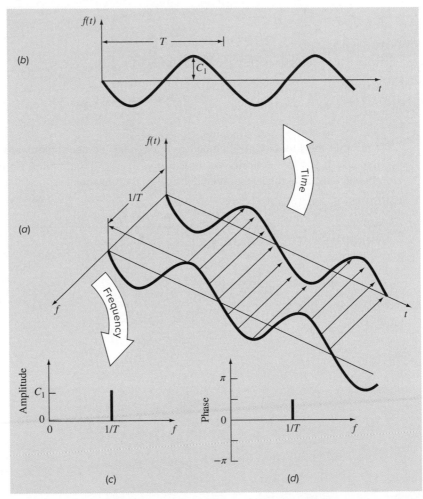

In this plot, the magnitude or amplitude of the curve $f(t)$ is the dependent variable, and time t and frequency $f = \omega_0/2\pi$ are the independent variables. Thus, the amplitude and the time axes form a *time plane*, and the amplitude and the frequency axes form a *frequency plane*. The sinusoid can, therefore, be conceived of as existing a distance $1/T$ out along the frequency axis and running parallel to the time axes. Consequently, when we speak about the behavior of the sinusoid in the time domain, we mean the projection of the curve onto the time plane (Fig. 16.6b). Similarly, the behavior in the frequency domain is merely its projection onto the frequency plane.

As in Fig. 16.6c, this projection is a measure of the sinusoid's maximum positive amplitude C_1. The full peak-to-peak swing is unnecessary because of the symmetry. Together with the location $1/T$ along the frequency axis, Fig. 16.6c now defines the amplitude and frequency of the sinusoid. This is enough information to reproduce the shape and size of the curve in the time domain. However, one more parameter—namely, the phase angle—is required to position the curve relative to $t = 0$. Consequently, a phase diagram, as shown in Fig. 16.6d, must also be included. The phase angle is determined as the distance (in radians) from zero to the point at which the positive peak occurs. If the peak occurs after zero, it is said to be delayed (recall our discussion of lags and leads in Sec. 16.1), and by convention, the phase angle is given a negative sign. Conversely, a peak before zero is said to be advanced and the phase angle is positive. Thus, for Fig. 16.6, the peak leads zero and the phase angle is plotted as $+\pi/2$. Figure 16.7 depicts some other possibilities.

We can now see that Fig. 16.6c and d provide an alternative way to present or summarize the pertinent features of the sinusoid in Fig. 16.6a. They are referred to as *line spectra*. Admittedly, for a single sinusoid they are not very interesting. However, when applied to a more complicated situation—say, a Fourier series—their true power and value is revealed. For example, Fig. 16.8 shows the amplitude and phase line spectra for the square-wave function from Example 16.2.

Such spectra provide information that would not be apparent from the time domain. This can be seen by contrasting Fig. 16.4 and Fig. 16.8. Figure 16.4 presents two alternative time domain perspectives. The first, the original square wave, tells us nothing about the sinusoids that comprise it. The alternative is to display these sinusoids—that is, $(4/\pi)\cos(\omega_0 t)$, $-(4/3\pi)\cos(3\omega_0 t)$, $(4/5\pi)\cos(5\omega_0 t)$, etc. This alternative does not provide an adequate visualization of the structure of these harmonics. In contrast, Fig. 16.8a and b provide a graphic display of this structure. As such, the line spectra represent "fingerprints" that can help us to characterize and understand a complicated waveform. They are particularly valuable for nonidealized cases where they sometimes allow us to discern structure in otherwise obscure signals. In the next section, we will describe the Fourier transform that will allow us to extend such analyses to nonperiodic waveforms.

16.4   FOURIER INTEGRAL AND TRANSFORM

Although the Fourier series is a useful tool for investigating periodic functions, there are many waveforms that do not repeat themselves regularly. For example, a lightning bolt occurs only once (or at least it will be a long time until it occurs again), but it will cause

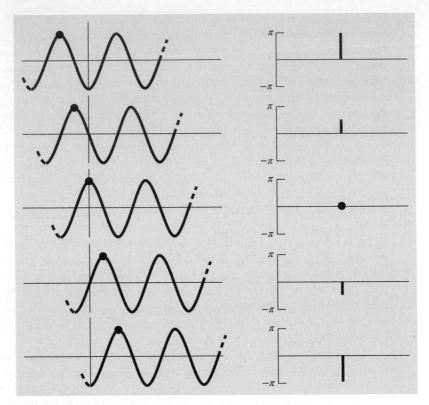

FIGURE 16.7
Various phases of a sinusoid showing the associated phase line spectra.

interference with receivers operating on a broad range of frequencies—for example, TVs, radios, and shortwave receivers. Such evidence suggests that a nonrecurring signal such as that produced by lightning exhibits a continuous frequency spectrum. Because such phenomena are of great interest to engineers, an alternative to the Fourier series would be valuable for analyzing these aperiodic waveforms.

The *Fourier integral* is the primary tool available for this purpose. It can be derived from the exponential form of the Fourier series [Eqs. (16.22) and (16.23)]. The transition from a periodic to a nonperiodic function can be effected by allowing the period to approach infinity. In other words, as T becomes infinite, the function never repeats itself and thus becomes aperiodic. If this is allowed to occur, it can be demonstrated (e.g., Van Valkenburg, 1974; Hayt and Kemmerly, 1986) that the Fourier series reduces to

$$f(t) = \frac{1}{2\pi} \int_{-\infty}^{\infty} F(\omega)e^{i\omega t}\, d\omega \qquad (16.24)$$

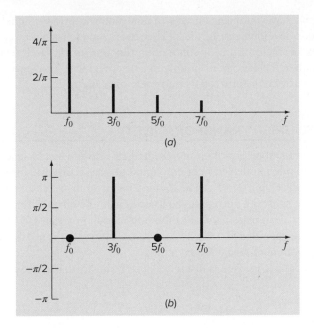

FIGURE 16.8
(*a*) Amplitude and (*b*) phase line spectra for the square wave from Fig. 16.4.

and the coefficients become a continuous function of the frequency variable ω, as in

$$F(\omega) = \int_{-\infty}^{\infty} f(t)e^{-i\omega t}\, dt \tag{16.25}$$

The function $F(\omega)$, as defined by Eq. (16.25), is called the *Fourier integral* of $f(t)$. In addition, Eqs. (16.24) and (16.25) are collectively referred to as the *Fourier transform pair*. Thus, along with being called the Fourier integral, $F(\omega)$ is also called the *Fourier transform* of $f(t)$. In the same spirit, $f(t)$, as defined by Eq. (16.24), is referred to as the *inverse Fourier transform* of $F(\omega)$. Thus, the pair allows us to transform back and forth between the time and the frequency domains for an aperiodic signal.

The distinction between the Fourier series and transform should now be quite clear. The major difference is that each applies to a different class of functions—the series to periodic and the transform to nonperiodic waveforms. Beyond this major distinction, the two approaches differ in how they move between the time and the frequency domains. The Fourier series converts a continuous, periodic time-domain function to frequency-domain magnitudes at discrete frequencies. In contrast, the Fourier transform converts a continuous time-domain function to a continuous frequency-domain function. Thus, the discrete frequency spectrum generated by the Fourier series is analogous to a continuous frequency spectrum generated by the Fourier transform.

Now that we have introduced a way to analyze an aperiodic signal, we will take the final step in our development. In the next section, we will acknowledge the fact that a signal is rarely characterized as a continuous function of the sort needed to implement Eq. (16.25). Rather, the data are invariably in a discrete form. Thus, we will now show how to compute a Fourier transform for such discrete measurements.

16.5   DISCRETE FOURIER TRANSFORM (DFT)

In engineering, functions are often represented by a finite set of discrete values. Additionally, data are often collected in or converted to such a discrete format. As depicted in Fig. 16.9, an interval from 0 to T can be divided into n equispaced subintervals with widths of $\Delta t = T/n$. The subscript j is employed to designate the discrete times at which samples are taken. Thus, f_j designates a value of the continuous function $f(t)$ taken at t_j. Note that the data points are specified at $j = 0, 1, 2, \ldots, n - 1$. A value is not included at $j = n$. (See Ramirez, 1985, for the rationale for excluding f_n.)

For the system in Fig. 16.9, a discrete Fourier transform can be written as

$$F_k = \sum_{j=0}^{n-1} f_j e^{-ik\omega_0 j} \qquad \text{for } k = 0 \text{ to } n - 1 \tag{16.26}$$

and the inverse Fourier transform as

$$f_j = \frac{1}{n} \sum_{k=0}^{n-1} F_k e^{ik\omega_0 j} \qquad \text{for } j = 0 \text{ to } n - 1 \tag{16.27}$$

where $\omega_0 = 2\pi/n$.

FIGURE 16.9
The sampling points of the discrete Fourier series.

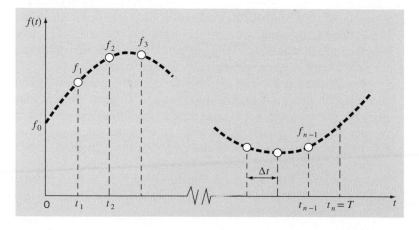

Equations (16.26) and (16.27) represent the discrete analogs of Eqs. (16.25) and (16.24), respectively. As such, they can be employed to compute both a direct and an inverse Fourier transform for discrete data. Note that the factor $1/n$ in Eq. (16.27) is merely a scale factor that can be included in either Eq. (16.26) or (16.27), but not both. For example, if it is shifted to Eq. (16.26), the first coefficient F_0 (which is the analog of the constant a_0) is equal to the arithmetic mean of the samples.

Before proceeding, several other aspects of the DFT bear mentioning. The highest frequency that can be measured in a signal, called the *Nyquist frequency*, is half the sampling frequency. Periodic variations that occur more rapidly than the shortest sampled time interval cannot be detected. The lowest frequency you can detect is the inverse of the total sample length.

As an example, suppose that you take 100 samples of data ($n = 100$ samples) at a sample frequency of $f_s = 1000$ Hz (i.e., 1000 samples per second). This means that the sample interval is

$$\Delta t = \frac{1}{f_s} = \frac{1}{1000 \text{ samples/s}} = 0.001 \text{ s/sample}$$

The total sample length is

$$t_n = \frac{n}{f_s} = \frac{100 \text{ samples}}{1000 \text{ samples/s}} = 0.1 \text{ s}$$

and the frequency increment is

$$\Delta f = \frac{f_s}{n} = \frac{1000 \text{ samples/s}}{100 \text{ samples}} = 10 \text{ Hz}$$

The Nyquist frequency is

$$f_{max} = 0.5 f_s = 0.5(1000 \text{ Hz}) = 500 \text{ Hz}$$

and the lowest detectable frequency is

$$f_{min} = \frac{1}{0.1 \text{ s}} = 10 \text{ Hz}$$

Thus, for this example, the DFT could detect signals with periods from $1/500 = 0.002$ s up to $1/10 = 0.1$ s.

16.5.1 Fast Fourier Transform (FFT)

Although an algorithm can be developed to compute the DFT based on Eq. (16.26), it is computationally burdensome because n^2 operations are required. Consequently, for data samples of even moderate size, the direct determination of the DFT can be extremely time consuming.

FIGURE 16.10
Plot of number of operations versus sample size for the standard DFT and the FFT.

The *fast Fourier transform*, or *FFT*, is an algorithm that has been developed to compute the DFT in an extremely economical fashion. Its speed stems from the fact that it utilizes the results of previous computations to reduce the number of operations. In particular, it exploits the periodicity and symmetry of trigonometric functions to compute the transform with approximately $n \log_2 n$ operations (Fig. 16.10). Thus, for $n = 50$ samples, the FFT is about 10 times faster than the standard DFT. For $n = 1000$, it is about 100 times faster.

The first FFT algorithm was developed by Gauss in the early nineteenth century (Heideman et al., 1984). Other major contributions were made by Runge, Danielson, Lanczos, and others in the early twentieth century. However, because discrete transforms often took days to weeks to calculate by hand, they did not attract broad interest prior to the development of the modern digital computer.

In 1965, J. W. Cooley and J. W. Tukey published a key paper in which they outlined an algorithm for calculating the FFT. This scheme, which is similar to those of Gauss and other earlier investigators, is called the Cooley-Tukey algorithm. Today, there are a host of other approaches that are offshoots of this method. As described next, MATLAB offers a function called fft that employs such efficient algorithms to compute the DFT.

16.5.2 MATLAB Function: fft

MATLAB's fft function provides an efficient way to compute the DFT. A simple representation of its syntax is

```
F = fft(f, n)
```

where F = a vector containing the DFT, and f = a vector containing the signal. The parameter n, which is optional, indicates that the user wants to implement an *n*-point FFT. If f has less than n points, it is padded with zeros and truncated if it has more.

Note that the elements in F are sequenced in what is called *reverse-wrap-around order*. The first half of the values are the positive frequencies (starting with the constant) and the second half are the negative frequencies. Thus, if $n = 8$, the order is 0, 1, 2, 3, 4, −3, −2, −1. The following example illustrates the function's use to calculate the DFT of a simple sinusoid.

EXAMPLE 16.3 Computing the DFT of a Simple Sinusoid with MATLAB

Problem Statement. Apply the MATLAB fft function to determine the discrete Fourier transform for a simple sinusoid:

$$f(t) = 5 + \cos(2\pi(12.5)t) + \sin(2\pi(18.75)t)$$

Generate 8 equispaced points with $\Delta t = 0.02$ s. Plot the result versus frequency.

Solution. Before generating the DFT, we can compute a number of quantities. The sampling frequency is

$$f_s = \frac{1}{\Delta t} = \frac{1}{0.02 \text{ s}} = 50 \text{ Hz}$$

The total sample length is

$$t_n = \frac{n}{f_s} = \frac{8 \text{ samples}}{50 \text{ samples/s}} = 0.16 \text{ s}$$

The Nyquist frequency is

$$f_{max} = 0.5 f_s = 0.5(50 \text{ Hz}) = 25 \text{ Hz}$$

and the lowest detectable frequency is

$$f_{min} = \frac{1}{0.16 \text{ s}} = 6.25 \text{ Hz}$$

Thus, the analysis can detect signals with periods from $1/25 = 0.04$ s up to $1/6.25 = 0.16$ s. So we should be able to detect both the 12.5 and 18.75 Hz signals.

The following MATLAB statements can be used to generate and plot the sample (Fig. 16.11a):

```
>> clc
>> n=8; dt=0.02; fs=1/dt; T = 0.16;
>> tspan=(0:n-1)/fs;
>> y=5+cos(2*pi*12.5*tspan)+sin(2*pi*18.75*tspan);
```

FIGURE 16.11
Results of computing a DFT with MATLAB's **fft** function: (*a*) the sample; and
plots of the (*b*) real and (*c*) imaginary parts of the DFT versus frequency.

```
>> subplot(3,1,1);
>> plot(tspan,y,'-ok','linewidth',2,'MarkerFaceColor','black');
>> title('(a) f(t) versus time (s)');
```

As was mentioned at the beginning of Sec. 16.5, notice that tspan omits the last point.
The fft function can be used to compute the DFT and display the results

```
>> Y=fft(y)/n;
>> Y'
```

We have divided the transform by n in order that the first coefficient is equal to the arithmetic mean of the samples. When this code is executed, the results are displayed as

```
ans =
    5.0000
    0.0000 - 0.0000i
    0.5000
   -0.0000 + 0.5000i
        0
   -0.0000 - 0.5000i
    0.5000
    0.0000 + 0.0000i
```

Notice that the first coefficient corresponds to the signal's mean value. In addition, because of the *reverse-wrap-around order*, the results can be interpreted as in the following table:

Index	k	Frequency	Period	Real	Imaginary
1	0	constant		5	0
2	1	6.25	0.16	0	0
3	2	12.5	0.08	0.5	0
4	3	18.75	0.053333	0	0.5
5	4	25	0.04	0	0
6	−3	31.25	0.032	0	−0.5
7	−2	37.5	0.026667	0.5	0
8	−1	43.75	0.022857	0	0

Notice that the fft has detected the 12.5- and 18.75-Hz signals. In addition, we have highlighted the Nyquist frequency to indicate that the values below it in the table are redundant. That is, they are merely reflections of the results below the Nyquist frequency.

  If we remove the constant value, we can plot both the real and imaginary parts of the DFT versus frequency

```
>> nyquist=fs/2;fmin=1/T;
>> f = linspace(fmin,nyquist,n/2);
>> Y(1)=[];YP=Y(1:n/2);
>> subplot(3,1,2)
>> stem(f,real(YP),'linewidth',2,'MarkerFaceColor','blue')
>> grid;title('(b) Real component versus frequency')
>> subplot(3,1,3)
>> stem(f,imag(YP),'linewidth',2,'MarkerFaceColor','blue')
>> grid;title('(b) Imaginary component versus frequency')
>> xlabel('frequency (Hz)')
```

As expected (recall Fig. 16.7), a positive peak occurs for the cosine at 12.5 Hz (Fig. 16.11*b*), and a negative peak occurs for the sine at 18.75 Hz (Fig. 16.11*c*).

16.6 THE POWER SPECTRUM

Beyond amplitude and phase spectra, power spectra provide another useful way to discern the underlying harmonics of seemingly random signals. As the name implies, it derives from the analysis of the power output of electrical systems. In terms of the DFT, a *power spectrum* consists of a plot of the power associated with each frequency component versus frequency. The power can be computed by summing the squares of the Fourier coefficients:

$$P_k = |\tilde{c}_k|^2$$

where P_k is the power associated with each frequency $k\omega_0$.

EXAMPLE   16.4 Computing the Power Spectrum with MATLAB

Problem Statement. Compute the power spectrum for the simple sinusoid for which the DFT was computed in Example 16.3.

Solution. The following script can be developed to compute the power spectrum:

```
% compute the DFT
clc;clf
n=8; dt=0.02;
fs=1/dt;tspan=(0:n-1)/fs;
y=5+cos(2*pi*12.5*tspan)+sin(2*pi*18.75*tspan);
Y=fft(y)/n;
f = (0:n-1)*fs/n;
Y(1)=[];f(1)=[];
% compute and display the power spectrum
nyquist=fs/2;
f = (1:n/2)/(n/2)*nyquist;
Pyy = abs(Y(1:n/2)).^2;
stem(f,Pyy,'linewidth',2,'MarkerFaceColor','blue')
title('Power spectrum')
xlabel('Frequency (Hz)');ylim([0 0.3])
```

As indicated, the first section merely computes the DFT with the pertinent statements from Example 16.3. The second section then computes and displays the power spectrum. As in Fig. 16.12, the resulting graph indicates that peaks occur at both 12.5 and 18.75 Hz as expected.

FIGURE 16.12
Power spectrum for a simple sinusoidal function with frequencies of 12.5 and 18.75 Hz.

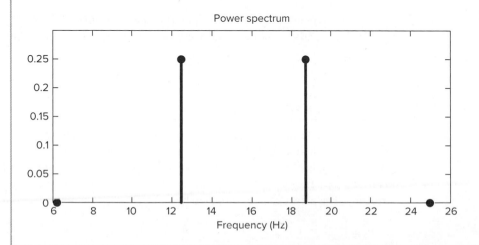

16.7 CASE STUDY   SUNSPOTS

Background. In 1848, Johann Rudolph Wolf devised a method for quantifying solar activity by counting the number of individual spots and groups of spots on the sun's surface. He computed a quantity, now called a *Wolf sunspot number*, by adding 10 times the number of groups plus the total count of individual spots. As in Fig. 16.13, the data set for the sunspot number extends back to 1700. On the basis of the early historical records, Wolf determined the cycle's length to be 11.1 years. Use a Fourier analysis to confirm this result by applying an FFT to the data.

Solution. The data for year and sunspot number are contained in a MATLAB file, sunspot.dat. The following statements load the file and assign the year and number information to vectors of the same name:

```
>> load sunspot.dat
>> year = sunspot(:,1);number = sunspot(:,2);
```

Before applying the Fourier analysis, it is noted that the data seem to exhibit an upward linear trend (Fig. 16.13). MATLAB can be used to remove this trend:

```
>> n = length(number);
>> a = polyfit(year,number,1);
>> lineartrend = polyval(a,year);
>> ft = number-lineartrend;
```

Next, the fft function is employed to generate the DFT

```
F = fft(ft);
```

The power spectrum can then be computed and plotted

```
fs = 1;
f = (0:n/2)*fs/n;
pow = abs(F(1:n/2+1)).^2;
```

FIGURE 16.13
Plot of Wolf sunspot number versus year. The dashed line indicates a mild, upward linear trend.

16.7 CASE STUDY continued

FIGURE 16.14
Power spectrum for Wolf sunspot number versus year.

```
plot(f,pow)
xlabel('Frequency (cycles/year)'); ylabel('Power')
title('Power versus frequency')
```

The result, as shown in Fig. 16.14, indicates a peak at a frequency of about 0.0915 cycles/yr. This corresponds to a period of $1/0.0915 = 10.93$ years. Thus, the Fourier analysis is consistent with Wolf's estimate of 11 years.

PROBLEMS

16.1 The following equation describes the variations of temperature of a tropical lake:

$$T(t) = 12.8 + 4 \cos\left(\tfrac{2\pi}{365}t\right) + 3 \sin\left(\tfrac{2\pi}{365}t\right)$$

What is **(a)** the mean temperature, **(b)** the amplitude, and **(c)** the period?

16.2 The temperature in a pond varies sinusoidally over the course of a year. Use linear least-squares regression to fit Eq. (16.11) to the following data. Use your fit to determine the mean, amplitude, and date of maximum temperature. Note that the period is 365 d.

t, d	15	45	75	105	135	165	225	255	285	315	345
$T, °C$	3.4	4.7	8.5	11.7	16	18.7	19.7	17.1	12.7	7.7	5.1

16.3 The pH in a reactor varies sinusoidally over the course of a day. Use least-squares regression to fit Eq. (16.11) to the following data. Use your fit to determine the mean, amplitude, and time of maximum pH. Note that the period is 24 hr

Time, hr	0	2	4	5	7	9
pH	7.6	7.2	7	6.5	7.5	7.2

Time, hr	12	15	20	22	24
pH	8.9	9.1	8.9	7.9	7

16.4 The solar radiation for Tucson, Arizona, has been tabulated as

Time, mo	J	F	M	A	M	J
Radiation, W/m²	144	188	245	311	351	359
Time, mo	J	A	S	O	N	D
Radiation, W/m²	308	287	260	211	159	131

Assuming each month is 30 days long, fit a sinusoid to these data. Use the resulting equation to predict the radiation in mid-August.

16.5 The average values of a function can be determined by

$$\bar{f} = \frac{\int_0^t f(t)\, dt}{t}$$

Use this relationship to verify the results of Eq. (16.13).

16.6 In electric circuits, it is common to see current behavior in the form of a square wave as shown in Fig. P16.6 (notice that square wave differs from the one described in Example 16.2). Solving for the Fourier series from

$$f(t) = \begin{cases} A_0 & 0 \le t \le T/2 \\ -A_0 & T/2 \le t \le T \end{cases}$$

the Fourier series can be represented as

$$f(t) = \sum_{n=1}^{\infty} \left(\frac{4A_0}{(2n-1)\pi} \right) \sin\left(\frac{2\pi(2n-1)t}{T} \right)$$

Develop a MATLAB function to generate a plot of the first n terms of the Fourier series individually, as well as the sum of these six terms. Design your function so that it plots the curves from $t = 0$ to $4T$. Use thin dotted red lines for the individual terms and a bold black solid line for the

FIGURE P16.6

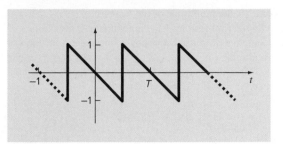

FIGURE P16.7
A sawtooth wave.

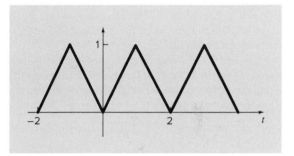

FIGURE P16.8
A triangular wave.

summation (i.e., 'k-','linewidth',2). The function's first line should be

```
function [t,f] = FourierSquare(A0,T,n)
```

Let $A_0 = 1$ and $T = 0.25$ s.

16.7 Use a continuous Fourier series to approximate the sawtooth wave in Fig. P16.7. Plot the first four terms along with the summation. In addition, construct amplitude and phase line spectra for the first four terms.

16.8 Use a continuous Fourier series to approximate the triangular wave form in Fig. P16.8. Plot the first four terms along with the summation. In addition, construct amplitude and phase line spectra for the first four terms.

16.9 Use the *Maclaurin series expansions* for e^x, cos x, and sin x to prove Euler's formula [Eq. (16.21)].

16.10 A half-wave rectifier can be characterized by

$$C_1 = \left[\frac{1}{\pi} + \frac{1}{2}\sin t - \frac{2}{3\pi}\cos 2t - \frac{2}{15\pi}\cos 4t \right. \\ \left. - \frac{2}{35\pi}\cos 6t - \cdots \right]$$

where C_1 is the amplitude of the wave.

(a) Plot the first four terms along with the summation.

(b) Construct amplitude and phase line spectra for the first four terms.

16.11 Duplicate Example 16.3, but for 64 points sampled at a rate of $\Delta t = 0.01$ s from the function

$$f(t) = \cos[2\pi(12.5)t] + \cos[2\pi(25)t]$$

Use fft to generate a DFT of these values and plot the results.

16.12 Use MATLAB to generate 64 points from the function

$$f(t) = \cos(10t) + \sin(3t)$$

from $t = 0$ to 2π. Add a random component to the signal with the function randn. Use fft to generate a DFT of these values and plot the results.

16.13 Use MATLAB to generate 32 points for the sinusoid depicted in Fig. 16.2 from $t = 0$ to 6 s. Compute the DFT and create subplots of **(a)** the original signal, **(b)** the real part, and **(c)** the imaginary part of the DFT versus frequency.

16.14 Use the fft function to compute a DFT for the triangular wave from Prob. 16.8. Sample the wave from $t = 0$ to $4T$ using 128 sample points.

16.15 Develop an M-file function that uses the fft function to generate a power spectrum plot. Use it to solve Prob. 16.11.

16.16 Use the fft function to compute the DFT for the following function:

$$f(t) = 1.5 + 1.8\cos(2\pi(12)t) + 0.8\sin(2\pi(20)t) \\ - 1.25\cos(2\pi(28)t)$$

Take $n = 64$ samples with a sampling frequency of $f_s = 128$ samples/s. Have your script compute values of Δt, t_n, Δf, f_{min}, and f_{max}. As illustrated in Examples 16.3 and 16.4, have your script generate plots as in Fig. 16.11 and Fig. 16.12.

16.17 If you take 128 samples of data ($n = 128$ samples) with a total sample length of $t_n = 0.4$ s, compute the following: **(a)** the sample frequency, f_s (sample/s); **(b)** the sample interval, Δt (s/sample); **(c)** the Nyquist frequency, f_{max} (Hz); **(d)** the minimum frequency, f_{min} (Hz).

17

Polynomial Interpolation

CHAPTER OBJECTIVES

The primary objective of this chapter is to introduce you to polynomial interpolation. Specific objectives and topics covered are

- Recognizing that evaluating polynomial coefficients with simultaneous equations is an ill-conditioned problem.
- Knowing how to evaluate polynomial coefficients and interpolate with MATLAB's `polyfit` and `polyval` functions.
- Knowing how to perform an interpolation with Newton's polynomial.
- Knowing how to perform an interpolation with a Lagrange polynomial.
- Knowing how to solve an inverse interpolation problem by recasting it as a roots problem.
- Appreciating the dangers of extrapolation.
- Recognizing that higher-order polynomials can manifest large oscillations.

YOU'VE GOT A PROBLEM

If we want to improve the velocity prediction for the free-falling bungee jumper, we might expand our model to account for other factors beyond mass and the drag coefficient. As was previously mentioned in Sec. 1.4, the drag coefficient can itself be formulated as a function of other factors such as the area of the jumper and characteristics such as the air's density and viscosity.

Air density and viscosity are commonly presented in tabular form as a function of temperature. For example, Table 17.1 is reprinted from a popular fluid mechanics textbook (White, 1999).

Suppose that you desired the density at a temperature not included in the table. In such a case, you would have to interpolate. That is, you would have to estimate the value

TABLE 17.1 Density (ρ), dynamic viscosity (μ), and kinematic viscosity (v) as a function of temperature (T) at 1 atm as reported by White (1999).

T, °C	ρ, kg/m^3	μ, N · s/m^2	v, m^2/s
−40	1.52	1.51×10^{-5}	0.99×10^{-5}
0	1.29	1.71×10^{-5}	1.33×10^{-5}
20	1.20	1.80×10^{-5}	1.50×10^{-5}
50	1.09	1.95×10^{-5}	1.79×10^{-5}
100	0.946	2.17×10^{-5}	2.30×10^{-5}
150	0.835	2.38×10^{-5}	2.85×10^{-5}
200	0.746	2.57×10^{-5}	3.45×10^{-5}
250	0.675	2.75×10^{-5}	4.08×10^{-5}
300	0.616	2.93×10^{-5}	4.75×10^{-5}
400	0.525	3.25×10^{-5}	6.20×10^{-5}
500	0.457	3.55×10^{-5}	7.77×10^{-5}

at the desired temperature based on the densities that bracket it. The simplest approach is to determine the equation for the straight line connecting the two adjacent values and use this equation to estimate the density at the desired intermediate temperature. Although such *linear interpolation* is perfectly adequate in many cases, error can be introduced when the data exhibit significant curvature. In this chapter, we will explore a number of different approaches for obtaining adequate estimates for such situations.

17.1 INTRODUCTION TO INTERPOLATION

You will frequently have occasion to estimate intermediate values between precise data points. The most common method used for this purpose is polynomial interpolation. The general formula for an $(n - 1)$th-order polynomial can be written as

$$f(x) = a_1 + a_2 x + a_3 x^2 + \cdots + a_n x^{n-1} \tag{17.1}$$

For n data points, there is one and only one polynomial of order $(n - 1)$ that passes through all the points. For example, there is only one straight line (i.e., a first-order polynomial) that connects two points (Fig. 17.1a). Similarly, only one parabola connects a set of three points (Fig. 17.1b). *Polynomial interpolation* consists of determining the unique $(n - 1)$th-order polynomial that fits n data points. This polynomial then provides a formula to compute intermediate values.

Before proceeding, we should note that MATLAB represents polynomial coefficients in a different manner than Eq. (17.1). Rather than using increasing powers of x, it uses decreasing powers as in

$$f(x) = p_1 x^{n-1} + p_2 x^{n-2} + \cdots + p_{n-1} x + p_n \tag{17.2}$$

To be consistent with MATLAB, we will adopt this scheme in the following section.

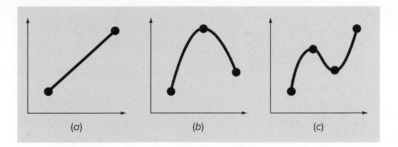

FIGURE 17.1
Examples of interpolating polynomials: (*a*) first-order (linear) connecting two points,
(*b*) second-order (quadratic or parabolic) connecting three points, and (*c*) third-order (cubic)
connecting four points.

17.1.1 Determining Polynomial Coefficients

A straightforward way for computing the coefficients of Eq. (17.2) is based on the fact that
n data points are required to determine the n coefficients. As in the following example, this
allows us to generate n linear algebraic equations that we can solve simultaneously for the
coefficients.

EXAMPLE 17.1   Determining Polynomial Coefficients with Simultaneous Equations

Problem Statement. Suppose that we want to determine the coefficients of the parabola,
$f(x) = p_1 x^2 + p_2 x + p_3$, that passes through the last three density values from Table 17.1:

$$x_1 = 300 \quad f(x_1) = 0.616$$
$$x_2 = 400 \quad f(x_2) = 0.525$$
$$x_3 = 500 \quad f(x_3) = 0.457$$

Each of these pairs can be substituted into Eq. (17.2) to yield a system of three equations:

$$0.616 = p_1(300)^2 + p_2(300) + p_3$$
$$0.525 = p_1(400)^2 + p_2(400) + p_3$$
$$0.457 = p_1(500)^2 + p_2(500) + p_3$$

or in matrix form:

$$\begin{bmatrix} 90,000 & 300 & 1 \\ 160,000 & 400 & 1 \\ 250,000 & 500 & 1 \end{bmatrix} \begin{Bmatrix} p_1 \\ p_2 \\ p_3 \end{Bmatrix} = \begin{Bmatrix} 0.616 \\ 0.525 \\ 0.457 \end{Bmatrix}$$

Thus, the problem reduces to solving three simultaneous linear algebraic equations for
the three unknown coefficients. A simple MATLAB session can be used to obtain the

solution:

```
>> format long
>> A = [90000 300 1;160000 400 1;250000 500 1];
>> b = [0.616 0.525 0.457]';
>> p = A\b

p =
    0.00000115000000
   -0.00171500000000
    1.02700000000000
```

Thus, the parabola that passes exactly through the three points is

$$f(x) = 0.00000115x^2 - 0.001715x + 1.027$$

This polynomial then provides a means to determine intermediate points. For example, the value of density at a temperature of 350 °C can be calculated as

$$f(350) = 0.00000115(350)^2 - 0.001715(350) + 1.027 = 0.567625$$

Although the approach in Example 17.1 provides an easy way to perform interpolation, it has a serious deficiency. To understand this flaw, notice that the coefficient matrix in Example 17.1 has a decided structure. This can be seen clearly by expressing it in general terms:

$$\begin{bmatrix} x_1^2 & x_1 & 1 \\ x_2^2 & x_2 & 1 \\ x_3^2 & x_3 & 1 \end{bmatrix} \begin{Bmatrix} p_1 \\ p_2 \\ p_3 \end{Bmatrix} = \begin{Bmatrix} f(x_1) \\ f(x_2) \\ f(x_3) \end{Bmatrix} \tag{17.3}$$

Coefficient matrices of this form are referred to as *Vandermonde matrices*. Such matrices are very ill-conditioned. That is, their solutions are very sensitive to round off errors. This can be illustrated by using MATLAB to compute the condition number for the coefficient matrix from Example 17.1 as

```
>> cond(A)

ans =

   5.8932e+006
```

This condition number, which is quite large for a 3×3 matrix, implies that about six digits of the solution would be questionable. The ill-conditioning becomes even worse as the number of simultaneous equations becomes larger.

As a consequence, there are alternative approaches that do not manifest this shortcoming. In this chapter, we will also describe two alternatives that are well-suited for computer implementation: the Newton and the Lagrange polynomials. Before doing this, however, we will first briefly review how the coefficients of the interpolating polynomial can be estimated directly with MATLAB's built-in functions.

17.1.2 MATLAB Functions: polyfit and polyval

Recall from Sec. 14.5.2, that the polyfit function can be used to perform polynomial regression. In such applications, the number of data points is greater than the number of coefficients being estimated. Consequently, the least-squares fit line does not necessarily pass through any of the points, but rather follows the general trend of the data.

For the case where the number of data points equals the number of coefficients, poly-fit performs interpolation. That is, it returns the coefficients of the polynomial that pass directly through the data points. For example, it can be used to determine the coefficients of the parabola that passes through the last three density values from Table 17.1:

```
>> format long
>> T = [300 400 500];
>> density = [0.616 0.525 0.457];
>> p = polyfit(T,density,2)

p =
    0.00000115000000   -0.00171500000000   1.02700000000000
```

We can then use the polyval function to perform an interpolation as in

```
>> d = polyval(p,350)

d =
    0.56762500000000
```

These results agree with those obtained previously in Example 17.1 with simultaneous equations.

17.2   NEWTON INTERPOLATING POLYNOMIAL

There are a variety of alternative forms for expressing an interpolating polynomial beyond the familiar format of Eq. (17.2). Newton's interpolating polynomial is among the most popular and useful forms. Before presenting the general equation, we will introduce the first- and second-order versions because of their simple visual interpretation.

17.2.1 Linear Interpolation

The simplest form of interpolation is to connect two data points with a straight line. This technique, called *linear interpolation,* is depicted graphically in Fig. 17.2. Using similar triangles,

$$\frac{f_1(x) - f(x_1)}{x - x_1} = \frac{f(x_2) - f(x_1)}{x_2 - x_1} \tag{17.4}$$

which can be rearranged to yield

$$f_1(x) = f(x_1) + \frac{f(x_2) - f(x_1)}{x_2 - x_1}(x - x_1) \tag{17.5}$$

which is the *Newton linear-interpolation formula.* The notation $f_1(x)$ designates that this is a first-order interpolating polynomial. Notice that besides representing the slope of the line connecting the points, the term $[f(x_2) - f(x_1)]/(x_2 - x_1)$ is a finite-difference approximation

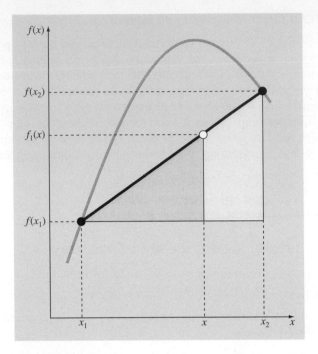

FIGURE 17.2
Graphical depiction of linear interpolation. The shaded areas indicate the similar triangles
used to derive the Newton linear-interpolation formula [Eq. (17.5)].

of the first derivative [recall Eq. (4.20)]. In general, the smaller the interval between the
data points, the better the approximation. This is due to the fact that, as the interval decreases, a continuous function will be better approximated by a straight line. This characteristic is demonstrated in the following example.

EXAMPLE 17.2   Linear Interpolation

Problem Statement.   Estimate the natural logarithm of 2 using linear interpolation. First,
perform the computation by interpolating between $\ln 1 = 0$ and $\ln 6 = 1.791759$. Then,
repeat the procedure, but use a smaller interval from $\ln 1$ to $\ln 4$ (1.386294). Note that the
true value of $\ln 2$ is 0.6931472.

Solution.   We use Eq. (17.5) from $x_1 = 1$ to $x_2 = 6$ to give

$$f_1(2) = 0 + \frac{1.791759 - 0}{6 - 1}(2 - 1) = 0.3583519$$

which represents an error of $\varepsilon_t = 48.3\%$. Using the smaller interval from $x_1 = 1$ to $x_2 = 4$ yields

$$f_1(2) = 0 + \frac{1.386294 - 0}{4 - 1}(2 - 1) = 0.4620981$$

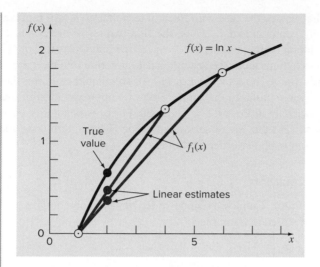

FIGURE 17.3
Two linear interpolations to estimate ln 2. Note how the smaller interval provides a better estimate.

Thus, using the shorter interval reduces the percent relative error to $\varepsilon_t = 33.3\%$. Both interpolations are shown in Fig. 17.3, along with the true function.

17.2.2 Quadratic Interpolation

The error in Example 17.2 resulted from approximating a curve with a straight line. Consequently, a strategy for improving the estimate is to introduce some curvature into the line connecting the points. If three data points are available, this can be accomplished with a second-order polynomial (also called a quadratic polynomial or a parabola). A particularly convenient form for this purpose is

$$f_2(x) = b_1 + b_2(x - x_1) + b_3(x - x_1)(x - x_2) \tag{17.6}$$

A simple procedure can be used to determine the values of the coefficients. For b_1, Eq. (17.6) with $x = x_1$ can be used to compute

$$b_1 = f(x_1) \tag{17.7}$$

Equation (17.7) can be substituted into Eq. (17.6), which can be evaluated at $x = x_2$ for

$$b_2 = \frac{f(x_2) - f(x_1)}{x_2 - x_1} \tag{17.8}$$

Finally, Eqs. (17.7) and (17.8) can be substituted into Eq. (17.6), which can be evaluated at $x = x_3$ and solved (after some algebraic manipulations) for

$$b_3 = \frac{\dfrac{f(x_3) - f(x_2)}{x_3 - x_2} - \dfrac{f(x_2) - f(x_1)}{x_2 - x_1}}{x_3 - x_1} \tag{17.9}$$

Notice that, as was the case with linear interpolation, b_2 still represents the slope of the line connecting points x_1 and x_2. Thus, the first two terms of Eq. (17.6) are equivalent to linear interpolation between x_1 and x_2, as specified previously in Eq. (17.5). The last term, $b_3(x - x_1)(x - x_2)$, introduces the second-order curvature into the formula.

Before illustrating how to use Eq. (17.6), we should examine the form of the coefficient b_3. It is very similar to the finite-difference approximation of the second derivative introduced previously in Eq. (4.27). Thus, Eq. (17.6) is beginning to manifest a structure that is very similar to the Taylor series expansion. That is, terms are added sequentially to capture increasingly higher-order curvature.

EXAMPLE 17.3   Quadratic Interpolation

Problem Statement.   Employ a second-order Newton polynomial to estimate ln 2 with the same three points used in Example 17.2:

$$x_1 = 1 \qquad f(x_1) = 0$$

$$x_2 = 4 \qquad f(x_2) = 1.386294$$

$$x_3 = 6 \qquad f(x_3) = 1.791759$$

Solution.   Applying Eq. (17.7) yields

$$b_1 = 0$$

Equation (17.8) gives

$$b_2 = \frac{1.386294 - 0}{4 - 1} = 0.4620981$$

FIGURE 17.4
The use of quadratic interpolation to estimate ln 2. The linear interpolation from $x = 1$ to 4 is also included for comparison.

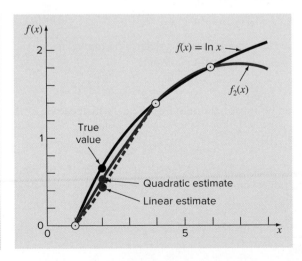

and Eq. (17.9) yields

$$b_3 = \frac{\dfrac{1.791759 - 1.386294}{6 - 4} - 0.4620981}{6 - 1} = -0.0518731$$

Substituting these values into Eq. (17.6) yields the quadratic formula

$$f_2(x) = 0 + 0.4620981(x - 1) - 0.0518731(x - 1)(x - 4)$$

which can be evaluated at $x = 2$ for $f_2(2) = 0.5658444$, which represents a relative error of $\varepsilon_t = 18.4\%$. Thus, the curvature introduced by the quadratic formula (Fig. 17.4) improves the interpolation compared with the result obtained using straight lines in Example 17.2 and Fig. 17.3.

17.2.3 General Form of Newton's Interpolating Polynomials

The preceding analysis can be generalized to fit an $(n - 1)$th-order polynomial to n data points. The $(n - 1)$th-order polynomial is

$$f_{n-1}(x) = b_1 + b_2(x - x_1) + \cdots + b_n(x - x_1)(x - x_2) \cdots (x - x_{n-1}) \tag{17.10}$$

As was done previously with linear and quadratic interpolation, data points can be used to evaluate the coefficients $b_1, b_2, \ldots, b_n$. For an $(n - 1)$th-order polynomial, n data points are required: $[x_1, f(x_1)], [x_2, f(x_2)], \ldots, [x_n, f(x_n)]$. We use these data points and the following equations to evaluate the coefficients:

$$b_1 = f(x_1) \tag{17.11}$$

$$b_2 = f[x_2, x_1] \tag{17.12}$$

$$b_3 = f[x_3, x_2, x_1] \tag{17.13}$$

$$\vdots$$

$$b_n = f[x_n, x_{n-1}, \ldots, x_2, x_1] \tag{17.14}$$

where the bracketed function evaluations are finite divided differences. For example, the first finite divided difference is represented generally as

$$f[x_i, x_j] = \frac{f(x_i) - f(x_j)}{x_i - x_j} \tag{17.15}$$

The second finite divided difference, which represents the difference of two first divided differences, is expressed generally as

$$f[x_i, x_j, x_k] = \frac{f[x_i, x_j] - f[x_j, x_k]}{x_i - x_k} \tag{17.16}$$

Similarly, the nth finite divided difference is

$$f[x_n, x_{n-1}, \ldots, x_2, x_1] = \frac{f[x_n, x_{n-1}, \ldots, x_2] - f[x_{n-1}, x_{n-2}, \ldots, x_1]}{x_n - x_1} \tag{17.17}$$

x_i	$f(x_i)$	First	Second	Third
x_1	$f(x_1)$	$f[x_2, x_1]$	$f[x_3, x_2, x_1]$	$f[x_4, x_3, x_2, x_1]$
x_2	$f(x_2)$	$f[x_3, x_2]$	$f[x_4, x_3, x_2]$	
x_3	$f(x_3)$	$f[x_4, x_3]$		
x_4	$f(x_4)$			

FIGURE 17.5
Graphical depiction of the recursive nature of finite divided differences. This representation is referred to as a divided difference table.

These differences can be used to evaluate the coefficients in Eqs. (17.11) through (17.14), which can then be substituted into Eq. (17.10) to yield the general form of Newton's interpolating polynomial:

$$f_{n-1}(x) = f(x_1) + (x - x_1)f[x_2, x_1] + (x - x_1)(x - x_2)f[x_3, x_2, x_1]$$
$$+ \cdots + (x - x_1)(x - x_2) \cdots (x - x_{n-1})f[x_n, x_{n-1}, \ldots, x_2, x_1] \qquad (17.18)$$

We should note that it is not necessary that the data points used in Eq. (17.18) be equally spaced or that the abscissa values necessarily be in ascending order, as illustrated in the following example. However, the points should be ordered so that they are centered around and as close as possible to the unknown. Also, notice how Eqs. (17.15) through (17.17) are recursive—that is, higher-order differences are computed by taking differences of lower-order differences (Fig. 17.5). This property will be exploited when we develop an efficient M-file to implement the method.

EXAMPLE 17.4   Newton Interpolating Polynomial

Problem Statement.   In Example 17.3, data points at $x_1 = 1$, $x_2 = 4$, and $x_3 = 6$ were used to estimate ln 2 with a parabola. Now, adding a fourth point [$x_4 = 5$; $f(x_4) = 1.609438$], estimate ln 2 with a third-order Newton's interpolating polynomial.

Solution.   The third-order polynomial, Eq. (17.10) with $n = 4$, is

$$f_3(x) = b_1 + b_2(x - x_1) + b_3(x - x_1)(x - x_2) + b_4(x - x_1)(x - x_2)(x - x_3)$$

The first divided differences for the problem are [Eq. (17.15)]

$$f[x_2, x_1] = \frac{1.386294 - 0}{4 - 1} = 0.4620981$$

$$f[x_3, x_2] = \frac{1.791759 - 1.386294}{6 - 4} = 0.2027326$$

$$f[x_4, x_3] = \frac{1.609438 - 1.791759}{5 - 6} = 0.1823216$$

The second divided differences are [Eq. (17.16)]

$$f[x_3, x_2, x_1] = \frac{0.2027326 - 0.4620981}{6 - 1} = -0.05187311$$

$$f[x_4, x_3, x_2] = \frac{0.1823216 - 0.2027326}{5 - 4} = -0.02041100$$

The third divided difference is [Eq. (17.17) with $n = 4$]

$$f[x_4, x_3, x_2, x_1] = \frac{-0.02041100 - (-0.05187311)}{5 - 1} = 0.007865529$$

Thus, the divided difference table is

x_i	$f(x_i)$	First	Second	Third
1	0	0.4620981	−0.05187311	0.007865529
4	1.386294	0.2027326	−0.02041100	
6	1.791759	0.1823216		
5	1.609438			

The results for $f(x_1), f[x_2, x_1], f[x_3, x_2, x_1]$, and $f[x_4, x_3, x_2, x_1]$ represent the coefficients b_1, b_2, b_3, and b_4, respectively, of Eq. (17.10). Thus, the interpolating cubic is

$$f_3(x) = 0 + 0.4620981(x - 1) - 0.05187311(x - 1)(x - 4)$$
$$+ 0.007865529(x - 1)(x - 4)(x - 6)$$

which can be used to evaluate $f_3(2) = 0.6287686$, which represents a relative error of $\varepsilon_t = 9.3\%$. The complete cubic polynomial is shown in Fig. 17.6.

FIGURE 17.6
The use of cubic interpolation to estimate ln 2.

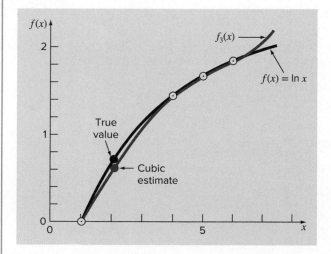

17.2.4 MATLAB M-file: Newtint

It is straightforward to develop an M-file to implement Newton interpolation. As in Fig. 17.7, the first step is to compute the finite divided differences and store them in an array. The differences are then used in conjunction with Eq. (17.18) to perform the interpolation.

An example of a session using the function would be to duplicate the calculation we just performed in Example 17.3:

```
>> format long
>> x = [1 4 6 5]';
```

FIGURE 17.7
An M-file to implement Newton interpolation.

```
function yint = Newtint(x,y,xx)
% Newtint: Newton interpolating polynomial
% yint = Newtint(x,y,xx): Uses an (n - 1)-order Newton
%    interpolating polynomial based on n data points (x, y)
%    to determine a value of the dependent variable (yint)
%    at a given value of the independent variable, xx.
% input:
%    x = independent variable
%    y = dependent variable
%    xx = value of independent variable at which
%         interpolation is calculated
% output:
%    yint = interpolated value of dependent variable

% compute the finite divided differences in the form of a
% difference table
n = length(x);
if length(y)~=n, error('x and y must be same length'); end
b = zeros(n,n);
% assign dependent variables to the first column of b.
b(:,1) = y(:);  % the (:) ensures that y is a column vector.
for j = 2:n
  for i = 1:n-j+1
    b(i,j) = (b(i+1,j-1)-b(i,j-1))/(x(i+j-1)-x(i));
  end
end
% use the finite divided differences to interpolate
xt = 1;
yint = b(1,1);
for j = 1:n-1
  xt = xt*(xx-x(j));
  yint = yint+b(1,j+1)*xt;
end
```

```
>> y = log(x);
>> Newtint(x,y,2)

ans =
   0.62876857890841
```

17.3   LAGRANGE INTERPOLATING POLYNOMIAL

Suppose we formulate a linear interpolating polynomial as the weighted average of the two values that we are connecting by a straight line:

$$f(x) = L_1 f(x_1) + L_2 f(x_2) \tag{17.19}$$

where the L's are the weighting coefficients. It is logical that the first weighting coefficient is the straight line that is equal to 1 at x_1 and 0 at x_2:

$$L_1 = \frac{x - x_2}{x_1 - x_2}$$

Similarly, the second coefficient is the straight line that is equal to 1 at x_2 and 0 at x_1:

$$L_2 = \frac{x - x_1}{x_2 - x_1}$$

Substituting these coefficients into Eq. (17.19) yields the straight line that connects the points (Fig. 17.8):

$$f_1(x) = \frac{x - x_2}{x_1 - x_2} f(x_1) + \frac{x - x_1}{x_2 - x_1} f(x_2) \tag{17.20}$$

where the nomenclature $f_1(x)$ designates that this is a first-order polynomial. Equation (17.20) is referred to as the *linear Lagrange interpolating polynomial.*

The same strategy can be employed to fit a parabola through three points. For this case three parabolas would be used with each one passing through one of the points and equaling zero at the other two. Their sum would then represent the unique parabola that connects the three points. Such a second-order Lagrange interpolating polynomial can be written as

$$f_2(x) = \frac{(x - x_2)(x - x_3)}{(x_1 - x_2)(x_1 - x_3)} f(x_1) + \frac{(x - x_1)(x - x_3)}{(x_2 - x_1)(x_2 - x_3)} f(x_2)$$

$$+ \frac{(x - x_1)(x - x_2)}{(x_3 - x_1)(x_3 - x_2)} f(x_3) \tag{17.21}$$

Notice how the first term is equal to $f(x_1)$ at x_1 and is equal to zero at x_2 and x_3. The other terms work in a similar fashion.

Both the first- and second-order versions as well as higher-order Lagrange polynomials can be represented concisely as

$$f_{n-1}(x) = \sum_{i=1}^{n} L_i(x) f(x_i) \tag{17.22}$$

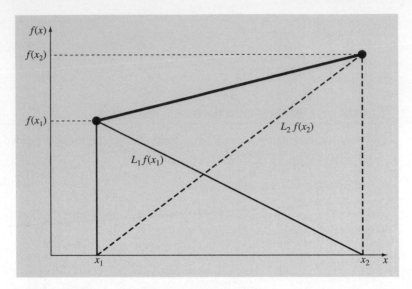

FIGURE 17.8
A visual depiction of the rationale behind Lagrange interpolating polynomials. The figure shows the first-order case. Each of the two terms of Eq. (17.20) passes through one of the points and is zero at the other. The summation of the two terms must, therefore, be the unique straight line that connects the two points.

where

$$L_i(x) = \prod_{\substack{j=1 \\ j \neq i}}^{n} \frac{x - x_j}{x_i - x_j} \tag{17.23}$$

where n = the number of data points and $\prod$ designates the "product of."

EXAMPLE 17.5   Lagrange Interpolating Polynomial

Problem Statement. Use a Lagrange interpolating polynomial of the first and second order to evaluate the density of unused motor oil at $T = 15\ ^\circ C$ based on the following data:

$$x_1 = 0 \qquad f(x_1) = 3.85$$

$$x_2 = 20 \qquad f(x_2) = 0.800$$

$$x_3 = 40 \qquad f(x_3) = 0.212$$

Solution. The first-order polynomial [Eq. (17.20)] can be used to obtain the estimate at $x = 15$:

$$f_1(x) = \frac{15 - 20}{0 - 20}\, 3.85 + \frac{15 - 0}{20 - 0}\, 0.800 = 1.5625$$

In a similar fashion, the second-order polynomial is developed as [Eq. (17.21)]

$$f_2(x) = \frac{(15-20)(15-40)}{(0-20)(0-40)}\,3.85 + \frac{(15-0)(15-40)}{(20-0)(20-40)}\,0.800$$

$$+ \frac{(15-0)(15-20)}{(40-0)(40-20)}\,0.212 = 1.3316875$$

17.3.1 MATLAB M-file: Lagrange

It is straightforward to develop an M-file based on Eqs. (17.22) and (17.23). As in Fig. 17.9, the function is passed two vectors containing the independent (x) and the dependent (y) variables. It is also passed the value of the independent variable where you want to interpolate (xx). The order of the polynomial is based on the length of the x vector that is passed. If n values are passed, an $(n-1)$th order polynomial is fit.

FIGURE 17.9
An M-file to implement Lagrange interpolation.

```
function yint = Lagrange(x,y,xx)
% Lagrange: Lagrange interpolating polynomial
%   yint = Lagrange(x,y,xx): Uses an (n - 1)-order
%     Lagrange interpolating polynomial based on n data points
%     to determine a value of the dependent variable (yint) at
%     a given value of the independent variable, xx.
% input:
%   x = independent variable
%   y = dependent variable
%   xx = value of independent variable at which the
%          interpolation is calculated
% output:
%   yint = interpolated value of dependent variable

n = length(x);
if length(y)~=n, error('x and y must be same length'); end
s = 0;
for i = 1:n
  product = y(i);
  for j = 1:n
    if i ~= j
      product = product*(xx-x(j))/(x(i)-x(j));
    end
  end
  s = s+product;
end
yint = s;
```

An example of a session using the function would be to predict the density of air at 1 atm pressure at a temperature of 15 °C based on the first four values from Table 17.1. Because four values are passed to the function, a third-order polynomial would be implemented by the Lagrange function to give:

```
>> format long
>> T = [-40 0 20 50];
>> d = [1.52 1.29 1.2 1.09];
>> density = Lagrange(T,d,15)

density =
    1.22112847222222
```

17.4 INVERSE INTERPOLATION

As the nomenclature implies, the $f(x)$ and x values in most interpolation contexts are the dependent and independent variables, respectively. As a consequence, the values of the x's are typically uniformly spaced. A simple example is a table of values derived for the function $f(x) = 1/x$:

x	1	2	3	4	5	6	7
$f(x)$	1	0.5	0.3333	0.25	0.2	0.1667	0.1429

Now suppose that you must use the same data, but you are given a value for $f(x)$ and must determine the corresponding value of x. For instance, for the data above, suppose that you were asked to determine the value of x that corresponded to $f(x) = 0.3$. For this case, because the function is available and easy to manipulate, the correct answer can be determined directly as $x = 1/0.3 = 3.3333$.

Such a problem is called *inverse interpolation*. For a more complicated case, you might be tempted to switch the $f(x)$ and x values [i.e., merely plot x versus $f(x)$] and use an approach like Newton or Lagrange interpolation to determine the result. Unfortunately, when you reverse the variables, there is no guarantee that the values along the new abscissa [the $f(x)$'s] will be evenly spaced. In fact, in many cases, the values will be "telescoped." That is, they will have the appearance of a logarithmic scale with some adjacent points bunched together and others spread out widely. For example, for $f(x) = 1/x$ the result is

$f(x)$	0.1429	0.1667	0.2	0.25	0.3333	0.5	1
x	7	6	5	4	3	2	1

Such nonuniform spacing on the abscissa often leads to oscillations in the resulting interpolating polynomial. This can occur even for lower-order polynomials. An alternative strategy is to fit an nth-order interpolating polynomial, $f_n(x)$, to the original data [i.e., with $f(x)$ versus x]. In most cases, because the x's are evenly spaced, this polynomial will not be ill-conditioned. The answer to your problem then amounts to finding the value of x that makes this polynomial equal to the given $f(x)$. Thus, the interpolation problem reduces to a roots problem!

For example, for the problem just outlined, a simple approach would be to fit a quadratic polynomial to the three points: (2, 0.5), (3, 0.3333), and (4, 0.25). The result would be

$$f_2(x) = 0.041667x^2 - 0.375x + 1.08333$$

The answer to the inverse interpolation problem of finding the x corresponding to $f(x) = 0.3$ would therefore involve determining the root of

$$0.3 = 0.041667x^2 - 0.375x + 1.08333$$

For this simple case, the quadratic formula can be used to calculate

$$x = \frac{0.375 \pm \sqrt{(-0.375)^2 - 4(0.041667)0.78333}}{2(0.041667)} = \frac{5.704158}{3.295842}$$

Thus, the second root, 3.296, is a good approximation of the true value of 3.333. If additional accuracy were desired, a third- or fourth-order polynomial along with one of the root-location methods from Chaps. 5 or 6 could be employed.

17.5   EXTRAPOLATION AND OSCILLATIONS

Before leaving this chapter, there are two issues related to polynomial interpolation that must be addressed. These are extrapolation and oscillations.

17.5.1 Extrapolation

Extrapolation is the process of estimating a value of $f(x)$ that lies outside the range of the known base points, $x_1, x_2, \ldots, x_n$. As depicted in Fig. 17.10, the open-ended nature of

FIGURE 17.10
Illustration of the possible divergence of an extrapolated prediction. The extrapolation is based on fitting a parabola through the first three known points.

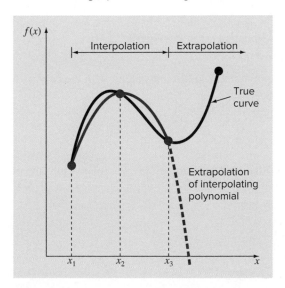

extrapolation represents a step into the unknown because the process extends the curve beyond the known region. As such, the true curve could easily diverge from the prediction. Extreme care should, therefore, be exercised whenever a case arises where one must extrapolate.

EXAMPLE 17.6   Dangers of Extrapolation

Problem Statement. This example is patterned after one originally developed by Forsythe, Malcolm, and Moler.[1] The population in millions of the United States from 1920 to 2000 can be tabulated as

Date	1920	1930	1940	1950	1960	1970	1980	1990	2000
Population	106.46	123.08	132.12	152.27	180.67	205.05	227.23	249.46	281.42

Fit a seventh-order polynomial to the first 8 points (1920 to 1990). Use it to compute the population in 2000 by extrapolation and compare your prediction with the actual result.

Solution. First, the data can be entered as

```
>> t = [1920:10:1990];
>> pop = [106.46 123.08 132.12 152.27 180.67 205.05 227.23
          249.46];
```

The polyfit function can be used to compute the coefficients

```
>> p = polyfit(t,pop,7)
```

However, when this is implemented, the following message is displayed:

```
Warning: Polynomial is badly conditioned. Remove repeated data points or try centering
         and scaling as described in HELP POLYFIT.
```

We can follow MATLAB's suggestion by scaling and centering the data values as in

```
>> ts = (t - 1955)/35;
```

Now polyfit works without an error message:

```
>> p = polyfit(ts,pop,7);
```

We can then use the polynomial coefficients along with the polyval function to predict the population in 2000 as

```
>> polyval(p,(2000-1955)/35)

ans =
   175.0800
```

which is much lower that the true value of 281.42. Insight into the problem can be gained by generating a plot of the data and the polynomial,

```
>> tt = linspace(1920,2000);
>> pp = polyval(p,(tt-1955)/35);
>> plot(t,pop,'o',tt,pp)
```

[1] Cleve Moler is one of the founders of The MathWorks, Inc., the makers of MATLAB.

FIGURE 17.11
Use of a seventh-order polynomial to make a prediction of U.S. population in 2000 based on data from 1920 through 1990.

As in Fig. 17.11, the result indicates that the polynomial seems to fit the data nicely from 1920 to 1990. However, once we move beyond the range of the data into the realm of extrapolation, the seventh-order polynomial plunges to the erroneous prediction in 2000.

17.5.2 Oscillations

Although "more is better" in many contexts, it is absolutely not true for polynomial interpolation. Higher-order polynomials tend to be very ill-conditioned—that is, they tend to be highly sensitive to round off error. The following example illustrates this point nicely.

EXAMPLE 17.7   Dangers of Higher-Order Polynomial Interpolation

Problem Statement. In 1901, Carl Runge published a study on the dangers of higher-order polynomial interpolation. He looked at the following simple-looking function:

$$f(x) = \frac{1}{1 + 25x^2} \tag{17.24}$$

which is now called *Runge's function*. He took equidistantly spaced data points from this function over the interval [−1, 1]. He then used interpolating polynomials of increasing order and found that as he took more points, the polynomials and the original curve differed considerably. Further, the situation deteriorated greatly as the order was increased. Duplicate Runge's result by using the polyfit and polyval functions to fit fourth- and tenth-order polynomials to 5 and 11 equally spaced points generated with Eq. (17.24). Create plots of your results along with the sampled values and the complete Runge's function.

Solution. The five equally spaced data points can be generated as in

```
>> x = linspace(-1,1,5);
>> y = 1./(1+25*x.^2);
```

Next, a more finally spaced vector of xx values can be computed so that we can create a smooth plot of the results:

```
>> xx = linspace(-1,1);
```

Recall that linspace automatically creates 100 points if the desired number of points is not specified. The polyfit function can be used to generate the coefficients of the fourth-order polynomial, and the polval function can be used to generate the polynomial interpolation at the finely spaced values of xx:

```
>> p = polyfit(x,y,4);
>> y4 = polyval(p,xx);
```

Finally, we can generate values for Runge's function itself and plot them along with the polynomial fit and the sampled data:

```
>> yr = 1./(1+25*xx.^2);
>> plot(x,y,'o',xx,y4,xx,yr,'--')
```

As in Fig. 17.12, the polynomial does a poor job of following Runge's function.

Continuing with the analysis, the tenth-order polynomial can be generated and plotted with

```
>> x = linspace(-1,1,11);
>> y = 1./(1+25*x.^2);
```

FIGURE 17.12
Comparison of Runge's function (dashed line) with a fourth-order polynomial fit to 5 points sampled from the function.

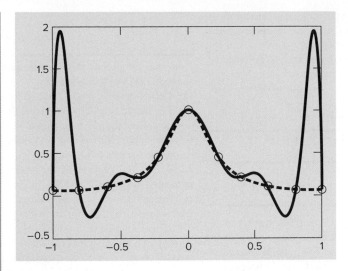

FIGURE 17.13
Comparison of Runge's function (dashed line) with a tenth-order polynomial fit to 11 points sampled from the function.

```
>> p = polyfit(x,y,10);
>> y10 = polyval(p,xx);
>> plot(x,y,'o',xx,y10,xx,yr,'--')
```

As in Fig. 17.13, the fit has gotten even worse, particularly at the ends of the interval!

Although there may be certain contexts where higher-order polynomials are necessary, they are usually to be avoided. In most engineering and scientific contexts, lower-order polynomials of the type described in this chapter can be used effectively to capture the curving trends of data without suffering from oscillations.

PROBLEMS

17.1 The following data come from a table that was measured with high precision. Use the best numerical method (for this type of problem) to determine y at $x = 3.5$. Note that a polynomial will yield an exact value. Your solution should prove that your result is exact.

x	0	1.8	5	6	8.2	9.2	12
y	26	16.415	5.375	3.5	2.015	2.54	8

17.2 Use Newton's interpolating polynomial to determine y at $x = 3.5$ to the best possible accuracy. Compute the finite divided differences as in Fig. 17.5, and order your points to attain optimal accuracy and convergence. That is, the points should be centered around and as close as possible to the unknown.

x	0	1	2.5	3	4.5	5	6
y	2	5.4375	7.3516	7.5625	8.4453	9.1875	12

17.3 Use Newton's interpolating polynomial to determine y at $x = 8$ to the best possible accuracy. Compute the finite divided differences as in Fig. 17.5, and order your points to attain optimal accuracy and convergence. That is, the points should be centered around and as close as possible to the unknown.

x	0	1	2	5.5	11	13	16	18
y	0.5	3.134	5.3	9.9	10.2	9.35	7.2	6.2

17.4 Given the data

x	1	2	2.5	3	4	5
$f(x)$	0	5	7	6.5	2	0

(a) Calculate $f(3.4)$ using Newton's interpolating polynomials of order 1 through 3. Choose the sequence of the points for your estimates to attain the best possible accuracy. That is, the points should be centered around and as close as possible to the unknown.
(b) Repeat (a) but use the Lagrange polynomial.

17.5 Given the data

x	1	2	3	5	6
$f(x)$	4.75	4	5.25	19.75	36

Calculate $f(4)$ using Newton's interpolating polynomials of order 1 through 4. Choose your base points to attain good accuracy. That is, the points should be centered around and as close as possible to the unknown. What do your results indicate regarding the order of the polynomial used to generate the data in the table?

17.6 Repeat Prob. 17.5 using the Lagrange polynomial of order 1 through 3.

17.7 Table P15.5 lists values for dissolved oxygen concentration in water as a function of temperature and chloride concentration.
(a) Use quadratic and cubic interpolation to determine the oxygen concentration for $T = 12\ °C$ and $c = 10$ g/L.
(b) Use linear interpolation to determine the oxygen concentration for $T = 12\ °C$ and $c = 15$ g/L.
(c) Repeat (b) but use quadratic interpolation.

17.8 Employ inverse interpolation using a cubic interpolating polynomial and bisection to determine the value of x that corresponds to $f(x) = 1.7$ for the following tabulated data:

x	1	2	3	4	5	6	7
$f(x)$	3.6	1.8	1.2	0.9	0.72	1.5	0.51429

17.9 Employ inverse interpolation to determine the value of x that corresponds to $f(x) = 0.93$ for the following tabulated data:

x	0	1	2	3	4	5
$f(x)$	0	0.5	0.8	0.9	0.941176	0.961538

Note that the values in the table were generated with the function $f(x) = x^2/(1 + x^2)$.
(a) Determine the correct value analytically.
(b) Use quadratic interpolation and the quadratic formula to determine the value numerically.
(c) Use cubic interpolation and bisection to determine the value numerically.

17.10 Use the portion of the given steam table for superheated water at 200 MPa to find (a) the corresponding entropy s for a specific volume v of 0.118 with linear interpolation, (b) the same corresponding entropy using quadratic interpolation, and (c) the volume corresponding to an entropy of 6.45 using inverse interpolation.

v, m³/kg	0.10377	0.11144	0.12547
s, kJ/(kg K)	6.4147	6.5453	6.7664

17.11 The following data for the density of nitrogen gas versus temperature come from a table that was measured with high precision. Use first- through fifth-order polynomials to estimate the density at a temperature of 330 K. What is your best estimate? Employ this best estimate and inverse interpolation to determine the corresponding temperature.

T, K	200	250	300	350	400	450
Density, kg/m³	1.708	1.367	1.139	0.967	0.854	0.759

17.12 Ohm's law states that the voltage drop V across an ideal resistor is linearly proportional to the current i flowing through the resister as in $V = i R$, where R is the resistance. However, real resistors may not always obey Ohm's law. Suppose that you performed some very precise experiments to measure the voltage drop and corresponding current for a resistor. The following results suggest a curvilinear relationship rather than the straight line represented by Ohm's law:

i	-2	-1	-0.5	0.5	1	2
V	-637	-96.5	-20.5	20.5	96.5	637

To quantify this relationship, a curve must be fit to the data. Because of measurement error, regression would typically be the preferred method of curve fitting for analyzing such experimental data. However, the smoothness of the relationship, as well as the precision of the experimental methods, suggests that interpolation might be appropriate. Use a fifth-order interpolating polynomial to fit the data and compute V for $i = 0.10$.

17.13 Bessel functions often arise in advanced engineering analyses such as the study of electric fields. Here are some selected values for the zero-order Bessel function of the first kind

x	1.8	2.0	2.2	2.4	2.6
$J_1(x)$	0.5815	0.5767	0.5560	0.5202	0.4708

Estimate $J_1(2.1)$ using third- and fourth-order interpolating polynomials. Determine the percent relative error for each case based on the true value, which can be determined with MATLAB's built-in function `besselj`.

17.14 Repeat Example 17.6 but using first-, second-, third-, and fourth-order interpolating polynomials to predict the population in 2000 based on the most recent data. That is, for the linear prediction use the data from 1980 and 1990, for the quadratic prediction use the data from 1970, 1980, and 1990, and so on. Which approach yields the best result?

17.15 The specific volume of a superheated steam is listed in steam tables for various temperatures.

T, °C	370	382	394	406	418
v, L/kg	5.9313	7.5838	8.8428	9.796	10.5311

Determine v at $T = 400$ °C.

17.16 The vertical stress σ_z under the corner of a rectangular area subjected to a uniform load of intensity q is given by the solution of Boussinesq's equation:

$$\sigma = \frac{q}{4\pi} \left[\frac{2mn\sqrt{m^2 + n^2 + 1}}{m^2 + n^2 + 1 + m^2n^2} \frac{m^2 + n^2 + 2}{m^2 + n^2 + 1} \right.$$
$$\left. + \sin^{-1} \left(\frac{2mn\sqrt{m^2 + n^2 + 1}}{m^2 + n^2 + 1 + m^2n^2} \right) \right]$$

Because this equation is inconvenient to solve manually, it has been reformulated as

$$\sigma_z = q f_z(m, n)$$

where $f_z(m, n)$ is called the influence value, and m and n are dimensionless ratios, with $m = a/z$ and $n = b/z$ and a and b are defined in Fig. P17.16. The influence value is then tabulated, a portion of which is given in Table P17.16. If $a = 4.6$ and $b = 14$, use a third-order interpolating polynomial to compute σ_z at a depth 10 m below the corner of a rectangular footing that is subject to a total load

FIGURE P17.16

of 100 t (metric tons). Express your answer in tonnes per square meter. Note that q is equal to the load per area.

TABLE P17.16

m	$n = 1.2$	$n = 1.4$	$n = 1.6$
0.1	0.02926	0.03007	0.03058
0.2	0.05733	0.05894	0.05994
0.3	0.08323	0.08561	0.08709
0.4	0.10631	0.10941	0.11135
0.5	0.12626	0.13003	0.13241
0.6	0.14309	0.14749	0.15027
0.7	0.15703	0.16199	0.16515
0.8	0.16843	0.17389	0.17739

17.17 You measure the voltage drop V across a resistor for a number of different values of current i. The results are

i	0.5	1.5	2.5	3.0	4.0
V	−0.45	−0.6	0.70	1.88	6.0

Use first- through fourth-order polynomial interpolation to estimate the voltage drop for $i = 2.3$. Interpret your results.

17.18 The current in a wire is measured with great precision as a function of time:

t	0	0.250	0.500	0.750	1.000
i	0	6.24	7.75	4.85	0.0000

Determine i at $t = 0.23$.

17.19 The acceleration due to gravity at an altitude y above the surface of the earth is given by

y, m	0	30,000	60,000	90,000	120,000
g, m/s²	9.8100	9.7487	9.6879	9.6278	9.5682

Compute g at $y = 55,000$ m.

17.20 Temperatures are measured at various points on a heated plate (Table P17.20). Estimate the temperature at **(a)** $x = 4$, $y = 3.2$ and **(b)** $x = 4.3$, $y = 2.7$.

TABLE P17.20 Temperatures (°C) at various points on a square heated plate.

	$x = 0$	$x = 2$	$x = 4$	$x = 6$	$x = 8$
$y = 0$	100.00	90.00	80.00	70.00	60.00
$y = 2$	85.00	64.49	53.50	48.15	50.00
$y = 4$	70.00	48.90	38.43	35.03	40.00
$y = 6$	55.00	38.78	30.39	27.07	30.00
$y = 8$	40.00	35.00	30.00	25.00	20.00

17.21 Use the portion of the given steam table for superheated H_2O at 200 MPa to **(a)** find the corresponding entropy s for a specific volume v of 0.108 m³/kg with linear interpolation, **(b)** find the same corresponding entropy using quadratic interpolation, and **(c)** find the volume corresponding to an entropy of 6.6 using inverse interpolation.

v (m³/kg)	0.10377	0.11144	0.12540
s (kJ/kg·K)	6.4147	6.5453	6.7664

17.22 Develop an M-file function, that uses `polyfit` and `polyval` for polynomial interpolation. Here is the script you can use to test your function

```
clear,clc,clf,format compact
x=[1 2 4 8];
fx=@(x) 10*exp(-0.2*x);
y=fx(x);
yint=polyint(x,y,3)
ytrue=fx(3)
et=abs((ytrue-yint)/ytrue)*100.
```

17.23 The following data comes from a table that was measured with high precision. Use the Newton interpolating polynomial to determine y at $x = 3.5$. Properly order *all the points* and then develop a divided difference table to compute the derivatives. Note that a polynomial will yield an exact value. Your solution should prove that your result is exact.

x	0	1	2.5	3	4.5	5	6
y	26	15.5	5.375	3.5	2.375	3.5	8

17.24 The following data are measured precisely:

T	2	2.1	2.2	2.7	3	3.4
Z	6	7.752	10.256	36.576	66	125.168

(a) Use Newton interpolating polynomials to determine z at $t = 2.5$. Make sure that you order your points to attain the most accurate results. What do your results tell you regarding the order of the polynomial used to generate the data?

(b) Use a third-order Lagrange interpolating polynomial to determine y at $t = 2.5$.

17.25 The following data for the density of water versus temperature come from a table that was measured with high precision. Use inverse interpolation to determine the temperature corresponding to a density of 0.999245 g/cm³. Base your estimate on a third-order interpolating polynomial (Even though you're doing this problem by hand, feel free to use the MATLAB `polyfit` function to determine the polynomial.) Determine the root with the Newton-Raphson method (by hand) with an initial guess of $T = 14$ °C. Be careful regarding roundoff errors.

T, °C	0	4	8	12	16
Density, g/cm³	0.99987	1	0.99988	0.99952	0.99897

18

Splines and Piecewise Interpolation

CHAPTER OBJECTIVES

The primary objective of this chapter is to introduce you to splines. Specific objectives and topics covered are

- Understanding that splines minimize oscillations by fitting lower-order polynomials to data in a piecewise fashion.
- Knowing how to develop code to perform a table lookup.
- Recognizing why cubic polynomials are preferable to quadratic and higher-order splines.
- Understanding the conditions that underlie a cubic spline fit.
- Understanding the differences between natural, clamped, and not-a-knot end conditions.
- Knowing how to fit a spline to data with MATLAB's built-in functions.
- Understanding how multidimensional interpolation is implemented with MATLAB.

18.1 INTRODUCTION TO SPLINES

In Chap. 17 $(n − 1)$th-order polynomials were used to interpolate between n data points. For example, for eight points, we can derive a perfect seventh-order polynomial. This curve would capture all the meanderings (at least up to and including seventh derivatives) suggested by the points. However, there are cases where these functions can lead to errone-ous results because of round off error and oscillations. An alternative approach is to apply lower-order polynomials in a piecewise fashion to subsets of data points. Such connecting polynomials are called *spline functions*.

For example, third-order curves employed to connect each pair of data points are called *cubic splines*. These functions can be constructed so that the connections between adjacent cubic equations are visually smooth. On the surface, it would seem that the third-order

453

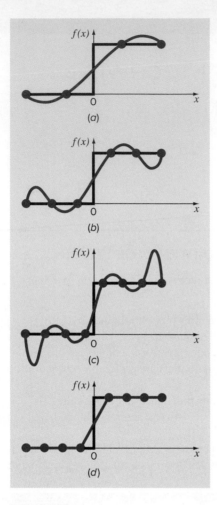

FIGURE 18.1
A visual representation of a situation where splines are superior to higher-order interpolating polynomials. The function to be fit undergoes an abrupt increase at $x = 0$. Parts (a) through (c) indicate that the abrupt change induces oscillations in interpolating polynomials. In contrast, because it is limited to straight-line connections, a linear spline (d) provides a much more acceptable approximation.

approximation of the splines would be inferior to the seventh-order expression. You might wonder why a spline would ever be preferable.

Figure 18.1 illustrates a situation where a spline performs better than a higher-order polynomial. This is the case where a function is generally smooth but undergoes an abrupt change somewhere along the region of interest. The step increase depicted in Fig. 18.1 is an extreme example of such a change and serves to illustrate the point.

Figure 18.1a through c illustrates how higher-order polynomials tend to swing through wild oscillations in the vicinity of an abrupt change. In contrast, the spline also connects the points, but because it is limited to lower-order changes, the oscillations are kept to a

FIGURE 18.2
The drafting technique of using a spline to draw smooth curves through a series of points.
Notice how, at the end points, the spline straightens out. This is called a "natural" spline.

minimum. As such, the spline usually provides a superior approximation of the behavior of functions that have local, abrupt changes.

The concept of the spline originated from the drafting technique of using a thin, flexible strip (called a *spline*) to draw smooth curves through a set of points. The process is depicted in Fig. 18.2 for a series of five pins (data points). In this technique, the drafter places paper over a wooden board and hammers nails or pins into the paper (and board) at the location of the data points. A smooth cubic curve results from interweaving the strip between the pins. Hence, the name "cubic spline" has been adopted for polynomials of this type.

In this chapter, simple linear functions will first be used to introduce some basic concepts and issues associated with spline interpolation. Then we derive an algorithm for fitting quadratic splines to data. This is followed by material on the cubic spline, which is the most common and useful version in engineering and science. Finally, we describe MATLAB's capabilities for piecewise interpolation including its ability to generate splines.

18.2   LINEAR SPLINES

The notation used for splines is displayed in Fig. 18.3. For n data points ($i = 1, 2, \ldots, n$), there are $n - 1$ intervals. Each interval i has its own spline function, $s_i(x)$. For linear splines, each function is merely the straight line connecting the two points at each end of the interval, which is formulated as

$$s_i(x) = a_i + b_i(x - x_i) \tag{18.1}$$

where a_i is the intercept, which is defined as

$$a_i = f_i \tag{18.2}$$

and b_i is the slope of the straight line connecting the points:

$$b_i = \frac{f_{i+1} - f_i}{x_{i+1} - x_i} \tag{18.3}$$

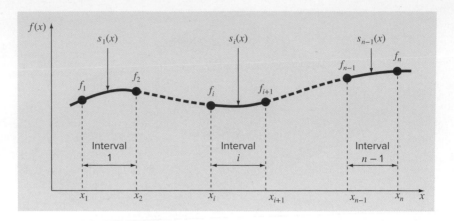

FIGURE 18.3
Notation used to derive splines. Notice that there are $n - 1$ intervals and n data points.

where f_i is shorthand for $f(x_i)$. Substituting Eqs. (18.1) and (18.2) into Eq. (18.3) gives

$$s_i(x) = f_i + \frac{f_{i+1} - f_i}{x_{i+1} - x_i}(x - x_i) \tag{18.4}$$

These equations can be used to evaluate the function at any point between x_1 and x_n by first locating the interval within which the point lies. Then the appropriate equation is used to determine the function value within the interval. Inspection of Eq. (18.4) indicates that the linear spline amounts to using Newton's first-order polynomial [Eq. (17.5)] to interpolate within each interval.

EXAMPLE 18.1   First-Order Splines

Problem Statement.   Fit the data in Table 18.1 with first-order splines. Evaluate the function at $x = 5$.

TABLE 18.1   Data to be fit with spline functions.

i	x_i	f_i
1	3.0	2.5
2	4.5	1.0
3	7.0	2.5
4	9.0	0.5

Solution.   The data can be substituted into Eq. (18.4) to generate the linear spline functions. For example, for the second interval from $x = 4.5$ to $x = 7$, the function is

$$s_2(x) = 1.0 + \frac{2.5 - 1.0}{7.0 - 4.5}(x - 4.5)$$

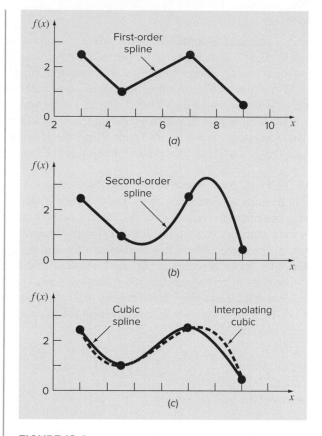

FIGURE 18.4
Spline fits of a set of four points. (*a*) Linear spline, (*b*) quadratic spline, and (*c*) cubic spline, with a cubic interpolating polynomial also plotted.

The equations for the other intervals can be computed, and the resulting first-order splines are plotted in Fig. 18.4*a*. The value at $x = 5$ is 1.3.

$$s_2(x) = 1.0 + \frac{2.5 - 1.0}{7.0 - 4.5}(5 - 4.5) = 1.3$$

Visual inspection of Fig. 18.4*a* indicates that the primary disadvantage of first-order splines is that they are not smooth. In essence, at the data points where two splines meet (called a *knot*), the slope changes abruptly. In formal terms, the first derivative of the function is discontinuous at these points. This deficiency is overcome by using higher-order polynomial splines that ensure smoothness at the knots by equating derivatives at these points, as will be discussed subsequently. Before doing that, the following section provides an application where linear splines are useful.

18.2.1 Table Lookup

A table lookup is a common task that is frequently encountered in engineering and science computer applications. It is useful for performing repeated interpolations from a table of independent and dependent variables. For example, suppose that you would like to set up an M-file that would use linear interpolation to determine air density at a particular temperature based on the data from Table 17.1. One way to do this would be to pass the M-file the temperature at which you want the interpolation to be performed along with the two adjoining values. A more general approach would be to pass in vectors containing all the data and have the M-file determine the bracket. This is called a *table lookup*.

Thus, the M-file would perform two tasks. First, it would search the independent variable vector to find the interval containing the unknown. Then it would perform the linear interpolation using one of the techniques described in this chapter or in Chap. 17.

For ordered data, there are two simple ways to find the interval. The first is called a *sequential search*. As the name implies, this method involves comparing the desired value with each element of the vector in sequence until the interval is located. For data in ascending order, this can be done by testing whether the unknown is less than the value being assessed. If so, we know that the unknown falls between this value and the previous one that we examined. If not, we move to the next value and repeat the comparison. Here is a simple M-file that accomplishes this objective:

```
function yi = TableLook(x, y, xx)

n = length(x);
if xx < x(1) | xx > x(n)
  error('Interpolation outside range')
end
% sequential search
i = 1;
while(1)
   if xx <= x(i + 1), break, end
   i = i + 1;
end
% linear interpolation
yi = y(i) + (y(i+1)-y(i))/(x(i+1)-x(i))*(xx-x(i));
```

The table's independent variables are stored in ascending order in the array x and the dependent variables stored in the array y. Before searching, an error trap is included to ensure that the desired value xx falls within the range of the x's. A while . . . break loop compares the value at which the interpolation is desired, xx, to determine whether it is less than the value at the top of the interval, x(i+1). For cases where xx is in the second interval or higher, this will not test true at first. In this case the counter i is incremented by one so that on the next iteration, xx is compared with the value at the top of the second interval. The loop is repeated until the xx is less than or equal to the interval's upper bound, in which case the loop is exited. At this point, the interpolation can be performed simply as shown.

For situations for which there are lots of data, the sequential sort is inefficient because it must search through all the preceding points to find values. In these cases, a simple alternative is the *binary search*. Here is an M-file that performs a binary search followed by linear interpolation:

```
function yi = TableLookBin(x, y, xx)

n = length(x);
if xx < x(1) | xx > x(n)
  error('Interpolation outside range')
end
% binary search
iL = 1; iU = n;
while (1)
  if iU - iL <= 1, break, end
  iM = fix((iL + iU) / 2);
  if x(iM) < xx
    iL = iM;
  else
    iU = iM;
  end
end
% linear interpolation
yi = y(iL) + (y(iL+1)-y(iL))/(x(iL+1)-x(iL))*(xx - x(iL));
```

The approach is akin to the bisection method for root location. Just as in bisection, the index at the midpoint iM is computed as the average of the first or "lower" index iL = 1 and the last or "upper" index iU = n. The unknown xx is then compared with the value of x at the midpoint x(iM) to assess whether it is in the lower half of the array or in the upper half. Depending on where it lies, either the lower or upper index is redefined as being the middle index. The process is repeated until the difference between thc upper and the lower index is less than or equal to zero. At this point, the lower index lies at the lower bound of the interval containing xx, the loop terminates, and the linear interpolation is performed.

Here is a MATLAB session illustrating how the binary search function can be applied to calculate the air density at 350 °C based on the data from Table 17.1. The sequential search would be similar.

```
>> T = [-40 0 20 50 100 150 200 250 300 400 500];
>> density = [1.52 1.29 1.2 1.09 .946 .935 .746 .675 .616... .525 .457];
>> TableLookBin(T,density,350)

ans =
    0.5705
```

This result can be verified by the hand calculation:

$$f(350) = 0.616 + \frac{0.525 - 0.616}{400 - 300}(350 - 300) = 0.5705$$

18.3 QUADRATIC SPLINES

To ensure that the *n*th derivatives are continuous at the knots, a spline of at least $n + 1$ order must be used. Third-order polynomials or cubic splines that ensure continuous first and second derivatives are most frequently used in practice. Although third and higher derivatives can be discontinuous when using cubic splines, they usually cannot be detected visually and consequently are ignored.

Because the derivation of cubic splines is somewhat involved, we have decided to first illustrate the concept of spline interpolation using second-order polynomials. These "quadratic splines" have continuous first derivatives at the knots. Although quadratic splines are not of practical importance, they serve nicely to demonstrate the general approach for developing higher-order splines.

The objective in quadratic splines is to derive a second-order polynomial for each interval between data points. The polynomial for each interval can be represented generally as

$$s_i(x) = a_i + b_i(x - x_i) + c_i(x - x_i)^2 \tag{18.5}$$

where the notation is as in Fig. 18.3. For n data points ($i = 1, 2, \ldots, n$), there are $n - 1$ intervals and, consequently, $3(n - 1)$ unknown constants (the a's, b's, and c's) to evaluate. Therefore, $3(n - 1)$ equations or conditions are required to evaluate the unknowns. These can be developed as follows:

1. The function must pass through all the points. This is called a *continuity condition*. It can be expressed mathematically as

$$f_i = a_i + b_i(x_i - x_i) + c_i(x_i - x_i)^2$$

which simplifies to

$$a_i = f_i \tag{18.6}$$

Therefore, the constant in each quadratic must be equal to the value of the dependent variable at the beginning of the interval. This result can be incorporated into Eq. (18.5):

$$s_i(x) = f_i + b_i(x - x_i) + c_i(x - x_i)^2$$

Note that because we have determined one of the coefficients, the number of conditions to be evaluated has now been reduced to $2(n - 1)$.

2. The function values of adjacent polynomials must be equal at the knots. This condition can be written for knot $i + 1$ as

$$f_i + b_i(x_{i+1} - x_i) + c_i(x_{i+1} - x_i)^2 = f_{i+1} + b_{i+1}(x_{i+1} - x_{i+1}) + c_{i+1}(x_{i+1} - x_{i+1})^2 \tag{18.7}$$

This equation can be simplified mathematically by defining the width of the ith interval as

$$h_i = x_{i+1} - x_i$$

Thus, Eq. (18.7) simplifies to

$$f_i + b_i h_i + c_i h_i^2 = f_{i+1} \tag{18.8}$$

This equation can be written for the nodes, $i = 1, \ldots, n - 1$. Since this amounts to $n - 1$ conditions, it means that there are $2(n - 1) - (n - 1) = n - 1$ remaining conditions.

3. The first derivatives at the interior nodes must be equal. This is an important condition, because it means that adjacent splines will be joined smoothly, rather than in the jagged fashion that we saw for the linear splines. Equation (18.5) can be differentiated to yield

$$s_i'(x) = b_i + 2c_i(x - x_i)$$

The equivalence of the derivatives at an interior node, $i + 1$ can therefore be written as

$$b_i + 2c_i h_i = b_{i+1} \qquad (18.9)$$

Writing this equation for all the interior nodes amounts to $n - 2$ conditions. This means that there is $n - 1 - (n - 2) = 1$ remaining condition. Unless we have some additional information regarding the functions or their derivatives, we must make an arbitrary choice to successfully compute the constants. Although there are a number of different choices that can be made, we select the following condition.

4. Assume that the second derivative is zero at the first point. Because the second derivative of Eq. (18.5) is $2c_i$, this condition can be expressed mathematically as

$$c_1 = 0$$

The visual interpretation of this condition is that the first two points will be connected by a straight line.

EXAMPLE 18.2 Quadratic Splines

Problem Statement. Fit quadratic splines to the same data employed in Example 18.1 (Table 18.1). Use the results to estimate the value at $x = 5$.

Solution. For the present problem, we have four data points and $n = 3$ intervals. Therefore, after applying the continuity condition and the zero second-derivative condition, this means that $2(4 - 1) - 1 = 5$ conditions are required. Equation (18.8) is written for $i = 1$ through 3 (with $c_1 = 0$) to give

$$f_1 + b_1 h_1 = f_2$$
$$f_2 + b_2 h_2 + c_2 h_2^2 = f_3$$
$$f_3 + b_3 h_3 + c_3 h_3^2 = f_4$$

Continuity of derivatives, Eq. (18.9), creates an additional $3 - 1 = 2$ conditions (again, recall that $c_1 = 0$):

$$b_1 = b_2$$
$$b_2 + 2c_2 h_2 = b_3$$

The necessary function and interval width values are

$$f_1 = 2.5 \qquad\qquad h_1 = 4.5 - 3.0 = 1.5$$
$$f_2 = 1.0 \qquad\qquad h_2 = 7.0 - 4.5 = 2.5$$
$$f_3 = 2.5 \qquad\qquad h_3 = 9.0 - 7.0 = 2.0$$
$$f_4 = 0.5$$

These values can be substituted into the conditions which can be expressed in matrix form as

$$\begin{bmatrix} 1.5 & 0 & 0 & 0 & 0 \\ 0 & 2.5 & 6.25 & 0 & 0 \\ 0 & 0 & 0 & 2 & 4 \\ 1 & -1 & 0 & 0 & 0 \\ 0 & 1 & 5 & -1 & 0 \end{bmatrix} \begin{Bmatrix} b_1 \\ b_2 \\ c_2 \\ b_3 \\ c_3 \end{Bmatrix} = \begin{Bmatrix} -1.5 \\ 1.5 \\ -2 \\ 0 \\ 0 \end{Bmatrix}$$

These equations can be solved using MATLAB with the results:

$$b_1 = -1$$
$$b_2 = -1 \qquad\qquad c_2 = 0.64$$
$$b_3 = 2.2 \qquad\qquad c_3 = -1.6$$

These results, along with the values for the a's [Eq. (18.6)], can be substituted into the original quadratic equations to develop the following quadratic splines for each interval:

$$s_1(x) = 2.5 - (x - 3)$$
$$s_2(x) = 1.0 - (x - 4.5) + 0.64(x - 4.5)^2$$
$$s_3(x) = 2.5 + 2.2(x - 7.0) - 1.6(x - 7.0)^2$$

Because $x = 5$ lies in the second interval, we use s_2 to make the prediction,

$$s_2(5) = 1.0 - (5 - 4.5) + 0.64(5 - 4.5)^2 = 0.66$$

The total quadratic spline fit is depicted in Fig. 18.4b. Notice that there are two shortcomings that detract from the fit: (1) the straight line connecting the first two points and (2) the spline for the last interval seems to swing too high. The cubic splines in the next section do not exhibit these shortcomings and, as a consequence, are better methods for spline interpolation.

18.4 CUBIC SPLINES

As stated at the beginning of the previous section, cubic splines are most frequently used in practice. The shortcomings of linear and quadratic splines have already been discussed. Quartic or higher-order splines are not used because they tend to exhibit the instabilities inherent in higher-order polynomials. Cubic splines are preferred because they provide the simplest representation that exhibits the desired appearance of smoothness.

The objective in cubic splines is to derive a third-order polynomial for each interval between knots as represented generally by

$$s_i(x) = a_i + b_i(x - x_i) + c_i(x - x_i)^2 + d_i(x - x_i)^3 \tag{18.10}$$

Thus, for n data points ($i = 1, 2, \ldots, n$), there are $n - 1$ intervals and $4(n - 1)$ unknown coefficients to evaluate. Consequently, $4(n - 1)$ conditions are required for their evaluation.

The first conditions are identical to those used for the quadratic case. That is, they are set up so that the functions pass through the points and that the first derivatives at the knots are equal. In addition to these, conditions are developed to ensure that the second derivatives at the knots are also equal. This greatly enhances the fit's smoothness.

After these conditions are developed, two additional conditions are required to obtain the solution. This is a much nicer outcome than occurred for quadratic splines where we needed to specify a single condition. In that case, we had to arbitrarily specify a zero second derivative for the first interval, hence making the result asymmetric. For cubic splines, we are in the advantageous position of needing two additional conditions and can, therefore, apply them evenhandedly at both ends.

For cubic splines, these last two conditions can be formulated in several different ways. A very common approach is to assume that the second derivatives at the first and last knots are equal to zero. The visual interpretation of these conditions is that the function becomes a straight line at the end nodes. Specification of such an end condition leads to what is termed a "natural" spline. It is given this name because the drafting spline naturally behaves in this fashion (Fig. 18.2).

There are a variety of other end conditions that can be specified. Two of the more popular are the clamped condition and the not-a-knot conditions. We will describe these options in Sec. 18.4.2. For the following derivation, we will limit ourselves to natural splines.

Once the additional end conditions are specified, we would have the $4(n - 1)$ conditions needed to evaluate the $4(n - 1)$ unknown coefficients. Whereas it is certainly possible to develop cubic splines in this fashion, we will present an alternative approach that requires the solution of only $n - 1$ equations. Further, the simultaneous equations will be tridiagonal and hence can be solved very efficiently. Although the derivation of this approach is less straightforward than for quadratic splines, the gain in efficiency is well worth the effort.

18.4.1 Derivation of Cubic Splines

As was the case with quadratic splines, the first condition is that the spline must pass through all the data points:

$$f_i = a_i + b_i(x_i - x_i) + c_i(x_i - x_i)^2 + d_i(x_i - x_i)^3$$

which simplifies to

$$a_i = f_i \qquad\qquad (18.11)$$

Therefore, the constant in each cubic must be equal to the value of the dependent variable at the beginning of the interval. This result can be incorporated into Eq. (18.10):

$$s_i(x) = f_i + b_i(x - x_i) + c_i(x - x_i)^2 + d_i(x - x_i)^3 \qquad\qquad (18.12)$$

Next, we will apply the condition that each of the cubics must join at the knots. For knot $i + 1$, this can be represented as

$$f_i + b_i h_i + c_i h_i^2 + d_i h_i^3 = f_{i+1} \qquad\qquad (18.13)$$

where

$$h_i = x_{i+1} - x_i$$

The first derivatives at the interior nodes must be equal. Equation (18.12) is differentiated to yield

$$s_i'(x) = b_i + 2c_i(x - x_i) + 3d_i(x - x_i)^2 \qquad (18.14)$$

The equivalence of the derivatives at an interior node, $i + 1$ can therefore be written as

$$b_i + 2c_i h_i + 3d_i h_i^2 = b_{i+1} \qquad (18.15)$$

The second derivatives at the interior nodes must also be equal. Equation (18.14) can be differentiated to yield

$$s_i''(x) = 2c_i + 6d_i(x - x_i) \qquad (18.16)$$

The equivalence of the second derivatives at an interior node, $i + 1$ can therefore be written as

$$c_i + 3d_i h_i = c_{i+1} \qquad (18.17)$$

Next, we can solve Eq. (18.17) for d_i:

$$d_i = \frac{c_{i+1} - c_i}{3h_i} \qquad (18.18)$$

This can be substituted into Eq. (18.13) to give

$$f_i + b_i h_i + \frac{h_i^2}{3}(2c_i + c_{i+1}) = f_{i+1} \qquad (18.19)$$

Equation (18.18) can also be substituted into Eq. (18.15) to give

$$b_{i+1} = b_i + h_i(c_i + c_{i+1}) \qquad (18.20)$$

Equation (18.19) can be solved for

$$b_i = \frac{f_{i+1} - f_i}{h_i} - \frac{h_i}{3}(2c_i + c_{i+1}) \qquad (18.21)$$

The index of this equation can be reduced by 1:

$$b_{i-1} = \frac{f_i - f_{i-1}}{h_{i-1}} - \frac{h_{i-1}}{3}(2c_{i-1} + c_i) \qquad (18.22)$$

The index of Eq. (18.20) can also be reduced by 1:

$$b_i = b_{i-1} + h_{i-1}(c_{i-1} + c_i) \qquad (18.23)$$

Equations (18.21) and (18.22) can be substituted into Eq. (18.23) and the result simplified to yield

$$h_{i-1}c_{i-1} + 2(h_{i-1} + h_i)c_i + h_i c_{i+1} = 3\frac{f_{i+1} - f_i}{h_i} - 3\frac{f_i - f_{i-1}}{h_{i-1}} \qquad (18.24)$$

This equation can be made a little more concise by recognizing that the terms on the right-hand side are finite differences [recall Eq. (17.15)]:

$$f[x_i, x_j] = \frac{f_i - f_j}{x_i - x_j}$$

Therefore, Eq. (18.24) can be written as

$$h_{i-1}c_{i-1} + 2(h_{i-1} + h_i)c_i + h_i c_{i+1} = 3(f[x_{i+1}, x_i] - f[x_i, x_{i-1}]) \tag{18.25}$$

Equation (18.25) can be written for the interior knots, $i = 2, 3, \ldots, n-2$, which results in $n-3$ simultaneous tridiagonal equations with $n-1$ unknown coefficients, $c_1, c_2, \ldots, c_{n-1}$. Therefore, if we have two additional conditions, we can solve for the c's. Once this is done, Eqs. (18.21) and (18.18) can be used to determine the remaining coefficients, b and d.

As stated previously, the two additional end conditions can be formulated in a number of ways. One common approach, the natural spline, assumes that the second derivatives at the end knots are equal to zero. To see how these can be integrated into the solution scheme, the second derivative at the first node [Eq. (18.16)] can be set to zero as in

$$s_1''(x_1) = 0 = 2c_1 + 6d_1(x_1 - x_1)$$

Thus, this condition amounts to setting c_1 equal to zero.

The same evaluation can be made at the last node:

$$s_{n-1}''(x_n) = 0 = 2c_{n-1} + 6d_{n-1}h_{n-1} \tag{18.26}$$

Recalling Eq. (18.17), we can conveniently define an extraneous parameter c_n, in which case Eq. (18.26) becomes

$$c_{n-1} + 3d_{n-1}h_{n-1} = c_n = 0$$

Thus, to impose a zero second derivative at the last node, we set $c_n = 0$.

The final equations can now be written in matrix form as

$$\begin{bmatrix} 1 & & & & & \\ h_1 & 2(h_1 + h_2) & h_2 & & & \\ & \ddots & \ddots & \ddots & & \\ & & h_{n-2} & 2(h_{n-2} + h_{n-1}) & h_{n-1} \\ & & & & 1 \end{bmatrix} \begin{Bmatrix} c_1 \\ c_2 \\ \vdots \\ c_{n-1} \\ c_n \end{Bmatrix}$$

$$= \begin{Bmatrix} 0 \\ 3(f[x_3, x_2] - f[x_2, x_1]) \\ \vdots \\ 3(f[x_n, x_{n-1}] - f[x_{n-1}, x_{n-2}]) \\ 0 \end{Bmatrix} \tag{18.27}$$

As shown, the system is tridiagonal and hence efficient to solve.

EXAMPLE 18.3　Natural Cubic Splines

Problem Statement. Fit cubic splines to the same data used in Examples 18.1 and 18.2 (Table 18.1). Utilize the results to estimate the value at $x = 5$.

Solution. The first step is to employ Eq. (18.27) to generate the set of simultaneous equations that will be utilized to determine the c coefficients:

$$
\begin{bmatrix}
1 & & & \\
h_1 & 2(h_1 + h_2) & h_2 & \\
& h_2 & 2(h_2 + h_3) & h_3 \\
& & & 1
\end{bmatrix}
\begin{Bmatrix} c_1 \\ c_2 \\ c_3 \\ c_4 \end{Bmatrix}
=
\begin{Bmatrix}
0 \\
3(f[x_3, x_2] - f[x_2, x_1]) \\
3(f[x_4, x_3] - f[x_3, x_2]) \\
0
\end{Bmatrix}
$$

The necessary function and interval width values are

$$
\begin{aligned}
f_1 &= 2.5 & h_1 &= 4.5 - 3.0 = 1.5 \\
f_2 &= 1.0 & h_2 &= 7.0 - 4.5 = 2.5 \\
f_3 &= 2.5 & h_3 &= 9.0 - 7.0 = 2.0 \\
f_4 &= 0.5
\end{aligned}
$$

These can be substituted to yield

$$
\begin{bmatrix}
1 & & & \\
1.5 & 8 & 2.5 & \\
& 2.5 & 9 & 2 \\
& & & 1
\end{bmatrix}
\begin{Bmatrix} c_1 \\ c_2 \\ c_3 \\ c_4 \end{Bmatrix}
=
\begin{Bmatrix} 0 \\ 4.8 \\ -4.8 \\ 0 \end{Bmatrix}
$$

These equations can be solved using MATLAB with the results:

$$
\begin{aligned}
c_1 &= 0 & c_2 &= 0.839543726 \\
c_3 &= -0.766539924 & c_4 &= 0
\end{aligned}
$$

Equations (18.21) and (18.18) can be used to compute the b's and d's

$$
\begin{aligned}
b_1 &= -1.419771863 & d_1 &= 0.186565272 \\
b_2 &= -0.160456274 & d_2 &= -0.214144487 \\
b_3 &= 0.022053232 & d_3 &= 0.127756654
\end{aligned}
$$

These results, along with the values for the a's [Eq. (18.11)], can be substituted into Eq. (18.10) to develop the following cubic splines for each interval:

$$
\begin{aligned}
s_1(x) &= 2.5 - 1.419771863(x - 3) + 0.186565272(x - 3)^3 \\
s_2(x) &= 1.0 - 0.160456274(x - 4.5) + 0.839543726(x - 4.5)^2 \\
&\quad - 0.214144487(x - 4.5)^3 \\
s_3(x) &= 2.5 + 0.022053232(x - 7.0) - 0.766539924(x - 7.0)^2 \\
&\quad + 0.127756654(x - 7.0)^3
\end{aligned}
$$

The three equations can then be employed to compute values within each interval. For example, the value at $x = 5$, which falls within the second interval, is calculated as

$$
\begin{aligned}
s_2(5) &= 1.0 - 0.160456274(5 - 4.5) + 0.839543726(5 - 4.5)^2 - 0.214144487(5 - 4.5)^3 \\
&= 1.102889734.
\end{aligned}
$$

The total cubic spline fit is depicted in Fig. 18.4c.

The results of Examples 18.1 through 18.3 are summarized in Fig. 18.4. Notice the progressive improvement of the fit as we move from linear to quadratic to cubic splines. We have also superimposed a cubic interpolating polynomial on Fig. 18.4c. Although the cubic spline consists of a series of third-order curves, the resulting fit differs from that obtained using the third-order polynomial. This is due to the fact that the natural spline requires zero second derivatives at the end knots, whereas the cubic polynomial has no such constraint.

18.4.2 End Conditions

Although its graphical basis is appealing, the natural spline is only one of several end conditions that can be specified for splines. Two of the most popular are

- *Clamped End Condition.* This option involves specifying the first derivatives at the first and last nodes. This is sometimes called a "clamped" spline because it is what occurs when you clamp the end of a drafting spline so that it has a desired slope. For example, if zero first derivatives are specified, the spline will level off or become horizontal at the ends.
- *"Not-a-Knot" End Condition.* A third alternative is to force continuity of the third derivative at the second and the next-to-last knots. Since the spline already specifies that the function value and its first and second derivatives are equal at these knots, specifying continuous third derivatives means that the same cubic functions will apply to each of the first and last two adjacent segments. Since the first internal knots no longer represent the junction of two different cubic functions, they are no longer true knots. Hence, this case is referred to as the *"not-a-knot" condition.* It has the additional property that for four points, it yields the same result as is obtained using an ordinary cubic interpolating polynomial of the sort described in Chap. 17.

These conditions can be readily applied by using Eq. (18.25) for the interior knots, $i = 2, 3, \ldots, n - 2$, and using first (1) and last equations $(n - 1)$ as written in Table 18.2.

Figure 18.5 shows a comparison of the three end conditions as applied to fit the data from Table 18.1. The clamped case is set up so that the derivatives at the ends are equal to zero.

As expected, the spline fit for the clamped case levels off at the ends. In contrast, the natural and not-a-knot cases follow the trend of the data points more closely. Notice how the natural spline tends to straighten out as would be expected because the second derivatives go to zero at the ends. Because it has nonzero second derivatives at the ends, the not-a-knot exhibits more curvature.

TABLE 18.2 The first and last equations needed to specify some commonly used end conditions for cubic splines.

Condition	First and Last Equations
Natural	$c_1 = 0, c_n = 0$
Clamped (where f_1' and f_n' are the specified first derivatives at the first and last nodes, respectively)	$2h_1 c_1 + h_1 c_2 = 3 f[x_2, x_1] - 3 f_1'$ $h_{n-1} c_{n-1} + 2h_{n-1} c_n = 3 f_n' - 3 f[x_n, x_{n-1}]$
Not-a-knot	$h_2 c_1 - (h_1 + h_2) c_2 + h_1 c_3 = 0$ $h_{n-1} c_{n-2} - (h_{n-2} + h_{n-1}) c_{n-1} + h_{n-2} c_n = 0$

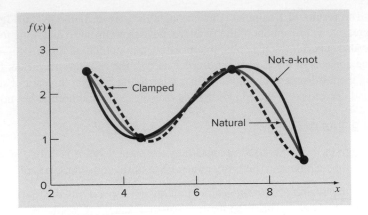

FIGURE 18.5
Comparison of the clamped (with zero first derivatives), not-a-knot, and natural splines for the data from Table 18.1.

18.5 PIECEWISE INTERPOLATION IN MATLAB

MATLAB has several built-in functions to implement piecewise interpolation. The spline function performs cubic spline interpolation as described in this chapter. The pchip function implements piecewise cubic Hermite interpolation. The interp1 function can also implement spline and Hermite interpolation, but can also perform a number of other types of piecewise interpolation.

18.5.1 MATLAB Function: spline

Cubic splines can be easily computed with the built-in MATLAB function, spline. It has the general syntax

$$yy = \text{spline}(x, y, xx) \tag{18.28}$$

where x and y = vectors containing the values that are to be interpolated, and yy = a vector containing the results of the spline interpolation as evaluated at the points in the vector xx.

By default, spline uses the not-a-knot condition. However, if y contains two more values than x has entries, then the first and last value in y are used as the derivatives at the end points. Consequently, this option provides the means to implement the clamped-end condition.

EXAMPLE 18.4 Splines in MATLAB

Problem Statement. Runge's function is a notorious example of a function that cannot be fit well with polynomials (recall Example 17.7):

$$f(x) = \frac{1}{1 + 25x^2}$$

Use MATLAB to fit nine equally spaced data points sampled from this function in the interval $[-1, 1]$. Employ **(a)** a not-a-knot spline and **(b)** a clamped spline with end slopes of $f_1' = 1$ and $f_{n-1}' = -4$.

Solution. **(a)** The nine equally spaced data points can be generated as in

```
>> x = linspace(-1,1,9);
>> y = 1./(1+25*x.^2);
```

Next, a more finely spaced vector of values can be generated so that we can create a smooth plot of the results as generated with the spline function:

```
>> xx = linspace(-1,1);
>> yy = spline(x,y,xx);
```

Recall that linspace automatically creates 100 points if the desired number of points are not specified. Finally, we can generate values for Runge's function itself and display them along with the spline fit and the original data:

```
>> yr = 1./(1+25*xx.^2);
>> plot(x,y,'o',xx,yy,xx,yr,'--')
```

As in Fig. 18.6, the not-a-knot spline does a nice job of following Runge's function without exhibiting wild oscillations between the points.

(b) The clamped condition can be implemented by creating a new vector yc that has the desired first derivatives as its first and last elements. The new vector can then be used to

FIGURE 18.6
Comparison of Runge's function (dashed line) with a 9-point not-a-knot spline fit generated with MATLAB (solid line).

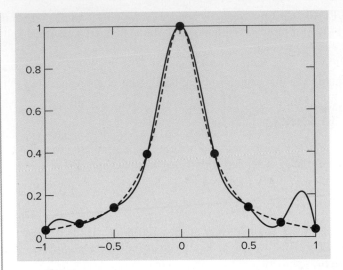

FIGURE 18.7
Comparison of Runge's function (dashed line) with a 9-point clamped end spline fit generated with MATLAB (solid line). Note that first derivatives of 1 and −4 are specified at the left and right boundaries, respectively.

generate and plot the spline fit:

```
>> yc = [1 y -4];
>> yyc = spline(x,yc,xx);
>> plot(x,y,'o',xx,yyc,xx,yr,'--')
```

As in Fig. 18.7, the clamped spline now exhibits some oscillations because of the artificial slopes that we have imposed at the boundaries. In other examples, where we have knowledge of the true first derivatives, the clamped spline tends to improve the fit.

18.5.2 MATLAB Function: interp1

The built-in function interp1 provides a handy means to implement a number of different types of piecewise one-dimensional interpolation. It has the general syntax

$$yi = \text{interp1}(x, y, xi, \text{'}method\text{'})$$

where x and y = vectors containing values that are to be interpolated, yi = a vector containing the results of the interpolation as evaluated at the points in the vector xi, and 'method' = the desired method. The various methods are

- 'nearest'—nearest neighbor interpolation. This method sets the value of an interpolated point to the value of the nearest existing data point. Thus, the interpolation looks like a series of plateaus, which can be thought of as zero-order polynomials.
- 'linear'—linear interpolation. This method uses straight lines to connect the points.

- 'spline'—piecewise cubic spline interpolation. This is identical to the spline function.
- 'pchip' and 'cubic'—piecewise cubic Hermite interpolation.

If the 'method' argument is omitted, the default is linear interpolation.

The pchip option (short for "*piecewise cubic Hermite interpolation*") merits more discussion. As with cubic splines, pchip uses cubic polynomials to connect data points with continuous first derivatives. However, it differs from cubic splines in that the second derivatives are not necessarily continuous. Further, the first derivatives at the knots will not be the same as for cubic splines. Rather, they are expressly chosen so that the interpolation is "shape preserving." That is, the interpolated values do not tend to overshoot the data points as can sometimes happen with cubic splines.

Therefore, there are trade-offs between the spline and the pchip options. The results of using spline will generally appear smoother because the human eye can detect discontinuities in the second derivative. In addition, it will be more accurate if the data are values of a smooth function. On the other hand, pchip has no overshoots and less oscillation if the data are not smooth. These trade-offs, as well as those involving the other options, are explored in the following example.

EXAMPLE 18.5   Trade-Offs Using interp1

Problem Statement. You perform a test drive on an automobile where you alternately accelerate the automobile and then hold it at a steady velocity. Note that you never decelerate during the experiment. The time series of spot measurements of velocity can be tabulated as

t	0	20	40	56	68	80	84	96	104	110
v	0	20	20	38	80	80	100	100	125	125

Use MATLAB's interp1 function to fit these data with (a) linear interpolation, (b) nearest neighbor, (c) cubic spline with not-a-knot end conditions, and (d) piecewise cubic Hermite interpolation.

Solution. (a) The data can be entered, fit with linear interpolation, and plotted with the following commands:

```
>> t = [0 20 40 56 68 80 84 96 104 110];
>> v = [0 20 20 38 80 80 100 100 125 125];
>> tt = linspace(0,110);
>> vl = interp1(t,v,tt);
>> plot(t,v,'o',tt,vl)
```

The results (Fig. 18.8*a*) are not smooth, but do not exhibit any overshoot.

(b) The commands to implement and plot the nearest neighbor interpolation are

```
>> vn = interp1(t,v,tt,'nearest');
>> plot(t,v,'o',tt,vn)
```

FIGURE 18.8
Use of several options of the `interp1` function to perform piecewise polynomial interpolation on a velocity time series for an automobile.

As in Fig. 18.8*b*, the results look like a series of plateaus. This option is neither a smooth nor an accurate depiction of the underlying process.

(c) The commands to implement the cubic spline are

```
>> vs = interp1(t,v,tt,'spline');
>> plot(t,v,'o',tt,vs)
```

These results (Fig. 18.8*c*) are quite smooth. However, severe overshoot occurs at several locations. This makes it appear that the automobile decelerated several times during the experiment.

(d) The commands to implement the piecewise cubic Hermite interpolation are

```
>> vh = interp1(t,v,tt,'pchip');
>> plot(t,v,'o',tt,vh)
```

For this case, the results (Fig. 18.8*d*) are physically realistic. Because of its shape-preserving nature, the velocities increase monotonically and never exhibit deceleration. Although the result is not as smooth as for the cubic splines, continuity of the first derivatives at the knots makes the transitions between points more gradual and hence more realistic.

18.6 MULTIDIMENSIONAL INTERPOLATION

The interpolation methods for one-dimensional problems can be extended to multidimensional interpolation. In this section, we will describe the simplest case of two-dimensional interpolation in Cartesian coordinates. In addition, we will describe MATLAB's capabilities for multidimensional interpolation.

18.6.1 Bilinear Interpolation

Two-dimensional interpolation deals with determining intermediate values for functions of two variables $z = f(x_i, y_i)$. As depicted in Fig. 18.9, we have values at four points: $f(x_1, y_1)$, $f(x_2, y_1)$, $f(x_1, y_2)$, and $f(x_2, y_2)$. We want to interpolate between these points

FIGURE 18.9
Graphical depiction of two-dimensional bilinear interpolation where an intermediate value (filled circle) is estimated based on four given values (open circles).

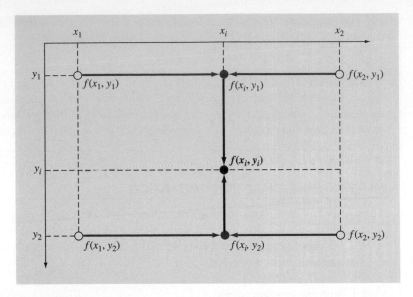

FIGURE 18.10
Two-dimensional bilinear interpolation can be implemented by first applying one-dimensional linear interpolation along the x dimension to determine values at x_i. These values can then be used to linearly interpolate along the y dimension to yield the final result at x_i, y_i.

to estimate the value at an intermediate point $f(x_i, y_i)$. If we use a linear function, the result is a plane connecting the points as in Fig. 18.9. Such functions are called *bilinear*.

A simple approach for developing the bilinear function is depicted in Fig. 18.10. First, we can hold the y value fixed and apply one-dimensional linear interpolation in the x direction. Using the Lagrange form, the result at (x_i, y_1) is

$$f(x_i, y_1) = \frac{x_i - x_2}{x_1 - x_2} f(x_1, y_1) + \frac{x_i - x_1}{x_2 - x_1} f(x_2, y_1) \tag{18.29}$$

and at (x_i, y_2) is

$$f(x_i, y_2) = \frac{x_i - x_2}{x_1 - x_2} f(x_1, y_2) + \frac{x_i - x_1}{x_2 - x_1} f(x_2, y_2) \tag{18.30}$$

These points can then be used to linearly interpolate along the y dimension to yield the final result:

$$f(x_i, y_i) = \frac{y_i - y_2}{y_1 - y_2} f(x_i, y_1) + \frac{y_i - y_1}{y_2 - y_1} f(x_i, y_2) \tag{18.31}$$

A single equation can be developed by substituting Eqs. (18.29) and (18.30) into Eq. (18.31) to give

$$f(x_i, y_i) = \frac{x_i - x_2}{x_1 - x_2} \frac{y_i - y_2}{y_1 - y_2} f(x_1, y_1) + \frac{x_i - x_1}{x_2 - x_1} \frac{y_i - y_2}{y_1 - y_2} f(x_2, y_1)$$

$$+ \frac{x_i - x_2}{x_1 - x_2} \frac{y_i - y_1}{y_2 - y_1} f(x_1, y_2) + \frac{x_i - x_1}{x_2 - x_1} \frac{y_i - y_1}{y_2 - y_1} f(x_2, y_2) \tag{18.32}$$

EXAMPLE 18.6 Bilinear Interpolation

Problem Statement. Suppose you have measured temperatures at a number of coordinates on the surface of a rectangular heated plate:

$$T(2, 1) = 60 \qquad\qquad T(9, 1) = 57.5$$
$$T(2, 6) = 55 \qquad\qquad T(9, 6) = 70$$

Use bilinear interpolation to estimate the temperature at $x_i = 5.25$ and $y_i = 4.8$.

Solution. Substituting these values into Eq. (18.32) gives

$$f(5.25, 4.8) = \frac{5.25 - 9}{2 - 9}\frac{4.8 - 6}{1 - 6} 60 + \frac{5.25 - 2}{9 - 2}\frac{4.8 - 6}{1 - 6} 57.5$$

$$+ \frac{5.25 - 9}{2 - 9}\frac{4.8 - 1}{6 - 1} 55 + \frac{5.25 - 2}{9 - 2}\frac{4.8 - 1}{6 - 1} 70 = 61.2143$$

18.6.2 Multidimensional Interpolation in MATLAB

MATLAB has two built-in functions for two- and three-dimensional piecewise interpolation: interp2 and interp3. As you might expect from their names, these functions operate in a similar fashion to interp1 (Sec. 18.5.2). For example, a simple representation of the syntax of interp2 is

```
zi = interp2(x, y, z, xi, yi, 'method')
```

where x and y = matrices containing the coordinates of the points at which the values in the matrix z are given, zi = a matrix containing the results of the interpolation as evaluated at the points in the matrices xi and yi, and method = the desired method. Note that the methods are identical to those used by interp1; that is, linear, nearest, spline, and cubic.

As with interp1, if the method argument is omitted, the default is linear interpolation. For example, interp2 can be used to make the same evaluation as in Example 18.6 as

```
>> x=[2 9];
>> y=[1 6];
>> z=[60 57.5;55 70];
>> interp2(x,y,z,5.25,4.8)

ans =
    61.2143
```

18.7 CASE STUDY   HEAT TRANSFER

Background. Lakes in the temperate zone can become thermally stratified during the summer. As depicted in Fig. 18.11, warm, buoyant water near the surface overlies colder, denser bottom water. Such stratification effectively divides the lake vertically into two layers: the *epilimnion* and the *hypolimnion*, separated by a plane called the *thermocline*.

Thermal stratification has great significance for environmental engineers and scientists studying such systems. In particular, the thermocline greatly diminishes mixing between the two layers. As a result, decomposition of organic matter can lead to severe depletion of oxygen in the isolated bottom waters.

The location of the thermocline can be defined as the inflection point of the temperature-depth curve—that is, the point at which $d^2T/dz^2 = 0$. It is also the point at which the absolute value of the first derivative or gradient is a maximum.

The temperature gradient is important in its own right because it can be used in conjunction with Fourier's law to determine the heat flux across the thermocline:

$$J = -D\rho C \frac{dT}{dz} \qquad (18.33)$$

where J = heat flux [cal/(cm^2 · s)], α = an eddy diffusion coefficient (cm^2/s), ρ = density ($\cong 1$ g/cm^3), and C = specific heat [$\cong 1$ cal/(g · C)].

In this case study, natural cubic splines are employed to determine the thermocline depth and temperature gradient for Platte Lake, Michigan (Table 18.3). The latter is also used to determine the heat flux for the case where $\alpha = 0.01$ cm^2/s.

FIGURE 18.11
Temperature versus depth during summer for Platte Lake, Michigan.

TABLE 18.3   Temperature versus depth during summer for Platte Lake, Michigan.

z, m	0	2.3	4.9	9.1	13.7	18.3	22.9	27.2
T, °C	22.8	22.8	22.8	20.6	13.9	11.7	11.1	11.1

Solution. As just described, we want to use natural spline end conditions to perform this analysis. Unfortunately, because it uses not-a-knot end conditions, the built-in MATLAB spline function does not meet our needs. Further, the spline function does not return the first and second derivatives we require for our analysis.

However, it is not difficult to develop our own M-file to implement a natural spline and return the derivatives. Such a code is shown in Fig. 18.12. After some preliminary error trapping, we set up and solve Eq. (18.27) for the second-order coefficients (c). Notice how

FIGURE 18.12
M-file to determine intermediate values and derivatives with a natural spline. Note that the diff function employed for error trapping is described in Sec. 21.7.1.

```
function [yy,dy,d2] = natspline(x,y,xx)
% natspline: natural spline with differentiation
%    [yy,dy,d2] = natspline(x,y,xx): uses a natural cubic spline
%    interpolation to find yy, the values of the underlying function
%    y at the points in the vector xx. The vector x specifies the
%    points at which the data y is given.
% input:
%    x = vector of independent variables
%    y = vector of dependent variables
%    xx = vector of desired values of dependent variables
% output:
%    yy = interpolated values at xx
%    dy = first derivatives at xx
%    d2 = second derivatives at xx
n = length(x);
if length(y)~=n, error('x and y must be same length'); end
if any(diff(x)<=0),error('x not strictly ascending'),end
m = length(xx);
b = zeros(n,n);
aa(1,1) = 1; aa(n,n) = 1;   %set up Eq. 18.27
bb(1)=0; bb(n)=0;
for i = 2:n-1
  aa(i,i-1) = h(x, i - 1);
  aa(i,i) = 2 * (h(x, i - 1) + h(x, i));
  aa(i,i+1) = h(x, i);
  bb(i) = 3 * (fd(i + 1, i, x, y) - fd(i, i - 1, x, y));
end
c=aa\bb';  %solve for c coefficients
for i = 1:n - 1   %solve for a, b and d coefficients
  a(i) = y(i);
  b(i) = fd(i + 1, i, x, y) - h(x, i) / 3 * (2 * c(i) + c(i + 1));
  d(i) = (c(i + 1) - c(i)) / 3 / h(x, i);
end
```

(continued)

```
for i = 1:m   %perform interpolations at desired values
  [yy(i),dy(i),d2(i)] = SplineInterp(x, n, a, b, c, d, xx(i));
end
end
function hh = h(x, i)
hh = x(i + 1) - x(i);
end
function fdd = fd(i, j, x, y)
fdd = (y(i) - y(j)) / (x(i) - x(j));
end
function [yyy,dyy,d2y]=SplineInterp(x, n, a, b, c, d, xi)
for ii = 1:n - 1
  if xi >= x(ii) - 0.000001 & xi <= x(ii + 1) + 0.000001
    yyy=a(ii)+b(ii)*(xi-x(ii))+c(ii)*(xi-x(ii))^2+d(ii)...
                                      *(xi-x(ii))^3;
    dyy=b(ii)+2*c(ii)*(xi-x(ii))+3*d(ii)*(xi-x(ii))^2;
    d2y=2*c(ii)+6*d(ii)*(xi-x(ii));
    break
  end
end
end
```

FIGURE 18.12 *(Continued)*

we use two subfunctions, h and fd, to compute the required finite differences. Once Eq. (18.27) is set up, we solve for the c's with back division. A loop is then employed to generate the other coefficients (a, b, and d).

At this point, we have all we need to generate intermediate values with the cubic equation:

$$f(x) = a_i + b_i (x - x_i) + c_i (x - x_i)^2 + d_i (x - x_i)^3$$

We can also determine the first and second derivatives by differentiating this equation twice to give

$$f'(x) = b_i + 2c_i (x - x_i) + 3d_i (x - x_i)^2$$

$$f''(x) = 2c_i + 6d_i (x - x_i)$$

As in Fig. 18.12, these equations can then be implemented in another subfunction, SplineInterp, to determine the values and the derivatives at the desired intermediate values.

Here is a script file that uses the natspline function to generate the spline and create plots of the results:

```
z = [0 2.3 4.9 9.1 13.7 18.3 22.9 27.2];
T=[22.8 22.8 22.8 20.6 13.9 11.7 11.1 11.1];
zz = linspace(z(1),z(length(z)));
```

18.7 CASE STUDY   continued

```
[TT,dT,dT2] = natspline(z,T,zz);
subplot(1,3,1),plot(T,z,'o',TT,zz)
title('(a) T'),legend('data','T')
set(gca,'YDir','reverse'),grid
subplot(1,3,2),plot(dT,zz)
title('(b) dT/dz')
set(gca,'YDir','reverse'),grid
subplot(1,3,3),plot(dT2,zz)
title('(c) d2T/dz2')
set(gca,'YDir','reverse'),grid
```

As in Fig. 18.13, the thermocline appears to be located at a depth of about 11.5 m. We can use root location (zero second derivative) or optimization methods (minimum first derivative) to refine this estimate. The result is that the thermocline is located at 11.35 m where the gradient is −1.61 °C/m.

FIGURE 18.13
Plots of (a) temperature, (b) gradient, and (c) second derivative versus depth (m) generated with the cubic spline program. The thermocline is located at the inflection point of the temperature-depth curve.

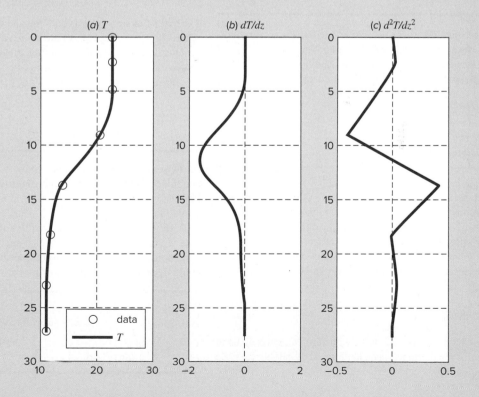

18.7 CASE STUDY   continued

The gradient can be used to compute the heat flux across the thermocline with Eq. (18.33):

$$J = -0.01 \frac{cm^2}{s} \times 1 \frac{g}{cm^3} \times 1 \frac{cal}{g \cdot {}^\circ C} \times \left(-1.61 \frac{{}^\circ C}{m}\right) \times \frac{1\,m}{100\,cm} \times \frac{86{,}400\,s}{d} = 13.9 \frac{cal}{cm^2 \cdot d}$$

The foregoing analysis demonstrates how spline interpolation can be used for engineering and scientific problem solving. However, it also is an example of numerical differentiation. As such, it illustrates how numerical approaches from different areas can be used in tandem for problem solving. We will be describing the topic of numerical differentiation in detail in Chap. 21.

PROBLEMS

18.1 Given the data

x	1	2	2.5	3	4	5
$f(x)$	1	5	7	8	2	1

Fit these data with (a) a cubic spline with natural end conditions, (b) a cubic spline with not-a-knot end conditions, and (c) piecewise cubic Hermite interpolation.

18.2 A reactor is thermally stratified as in the following table:

Depth, m	0	0.5	1	1.5	2	2.5	3
Temperature, °C	70	70	55	22	13	10	10

Based on these temperatures, the tank can be idealized as two zones separated by a strong temperature gradient or *thermocline*. The depth of the thermocline can be defined as the inflection point of the temperature-depth curve—that is, the point at which $d^2T/dz^2 = 0$. At this depth, the heat flux from the surface to the bottom layer can be computed with Fourier's law:

$$J = -k \frac{dT}{dz}$$

Use a clamped cubic spline fit with zero end derivatives to determine the thermocline depth. If $k = 0.01$ cal/ (s · cm · °C) compute the flux across this interface.

18.3 The following is the built-in humps function that MATLAB uses to demonstrate some of its numerical capabilities:

$$f(x) = \frac{1}{(x - 0.3)^2 + 0.01} + \frac{1}{(x - 0.9)^2 + 0.04} - 6$$

The humps function exhibits both flat and steep regions over a relatively short x range. Here are some values that have been generated at intervals of 0.1 over the range from $x = 0$ to 1:

x	0	0.1	0.2	0.3	0.4	0.5
$f(x)$	5.176	15.471	45.887	96.500	47.448	19.000

x	0.6	0.7	0.8	0.9	1
$f(x)$	11.692	12.382	17.846	21.703	16.000

Fit these data with a (a) cubic spline with not-a-knot end conditions and (b) piecewise cubic Hermite interpolation. In both cases, create a plot comparing the fit with the exact humps function.

18.4 Develop a plot of a cubic spline fit of the following data with (a) natural end conditions and (b) not-a-knot end conditions. In addition, develop a plot using (c) piecewise cubic Hermite interpolation (pchip).

x	0	100	200	400
$f(x)$	0	0.82436	1.00000	0.73576

x	600	800	1000
$f(x)$	0.40601	0.19915	0.09158

In each case, compare your plot with the following equation which was used to generate the data:

$$f(x) = \frac{x}{200}\, e^{-x/200+1}$$

18.5 The following data are sampled from the step function depicted in Fig. 18.1:

x	−1	−0.6	−0.2	0.2	0.6	1
$f(x)$	0	0	0	1	1	1

Fit these data with a **(a)** cubic spline with not-a-knot end conditions, **(b)** cubic spline with zero-slope clamped end conditions, and **(c)** piecewise cubic Hermite interpolation. In each case, create a plot comparing the fit with the step function.

18.6 Develop an M-file to compute a cubic spline fit with natural end conditions. Test your code by using it to duplicate Example 18.3.

18.7 The following data were generated with the fifth-order polynomial: $f(x) = 0.0185x^5 - 0.444x^4 + 3.9125x^3 - 15.456x^2 + 27.069x - 14.1$:

x	1	3	5	6	7	9
$f(x)$	1.000	2.172	4.220	5.430	4.912	9.120

(a) Fit these data with a cubic spline with not-a-knot end conditions. Create a plot comparing the fit with the function. **(b)** Repeat **(a)** but use clamped end conditions where the end slopes are set at the exact values as determined by differentiating the function.

18.8 Bessel functions often arise in advanced engineering and scientific analyses such as the study of electric fields. These functions are usually not amenable to straightforward evaluation and, therefore, are often compiled in standard mathematical tables. For example,

x	1.8	2	2.2	2.4	2.6
$J_1(x)$	0.5815	0.5767	0.556	0.5202	0.4708

Estimate $J_1(2.1)$, **(a)** using an interpolating polynomial and **(b)** using cubic splines. Note that the true value is 0.5683.

18.9 The following data define the sea-level concentration of dissolved oxygen for fresh water as a function of temperature:

T, °C	0	8	16	24	32	40
o, mg/L	14.621	11.843	9.870	8.418	7.305	6.413

Use MATLAB to fit the data with **(a)** piecewise linear interpolation, **(b)** a fifth-order polynomial, and **(c)** a spline. Display the results graphically and use each approach to estimate $o(27)$. Note that the exact result is 7.986 mg/L.

18.10 **(a)** Use MATLAB to fit a cubic spline to the following data to determine y at $x = 1.5$:

x	0	2	4	7	10	12
y	20	20	12	7	6	6

(b) Repeat **(a)**, but with zero first derivatives at the end knots.

18.11 Runge's function is written as

$$f(x) = \frac{1}{1 + 25x^2}$$

Generate five equidistantly spaced values of this function over the interval: $[-1, 1]$. Fit these data with **(a)** a fourth-order polynomial, **(b)** a linear spline, and **(c)** a cubic spline. Present your results graphically.

18.12 Use MATLAB to generate eight points from the function

$$f(t) = \sin^2 t$$

from $t = 0$ to 2π. Fit these data using **(a)** cubic spline with not-a-knot end conditions, **(b)** cubic spline with derivative end conditions equal to the exact values calculated with differentiation, and **(c)** piecewise cubic Hermite interpolation. Develop plots of each fit as well as plots of the absolute error (E_t = approximation − true) for each.

18.13 The drag coefficient for spheres such as sporting balls is known to vary as a function of the *Reynolds* number Re, a dimensionless number that gives a measure of the ratio of inertial forces to viscous forces:

$$Re = \frac{\rho V D}{\mu}$$

where ρ = the fluid's density (kg/m³), V = its velocity (m/s), D = diameter (m), and μ = dynamic viscosity (N · s/m²). Although the relationship of drag to the Reynolds number is sometimes available in equation form, it is frequently tabulated. For example, the following table provides values for a smooth spherical ball:

Re ($\times 10^{-4}$)	2	5.8	16.8	27.2	29.9	33.9
C_D	0.52	0.52	0.52	0.5	0.49	0.44
Re ($\times 10^{-4}$)	36.3	40	46	60	100	200	400
C_D	0.18	0.074	0.067	0.08	0.12	0.16	0.19

(a) Develop a MATLAB function that employs an appropriate interpolation function to return a value of C_D as a function of the Reynolds number. The first line of the function should be

```
function CDout = Drag(ReCD,ReIn)
```

where ReCD = a 2-row matrix containing the table, ReIn = the Reynolds number at which you want to estimate the drag, and CDout = the corresponding drag coefficient.

(b) Write a script that uses the function developed in part (a) to generate a labeled plot of the drag force versus velocity (recall Sec. 1.4). Use the following parameter values for the script: $D = 22$ cm, $\rho = 1.3$ kg/m^3, and $\mu = 1.78 \times 10^{-5}$ Pa · s. Employ a range of velocities from 4 to 40 m/s for your plot.

18.14 The following function describes the temperature distribution on a rectangular plate for the range $-2 \leq x \leq 0$ and $0 \leq y \leq 3$

$$T = 2 + x - y + 2x^2 + 2xy + y^2$$

Develop a script to: (a) Generate a meshplot of this function using the MATLAB function surfc. Employ the linspace function with default spacing (i.e., 100 interior points) to generate the x and y values. (b) Use the MATLAB function interp2 with the default interpolation option ('linear') to compute the temperature at $x = -1.63$ and $y = 1.627$. Determine the percent relative error of your result. (c) Repeat (b), but with 'spline'. Note: for parts (b) and (c), employ the linspace function with 9 interior points.

18.15 The U.S. Standard Atmosphere specifies atmospheric properties as a function of altitude above sea level. The following table shows selected values of temperature, pressure, and density

Altitude (km)	T (°C)	p (atm)	ρ (kg/m^3)
−0.5	18.4	1.0607	1.2850
2.5	−1.1	0.73702	0.95697
6	−23.8	0.46589	0.66015
11	−56.2	0.22394	0.36481
20	−56.3	0.054557	0.088911
28	−48.5	0.015946	0.025076
50	−2.3	7.8721×10^{-4}	1.0269×10^{-3}
60	−17.2	2.2165×10^{-4}	3.0588×10^{-4}
80	−92.3	1.0227×10^{-5}	1.9992×10^{-5}
90	−92.3	1.6216×10^{-6}	3.1703×10^{-6}

Develop a MATLAB function, StdAtm, to determine values of the three properties for a given altitude. Base the function on the pchip option for interp1. If the user requests a value outside the range of altitudes, have the function display an error message and terminate the application. Use the following script as the starting point to create a 3-panel plot of altitude versus the properties as depicted in Fig. P18.15.

```
% Script to generate a plot of temperature, pressure
and density
% for the U.S. Standard Atmosphere
clc, clf
z=[-0.5 2.5 6 11 20 28 50 60 80 90];
T=[18.4 -1.1 -23.8 -56.2 -56.3 -48.5 -2.3 -17.2 -92.3
   -92.3];
p=[1.0607 0.73702 0.46589 0.22394 0.054557 0.015946 ...
   7.8721e-4 2.2165e-4 1.02275e-05 1.6216e-06];
rho=[1.285025 0.95697 0.6601525 0.364805 0.0889105 ...
   0.02507575 0.001026918 0.000305883 0.000019992
   3.1703e-06];
zint=[-0.5:0.1:90];
for i=1:length(zint)
 [Tint(i),pint(i),rint(i)]=StdAtm(z,T,p,rho,zint(i));
end

% Create plot

Te=StdAtm(z,T,p,rho,-1000);
```

18.16 Felix Baumgartner ascended to 39 km in a stratospheric balloon and made a free-fall jump rushing toward earth at supersonic speeds before parachuting to the ground. As he fell, his drag coefficient changed primarily because the air density changed. Recall from Chap. 1 that the terminal velocity, $v_{terminal}$ (m/s), of a free-falling object can be computed as

$$v_{terminal} = \sqrt{\frac{gm}{c_d}}$$

where g = gravitational acceleration (m/s^2), m = mass (kg), and c_d = drag coefficient (kg/m). The drag coefficient can be computed as

$$c_d = 0.5\rho A C_d$$

where ρ = the fluid density (kg/m^3), A = the projected area (m^2), and C_d = a dimensionless drag coefficient. Note that the gravitational acceleration, g (m/s^2), can be related to elevation by

$$g = 9.806412 - 0.003039734z$$

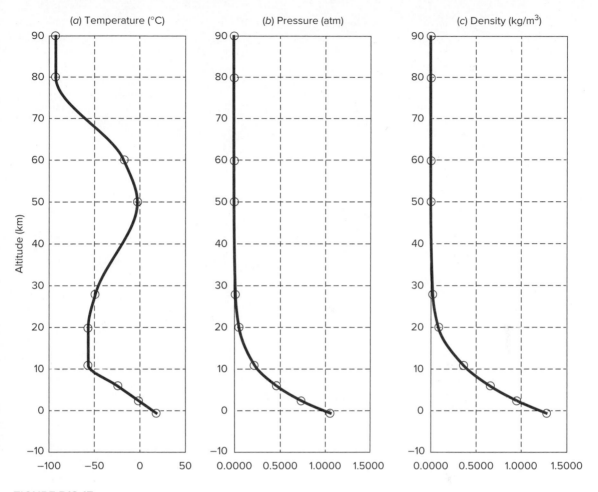

FIGURE P18.15

where z = elevation above the earth's surface (km) and the density of air, ρ (kg/m³), at various elevations can be tabulated as

z (km)	ρ (kg/m³)	z (km)	ρ (kg/m³)	z (km)	ρ (kg/m³)
−1	1.347	6	0.6601	25	0.04008
0	1.225	7	0.5900	30	0.01841
1	1.112	8	0.5258	40	0.003996
2	1.007	9	0.4671	50	0.001027
3	0.9093	10	0.4135	60	0.0003097
4	0.8194	15	0.1948	70	8.283×10^{-5}
5	0.7364	20	0.08891	80	1.846×10^{-5}

Assume that $m = 80$ kg, $A = 0.55$ m², and $C_d = 1.1$. Develop a MATLAB script to create a labeled plot of terminal velocity versus elevation for $z = [0:0.5:40]$. Use a spline to generate the required densities needed to construct the plot.

PART FIVE

Integration and Differentiation

5.1 OVERVIEW

In high school or during your first year of college, you were introduced to differential and integral calculus. There you learned techniques to obtain analytical or exact derivatives and integrals.

Mathematically, the *derivative* represents the rate of change of a dependent variable with respect to an independent variable. For example, if we are given a function $y(t)$ that specifies an object's position as a function of time, differentiation provides a means to determine its velocity, as in:

$$v(t) = \frac{d}{dt} y(t)$$

As in Fig. PT5.1a, the derivative can be visualized as the slope of a function.

Integration is the inverse of differentiation. Just as differentiation uses differences to quantify an instantaneous process, integration involves summing instantaneous information to give a total result over an interval. Thus, if we are provided with velocity as a function of time, integration can be used to determine the distance traveled:

$$y(t) = \int_0^t v(t) \, dt$$

As in Fig. PT5.1b, for functions lying above the abscissa, the integral can be visualized as the area under the curve of $v(t)$ from 0 to t. Consequently, just as a derivative can be thought of as a slope, an integral can be envisaged as a summation.

Because of the close relationship between differentiation and integration, we have opted to devote this part of the book to both processes. Among other things, this will provide the opportunity to highlight their similarities

485

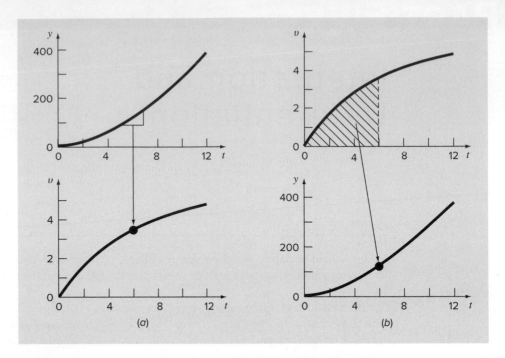

FIGURE PT5.1
The contrast between (*a*) differentiation and (*b*) integration.

and differences from a numerical perspective. In addition, the material will have relevance to the next part of the book where we will cover differential equations.

Although differentiation is taught before integration in calculus, we reverse their order in the following chapters. We do this for several reasons. First, we have already introduced you to the basics of numerical differentiation in Chap. 4. Second, in part because it is much less sensitive to roundoff errors, integration represents a more highly developed area of numerical methods. Finally, although numerical differentiation is not as widely employed, it does have great significance for the solution of differential equations. Hence, it makes sense to cover it as the last topic prior to describing differential equations in Part Six.

5.2 PART ORGANIZATION

Chapter 19 is devoted to the most common approaches for numerical integration—the *Newton-Cotes formulas*. These relationships are based on replacing a complicated function or tabulated data with a simple polynomial that is easy to integrate. Three of the most widely used Newton-Cotes formulas are discussed in detail: the *trapezoidal rule, Simpson's 1/3 rule,* and *Simpson's 3/8 rule.* All these formulas are designed for cases where the data to be integrated are evenly spaced. In addition, we also include a discussion of numerical integration of unequally spaced data. This is a very important topic because many real-world applications deal with data that are in this form.

All the above material relates to *closed integration,* where the function values at the ends of the limits of integration are known. At the end of Chap. 19, we present *open integration formulas,* where the integration limits extend beyond the range of the known data. Although they are not commonly used for definite integration, open integration formulas are presented here because they are utilized in the solution of ordinary differential equations in Part Six.

The formulations covered in Chap. 19 can be employed to analyze both tabulated data and equations. *Chapter 20* deals with two techniques that are expressly designed to integrate equations and functions: *Romberg integration* and *Gauss quadrature.* Computer algorithms are provided for both of these methods. In addition, *adaptive integration* is discussed.

In *Chap. 21,* we present additional information on *numerical differentiation* to supplement the introductory material from Chap. 4. Topics include *high-accuracy finite-difference formulas, Richardson extrapolation,* and the differentiation of unequally spaced data. The effect of errors on both numerical differentiation and integration is also discussed.

19

Numerical Integration Formulas

CHAPTER OBJECTIVES

The primary objective of this chapter is to introduce you to numerical integration. Specific objectives and topics covered are

- Recognizing that Newton-Cotes integration formulas are based on the strategy of replacing a complicated function or tabulated data with a polynomial that is easy to integrate.
- Knowing how to implement the following single application Newton-Cotes formulas:
    Trapezoidal rule
    Simpson's 1/3 rule
    Simpson's 3/8 rule
- Knowing how to implement the following composite Newton-Cotes formulas:
    Trapezoidal rule
    Simpson's 1/3 rule
- Recognizing that even-segment–odd-point formulas like Simpson's 1/3 rule achieve higher than expected accuracy.
- Knowing how to use the trapezoidal rule to integrate unequally spaced data.
- Understanding the difference between open and closed integration formulas.

YOU'VE GOT A PROBLEM

R ecall that the velocity of a free-falling bungee jumper as a function of time can be computed as

$$v(t) = \sqrt{\frac{gm}{c_d}} \tanh \left(\sqrt{\frac{gc_d}{m}} t \right) \tag{19.1}$$

Suppose that we would like to know the vertical distance z the jumper has fallen after a certain time t. This distance can be evaluated by integration:

$$z(t) = \int_0^t v(t)\, dt \tag{19.2}$$

Substituting Eq. (19.1) into Eq. (19.2) gives

$$z(t) = \int_0^t \sqrt{\frac{gm}{c_d}} \tanh\left(\sqrt{\frac{gc_d}{m}}\, t\right) dt \tag{19.3}$$

Thus, integration provides the means to determine the distance from the velocity. Calculus can be used to solve Eq. (19.3) for

$$z(t) = \frac{m}{c_d} \ln\left[\cosh\left(\sqrt{\frac{gc_d}{m}}\, t\right)\right] \tag{19.4}$$

Although a closed form solution can be developed for this case, there are other functions that cannot be integrated analytically. Further, suppose that there was some way to measure the jumper's velocity at various times during the fall. These velocities along with their associated times could be assembled as a table of discrete values. In this situation, it would also be possible to integrate the discrete data to determine the distance. In both these instances, numerical integration methods are available to obtain solutions. Chapters 19 and 20 will introduce you to some of these methods.

19.1   INTRODUCTION AND BACKGROUND

19.1.1 What Is Integration?

According to the dictionary definition, to integrate means "to bring together, as parts, into a whole; to unite; to indicate the total amount. . . ." Mathematically, definite integration is represented by

$$I = \int_a^b f(x)\, dx \tag{19.5}$$

which stands for the integral of the function $f(x)$ with respect to the independent variable x, evaluated between the limits $x = a$ to $x = b$.

As suggested by the dictionary definition, the "meaning" of Eq. (19.5) is the total value, or summation, of $f(x)\, dx$ over the range $x = a$ to b. In fact, the symbol $\int$ is actually a stylized capital S that is intended to signify the close connection between integration and summation.

Figure 19.1 represents a graphical manifestation of the concept. For functions lying above the x axis, the integral expressed by Eq. (19.5) corresponds to the area under the curve of $f(x)$ between $x = a$ and b.

Numerical integration is sometimes referred to as quadrature. This is an archaic term that originally meant the construction of a square having the same area as some curvilinear figure. Today, the term *quadrature* is generally taken to be synonymous with numerical definite integration.

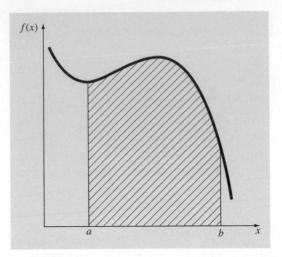

FIGURE 19.1
Graphical representation of the integral of $f(x)$ between the limits $x = a$ to b. The integral is equivalent to the area under the curve.

19.1.2 Integration in Engineering and Science

Integration has so many engineering and scientific applications that you were required to take integral calculus in your first year at college. Many specific examples of such applications could be given in all fields of engineering and science. A number of examples relate directly to the idea of the integral as the area under a curve. Figure 19.2 depicts a few cases where integration is used for this purpose.

Other common applications relate to the analogy between integration and summation. For example, a common application is to determine the mean of a continuous function. Recall that the mean of n discrete data points can be calculated by [Eq. (14.2)].

$$\text{Mean} = \frac{\sum_{i=1}^{n} y_i}{n} \tag{19.6}$$

where y_i are individual measurements. The determination of the mean of discrete points is depicted in Fig. 19.3a.

In contrast, suppose that y is a continuous function of an independent variable x, as depicted in Fig. 19.3b. For this case, there are an infinite number of values between a and b. Just as Eq. (19.6) can be applied to determine the mean of the discrete readings, you might also be interested in computing the mean or average of the continuous function $y = f(x)$ for the interval from a to b. Integration is used for this purpose, as specified by

$$\text{Mean} = \frac{\int_a^b f(x)dx}{b - a} \tag{19.7}$$

This formula has hundreds of engineering and scientific applications. For example, it is used to calculate the center of gravity of irregular objects in mechanical and civil engineering and to determine the root-mean-square current in electrical engineering.

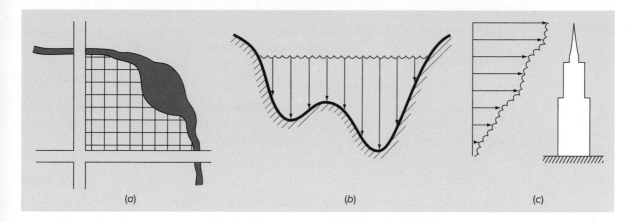

FIGURE 19.2
Examples of how integration is used to evaluate areas in engineering and scientific applications. (*a*) A surveyor might need to know the area of a field bounded by a meandering stream and two roads. (*b*) A hydrologist might need to know the cross-sectional area of a river. (*c*) A structural engineer might need to determine the net force due to a nonuniform wind blowing against the side of a skyscraper.

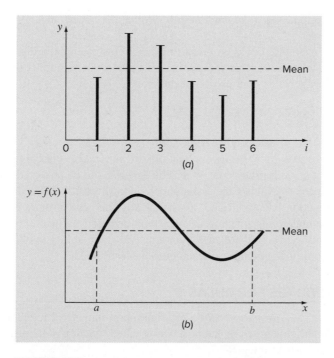

FIGURE 19.3
An illustration of the mean for (*a*) discrete and (*b*) continuous data.

Integrals are also employed by engineers and scientists to evaluate the total amount or quantity of a given physical variable. The integral may be evaluated over a line, an area, or a volume. For example, the total mass of chemical contained in a reactor is given as the product of the concentration of chemical and the reactor volume, or

$$\text{Mass} = \text{concentration} \times \text{volume}$$

where concentration has units of mass per volume. However, suppose that concentration varies from location to location within the reactor. In this case, it is necessary to sum the products of local concentrations c_i and corresponding elemental volumes ΔV_i:

$$\text{Mass} = \sum_{i=1}^{n} c_i \Delta V_i$$

where n is the number of discrete volumes. For the continuous case, where $c(x, y, z)$ is a known function and x, y, and z are independent variables designating position in Cartesian coordinates, integration can be used for the same purpose:

$$\text{Mass} = \iiint c(x, y, z)\, dx\, dy\, dz$$

or

$$\text{Mass} = \iiint_V c(V)\, dV$$

which is referred to as a *volume integral*. Notice the strong analogy between summation and integration.

Similar examples could be given in other fields of engineering and science. For example, the total rate of energy transfer across a plane where the flux (in calories per square centimeter per second) is a function of position is given by

$$\text{Flux} = \iint_A \text{flux}\, dA$$

which is referred to as an *areal integral,* where $A = $ area.

These are just a few of the applications of integration that you might face regularly in the pursuit of your profession. When the functions to be analyzed are simple, you will normally choose to evaluate them analytically. However, it is often difficult or impossible when the function is complicated, as is typically the case in more realistic examples. In addition, the underlying function is often unknown and defined only by measurement at discrete points. For both these cases, you must have the ability to obtain approximate values for integrals using numerical techniques as described next.

19.2 NEWTON-COTES FORMULAS

The *Newton Cotes formulas* are the most common numerical integration schemes. They are based on the strategy of replacing a complicated function or tabulated data with a polynomial that is easy to integrate:

$$I = \int_a^b f(x)\, dx \cong \int_a^b f_n(x)\, dx \tag{19.8}$$

where $f_n(x)$ = a polynomial of the form

$$f_n(x) = a_0 + a_1 x + \cdots + a_{n-1} x^{n-1} + a_n x^n \tag{19.9}$$

where n is the order of the polynomial. For example, in Fig. 19.4a, a first-order polynomial (a straight line) is used as an approximation. In Fig. 19.4b, a parabola is employed for the same purpose.

The integral can also be approximated using a series of polynomials applied piecewise to the function or data over segments of constant length. For example, in Fig. 19.5, three

FIGURE 19.4
The approximation of an integral by the area under (a) a straight line and (b) a parabola.

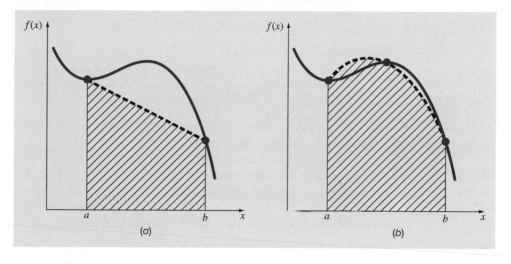

(a)   (b)

FIGURE 19.5
The approximation of an integral by the area under three straight-line segments.

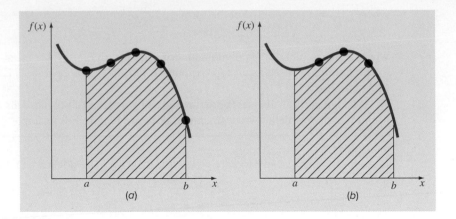

FIGURE 19.6
The difference between (*a*) closed and (*b*) open integration formulas.

straight-line segments are used to approximate the integral. Higher-order polynomials can be utilized for the same purpose.

Closed and open forms of the Newton-Cotes formulas are available. The *closed forms* are those where the data points at the beginning and end of the limits of integration are known (Fig. 19.6*a*). The *open forms* have integration limits that extend beyond the range of the data (Fig. 19.6*b*). This chapter emphasizes the closed forms. However, material on open Newton-Cotes formulas is briefly introduced in Sec. 19.7.

19.3   THE TRAPEZOIDAL RULE

The *trapezoidal rule* is the first of the Newton-Cotes closed integration formulas. It corresponds to the case where the polynomial in Eq. (19.8) is first-order:

$$I = \int_a^b \left[f(a) + \frac{f(b) - f(a)}{b - a}(x - a) \right] dx \tag{19.10}$$

The result of the integration is

$$I = (b - a)\frac{f(a) + f(b)}{2} \tag{19.11}$$

which is called the *trapezoidal rule*.

Geometrically, the trapezoidal rule is equivalent to approximating the area of the trapezoid under the straight line connecting $f(a)$ and $f(b)$ in Fig. 19.7. Recall from geometry that the formula for computing the area of a trapezoid is the height times the average of the bases. In our case, the concept is the same but the trapezoid is on its side. Therefore, the integral estimate can be represented as

$$I = \text{width} \times \text{average height} \tag{19.12}$$

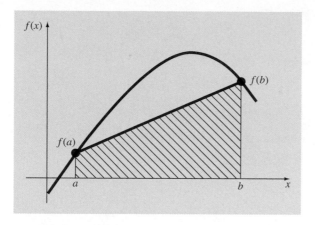

FIGURE 19.7
Graphical depiction of the trapezoidal rule.

or

$$I = (b - a) \times \text{average height} \tag{19.13}$$

where, for the trapezoidal rule, the average height is the average of the function values at the end points, or $[f(a) + f(b)]/2$.

All the Newton-Cotes closed formulas can be expressed in the general format of Eq. (19.13). That is, they differ only with respect to the formulation of the average height.

19.3.1 Error of the Trapezoidal Rule

When we employ the integral under a straight-line segment to approximate the integral under a curve, we obviously can incur an error that may be substantial (Fig. 19.8). An estimate for the local truncation error of a single application of the trapezoidal rule is

$$E_t = -\frac{1}{12} f''(\xi)(b - a)^3 \tag{19.14}$$

where ξ lies somewhere in the interval from a to b. Equation (19.14) indicates that if the function being integrated is linear, the trapezoidal rule will be exact because the second derivative of a straight line is zero. Otherwise, for functions with second- and higher-order derivatives (i.e., with curvature), some error can occur.

EXAMPLE 19.1   Single Application of the Trapezoidal Rule

Problem Statement. Use Eq. (19.11) to numerically integrate

$$f(x) = 0.2 + 25x - 200x^2 + 675x^3 - 900x^4 + 400x^5$$

from $a = 0$ to $b = 0.8$. Note that the exact value of the integral can be determined analytically to be 1.640533.

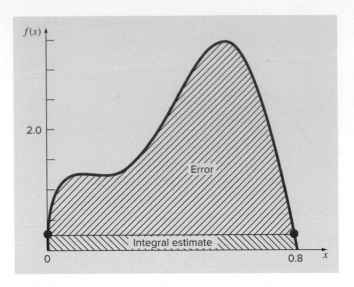

FIGURE 19.8
Graphical depiction of the use of a single application of the trapezoidal rule to approximate the integral of $f(x) = 0.2 + 25x - 200x^2 + 675x^3 - 900x^4 + 400x^5$ from $x = 0$ to 0.8.

Solution. The function values $f(0) = 0.2$ and $f(0.8) = 0.232$ can be substituted into Eq. (19.11) to yield

$$I = (0.8 - 0)\frac{0.2 + 0.232}{2} = 0.1728$$

which represents an error of $E_t = 1.640533 - 0.1728 = 1.467733$, which corresponds to a percent relative error of $\varepsilon_t = 89.5\%$. The reason for this large error is evident from the graphical depiction in Fig. 19.8. Notice that the area under the straight line neglects a significant portion of the integral lying above the line.

In actual situations, we would have no foreknowledge of the true value. Therefore, an approximate error estimate is required. To obtain this estimate, the function's second derivative over the interval can be computed by differentiating the original function twice to give

$$f''(x) = -400 + 4,050x - 10,800x^2 + 8,000x^3$$

The average value of the second derivative can be computed as [Eq. (19.7)]

$$\bar{f}''(x) = \frac{\int_0^{0.8}(-400 + 4,050x - 10,800x^2 + 8,000x^3)\,dx}{0.8 - 0} = -60$$

which can be substituted into Eq. (19.14) to yield

$$E_a = -\frac{1}{12}(-60)(0.8)^3 = 2.56$$

which is of the same order of magnitude and sign as the true error. A discrepancy does exist, however, because of the fact that for an interval of this size, the average second derivative is not necessarily an accurate approximation of $f''(\xi)$. Thus, we denote that the error is approximate by using the notation E_a, rather than exact by using E_t.

19.3.2 The Composite Trapezoidal Rule

One way to improve the accuracy of the trapezoidal rule is to divide the integration interval from a to b into a number of segments and apply the method to each segment (Fig. 19.9). The areas of individual segments can then be added to yield the integral for the entire interval. The resulting equations are called *composite,* or *multiple-segment, integration formulas.*

Figure 19.9 shows the general format and nomenclature we will use to characterize composite integrals. There are $n + 1$ equally spaced base points $(x_0, x_1, x_2, \ldots, x_n)$. Consequently, there are n segments of equal width:

$$h = \frac{b-a}{n} \tag{19.15}$$

If a and b are designated as x_0, and x_n, respectively, the total integral can be represented as

$$I = \int_{x_0}^{x_1} f(x)\, dx + \int_{x_1}^{x_2} f(x)\, dx + \cdots + \int_{x_{n-1}}^{x_n} f(x)\, dx$$

FIGURE 19.9
Composite trapezoidal rule.

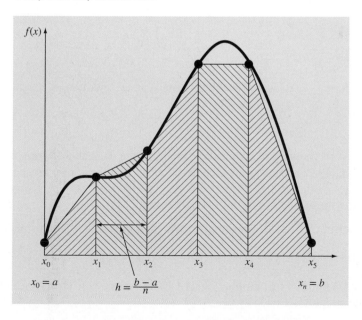

Substituting the trapezoidal rule for each integral yields

$$I = h\frac{f(x_0) + f(x_1)}{2} + h\frac{f(x_1) + f(x_2)}{2} + \cdots + h\frac{f(x_{n-1}) + f(x_n)}{2} \tag{19.16}$$

or, grouping terms:

$$I = \frac{h}{2}\left[f(x_0) + 2\sum_{i=1}^{n-1}f(x_i) + f(x_n)\right] \tag{19.17}$$

or, using Eq. (19.15) to express Eq. (19.17) in the general form of Eq. (19.13):

$$I = \underbrace{(b-a)}_{\text{Width}} \underbrace{\frac{f(x_0) + 2\sum_{i=1}^{n-1}f(x_i) + f(x_n)}{2n}}_{\text{Average height}} \tag{19.18}$$

Because the summation of the coefficients of $f(x)$ in the numerator divided by $2n$ is equal to 1, the average height represents a weighted average of the function values. According to Eq. (19.18), the interior points are given twice the weight of the two end points $f(x_0)$ and $f(x_n)$.

An error for the composite trapezoidal rule can be obtained by summing the individual errors for each segment to give

$$E_t = -\frac{(b-a)^3}{12n^3}\sum_{i=1}^{n}f''(\xi_i) \tag{19.19}$$

where $f''(\xi_i)$ is the second derivative at a point ξ_i located in segment i. This result can be simplified by estimating the mean or average value of the second derivative for the entire interval as

$$\bar{f}'' \cong \frac{\sum_{i=1}^{n}f''(\xi_i)}{n} \tag{19.20}$$

Therefore $\sum f''(\xi_i) \cong n\bar{f}''$ and Eq. (19.19) can be rewritten as

$$E_a = -\frac{(b-a)^3}{12n^2}\bar{f}'' \tag{19.21}$$

Thus, if the number of segments is doubled, the truncation error will be quartered. Note that Eq. (19.21) is an approximate error because of the approximate nature of Eq. (19.20).

EXAMPLE 19.2   Composite Application of the Trapezoidal Rule

Problem Statement. Use the two-segment trapezoidal rule to estimate the integral of

$$f(x) = 0.2 + 25x - 200x^2 + 675x^3 - 900x^4 + 400x^5$$

from $a = 0$ to $b = 0.8$. Employ Eq. (19.21) to estimate the error. Recall that the exact value of the integral is 1.640533.

Solution. For $n = 2$ $(h = 0.4)$:

$$f(0) = 0.2 \qquad f(0.4) = 2.456 \qquad f(0.8) = 0.232$$

$$I = 0.8 \frac{0.2 + 2(2.456) + 0.232}{4} = 1.0688$$

$$E_t = 1.640533 - 1.0688 = 0.57173 \qquad \varepsilon_t = 34.9\%$$

$$E_a = -\frac{0.8^3}{12(2)^2}(-60) = 0.64$$

where -60 is the average second derivative determined previously in Example 19.1.

The results of the previous example, along with three- through ten-segment applications of the trapezoidal rule, are summarized in Table 19.1. Notice how the error decreases as the number of segments increases. However, also notice that the rate of decrease is gradual. This is because the error is inversely related to the square of n [Eq. (19.21)]. Therefore, doubling the number of segments quarters the error. In subsequent sections we develop higher-order formulas that are more accurate and that converge more quickly on the true integral as the segments are increased. However, before investigating these formulas, we will first discuss how MATLAB can be used to implement the trapezoidal rule.

19.3.3 MATLAB M-file: trap

A simple algorithm to implement the composite trapezoidal rule can be written as in Fig. 19.10. The function to be integrated is passed into the M-file along with the limits of integration and the number of segments. A loop is then employed to generate the integral following Eq. (19.18).

TABLE 19.1   Results for the composite trapezoidal rule to estimate the integral of $f(x) = 0.2 + 25x - 200x^2 + 675x^3 - 900x^4 + 400x^5$ from $x = 0$ to 0.8. The exact value is 1.640533.

n	h	I	ε_t (%)
2	0.4	1.0688	34.9
3	0.2667	1.3695	16.5
4	0.2	1.4848	9.5
5	0.16	1.5399	6.1
6	0.1333	1.5703	4.3
7	0.1143	1.5887	3.2
8	0.1	1.6008	2.4
9	0.0889	1.6091	1.9
10	0.08	1.6150	1.6

```
function I = trap(func,a,b,n,varargin)
% trap: composite trapezoidal rule quadrature
%   I = trap(func,a,b,n,p1,p2,...):
%                     composite trapezoidal rule
% input:
%   func = name of function to be integrated
%   a, b = integration limits
%   n = number of segments (default = 100)
%   p1,p2,... = additional parameters used by func
% output:
%   I = integral estimate

if nargin<3,error('at least 3 input arguments required'),end
if ~(b>a),error('upper bound must be greater than lower'),end
if nargin<4|isempty(n),n=100;end
x = a; h = (b - a)/n;
s=func(a,varargin{:});
for i = 1 : n-1
  x = x + h;
  s = s + 2*func(x,varargin{:});
end
s = s + func(b,varargin{:});
I = (b - a) * s/(2*n);
```

FIGURE 19.10
M-file to implement the composite trapezoidal rule.

An application of the M-file can be developed to determine the distance fallen by the free-falling bungee jumper in the first 3 s by evaluating the integral of Eq. (19.3). For this example, assume the following parameter values: $g = 9.81$ m/s^2, $m = 68.1$ kg, and $c_d = 0.25$ kg/m. Note that the exact value of the integral can be computed with Eq. (19.4) as 41.94805.

The function to be integrated can be developed as an M-file or with an anonymous function,

```
>> v=@(t) sqrt(9.81*68.1/0.25)*tanh(sqrt(9.81*0.25/68.1)*t)

v =
    @(t) sqrt(9.81*68.1/0.25)*tanh(sqrt(9.81*0.25/68.1)*t)
```

First, let's evaluate the integral with a crude five-segment approximation:

```
>> format long
>> trap(v,0,3,5)

ans =
    41.86992959072735
```

As would be expected, this result has a relatively high true error of 18.6%. To obtain a more accurate result, we can use a very fine approximation based on 10,000 segments:

```
>> trap(v,0,3,10000)

x =
   41.94804999917528
```

which is very close to the true value.

19.4 SIMPSON'S RULES

Aside from applying the trapezoidal rule with finer segmentation, another way to obtain a more accurate estimate of an integral is to use higher-order polynomials to connect the points. For example, if there is an extra point midway between $f(a)$ and $f(b)$, the three points can be connected with a parabola (Fig. 19.11a). If there are two points equally spaced between $f(a)$ and $f(b)$, the four points can be connected with a third-order polynomial (Fig. 19.11b). The formulas that result from taking the integrals under these polynomials are called *Simpson's rules*.

19.4.1 Simpson's 1/3 Rule

Simpson's 1/3 rule corresponds to the case where the polynomial in Eq. (19.8) is second-order:

$$I = \int_{x_0}^{x_2} \left[\frac{(x - x_1)(x - x_2)}{(x_0 - x_1)(x_0 - x_2)} f(x_0) + \frac{(x - x_0)(x - x_2)}{(x_1 - x_0)(x_1 - x_2)} f(x_1) \right.$$

$$\left. + \frac{(x - x_0)(x - x_1)}{(x_2 - x_0)(x_2 - x_1)} f(x_2) \right] dx$$

FIGURE 19.11
(a) Graphical depiction of Simpson's 1/3 rule: It consists of taking the area under a parabola connecting three points. (b) Graphical depiction of Simpson's 3/8 rule: It consists of taking the area under a cubic equation connecting four points.

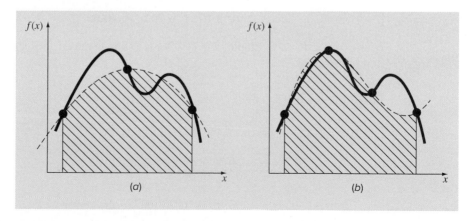

where a and b are designated as x_0 and x_2, respectively. The result of the integration is

$$I = \frac{h}{3}[f(x_0) + 4f(x_1) + f(x_2)] \qquad (19.22)$$

where, for this case, $h = (b - a)/2$. This equation is known as *Simpson's 1/3 rule*. The label "1/3" stems from the fact that h is divided by 3 in Eq. (19.22). Simpson's 1/3 rule can also be expressed using the format of Eq. (19.13):

$$I = (b - a)\frac{f(x_0) + 4f(x_1) + f(x_2)}{6} \qquad (19.23)$$

where $a = x_0$, $b = x_2$, and $x_1 = $ the point midway between a and b, which is given by $(a + b)/2$. Notice that, according to Eq. (19.23), the middle point is weighted by two-thirds and the two end points by one-sixth.

It can be shown that a single-segment application of Simpson's 1/3 rule has a truncation error of

$$E_t = -\frac{1}{90}h^5 f^{(4)}(\xi)$$

or, because $h = (b - a)/2$:

$$E_t = -\frac{(b-a)^5}{2880}f^{(4)}(\xi) \qquad (19.24)$$

where ξ lies somewhere in the interval from a to b. Thus, Simpson's 1/3 rule is more accurate than the trapezoidal rule. However, comparison with Eq. (19.14) indicates that it is more accurate than expected. Rather than being proportional to the third derivative, the error is proportional to the fourth derivative. Consequently, Simpson's 1/3 rule is third-order accurate even though it is based on only three points. In other words, it yields exact results for cubic polynomials even though it is derived from a parabola!

EXAMPLE 19.3   Single Application of Simpson's 1/3 Rule

Problem Statement. Use Eq. (19.23) to integrate

$$f(x) = 0.2 + 25x - 200x^2 + 675x^3 - 900x^4 + 400x^5$$

from $a = 0$ to $b = 0.8$. Employ Eq. (19.24) to estimate the error. Recall that the exact integral is 1.640533.

Solution. $n = 2(h = 0.4)$:

$$f(0) = 0.2 \qquad f(0.4) = 2.456 \qquad f(0.8) = 0.232$$

$$I = 0.8\frac{0.2 + 4(2.456) + 0.232}{6} = 1.367467$$

$$E_t = 1.640533 - 1.367467 = 0.2730667 \qquad \varepsilon_t = 16.6\%$$

which is approximately five times more accurate than for a single application of the trapezoidal rule (Example 19.1). The approximate error can be estimated as

$$E_a = -\frac{0.8^5}{2880}(-2400) = 0.2730667$$

where -2400 is the average fourth derivative for the interval. As was the case in Example 19.1, the error is approximate (E_a) because the average fourth derivative is generally not an exact estimate of $f^{(4)}(\xi)$. However, because this case deals with a fifth-order polynomial, the result matches exactly.

19.4.2 The Composite Simpson's 1/3 Rule

Just as with the trapezoidal rule, Simpson's rule can be improved by dividing the integration interval into a number of segments of equal width (Fig. 19.12). The total integral can be represented as

$$I = \int_{x_0}^{x_2} f(x)\,dx + \int_{x_2}^{x_4} f(x)\,dx + \cdots + \int_{x_{n-2}}^{x_n} f(x)\,dx \tag{19.25}$$

FIGURE 19.12
Composite Simpson's 1/3 rule. The relative weights are depicted above the function values. Note that the method can be employed only if the number of segments is even.

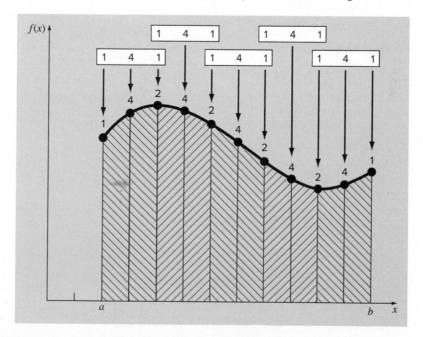

Substituting Simpson's 1/3 rule for each integral yields

$$I = 2h\frac{f(x_0) + 4f(x_1) + f(x_2)}{6} + 2h\frac{f(x_2) + 4f(x_3) + f(x_4)}{6}$$

$$+ \cdots + 2h\frac{f(x_{n-2}) + 4f(x_{n-1}) + f(x_n)}{6}$$

or, grouping terms and using Eq. (19.15):

$$I = (b - a)\frac{f(x_0) + 4\sum_{i=1,3,5}^{n-1} f(x_i) + 2\sum_{j=2,4,6}^{n-2} f(x_j) + f(x_n)}{3n} \tag{19.26}$$

Notice that, as illustrated in Fig. 19.12, an even number of segments must be utilized to implement the method. In addition, the coefficients "4" and "2" in Eq. (19.26) might seem peculiar at first glance. However, they follow naturally from Simpson's 1/3 rule. As illustrated in Fig. 19.12, the odd points represent the middle term for each application and hence carry the weight of four from Eq. (19.23). The even points are common to adjacent applications and hence are counted twice.

An error estimate for the composite Simpson's rule is obtained in the same fashion as for the trapezoidal rule by summing the individual errors for the segments and averaging the derivative to yield

$$E_a = -\frac{(b - a)^5}{180n^4}\bar{f}^{(4)} \tag{19.27}$$

where $\bar{f}^{(4)}$ is the average fourth derivative for the interval.

EXAMPLE 19.4   Composite Simpson's 1/3 Rule

Problem Statement. Use Eq. (19.26) with $n = 4$ to estimate the integral of

$$f(x) = 0.2 + 25x - 200x^2 + 675x^3 - 900x^4 + 400x^5$$

from $a = 0$ to $b = 0.8$. Employ Eq. (19.27) to estimate the error. Recall that the exact integral is 1.640533.

Solution. $n = 4 (h = 0.2)$:

$$f(0) = 0.2 \qquad f(0.2) = 1.288$$

$$f(0.4) = 2.456 \qquad f(0.6) = 3.464$$

$$f(0.8) = 0.232$$

From Eq. (19.26):

$$I = 0.8\frac{0.2 + 4(1.288 + 3.464) + 2(2.456) + 0.232}{12} = 1.623467$$

$$E_t = 1.640533 - 1.623467 = 0.017067 \qquad \varepsilon_t = 1.04\%$$

The estimated error [Eq. (19.27)] is

$$E_a = -\frac{(0.8)^5}{180(4)^4}(-2400) = 0.017067$$

which is exact (as was also the case for Example 19.3).

As in Example 19.4, the composite version of Simpson's 1/3 rule is considered superior to the trapezoidal rule for most applications. However, as mentioned previously, it is limited to cases where the values are equispaced. Further, it is limited to situations where there are an even number of segments and an odd number of points. Consequently, as discussed in Sec. 19.4.3, an odd-segment–even-point formula known as Simpson's 3/8 rule can be used in conjunction with the 1/3 rule to permit evaluation of both even and odd numbers of equispaced segments.

19.4.3 Simpson's 3/8 Rule

In a similar manner to the derivation of the trapezoidal and Simpson's 1/3 rule, a third-order Lagrange polynomial can be fit to four points and integrated to yield

$$I = \frac{3h}{8}[f(x_0) + 3f(x_1) + 3f(x_2) + f(x_3)]$$

where $h = (b - a)/3$. This equation is known as *Simpsons 3/8 rule* because h is multiplied by 3/8. It is the third Newton-Cotes closed integration formula. The 3/8 rule can also be expressed in the form of Eq. (19.13):

$$I = (b - a)\frac{f(x_0) + 3f(x_1) + 3f(x_2) + f(x_3)}{8} \tag{19.28}$$

Thus, the two interior points are given weights of three-eighths, whereas the end points are weighted with one-eighth. Simpson's 3/8 rule has an error of

$$E_t = -\frac{3}{80}h^5 f^{(4)}(\xi)$$

or, because $h = (b - a)/3$:

$$E_t = -\frac{(b - a)^5}{6480}f^{(4)}(\xi) \tag{19.29}$$

Because the denominator of Eq. (19.29) is larger than for Eq. (19.24), the 3/8 rule is somewhat more accurate than the 1/3 rule.

Simpson's 1/3 rule is usually the method of preference because it attains third-order accuracy with three points rather than the four points required for the 3/8 version. However, the 3/8 rule has utility when the number of segments is odd. For instance, in Example 19.4 we used Simpson's rule to integrate the function for four segments. Suppose that you desired an estimate for five segments. One option would be to use a composite version of the trapezoidal rule as was done in Example 19.2. This may not be advisable, however, because of the large truncation error associated with this method. An alternative would be to apply Simpson's 1/3 rule to the first two segments and Simpson's 3/8 rule to the last

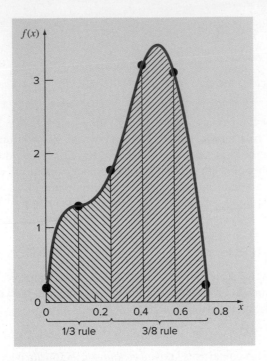

FIGURE 19.13
Illustration of how Simpson's 1/3 and 3/8 rules can be applied in tandem to handle multiple applications with odd numbers of intervals.

three (Fig. 19.13). In this way, we could obtain an estimate with third-order accuracy across the entire interval.

EXAMPLE 19.5   Simpson's 3/8 Rule

Problem Statement. **(a)** Use Simpson's 3/8 rule to integrate

$$f(x) = 0.2 + 25x - 200x^2 + 675x^3 - 900x^4 + 400x^5$$

from $a = 0$ to $b = 0.8$. **(b)** Use it in conjunction with Simpson's 1/3 rule to integrate the same function for five segments.

Solution. **(a)** A single application of Simpson's 3/8 rule requires four equally spaced points:

$f(0) = 0.2$ $f(0.2667) = 1.432724$

$f(0.5333) = 3.487177$ $f(0.8) = 0.232$

Using Eq. (19.28):

$$I = 0.8 \frac{0.2 + 3(1.432724 + 3.487177) + 0.232}{8} = 1.51917$$

(b) The data needed for a five-segment application ($h = 0.16$) are

$$f(0) = 0.2 \qquad\qquad f(0.16) = 1.296919$$

$$f(0.32) = 1.743393 \qquad f(0.48) = 3.186015$$

$$f(0.64) = 3.181929 \qquad f(0.80) = 0.232$$

The integral for the first two segments is obtained using Simpson's 1/3 rule:

$$I = 0.32 \frac{0.2 + 4(1.296919) + 1.743393}{6} = 0.3803237$$

For the last three segments, the 3/8 rule can be used to obtain

$$I = 0.48 \frac{1.743393 + 3(3.186015 + 3.181929) + 0.232}{8} = 1.264754$$

The total integral is computed by summing the two results:

$$I = 0.3803237 + 1.264754 = 1.645077$$

19.5   HIGHER-ORDER NEWTON-COTES FORMULAS

As noted previously, the trapezoidal rule and both of Simpson's rules are members of a family of integrating equations known as the Newton-Cotes closed integration formulas. Some of the formulas are summarized in Table 19.2 along with their truncation-error estimates.

Notice that, as was the case with Simpson's 1/3 and 3/8 rules, the five- and six-point formulas have the same order error. This general characteristic holds for the higher-point formulas and leads to the result that the even-segment–odd-point formulas (e.g., 1/3 rule and Boole's rule) are usually the methods of preference.

TABLE 19.2 Newton-Cotes closed integration formulas. The formulas are presented in the format of Eq. (19.13) so that the weighting of the data points to estimate the average height is apparent. The step size is given by $h = (b - a)/n$.

Segments (n)	Points	Name	Formula	Truncation Error
1	2	Trapezoidal rule	$(b - a)\dfrac{f(x_0) + f(x_1)}{2}$	$-(1/12)h^3 f''(\xi)$
2	3	Simpson's 1/3 rule	$(b - a)\dfrac{f(x_0) + 4f(x_1) + f(x_2)}{6}$	$-(1/90)h^5 f^{(4)}(\xi)$
3	4	Simpson's 3/8 rule	$(b - a)\dfrac{f(x_0) + 3f(x_1) + 3f(x_2) + f(x_3)}{8}$	$-(3/80)h^5 f^{(4)}(\xi)$
4	5	Boole's rule	$(b - a)\dfrac{7f(x_0) + 32 f(x_1) + 12 f(x_2) + 32 f(x_3) + 7 f(x_4)}{90}$	$-(8/945) h^7 f^{(6)}(\xi)$
5	6		$(b - a)\dfrac{19f(x_0) + 75 f(x_1) + 50 f(x_2) + 50 f(x_3) + 75 f(x_4) + 19 f(x_5)}{288}$	$-(275/12{,}096)h^7 f^{(6)}(\xi)$

However, it must also be stressed that, in engineering and science practice, the higher-order (i.e., greater than four-point) formulas are not commonly used. Simpson's rules are sufficient for most applications. Accuracy can be improved by using the composite version. Furthermore, when the function is known and high accuracy is required, methods such as Romberg integration or Gauss quadrature, described in Chap. 20, offer viable and attractive alternatives.

19.6 INTEGRATION WITH UNEQUAL SEGMENTS

To this point, all formulas for numerical integration have been based on equispaced data points. In practice, there are many situations where this assumption does not hold and we must deal with unequal-sized segments. For example, experimentally derived data are often of this type. For these cases, one method is to apply the trapezoidal rule to each segment and sum the results:

$$I = h_1 \frac{f(x_0) + f(x_1)}{2} + h_2 \frac{f(x_1) + f(x_2)}{2} + \cdots + h_n \frac{f(x_{n-1}) + f(x_n)}{2} \tag{19.30}$$

where h_i = the width of segment i. Note that this was the same approach used for the composite trapezoidal rule. The only difference between Eqs. (19.16) and (19.30) is that the h's in the former are constant.

EXAMPLE 19.6   Trapezoidal Rule with Unequal Segments

Problem Statement.   The information in Table 19.3 was generated using the same polynomial employed in Example 19.1. Use Eq. (19.30) to determine the integral for these data. Recall that the correct answer is 1.640533.

TABLE 19.3   Data for $f(x) = 0.2 + 25x - 200x^2 + 675x^3 - 900x^4 + 400x^5$, with unequally spaced values of x.

x	$f(x)$	x	$f(x)$
0.00	0.200000	0.44	2.842985
0.12	1.309729	0.54	3.507297
0.22	1.305241	0.64	3.181929
0.32	1.743393	0.70	2.363000
0.36	2.074903	0.80	0.232000
0.40	2.456000		

Solution.   Applying Eq. (19.30) yields

$$I = 0.12 \frac{0.2 + 1.309729}{2} + 0.10 \frac{1.309729 + 1.305241}{2}$$

$$+ \cdots + 0.10 \frac{2.363 + 0.232}{2} = 1.594801$$

which represents an absolute percent relative error of $\varepsilon_t = 2.8\%$.

19.6.1 MATLAB M-file: trapuneq

A simple algorithm to implement the trapezoidal rule for unequally spaced data can be written as in Fig. 19.14. Two vectors, x and y, holding the independent and dependent variables are passed into the M-file. Two error traps are included to ensure that (a) the two vectors are of the same length and (b) the x's are in ascending order.[1] A loop is employed to generate the integral. Notice that we have modified the subscripts from those of Eq. (19.30) to account for the fact that MATLAB does not allow zero subscripts in arrays.

An application of the M-file can be developed for the same problem that was solved in Example 19.6:

```
>> x = [0 .12 .22 .32 .36 .4 .44 .54 .64 .7 .8];
>> y = 0.2+25*x-200*x.^2+675*x.^3-900*x.^4+400*x.^5;
>> trapuneq(x,y)

ans =

    1.5948
```

which is identical to the result obtained in Example 19.6.

FIGURE 19.14
M-file to implement the trapezoidal rule for unequally spaced data.

```
function I = trapuneq(x,y)
% trapuneq: unequal spaced trapezoidal rule quadrature
%   I = trapuneq(x,y):
%   Applies the trapezoidal rule to determine the integral
%   for n data points (x, y) where x and y must be of the
%   same length and x must be monotonically ascending
% input:
%   x = vector of independent variables
%   y = vector of dependent variables
% output:
%   I = integral estimate

if nargin<2,error('at least 2 input arguments required'),end
if any(diff(x)<0),error('x not monotonically ascending'),end
n = length(x);
if length(y)~=n,error('x and y must be same length'); end
s = 0;
for k = 1:n-1
  s = s + (x(k+1)-x(k))*(y(k)+y(k+1))/2;
end
I = s;
```

---

[1] The diff function is described in Sec. 21.7.1.

19.6.2 MATLAB Functions: trapz and cumtrapz

MATLAB has a built-in function that evaluates integrals for data in the same fashion as the M-file we just presented in Fig. 19.14. It has the general syntax

```
z = trapz(x, y)
```

where the two vectors, x and y, hold the independent and dependent variables, respectively. Here is a simple MATLAB session that uses this function to integrate the data from Table 19.3:

```
>> x = [0 .12 .22 .32 .36 .4 .44 .54 .64 .7 .8];
>> y = 0.2+25*x-200*x.^2+675*x.^3-900*x.^4+400*x.^5;
>> trapz(x,y)

ans =

    1.5948
```

In addition, MATLAB has another function, cumtrapz, that computes the cumulative integral. A simple representation of its syntax is

```
z = cumtrapz(x, y)
```

where the two vectors, x and y, hold the independent and dependent variables, respectively, and $z =$ a vector whose elements $z(k)$ hold the integral from $x(1)$ to $x(k)$.

EXAMPLE 19.7   Using Numerical Integration to Compute Distance from Velocity

Problem Statement. As described at the beginning of this chapter, a nice application of integration is to compute the distance $z(t)$ of an object based on its velocity $v(t)$ as in [recall Eq. (19.2)]:

$$z(t) = \int_0^t v(t)\, dt$$

Suppose that we had measurements of velocity at a series of discrete unequally spaced times during free fall. Use Eq. (19.2) to synthetically generate such information for a 70-kg jumper with a drag coefficient of 0.275 kg/m. Incorporate some random error by rounding the velocities to the nearest integer. Then use cumtrapz to determine the distance fallen and compare the results to the analytical solution [Eq. (19.4)]. In addition, develop a plot of the analytical and computed distances along with velocity on the same graph.

Solution. Some unequally spaced times and rounded velocities can be generated as

```
>> format short g
>> t=[0 1 1.4 2 3 4.3 6 6.7 8];
>> g=9.81;m=70;cd=0.275;
>> v=round(sqrt(g*m/cd)*tanh(sqrt(g*cd/m)*t));
```

The distances can then be computed as

```
>> z=cumtrapz(t,v)

z=

    0  5  9.6  19.2  41.7  80.7  144.45  173.85  231.7
```

Thus, after 8 seconds, the jumper has fallen 231.7 m. This result is reasonably close to the analytical solution [Eq. (19.4)]:

$$z(t) = \frac{70}{0.275} \ln \left[\cosh \left(\sqrt{\frac{9.81(0.275)}{70}}\, 8 \right) \right] = 234.1$$

A graph of the numerical and analytical solutions along with both the exact and rounded velocities can be generated with the following commands:

```
>> ta=linspace(t(1),t(length(t)));
>> za=m/cd*log(cosh(sqrt(g*cd/m)*ta));
>> plot(ta,za,t,z,'o')
>> title('Distance versus time')
>> xlabel('t (s)'),ylabel('x (m)')
>> legend('analytical','numerical')
```

As in Fig. 19.15, the numerical and analytical results match fairly well.

FIGURE 19.15
Plot of distance versus time. The line was computed with the analytical solution, whereas the points were determined numerically with the cumtrapz function.

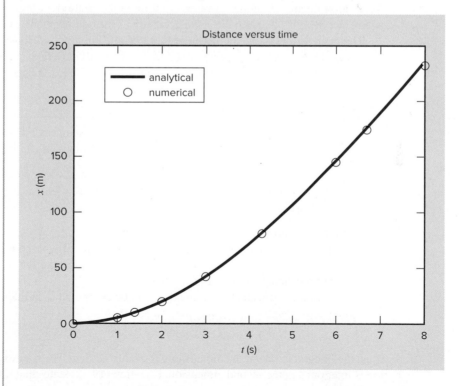

TABLE 19.4 Newton-Cotes open integration formulas. The formulas are presented in the format of Eq. (19.13) so that the weighting of the data points to estimate the average height is apparent. The step size is given by $h = (b - a)/n$.

Segments (n)	Points	Name	Formula	Truncation Error
2	1	Midpoint method	$(b - a) f(x_1)$	$(1/3) h^3 f''(\xi)$
3	2		$(b - a) \dfrac{f(x_1) + f(x_2)}{2}$	$(3/4) h^3 f''(\xi)$
4	3		$(b - a) \dfrac{2 f(x_1) - f(x_2) + 2 f(x_3)}{3}$	$(14/45) h^5 f^{(4)}(\xi)$
5	4		$(b - a) \dfrac{11 f(x_1) + f(x_2) + f(x_3) + 11 f(x_4)}{24}$	$(95/144) h^5 f^{(4)}(\xi)$
6	5		$(b - a) \dfrac{11 f(x_1) - 14 f(x_2) + 26 f(x_3) - 14 f(x_4) + 11 f(x_5)}{20}$	$(41/140) h^7 f^{(6)}(\xi)$

19.7 OPEN METHODS

Recall from Fig. 19.6b that open integration formulas have limits that extend beyond the range of the data. Table 19.4 summarizes the *Newton-Cotes open integration formulas*. The formulas are expressed in the form of Eq. (19.13) so that the weighting factors are evident. As with the closed versions, successive pairs of the formulas have the same-order error. The even-segment–odd-point formulas are usually the methods of preference because they require fewer points to attain the same accuracy as the odd-segment–even-point formulas.

The open formulas are not often used for definite integration. However, they have utility for analyzing improper integrals. In addition, they will have relevance to our discussion of methods for solving ordinary differential equations in Chaps. 22 and 23.

19.8 MULTIPLE INTEGRALS

Multiple integrals are widely used in engineering and science. For example, a general equation to compute the average of a two-dimensional function can be written as [recall Eq. (19.7)]

$$\bar{f} = \frac{\int_c^d \left(\int_a^b f(x, y) \, dx \right)}{(d - c)(b - a)} \tag{19.31}$$

The numerator is called a *double integral*.

The techniques discussed in this chapter (and Chap. 20) can be readily employed to evaluate multiple integrals. A simple example would be to take the double integral of a function over a rectangular area (Fig. 19.16).

Recall from calculus that such integrals can be computed as iterated integrals:

$$\int_c^d \left(\int_a^b f(x, y) \, dx \right) dy = \int_a^b \left(\int_c^d f(x, y) \, dy \right) dx \tag{19.32}$$

Thus, the integral in one of the dimensions is evaluated first. The result of this first integration is integrated in the second dimension. Equation (19.32) states that the order of integration is not important.

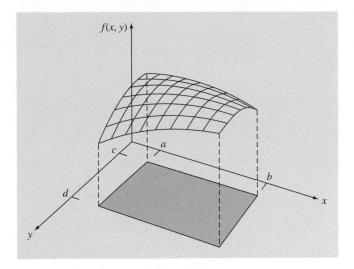

FIGURE 19.16
Double integral as the area under the function surface.

A numerical double integral would be based on the same idea. First, methods such as the composite trapezoidal or Simpson's rule would be applied in the first dimension with each value of the second dimension held constant. Then the method would be applied to integrate the second dimension. The approach is illustrated in the following example.

EXAMPLE 19.8   Using Double Integral to Determine Average Temperature

Problem Statement. Suppose that the temperature of a rectangular heated plate is described by the following function:

$$T(x, y) = 2xy + 2x - x^2 - 2y^2 + 72$$

If the plate is 8 m long (x dimension) and 6 m wide (y dimension), compute the average temperature.

Solution. First, let us merely use two-segment applications of the trapezoidal rule in each dimension. The temperatures at the necessary x and y values are depicted in Fig. 19.17. Note that a simple average of these values is 47.33. The function can also be evaluated analytically to yield a result of 58.66667.

To make the same evaluation numerically, the trapezoidal rule is first implemented along the x dimension for each y value. These values are then integrated along the y dimension to give the final result of 2544. Dividing this by the area yields the average temperature as $2544/(6 \times 8) = 53$.

Now we can apply a single-segment Simpson's 1/3 rule in the same fashion. This results in an integral of 2816 and an average of 58.66667, which is exact. Why does this occur? Recall that Simpson's 1/3 rule yielded perfect results for cubic polynomials. Since the highest-order term in the function is second order, the same exact result occurs for the present case.

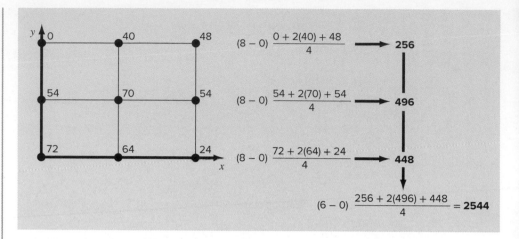

FIGURE 19.17
Numerical evaluation of a double integral using the two-segment trapezoidal rule.

For higher-order algebraic functions as well as transcendental functions, it would be necessary to use composite applications to attain accurate integral estimates. In addition, Chap. 20 introduces techniques that are more efficient than the Newton-Cotes formulas for evaluating integrals of given functions. These often provide a superior means to implement the numerical integrations for multiple integrals.

19.8.1 MATLAB Functions: integral2 and integral3

MATLAB has functions to implement both double (integral2) and triple (integral3) integration. A simple representation of the syntax for integral2 is

```
q = integral2(fun, xmin, xmax, ymin, ymax)
```

where q is the double integral of the function *fun* over the ranges from *xmin* to *xmax* and *ymin* to *ymax*.

Here is an example of how this function can be used to compute the double integral evaluated in Example 19.7:

```
>> q = integral2(@(x,y) 2*x*y+2*x-x.^2-2*y.^2+72,0,8,0,6)

q =
        2816
```

19.9 CASE STUDY COMPUTING WORK WITH NUMERICAL INTEGRATION

Background. The calculation of work is an important component of many areas of engineering and science. The general formula is

Work = force × distance

When you were introduced to this concept in high school physics, simple applications were presented using forces that remained constant throughout the displacement. For example, if a force of 10 N was used to pull a block a distance of 5 m, the work would be calculated as 50 J (1 joule = 1 N · m).

Although such a simple computation is useful for introducing the concept, realistic problem settings are usually more complex. For example, suppose that the force varies during the course of the calculation. In such cases, the work equation is reexpressed as

$$W = \int_{x_0}^{x_n} F(x)\, dx \tag{19.33}$$

where W = work (J), x_0 and x_n = the initial and final positions (m), respectively, and $F(x)$ = a force that varies as a function of position (N). If $F(x)$ is easy to integrate, Eq. (19.33) can be evaluated analytically. However, in a realistic problem setting, the force might not be expressed in such a manner. In fact, when analyzing measured data, the force might be available only in tabular form. For such cases, numerical integration is the only viable option for the evaluation.

Further complexity is introduced if the angle between the force and the direction of movement also varies as a function of position (Fig. 19.18). The work equation can be modified further to account for this effect, as in

$$W = \int_{x_0}^{x_n} F(x) \cos[\theta(x)]\, dx \tag{19.34}$$

Again, if $F(x)$ and $\theta(x)$ are simple functions, Eq. (19.34) might be solved analytically. However, as in Fig. 19.18, it is more likely that the functional relationship is complicated. For this situation, numerical methods provide the only alternative for determining the integral.

Suppose that you have to perform the computation for the situation depicted in Fig. 19.18. Although the figure shows the continuous values for $F(x)$ and $\theta(x)$, assume that, because of experimental constraints, you are provided with only discrete measurements at $x = 5$-m intervals (Table 19.5). Use single- and composite versions of the trapezoidal rule and Simpson's 1/3 and 3/8 rules to compute work for these data.

Solution. The results of the analysis are summarized in Table 19.6. A percent relative error ε_t was computed in reference to a true value of the integral of 129.52 that was estimated on the basis of values taken from Fig. 19.18 at 1-m intervals.

The results are interesting because the most accurate outcome occurs for the simple two-segment trapezoidal rule. More refined estimates using more segments, as well as Simpson's rules, yield less accurate results.

The reason for this apparently counterintuitive result is that the coarse spacing of the points is not adequate to capture the variations of the forces and angles. This is particularly

FIGURE 19.18
The case of a variable force acting on a block. For this case the angle, as well as the magnitude, of the force varies.

TABLE 19.5 Data for force $F(x)$ and angle $\theta(x)$ as a function of position x.

x, m	$F(x)$, N	θ, rad	$F(x)\cos\theta$
0	0.0	0.50	0.0000
5	9.0	1.40	1.5297
10	13.0	0.75	9.5120
15	14.0	0.90	8.7025
20	10.5	1.30	2.8087
25	12.0	1.48	1.0881
30	5.0	1.50	0.3537

19.9 CASE STUDY continued

TABLE 19.6 Estimates of work calculated using the trapezoidal rule and Simpson's rules. The percent relative error ε_t as computed in reference to a true value of the integral (129.52 Pa) that was estimated on the basis of values at 1-m intervals.

Technique	Segments	Work	ε_t, %
Trapezoidal rule	1	5.31	95.9
	2	133.19	2.84
	3	124.98	3.51
	6	119.09	8.05
Simpson's 1/3 rule	2	175.82	35.75
	6	117.13	9.57
Simpson's 3/8 rule	3	139.93	8.04

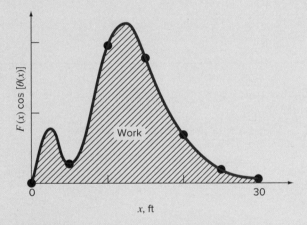

FIGURE 19.19
A continuous plot of $F(x) \cos [\theta (x)]$ versus position with the seven discrete points used to develop the numerical integration estimates in Table 19.6. Notice how the use of seven points to characterize this continuously varying function misses two peaks at $x = 2.5$ and 12.5 m.

evident in Fig. 19.19, where we have plotted the continuous curve for the product of $F(x)$ and $\cos [\theta(x)]$. Notice how the use of seven points to characterize the continuously varying function misses the two peaks at $x = 2.5$ and 12.5 m. The omission of these two points effectively limits the accuracy of the numerical integration estimates in Table 19.6. The fact that the two-segment trapezoidal rule yields the most accurate result is due to the chance positioning of the points for this particular problem (Fig. 19.20).

The conclusion to be drawn from Fig. 19.20 is that an adequate number of measurements must be made to accurately compute integrals. For the present case, if data were available at $F(2.5) \cos [\theta(2.5)] = 3.9007$ and $F(12.5) \cos [\theta (12.5)] = 11.3940$, we could

19.9 CASE STUDY continued

FIGURE 19.20
Graphical depiction of why the two-segment trapezoidal rule yields a good estimate of the integral for this particular case. By chance, the use of two trapezoids happens to lead to an even balance between positive and negative errors.

determine an improved integral estimate. For example, using the MATLAB trapz function, we could compute

```
>> x=[0 2.5 5 10 12.5 15 20 25 30];
>> y=[0 3.9007 1.5297 9.5120 11.3940 8.7025 2.8087 ...
                                    1.0881 0.3537];
>> trapz(x,y)

ans =
    132.6458
```

Including the two additional points yields an improved integral estimate of 132.6458 ($\varepsilon_t = 2.16\%$). Thus, the inclusion of the additional data incorporates the peaks that were missed previously and, as a consequence, lead to better results.

PROBLEMS

19.1 Derive Eq. (19.4) by integrating Eq. (19.3).
19.2 Evaluate the following integral:

$$\int_0^4 (1 - e^{-x})\,dx$$

(a) analytically, **(b)** single application of the trapezoidal rule, **(c)** composite trapezoidal rule with $n = 2$ and 4, **(d)** single application of Simpson's 1/3 rule, **(e)** composite Simpson's

1/3 rule with $n = 4$, **(f)** Simpson's 3/8 rule, and **(g)** composite Simpson's rule, with $n = 5$. For each of the numerical estimates **(b)** through **(g)**, determine the true percent relative error based on **(a)**.
19.3 Evaluate the following integral:

$$\int_0^{\pi/2} (8 + 4 \cos x)\,dx$$

(a) analytically, (b) single application of the trapezoidal rule, (c) composite trapezoidal rule with $n = 2$ and 4, (d) single application of Simpson's 1/3 rule, (e) composite Simpson's 1/3 rule with $n = 4$, (f) Simpson's 3/8 rule, and (g) composite Simpson's rule, with $n = 5$. For each of the numerical estimates (b) through (g), determine the true percent relative error based on (a).

19.4 Evaluate the following integral:

$$\int_{-2}^{4} (1 - x - 4x^3 + 2x^5)\,dx$$

(a) analytically, (b) single application of the trapezoidal rule, (c) composite trapezoidal rule with $n = 2$ and 4, (d) single application of Simpson's 1/3 rule, (e) Simpson's 3/8 rule, and (f) Boole's rule. For each of the numerical estimates (b) through (f), determine the true percent relative error based on (a).

19.5 The function

$$f(x) = e^{-x}$$

can be used to generate the following table of unequally spaced data:

x	0	0.1	0.3	0.5	0.7	0.95	1.2
$f(x)$	1	0.9048	0.7408	0.6065	0.4966	0.3867	0.3012

Evaluate the integral from $a = 0$ to $b = 1.2$ using (a) analytical means, (b) the trapezoidal rule, and (c) a combination of the trapezoidal and Simpson's rules wherever possible to attain the highest accuracy. For (b) and (c), compute the true percent relative error.

19.6 Evaluate the double integral

$$\int_{-2}^{2} \int_{0}^{4} (x^2 - 3y^2 + xy^3)\,dx\,dy$$

(a) analytically, (b) using the composite trapezoidal rule with $n = 2$, (c) using single applications of Simpson's 1/3 rule, and (d) using the integral2 function. For (b) and (c), compute the percent relative error.

19.7 Evaluate the triple integral

$$\int_{-4}^{4} \int_{0}^{6} \int_{-1}^{3} (x^3 - 2yz)\,dx\,dy\,dz$$

(a) analytically, (b) using single applications of Simpson's 1/3 rule, and (c) the integral3 function. For (b), compute the true percent relative error.

19.8 Determine the distance traveled from the following velocity data:

t	1	2	3.25	4.5	6	7	8	8.5	9	10
v	5	6	5.5	7	8.5	8	6	7	7	5

(a) Use the trapezoidal rule. In addition, determine the average velocity.
(b) Fit the data with a cubic equation using polynomial regression. Integrate the cubic equation to determine the distance.

19.9 Water exerts pressure on the upstream face of a dam as shown in Fig. P19.9. The pressure can be characterized by

$$p(z) = \rho g(D - z)$$

where $p(z) = $ pressure in pascals (or N/m²) exerted at an elevation z meters above the reservoir bottom; $\rho = $ density of water, which for this problem is assumed to be a constant 10^3 kg/m³; $g = $ acceleration due to gravity (9.81 m/s²); and $D = $ elevation (in m) of the water surface above the reservoir bottom. According to Eq. (P19.9), pressure increases linearly with depth, as depicted in Fig. P19.9a. Omitting atmospheric pressure (because it works against both sides of the dam face and essentially cancels out), the total force f_t can be determined by multiplying pressure times the area of the dam face (as shown in Fig. P19.9b). Because both pressure and area vary with elevation, the total force is obtained by evaluating

$$f_t = \int_{0}^{D} \rho g w(z)(D - z)\,dz$$

FIGURE P19.9
Water exerting pressure on the upstream face of a dam: (a) side view showing force increasing linearly with depth; (b) front view showing width of dam in meters.

where $w(z)$ = width of the dam face (m) at elevation z (Fig. P19.9b). The line of action can also be obtained by evaluating

$$d = \frac{\int_0^D \rho g z w(z)(D - z)\, dz}{\int_0^D \rho g w(z)(D - z)\, dz}$$

Use Simpson's rule to compute f_t and d.

19.10 The force on a sailboat mast can be represented by the following function:

$$f(z) = 200 \left(\frac{z}{5 + z} \right) e^{-2z/H}$$

where z = the elevation above the deck and H = the height of the mast. The total force F exerted on the mast can be determined by integrating this function over the height of the mast:

$$F = \int_0^H f(z)\, dz$$

The line of action can also be determined by integration:

$$d = \frac{\int_0^H z f(z)\, dz}{\int_0^H f(z)\, dz}$$

(a) Use the composite trapezoidal rule to compute F and d for the case where $H = 30$ ($n = 6$).
(b) Repeat (a), but use the composite Simpson's 1/3 rule.

19.11 A wind force distributed against the side of a skyscraper is measured as

Height l, m	0	30	60	90	120
Force, $F(l)$, N/m	0	340	1200	1550	2700

Height l, m	150	180	210	240
Force, $F(l)$, N/m	3100	3200	3500	3750

Compute the net force and the line of action due to this distributed wind.

19.12 An 11-m beam is subjected to a load, and the shear force follows the equation

$$V(x) = 5 + 0.25x^2$$

where V is the shear force, and x is length in distance along the beam. We know that $V = dM/dx$, and M is the bending moment. Integration yields the relationship

$$M = M_o + \int_0^x V\, dx$$

If M_o is zero and $x = 11$, calculate M using (a) analytical integration, (b) composite trapezoidal rule, and (c) composite Simpson's rules. For (b) and (c) use 1-m increments.

19.13 The total mass of a variable density rod is given by

$$m = \int_0^L \rho(x) A_c(x)\, dx$$

where m = mass, $\rho(x)$ = density, $A_c(x)$ = cross-sectional area, x = distance along the rod, and L = the total length of the rod. The following data have been measured for a 20-m length rod. Determine the mass in grams to the best possible accuracy.

x, m	0	4	6	8	12	16	20
ρ, g/cm³	4.00	3.95	3.89	3.80	3.60	3.41	3.30
A_c, cm²	100	103	106	110	120	133	150

19.14 A transportation engineering study requires that you determine the number of cars that pass through an intersection traveling during morning rush hour. You stand at the side of the road and count the number of cars that pass every 4 minutes at several times as tabulated below. Use the best numerical method to determine (a) the total number of cars that pass between 7:30 and 9:15 and (b) the rate of cars going through the intersection per minute. (Hint: Be careful with units.)

Time (hr)	7:30	7:45	8:00	8:15	8:45	9:15
Rate (cars per 4 min)	18	23	14	24	20	9

19.15 Determine the average value for the data in Fig. P19.15. Perform the integral needed for the average in the order shown by the following equation:

$$I = \int_{x_0}^{x_n} \left[\int_{y_0}^{y_m} f(x, y)\, dy \right] dx$$

19.16 Integration provides a means to compute how much mass enters or leaves a reactor over a specified time period, as in

$$M = \int_{t_1}^{t_2} Qc\, dt$$

where t_1 and t_2 = the initial and final times, respectively. This formula makes intuitive sense if you recall the analogy between integration and summation. Thus, the integral represents the summation of the product of flow times concentration to give the total mass entering or leaving from t_1 to t_2. Use numerical integration to evaluate this equation for the data listed below:

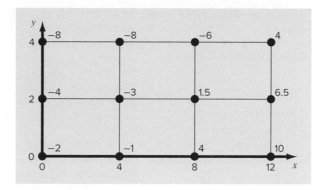

FIGURE P19.15

t, min	0	10	20	30	35	40	45	50
Q, m³/min	4	4.8	5.2	5.0	4.6	4.3	4.3	5.0
c, mg/m³	10	35	55	52	40	37	32	34

19.17 The cross-sectional area of a channel can be computed as

$$A_c = \int_0^B H(y)\, dy$$

where B = the total channel width (m), H = the depth (m), and y = distance from the bank (m). In a similar fashion, the average flow Q (m³/s) can be computed as

$$Q = \int_0^B U(y)\, H(y)\, dy$$

where U = water velocity (m/s). Use these relationships and a numerical method to determine A_c and Q for the following data:

y, m	0	2	4	5	6	9
H, m	0.5	1.3	1.25	1.8	1	0.25
U, m/s	0.03	0.06	0.05	0.13	0.11	0.02

19.18 The average concentration of a substance $\bar{c}$ (g/m³) in a lake where the area A_s(m²) varies with depth z(m) can be computed by integration:

$$\bar{c} = \frac{\int_0^Z c(z) A_s(z)\, dz}{\int_0^Z A_s(z)\, dz}$$

where Z = the total depth (m). Determine the average concentration based on the following data:

z, m	0	4	8	12	16
A, 10⁶ m²	9.8175	5.1051	1.9635	0.3927	0.0000
c, g/m³	10.2	8.5	7.4	5.2	4.1

19.19 As was done in Sec. 19.9, determine the work performed if a constant force of 1 N applied at an angle θ results in the following displacements. Use the cumtrapz function to determine the cumulative work and plot the result versus θ.

x, m	0	1	2.8	3.9	3.8	3.2	1.3
θ, deg	0	30	60	90	120	150	180

19.20 Compute work as described in Sec. 19.9, but use the following equations for $F(x)$ and $\theta(x)$:

$$F(x) = 1.6x - 0.045x^2$$
$$\theta(x) = -0.00055x^3 + 0.0123x^2 + 0.13x$$

The force is in Newtons and the angle is in radians. Perform the integration from $x = 0$ to 30 m.

19.21 As specified in the following table, a manufactured spherical particle has a density that varies as a function of the distance from its center ($r = 0$):

r, mm	0	0.12	0.24	0.36	0.49
ρ(g/cm³)	6	5.81	5.14	4.29	3.39

r, mm	0.62	0.79	0.86	0.93	1
ρ(g/cm³)	2.7	2.19	2.1	2.04	2

Use numerical integration to estimate the particle's mass (in g) and average density (in g/cm³).

19.22 As specified in the following table, the earth's density varies as a function of the distance from its center ($r = 0$):

r, km	0	1100	1500	2450	3400	3630
ρ(g/cm³)	13	12.4	12	11.2	9.7	5.7

r, km	4500	5380	6060	6280	6380
ρ(g/cm³)	5.2	4.7	3.6	3.4	3

Use numerical integration to estimate the earth's mass (in metric tonnes) and average density (in g/cm³). Develop vertically stacked subplots of (top) density versus radius, and (bottom) mass versus radius. Assume that the earth is a perfect sphere.

19.23 A spherical tank has a circular orifice in its bottom through which the liquid flows out (Fig. P19.23). The following data is collected for the flow rate through the orifice as a function of time:

t, s	0	500	1000	1500	2200	2900
Q, m³/hr	10.55	9.576	9.072	8.640	8.100	7.560

t, s	3600	4300	5200	6500	7000	7500
Q, m³/hr	7.020	6.480	5.688	4.752	3.348	1.404

Write a script with supporting functions **(a)** to estimate the volume of fluid (in liters) drained over the entire measurement period and **(b)** to estimate the liquid level in the tank at $t = 0$ s. Note that $r = 1.5$ m.

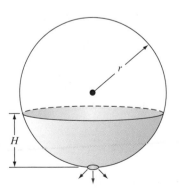

FIGURE P19.23

19.24 Develop an M-file function to implement the composite Simpson's 1/3 rule for equispaced data. Have the function print error messages and terminate (1) if the data is not equispaced or (2) if the input vectors holding the data are not of equal length. If there are only 2 data points, implement the trapezoidal rule. If there are an even number of data points n (i.e., an odd number of segments, $n - 1$), use Simpson's 3/8 rule for the final 3 segments.

19.25 During a storm a high wind blows along one side of a rectangular skyscraper as depicted in Fig. P19.25. As described in Prob. 19.9, use the best lower-order Newton-Cotes formulas (trapezoidal, Simpsons 1/3 and 3/8 rules) to determine **(a)** the force on the building in Newtons and **(b)** the line of force in meters.

19.26 The following data is provided for the velocity of an object as a function of time:

t, s	0	4	8	12	16	20	24	28	30
v, m/s	0	18	31	42	50	56	61	65	70

(a) Limiting yourself to trapezoidal rule and Simpson's 1/3 and 3/8 rules, make the best estimate of how far the object travels from $t = 0$ to 30 s?

(b) Employ the results of **(a)** to compute the average velocity.

FIGURE P19.25

19.27 The total mass of a variable density rod is given by

$$m = \int_0^L \rho(x) A_c(x)\, dx$$

where m = mass, $\rho(x)$ = density, $A_c(x)$ = cross-sectional area, and x = distance along the rod. The following data has been measured for a 10-m length rod:

x (m)	0	2	3	4	6	8	10
ρ (g/cm³)	4.00	3.95	3.89	3.80	3.60	3.41	3.30
A_c (cm²)	100	103	106	110	120	133	150

Determine the mass in grams to the best possible accuracy limiting yourself to trapezoidal rule and Simpson's 1/3 and 3/8 rules.

19.28 A gas is expanded in an engine cylinder, following the law

$$PV^{1.3} = c$$

The initial pressure is 2550 kPa and the final pressure is 210 kPa. If the volume at the end of expansion is 0.75 m³, compute the work done by the gas.

19.29 The pressure p and volume v of a given mass of gas are connected by the relation

$$(p + a/v^2)(v - b)$$

where a, b, and k are constants. Express p in terms of v, and write a script to compute the work done by the gas in expanding from an initial volume to a final volume. Test your solution with $a = 0.01$, $b = 0.001$, initial pressure and volume = 100 kPa and 1 m³, respectively, and final volume = 2 m³.

20

Numerical Integration of Functions

CHAPTER OBJECTIVES

The primary objective of this chapter is to introduce you to numerical methods for integrating given functions. Specific objectives and topics covered are

- Understanding how Richardson extrapolation provides a means to create a more accurate integral estimate by combining two less accurate estimates.
- Understanding how Gauss quadrature provides superior integral estimates by picking optimal abscissas at which to evaluate the function.
- Knowing how to use MATLAB's built-in function `integral` to integrate functions.

20.1 INTRODUCTION

In Chap. 19, we noted that functions to be integrated numerically will typically be of two forms: a table of values or a function. The form of the data has an important influence on the approaches that can be used to evaluate the integral. For tabulated information, you are limited by the number of points that are given. In contrast, if the function is available, you can generate as many values of $f(x)$ as are required to attain acceptable accuracy.

At face value, the composite Simpson's 1/3 rule might seem to be a reasonable tool for such problems. Although it is certainly adequate for many problems, there are more efficient methods that are available. This chapter is devoted to three such techniques, which capitalize on the ability to generate function values to develop efficient schemes for numerical integration.

The first technique is based on *Richardson extrapolation,* which is a method for combining two numerical integral estimates to obtain a third, more accurate value. The computational algorithm for implementing Richardson extrapolation in a highly efficient manner is called *Romberg integration.* This technique can be used to generate an integral estimate within a prespecified error tolerance.

The second method is called *Gauss quadrature.* Recall that, in Chap. 19, values of $f(x)$ for the Newton-Cotes formulas were determined at specified values of x. For example, if we used the trapezoidal rule to determine an integral, we were constrained to take the weighted average of $f(x)$ at the ends of the interval. Gauss-quadrature formulas employ x values that are positioned between the integration limits in such a manner that a much more accurate integral estimate results.

The third approach is called *adaptive quadrature.* This techniques applies composite Simpson's 1/3 rule to subintervals of the integration range in a way that allows error estimates to be computed. These error estimates are then used to determine whether more refined estimates are required for a subinterval. In this way, more refined segmentation is only used where it is necessary. A built-in MATLAB function that uses adaptive quadrature is illustrated.

20.2 ROMBERG INTEGRATION

Romberg integration is one technique that is designed to attain efficient numerical integrals of functions. It is quite similar to the techniques discussed in Chap. 19 in the sense that it is based on successive application of the trapezoidal rule. However, through mathematical manipulations, superior results are attained for less effort.

20.2.1 Richardson Extrapolation

Techniques are available to improve the results of numerical integration on the basis of the integral estimates themselves. Generally called *Richardson extrapolation,* these methods use two estimates of an integral to compute a third, more accurate approximation.

The estimate and the error associated with the composite trapezoidal rule can be represented generally as

$$I = I(h) + E(h)$$

where I = the exact value of the integral, $I(h)$ = the approximation from an n-segment application of the trapezoidal rule with step size $h = (b - a)/n$, and $E(h)$ = the truncation error. If we make two separate estimates using step sizes of h_1 and h_2 and have exact values for the error:

$$I(h_1) + E(h_1) = I(h_2) + E(h_2) \tag{20.1}$$

Now recall that the error of the composite trapezoidal rule can be represented approximately by Eq. (19.21) [with $n = (b - a)/h$]:

$$E \cong -\frac{b-a}{12} h^2 \bar{f}'' \tag{20.2}$$

If it is assumed that $\bar{f}''$ is constant regardless of step size, Eq. (20.2) can be used to determine that the ratio of the two errors will be

$$\frac{E(h_1)}{E(h_2)} \cong \frac{h_1^2}{h_2^2} \tag{20.3}$$

This calculation has the important effect of removing the term $\bar{f}''$ from the computation. In so doing, we have made it possible to utilize the information embodied by Eq. (20.2) without prior knowledge of the function's second derivative. To do this, we rearrange Eq. (20.3) to give

$$E(h_1) \cong E(h_2) \left(\frac{h_1}{h_2}\right)^2$$

which can be substituted into Eq. (20.1):

$$I(h_1) + E(h_2) \left(\frac{h_1}{h_2}\right)^2 = I(h_2) + E(h_2)$$

which can be solved for

$$E(h_2) = \frac{I(h_1) - I(h_2)}{1 - (h_1/h_2)^2}$$

Thus, we have developed an estimate of the truncation error in terms of the integral estimates and their step sizes. This estimate can then be substituted into

$$I = I(h_2) + E(h_2)$$

to yield an improved estimate of the integral:

$$I = I(h_2) + \frac{1}{(h_1/h_2)^2 - 1} [I(h_2) - I(h_1)] \tag{20.4}$$

It can be shown (Ralston and Rabinowitz, 1978) that the error of this estimate is $O(h^4)$. Thus, we have combined two trapezoidal rule estimates of $O(h^2)$ to yield a new estimate of $O(h^4)$. For the special case where the interval is halved ($h_2 = h_1/2$), this equation becomes

$$I = \frac{4}{3} I(h_2) - \frac{1}{3} I(h_1) \tag{20.5}$$

EXAMPLE 20.1   Richardson Extrapolation

Problem Statement.   Use Richardson extrapolation to evaluate the integral of $f(x) = 0.2 + 25x - 200x^2 + 675x^3 - 900x^4 + 400x^5$ from $a = 0$ to $b = 0.8$.

Solution.   Single and composite applications of the trapezoidal rule can be used to evaluate the integral:

Segments	h	Integral	ε_t
1	0.8	0.1728	89.5%
2	0.4	1.0688	34.9%
4	0.2	1.4848	9.5%

Richardson extrapolation can be used to combine these results to obtain improved estimates of the integral. For example, the estimates for one and two segments can be combined to yield

$$I = \frac{4}{3}(1.0688) - \frac{1}{3}(0.1728) = 1.367467$$

The error of the improved integral is $E_t = 1.640533 - 1.367467 = 0.273067(\varepsilon_t = 16.6\%)$, which is superior to the estimates upon which it was based.

In the same manner, the estimates for two and four segments can be combined to give

$$I = \frac{4}{3}(1.4848) - \frac{1}{3}(1.0688) = 1.623467$$

which represents an error of $E_t = 1.640533 - 1.623467 = 0.017067\,(\varepsilon_t = 1.0\%)$.

Equation (20.4) provides a way to combine two applications of the trapezoidal rule with error $O(h^2)$ to compute a third estimate with error $O(h^4)$. This approach is a subset of a more general method for combining integrals to obtain improved estimates. For instance, in Example 20.1, we computed two improved integrals of $O(h^4)$ on the basis of three trapezoidal rule estimates. These two improved integrals can, in turn, be combined to yield an even better value with $O(h^6)$. For the special case where the original trapezoidal estimates are based on successive halving of the step size, the equation used for $O(h^6)$ accuracy is

$$I = \frac{16}{15}I_m - \frac{1}{15}I_l \tag{20.6}$$

where I_m and I_l are the more and less accurate estimates, respectively. Similarly, two $O(h^6)$ results can be combined to compute an integral that is $O(h^8)$ using

$$I = \frac{64}{63}I_m - \frac{1}{63}I_l \tag{20.7}$$

EXAMPLE 20.2   Higher-Order Corrections

Problem Statement.   In Example 20.1, we used Richardson extrapolation to compute two integral estimates of $O(h^4)$. Utilize Eq. (20.6) to combine these estimates to compute an integral with $O(h^6)$.

Solution.   The two integral estimates of $O(h^4)$ obtained in Example 20.1 were 1.367467 and 1.623467. These values can be substituted into Eq. (20.6) to yield

$$I = \frac{16}{15}(1.623467) - \frac{1}{15}(1.367467) = 1.640533$$

which is the exact value of the integral.

20.2.2 The Romberg Integration Algorithm

Notice that the coefficients in each of the extrapolation equations [Eqs. (20.5), (20.6), and (20.7)] add up to 1. Thus, they represent weighting factors that, as accuracy increases, place relatively greater weight on the superior integral estimate. These formulations can be expressed in a general form that is well suited for computer implementation:

$$I_{j,k} = \frac{4^{k-1} I_{j+1,k-1} - I_{j,k-1}}{4^{k-1} - 1} \tag{20.8}$$

where $I_{j+1,k-1}$ and $I_{j,k-1}$ = the more and less accurate integrals, respectively, and $I_{j,k}$ = the improved integral. The index k signifies the level of the integration, where $k = 1$ corresponds to the original trapezoidal rule estimates, $k = 2$ corresponds to the $O(h^4)$ estimates, $k = 3$ to the $O(h^6)$, and so forth. The index j is used to distinguish between the more ($j + 1$) and the less (j) accurate estimates. For example, for $k = 2$ and $j = 1$, Eq. (20.8) becomes

$$I_{1,2} = \frac{4I_{2,1} - I_{1,1}}{3}$$

which is equivalent to Eq. (20.5).

The general form represented by Eq. (20.8) is attributed to Romberg, and its systematic application to evaluate integrals is known as *Romberg integration*. Figure 20.1 is a graphical depiction of the sequence of integral estimates generated using this approach. Each matrix corresponds to a single iteration. The first column contains the trapezoidal rule evaluations that are designated $I_{j,1}$, where $j = 1$ is for a single-segment application (step size is $b - a$), $j = 2$ is for a two-segment application [step size is $(b - a)/2$], $j = 3$ is for a four-segment application [step size is $(b - a)/4$], and so forth. The other columns of the matrix are generated by systematically applying Eq. (20.8) to obtain successively better estimates of the integral.

For example, the first iteration (Fig. 20.1a) involves computing the one- and two-segment trapezoidal rule estimates ($I_{1,1}$ and $I_{2,1}$). Equation (20.8) is then used to compute the element $I_{1,2} = 1.367467$, which has an error of $O(h^4)$.

FIGURE 20.1
Graphical depiction of the sequence of integral estimates generated using Romberg integration. (*a*) First iteration. (*b*) Second iteration. (*c*) Third iteration.

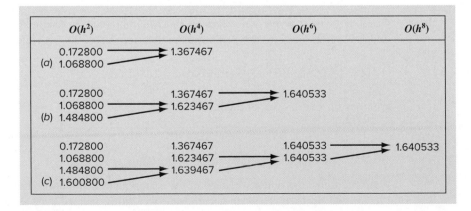

$O(h^2)$	$O(h^4)$	$O(h^6)$	$O(h^8)$
0.172800	1.367467		
(a) 1.068800			
0.172800	1.367467	1.640533	
1.068800	1.623467		
(b) 1.484800			
0.172800	1.367467	1.640533	1.640533
1.068800	1.623467	1.640533	
1.484800	1.639467		
(c) 1.600800			

Now, we must check to determine whether this result is adequate for our needs. As in other approximate methods in this book, a termination, or stopping, criterion is required to assess the accuracy of the results. One method that can be employed for the present purposes is

$$|\varepsilon_a| = \left| \frac{I_{1,k} - I_{2,k-1}}{I_{1,k}} \right| \times 100\% \tag{20.9}$$

where ε_a = an estimate of the percent relative error. Thus, as was done previously in other iterative processes, we compare the new estimate with a previous value. For Eq. (20.9), the previous value is the most accurate estimate from the previous level of integration (i.e., the $k-1$ level of integration with $j = 2$). When the change between the old and new values as represented by ε_a is below a prespecified error criterion ε_s, the computation is terminated. For Fig. 20.1a, this evaluation indicates the following percent change over the course of the first iteration:

$$|\varepsilon_a| = \left| \frac{1.367467 - 1.068800}{1.367467} \right| \times 100\% = 21.8\%$$

The object of the second iteration (Fig. 20.1b) is to obtain the $O(h^6)$ estimate—$I_{1,3}$. To do this, a four-segment trapezoidal rule estimate, $I_{3,1} = 1.4848$, is determined. Then it is combined with $I_{2,1}$ using Eq. (20.8) to generate $I_{2,2} = 1.623467$. The result is, in turn, combined with $I_{1,2}$ to yield $I_{1,3} = 1.640533$. Equation (20.9) can be applied to determine that this result represents a change of 1.0% when compared with the previous result $I_{2,2}$.

The third iteration (Fig. 20.1c) continues the process in the same fashion. In this case, an eight-segment trapezoidal estimate is added to the first column, and then Eq. (20.8) is applied to compute successively more accurate integrals along the lower diagonal. After only three iterations, because we are evaluating a fifth-order polynomial, the result ($I_{1,4} = 1.640533$) is exact.

Romberg integration is more efficient than the trapezoidal rule and Simpson's rules. For example, for determination of the integral as shown in Fig. 20.1, Simpson's 1/3 rule would require about a 48-segment application in double precision to yield an estimate of the integral to seven significant digits: 1.640533. In contrast, Romberg integration produces the same result based on combining one-, two-, four-, and eight-segment trapezoidal rules—that is, with only 15 function evaluations!

Figure 20.2 presents an M-file for Romberg integration. By using loops, this algorithm implements the method in an efficient manner. Note that the function uses another function trap to implement the composite trapezoidal rule evaluations (recall Fig. 19.10). Here is a MATLAB session showing how it can be used to determine the integral of the polynomial from Example 20.1:

```
>> f=@(x) 0.2+25*x-200*x^2+675*x^3-900*x^4+400*x^5;
>> romberg(f,0,0.8)

ans =
    1.6405
```

```
function [q,ea,iter]=romberg(func,a,b,es,maxit,varargin)
% romberg: Romberg integration quadrature
%   q = romberg(func,a,b,es,maxit,p1,p2,...):
%                       Romberg integration.
% input:
%   func = name of function to be integrated
%   a, b = integration limits
%   es = desired relative error (default = 0.000001%)
%   maxit = maximum allowable iterations (default = 30)
%   p1,p2,... = additional parameters used by func
% output:
%   q = integral estimate
%   ea = approximate relative error (%)
%   iter = number of iterations

if nargin<3,error('at least 3 input arguments required'),end
if nargin<4|isempty(es), es=0.000001;end
if nargin<5|isempty(maxit), maxit=50;end
n = 1;
I(1,1) = trap(func,a,b,n,varargin{:});
iter = 0;
while iter<maxit
  iter = iter+1;
  n = 2^iter;
  I(iter+1,1) = trap(func,a,b,n,varargin{:});
  for k = 2:iter+1
    j = 2+iter-k;
    I(j,k) = (4^(k-1)*I(j+1,k-1)-I(j,k-1))/(4^(k-1)-1);
  end
  ea = abs((I(1,iter+1)-I(2,iter))/I(1,iter+1))*100;
  if ea<=es, break; end
end
q = I(1,iter+1);
```

FIGURE 20.2
M-file to implement Romberg integration.

20.3 GAUSS QUADRATURE

In Chap. 19, we employed the Newton-Cotes equations. A characteristic of these formulas (with the exception of the special case of unequally spaced data) was that the integral estimate was based on evenly spaced function values. Consequently, the location of the base points used in these equations was predetermined or fixed.

For example, as depicted in Fig. 20.3a, the trapezoidal rule is based on taking the area under the straight line connecting the function values at the ends of the integration interval. The formula that is used to compute this area is

$$I \cong (b - a)\frac{f(a) + f(b)}{2} \tag{20.10}$$

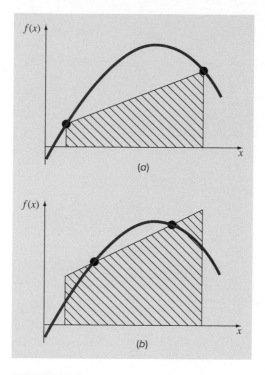

FIGURE 20.3
(a) Graphical depiction of the trapezoidal rule as the area under the straight line joining fixed end points. (b) An improved integral estimate obtained by taking the area under the straight line passing through two intermediate points. By positioning these points wisely, the positive and negative errors are better balanced, and an improved integral estimate results.

where a and $b =$ the limits of integration and $b - a =$ the width of the integration interval. Because the trapezoidal rule must pass through the end points, there are cases such as Fig. 20.3a where the formula results in a large error.

Now, suppose that the constraint of fixed base points was removed and we were free to evaluate the area under a straight line joining any two points on the curve. By positioning these points wisely, we could define a straight line that would balance the positive and negative errors. Hence, as in Fig. 20.3b, we would arrive at an improved estimate of the integral.

Gauss quadrature is the name for a class of techniques to implement such a strategy. The particular Gauss quadrature formulas described in this section are called *Gauss-Legendre* formulas. Before describing the approach, we will show how numerical integration formulas such as the trapezoidal rule can be derived using the method of undetermined coefficients. This method will then be employed to develop the Gauss-Legendre formulas.

20.3.1 Method of Undetermined Coefficients

In Chap. 19, we derived the trapezoidal rule by integrating a linear interpolating polynomial and by geometrical reasoning. The method of undetermined coefficients offers a third approach that also has utility in deriving other integration techniques such as Gauss quadrature.

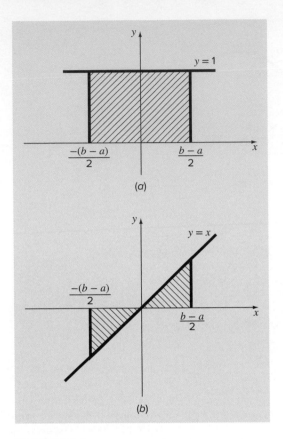

FIGURE 20.4
Two integrals that should be evaluated exactly by the trapezoidal rule: (a) a constant and (b) a straight line.

To illustrate the approach, Eq. (20.10) is expressed as

$$I \cong c_0 f(a) + c_1 f(b) \tag{20.11}$$

where the c's = constants. Now realize that the trapezoidal rule should yield exact results when the function being integrated is a constant or a straight line. Two simple equations that represent these cases are $y = 1$ and $y = x$ (Fig. 20.4). Thus, the following equalities should hold:

$$c_0 + c_1 = \int_{-(b-a)/2}^{(b-a)/2} 1 \, dx$$

and

$$-c_0 \frac{b-a}{2} + c_1 \frac{b-a}{2} = \int_{-(b-a)/2}^{(b-a)/2} x \, dx$$

or, evaluating the integrals,

$$c_0 + c_1 = b - a$$

and

$$-c_0 \frac{b-a}{2} + c_1 \frac{b-a}{2} = 0$$

These are two equations with two unknowns that can be solved for

$$c_0 = c_1 = \frac{b-a}{2}$$

which, when substituted back into Eq. (20.11), gives

$$I = \frac{b-a}{2} f(a) + \frac{b-a}{2} f(b)$$

which is equivalent to the trapezoidal rule.

20.3.2 Derivation of the Two-Point Gauss-Legendre Formula

Just as was the case for the previous derivation of the trapezoidal rule, the object of Gauss quadrature is to determine the coefficients of an equation of the form

$$I \cong c_0 f(x_0) + c_1 f(x_1) \tag{20.12}$$

where the c's = the unknown coefficients. However, in contrast to the trapezoidal rule that used fixed end points a and b, the function arguments x_0 and x_1 are not fixed at the end points, but are unknowns (Fig. 20.5). Thus, we now have a total of four unknowns that must be evaluated, and consequently, we require four conditions to determine them exactly.

Just as for the trapezoidal rule, we can obtain two of these conditions by assuming that Eq. (20.12) fits the integral of a constant and a linear function exactly. Then, to arrive at the other two conditions, we merely extend this reasoning by assuming that it also fits the integral of a parabolic ($y = x^2$) and a cubic ($y = x^3$) function. By doing this, we determine

FIGURE 20.5
Graphical depiction of the unknown variables x_0 and x_1 for integration by Gauss quadrature.

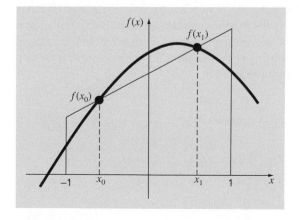

all four unknowns and in the bargain derive a linear two-point integration formula that is exact for cubics. The four equations to be solved are

$$c_0 + c_1 = \int_{-1}^{1} 1 \, dx = 2 \tag{20.13}$$

$$c_0 x_0 + c_1 x_1 = \int_{-1}^{1} x \, dx = 0 \tag{20.14}$$

$$c_0 x_0^2 + c_1 x_1^2 = \int_{-1}^{1} x^2 \, dx = \frac{2}{3} \tag{20.15}$$

$$c_0 x_0^3 + c_1 x_1^3 = \int_{-1}^{1} x^3 \, dx = 0 \tag{20.16}$$

Equations (20.13) through (20.16) can be solved simultaneously for the four unknowns. First, solve Eq. (20.14) for c_1 and substitute the result into Eq. (20.16), which can be solved for

$$x_0^2 = x_1^2$$

Since x_0 and x_1 cannot be equal, this means that $x_0 = -x_1$. Substituting this result into Eq. (20.14) yields $c_0 = c_1$. Consequently from Eq. (20.13) it follows that

$$c_0 = c_1 = 1$$

Substituting these results into Eq. (20.15) gives

$$x_0 = -\frac{1}{\sqrt{3}} = -0.5773503 \ldots$$

$$x_1 = \frac{1}{\sqrt{3}} = 0.5773503 \ldots$$

Therefore, the two-point Gauss-Legendre formula is

$$I = f\left(\frac{-1}{\sqrt{3}}\right) + f\left(\frac{1}{\sqrt{3}}\right) \tag{20.17}$$

Thus, we arrive at the interesting result that the simple addition of the function values at $x = -1/\sqrt{3}$ and $1/\sqrt{3}$ yields an integral estimate that is third-order accurate.

Notice that the integration limits in Eqs. (20.13) through (20.16) are from -1 to 1. This was done to simplify the mathematics and to make the formulation as general as possible. A simple change of variable can be used to translate other limits of integration into this form. This is accomplished by assuming that a new variable x_d is related to the original variable x in a linear fashion, as in

$$x = a_1 + a_2 x_d \tag{20.18}$$

If the lower limit, $x = a$, corresponds to $x_d = -1$, these values can be substituted into Eq. (20.18) to yield

$$a = a_1 + a_2(-1) \tag{20.19}$$

Similarly, the upper limit, $x = b$, corresponds to $x_d = 1$, to give

$$b = a_1 + a_2(1) \tag{20.20}$$

Equations (20.19) and (20.20) can be solved simultaneously for

$$a_1 = \frac{b+a}{2} \qquad \text{and} \qquad a_2 = \frac{b-a}{2} \tag{20.21}$$

which can be substituted into Eq. (20.18) to yield

$$x = \frac{(b+a) + (b-a)x_d}{2} \tag{20.22}$$

This equation can be differentiated to give

$$dx = \frac{b-a}{2} dx_d \tag{20.23}$$

Equations (20.22) and (20.23) can be substituted for x and dx, respectively, in the equation to be integrated. These substitutions effectively transform the integration interval without changing the value of the integral. The following example illustrates how this is done in practice.

EXAMPLE 20.3   Two-Point Gauss-Legendre Formula

Problem Statement. Use Eq. (20.17) to evaluate the integral of

$$f(x) = 0.2 + 25x - 200x^2 + 675x^3 - 900x^4 + 400x^5$$

between the limits $x = 0$ to 0.8. The exact value of the integral is 1.640533.

Solution. Before integrating the function, we must perform a change of variable so that the limits are from -1 to $+1$. To do this, we substitute $a = 0$ and $b = 0.8$ into Eqs. (20.22) and (20.23) to yield

$$x = 0.4 + 0.4x_d \qquad \text{and} \qquad dx = 0.4dx_d$$

Both of these can be substituted into the original equation to yield

$$\int_0^{0.8} (0.2 + 25x - 200x^2 + 675x^3 - 900x^4 + 400x^5) \, dx$$

$$= \int_{-1}^1 [0.2 + 25(0.4 + 0.4x_d) - 200(0.4 + 0.4x_d)^2 + 675(0.4 + 0.4x_d)^3$$

$$-900(0.4 + 0.4x_d)^4 + 400(0.4 + 0.4x_d)^5]0.4dx_d$$

Therefore, the right-hand side is in the form that is suitable for evaluation using Gauss quadrature. The transformed function can be evaluated at $x_d = -1/\sqrt{3}$ as 0.516741 and at $x_d = 1/\sqrt{3}$ as 1.305837. Therefore, the integral according to Eq. (20.17) is 0.516741 + 1.305837 = 1.822578, which represents a percent relative error of -11.1%. This result is comparable in magnitude to a four-segment application of the trapezoidal rule or a single application of Simpson's 1/3 and 3/8 rules. This latter result is to be expected because Simpson's rules are also third-order accurate. However, because of the clever choice of base points, Gauss quadrature attains this accuracy on the basis of only two function evaluations.

TABLE 20.1 Weighting factors and function arguments used in Gauss-Legendre formulas.

Points	Weighting Factors	Function Arguments	Truncation Error
1	$c_0 = 2$	$x_0 = 0.0$	$\cong f^{(2)}(\xi)$
2	$c_0 = 1$ $c_1 = 1$	$x_0 = -1/\sqrt{3}$ $x_1 = 1/\sqrt{3}$	$\cong f^{(4)}(\xi)$
3	$c_0 = 5/9$ $c_1 = 8/9$ $c_2 = 5/9$	$x_0 = -\sqrt{3/5}$ $x_1 = 0.0$ $x_2 = \sqrt{3/5}$	$\cong f^{(6)}(\xi)$
4	$c_0 = (18 - \sqrt{30})/36$ $c_1 = (18 + \sqrt{30})/36$ $c_2 = (18 + \sqrt{30})/36$ $c_3 = (18 - \sqrt{30})/36$	$x_0 = -\sqrt{525 + 70\sqrt{30}}/35$ $x_1 = -\sqrt{525 - 70\sqrt{30}}/35$ $x_2 = \sqrt{525 - 70\sqrt{30}}/35$ $x_3 = \sqrt{525 + 70\sqrt{30}}/35$	$\cong f^{(8)}(\xi)$
5	$c_0 = (322 - 13\sqrt{70})/900$ $c_1 = (322 + 13\sqrt{70})/900$ $c_2 = 128/225$ $c_3 = (322 + 13\sqrt{70})/900$ $c_4 = (322 - 13\sqrt{70})/900$	$x_0 = -\sqrt{245 + 14\sqrt{70}}/21$ $x_1 = -\sqrt{245 - 14\sqrt{70}}/21$ $x_2 = 0.0$ $x_3 = \sqrt{245 - 14\sqrt{70}}/21$ $x_4 = \sqrt{245 + 14\sqrt{70}}/21$	$\cong f^{(10)}(\xi)$
6	$c_0 = 0.171324492379170$ $c_1 = 0.360761573048139$ $c_2 = 0.467913934572691$ $c_3 = 0.467913934572691$ $c_4 = 0.360761573048131$ $c_5 = 0.171324492379170$	$x_0 = -0.932469514203152$ $x_1 = -0.661209386466265$ $x_2 = -0.238619186083197$ $x_3 = 0.238619186083197$ $x_4 = 0.661209386466265$ $x_5 = 0.932469514203152$	$\cong f^{(12)}(\xi)$

20.3.3 Higher-Point Formulas

Beyond the two-point formula described in the previous section, higher-point versions can be developed in the general form

$$I \cong c_0 f(x_0) + c_1 f(x_1) + \cdots + c_{n-1} f(x_{n-1}) \tag{20.24}$$

where n = the number of points. Values for c's and x's for up to and including the six-point formula are summarized in Table 20.1.

EXAMPLE 20.4   Three-Point Gauss-Legendre Formula

Problem Statement. Use the three-point formula from Table 20.1 to estimate the integral for the same function as in Example 20.3.

Solution. According to Table 20.1, the three-point formula is

$$I = 0.5555556 f(-0.7745967) + 0.8888889 f(0) + 0.5555556 f(0.7745967)$$

which is equal to

$$I = 0.2813013 + 0.8732444 + 0.4859876 = 1.640533$$

which is exact.

Because Gauss quadrature requires function evaluations at nonuniformly spaced points within the integration interval, it is not appropriate for cases where the function is unknown. Thus, it is not suited for engineering problems that deal with tabulated data. However, where the function is known, its efficiency can be a decided advantage. This is particularly true when numerous integral evaluations must be performed.

20.4 ADAPTIVE QUADRATURE

Although Romberg integration is more efficient than the composite Simpson's 1/3 rule, both use equally spaced points. This constraint does not take into account that some functions have regions of relatively abrupt changes where more refined spacing might be required. Hence, to achieve a desired accuracy, fine spacing must be applied everywhere even though it is only needed for the regions of sharp change. Adaptive quadrature methods remedy this situation by automatically adjusting the step size so that small steps are taken in regions of sharp variations and larger steps are taken where the function changes gradually.

20.4.1 MATLAB M-file: quadadapt

Adaptive quadrature methods accommodate the fact that many functions have regions of high variability along with other sections where change is gradual. They accomplish this by adjusting the step size so that small intervals are used in regions of rapid variations and larger intervals are used where the function changes gradually. Many of these techniques are based on applying the composite Simpson's 1/3 rule to subintervals in a fashion that is very similar to the way in which the composite trapezoidal rule was used in Richardson extrapolation. That is, the 1/3 rule is applied at two levels of refinement, and the difference between these two levels is used to estimate the truncation error. If the truncation error is acceptable, no further refinement is required, and the integral estimate for the subinterval is deemed acceptable. If the error estimate is too large, the step size is refined and the process repeated until the error falls to acceptable levels. The total integral is then computed as the summation of the integral estimates for the subintervals.

The theoretical basis of the approach can be illustrated for an interval $x = a$ to $x = b$ with a width of $h_1 = b - a$. A first estimate of the integral can be estimated with Simpson's 1/3 rule:

$$I(h_1) = \frac{h_1}{6} [f(a) + 4f(c) + f(b)] \tag{20.25}$$

where $c = (a + b)/2$.

As in Richardson extrapolation, a more refined estimate can be obtained by halving the step size. That is, by applying the composite Simpson's 1/3 rule with $n = 4$:

$$I(h_2) = \frac{h_2}{6} [f(a) + 4f(d) + 2f(c) + 4f(e) + f(b)] \tag{20.26}$$

where $d = (a + c)/2$, $e = (c + b)/2$, and $h_2 = h_1/2$.

Because both $I(h_1)$ and $I(h_2)$ are estimates of the same integral, their difference provides a measure of the error. That is,

$$E \cong I(h_2) - I(h_1) \tag{20.27}$$

In addition, the estimate and error associated with either application can be represented generally as

$$I = I(h) + E(h) \tag{20.28}$$

where I = the exact value of the integral, $I(h)$ = the approximation from an n-segment application of the Simpson's 1/3 rule with step size $h = (b - a)/n$, and $E(h)$ = the corresponding truncation error.

Using an approach similar to Richardson extrapolation, we can derive an estimate in the error of the more refined estimate $I(h_2)$ as a function of the difference between the two integral estimates:

$$E(h_2) = \frac{1}{15}[I(h_2) - I(h_1)] \tag{20.29}$$

The error can then be added to $I(h_2)$ to generate an even better estimate:

$$I = I(h_2) + \frac{1}{15}[I(h_2) - I(h_1)] \tag{20.30}$$

This result is equivalent to *Boole's rule* (Table 19.2).

The equations just developed can now be combined into an efficient algorithm. Figure 20.6 presents an M-file function that is based on an algorithm originally developed by Cleve Moler (2004).

The function consists of a main calling function quadadapt along with a recursive function qstep that actually performs the integration. The main calling function quadadapt is passed the function f and the integration limits a and b. After setting the tolerance, the function evaluations required for the initial application of Simpson's 1/3 rule [Eq. (20.25)] are computed. These values along with the integration limits are then passed to qstep. Within qstep, the remaining step sizes and function values are determined, and the two integral estimates [Eqs. (20.25) and (20.26)] are computed.

At this point, the error is estimated as the absolute difference between the integral estimates. Depending on the value of the error, two things can then happen:

1. If the error is less or equal to the tolerance (tol), Boole's rule is generated; the function terminates and passes back the result.
2. If the error is larger than the tolerance, qstep is invoked twice to evaluate each of the two subintervals of the current call.

The two recursive calls in the second step represent the real beauty of this algorithm. They just keep subdividing until the tolerance is met. Once this occurs, their results are passed back up the recursive path, combining with the other integral estimates along the way. The process ends when the final call is satisfied and the total integral is evaluated and returned to the main calling function.

It should be stressed that the algorithm in Fig. 20.6 is a stripped-down version of the integral function, which is the professional root-location function employed in MATLAB. Thus, it does not guard against failure such as cases where integrals do not exist. Nevertheless,

```
function q = quadadapt(f,a,b,tol,varargin)
% Evaluates definite integral of f(x) from a to b
if nargin < 4 | isempty(tol),tol = 1.e-6;end
c = (a + b)/2;
fa = feval(f,a,varargin{:});
fc = feval(f,c,varargin{:});
fb = feval(f,b,varargin{:});
q = quadstep(f, a, b, tol, fa, fc, fb, varargin{:});
end

function q = quadstep(f,a,b,tol,fa,fc,fb,varargin)
% Recursive subfunction used by quadadapt.
h = b - a; c = (a + b)/2;
fd = feval(f,(a+c)/2,varargin{:});
fe = feval(f,(c+b)/2,varargin{:});
q1 = h/6 * (fa + 4*fc + fb);
q2 = h/12 * (fa + 4*fd + 2*fc + 4*fe + fb);
if abs(q2 - q1) <= tol
  q  = q2 + (q2 - q1)/15;
else
  qa = quadstep(f, a, c, tol, fa, fd, fc, varargin{:});
  qb = quadstep(f, c, b, tol, fc, fe, fb, varargin{:});
  q  = qa + qb;
end
end
```

FIGURE 20.6
An M-file to implement an adaptive quadrature algorithm based on an algorithm originally developed by Cleve Moler (2004).

it works just fine for many applications, and certainly serves to illustrate how adaptive quadrature works. Here is a MATLAB session showing how quadadapt can be used to determine the integral of the polynomial from Example 20.1:

```
>> f=@(x) 0.2+25*x-200*x^2+675*x^3-900*x^4+400*x^5;
>> q = quadadapt(f,0,0.8)

q =
    1.640533333333336
```

20.4.2 MATLAB Function: integral

MATLAB has a function for implementing adaptive quadrature:

```
q = integral(fun, a, b)
```

where *fun* is the function to be integrated, and *a* and *b* = the integration bounds. It should be noted that array operators .*, ./ and .^ should be used in the definition of *fun*.

EXAMPLE 20.5   Adaptive Quadrature

Problem Statement. Use integral to integrate the following function:

$$f(x) = \frac{1}{(x-q)^2 + 0.01} + \frac{1}{(x-r)^2 + 0.04} - s$$

between the limits $x = 0$ to 1. Note that for $q = 0.3$, $r = 0.9$, and $s = 6$, this is the built-in humps function that MATLAB uses to demonstrate some of its numerical capabilities. The humps function exhibits both flat and steep regions over a relatively short x range. Hence, it is useful for demonstrating and testing a function like integral. Note that the humps function can be integrated analytically between the given limits to yield an exact integral of 29.85832539549867.

Solution. First, let's evaluate the integral using the built-in version of humps

```
>> format long
>> Q=integral(@(x) humps(x),0,1)

ans =
   29.85832612842764
```

Thus, the solution is correct to seven significant digits.

20.5 CASE STUDY   ROOT-MEAN-SQUARE CURRENT

Background. Because it results in efficient energy transmission, the current in an AC circuit is often in the form of a sine wave:

$$i = i_{peak} \sin(\omega t)$$

where i = the current (A = C/s), i_{peak} = the peak current (A), ω = the angular frequency (radians/s), and t = time (s). The angular frequency is related to the period $T(s)$ by $\omega = 2\pi/T$.

The power generated is related to the magnitude of the current. Integration can be used to determine the average current over one cycle:

$$\bar{i} = \frac{1}{T} \int_0^T i_{peak} \sin(\omega t) \, dt = \frac{i_{peak}}{T}(-\cos(2\pi) + \cos(0)) = 0$$

Despite the fact that the average is zero, such a current is capable of generating power. Therefore, an alternative to the average current must be derived.

To do this, electrical engineers and scientists determine the root mean square current i_{rms} (A), which is calculated as

$$i_{rms} = \sqrt{\frac{1}{T} \int_0^T i_{peak}^2 \sin^2(\omega t) \, dt} = \frac{i_{peak}}{\sqrt{2}} \tag{20.31}$$

Thus, as the name implies, the rms current is the square root of the mean of the squared current. Because $1/\sqrt{2} = 0.70707$, i_{rms} is equal to about 70% of the peak current for our assumed sinusoidal wave form.

This quantity has meaning because it is directly related to the average power absorbed by an element in an AC circuit. To understand this, recall that *Joule's law* states that the instantaneous power absorbed by a circuit element is equal to product of the voltage across it and the current through it:

$$P = iV \tag{20.32}$$

where P = the power (W = J/s), and V = voltage (V = J/C). For a resistor, *Ohm's law* states that the voltage is directly proportional to the current:

$$V = iR \tag{20.33}$$

where R = the resistance (Ω = V/A = J $\cdot$ s/C^2). Substituting Eq. (20.33) into (20.32) gives

$$P = i^2R \tag{20.34}$$

The average power can be determined by integrating Eq. (20.34) over a period with the result:

$$\bar{P} = i^2_{rms}\, R$$

Thus, the AC circuit generates the equivalent power as a DC circuit with a constant current of i_{rms}.

Now, although the simple sinusoid is widely employed, it is by no means the only waveform that is used. For some of these forms, such as triangular or square waves, the i_{rms} can be evaluated analytically with closed-form integration. However, some waveforms must be analyzed with numerical integration methods.

In this case study, we will calculate the root-mean-square current of a non sinusoidal wave form. We will use both the Newton-Cotes formulas from Chap. 19 as well as the approaches described in this chapter.

Solution. The integral that must be evaluated is

$$i^2_{rms} = \int_0^{1/2} (10e^{-t} \sin 2\pi t)^2 \, dt \tag{20.35}$$

For comparative purposes, the exact value of this integral to fifteen significant digits is 15.41260804810169.

Integral estimates for various applications of the trapezoidal rule and Simpson's 1/3 rule are listed in Table 20.2. Notice that Simpson's rule is more accurate than the trapezoidal rule. The value for the integral to seven significant digits is obtained using a 128-segment trapezoidal rule or a 32-segment Simpson's rule.

The M-file we developed in Fig. 20.2 can be used to evaluate the integral with Romberg integration:

```
>> format long
>> i2=@(t) (10*exp(-t).*sin(2*pi*t)).^2;
>> [q,ea,iter]=romberg(i2,0,.5)
```

20.5 CASE STUDY   continued

TABLE 20.2   Values for the integral calculated using Newton-Cotes formulas.

Technique	Segments	Integral	$\varepsilon_t(\%)$
Trapezoidal rule	1	0.0	100.0000
	2	15.163266493	1.6178
	4	15.401429095	0.0725
	8	15.411958360	4.22×10^{-3}
	16	15.412568151	2.59×10^{-4}
	32	15.412605565	1.61×10^{-5}
	64	15.412607893	1.01×10^{-6}
	128	15.412608038	6.28×10^{-8}
Simpson's 1/3 rule	2	20.217688657	31.1763
	4	15.480816629	0.4426
	8	15.415468115	0.0186
	16	15.412771415	1.06×10^{-3}
	32	15.412618037	6.48×10^{-5}

```
q =
   15.41260804288977
ea =
    1.480058787326946e-008
iter =
    5
```

Thus, with the default stopping criterion of es $= 1 \times 10^{-6}$, we obtain a result that is correct to over nine significant figures in five iterations. We can obtain an even better result if we impose a more stringent stopping criterion:

```
>> [q,ea,iter]=romberg(i2,0,.5,1e-15)

q =
   15.41260804810169
ea =
    0
iter =
    7
```

Gauss quadrature can also be used to make the same estimate. First, a change in variable is performed by applying Eqs. (20.22) and (20.23) to yield

$$t = \frac{1}{4} + \frac{1}{4} t_d \qquad dt = \frac{1}{4} dt_d$$

These relationships can be substituted into Eq. (20.35) to yield

$$i_{\text{rms}}^2 = \int_{-1}^{1} [10e^{-(0.25+0.25t_d)} \sin 2\pi(0.25 + 0.25t_d)]^2 \, 0.25 \, dt \qquad (20.36)$$

For the two-point Gauss-Legendre formula, this function is evaluated at $t_d = -1/\sqrt{3}$ and $1/\sqrt{3}$, with the results being 7.684096 and 4.313728, respectively. These values can be substituted into Eq. (20.17) to yield an integral estimate of 11.99782, which represents an error of $\varepsilon_t = 22.1\%$.

The three-point formula is (Table 20.1)

$$I = 0.5555556(1.237449) + 0.8888889(15.16327) + 0.5555556(2.684915) = 15.65755$$

which has $\varepsilon_t = 1.6\%$. The results of using the higher-point formulas are summarized in Table 20.3.

Finally, the integral can be evaluated with the built-in MATLAB function integral:

```
>> irms2=integral(i2,0,.5)

irms2 =
  15.412608049345090
```

We can now compute the i_{rms} by merely taking the square root of the integral. For example, using the result computed with integral, we get

```
>> irms=sqrt(irms2)

irms =
   3.925889459485796
```

This result could then be employed to guide other aspects of the design and operation of the circuit such as power dissipation computations.

As we did for the simple sinusoid in Eq. (20.31), an interesting calculation involves comparing this result with the peak current. Recognizing that this is an optimization problem, we can readily employ the fminbnd function to determine this value. Because we are looking for a maximum, we evaluate the negative of the function:

```
>> [tmax,imax]=fminbnd(@(t) -10*exp(-t).*sin(2*pi*t),0,.5)

tmax =
   0.22487940319321
imax =
  -7.886853873932577
```

A maximum current of 7.88685 A occurs at $t = 0.2249$ s. Hence, for this particular wave form, the root-mean-square value is about 49.8% of the maximum.

TABLE 20.3 Results of using various-point Gauss quadrature formulas to approximate the integral.

Points	Estimate	ε_t (%)
2	11.9978243	22.1
3	15.6575502	1.59
4	15.4058023	4.42×10^{-2}
5	15.4126391	2.01×10^{-4}
6	15.4126109	1.82×10^{-5}

PROBLEMS

20.1 Use Romberg integration to evaluate

$$I = \int_1^2 \left(x + \frac{1}{x}\right)^2 dx$$

to an accuracy of $\varepsilon_s = 0.5\%$. Your results should be presented in the format of Fig. 20.1. Use the analytical solution of the integral to determine the percent relative error of the result obtained with Romberg integration. Check that ε_t is less than ε_s.

20.2 Evaluate the following integral (a) analytically, (b) Romberg integration ($\varepsilon_s = 0.5\%$), (c) the three-point Gauss quadrature formula, and (d) MATLAB integral function:

$$I = \int_0^8 -0.055x^4 + 0.86x^3 - 4.2x^2 + 6.3x + 2 \; dx$$

20.3 Evaluate the following integral with (a) Romberg integration ($\varepsilon_s = 0.5\%$), (b) the two-point Gauss quadrature formula, and (c) MATLAB integral function:

$$I = \int_0^3 xe^{2x} dx$$

20.4 There is no closed form solution for the error function

$$\text{erf}(a) = \frac{2}{\sqrt{\pi}} \int_0^a e^{-x^2} dx$$

Use the (a) two-point and (b) three-point Gauss-Legendre formulas to estimate erf(1.5). Determine the percent relative error for each case based on the true value, which can be determined with MATLAB's built-in function erf.

20.5 The force on a sailboat mast can be represented by the following function:

$$F = \int_0^H 200 \left(\frac{z}{5+z}\right) e^{-2z/H} dz$$

where $z =$ the elevation above the deck and $H =$ the height of the mast. Compute F for the case where $H = 30$ using (a) Romberg integration to a tolerance of $\varepsilon_s = 0.5\%$, (b) the two-point Gauss-Legendre formula, and (c) the MATLAB integral function.

20.6 The root-mean-square current can be computed as

$$I_{RMS} = \sqrt{\frac{1}{T} \int_0^T i^2(t) \, dt}$$

For $T = 1$, suppose that $i(t)$ is defined as

$$i(t) = 8e^{-t/T} \sin\left(2\pi \frac{t}{T}\right) \qquad \text{for } 0 \le t \le T/2$$

$$i(t) = 0 \qquad \text{for } T/2 \le t \le T$$

Evaluate the I_{RMS} using (a) Romberg integration to a tolerance of 0.1%, (b) the two- and three-point Gauss-Legendre formulas, and (c) the MATLAB integral function.

20.7 The heat required, ΔH (cal), to induce a temperature change, $\Delta T(°C)$, of a material can be computed as

$$\Delta H = mC_p(T)\Delta T$$

where $m =$ mass (g), and $C_p(T) =$ heat capacity [cal/(g $\cdot$°C)]. The heat capacity increases with temperature, T (°C), according to

$$C_p(T) = 0.132 + 1.56 \times 10^{-4}T + 2.64 \times 10^{-7}T^2$$

Write a script that uses the integral function to generate a plot of ΔH versus ΔT for cases where $m = 1$ kg, the starting temperature is -100 °C, and ΔT ranges from 0 to 300 °C.

20.8 The amount of mass transported via a pipe over a period of time can be computed as

$$M = \int_{t_1}^{t_2} Q(t)c(t) \, dt$$

where $M =$ mass (mg), $t_1 =$ the initial time (min), $t_2 =$ the final time (min), $Q(t) =$ flow rate (m³/min), and $c(t) =$ concentration (mg/m³). The following functional representations define the temporal variations in flow and concentration:

$$Q(t) = 9 + 5 \cos^2(0.4t)$$

$$c(t) = 5e^{-0.5t} + 2e^{0.15t}$$

Determine the mass transported between $t_1 = 2$ and $t_2 = 8$ min with (a) Romberg integration to a tolerance of 0.1% and (b) the MATLAB integral function.

20.9 Evaluate the double integral

$$\int_{-2}^2 \int_0^4 (x^2 - 3y^2 + xy^3) \, dx \, dy$$

(a) analytically, and (b) with the integral2 function.

20.10 Compute work as described in Sec. 19.9, but use the following equations for $F(x)$ and $\theta(x)$:

$$F(x) = 1.6x - 0.045x^2$$

$$\theta(x) = -0.00055x^3 + 0.0123x^2 + 0.13x$$

The force is in newtons and the angle is in radians. Perform the integration from $x = 0$ to 30 m.

20.11 Perform the same computation as in Sec. 20.5, but for the current as specified by

$$i(t) = 6e^{-1.25t} \sin 2\pi t \qquad \text{for } 0 \le t \le T/2$$

$$i(t) = 0 \qquad \text{for } T/2 < t \le T$$

where $T = 1$ s.

20.12 Compute the power absorbed by an element in a circuit as described in Sec. 20.5, but for a simple sinusoidal current $i = \sin(2\pi t/T)$ where $T = 1$ s.

(a) Assume that Ohm's law holds and $R = 5 \, \Omega$.

(b) Assume that Ohm's law does not hold and that voltage and current are related by the following nonlinear relationship: $V = (5i - 1.25i^3)$.

20.13 Suppose that the current through a resistor is described by the function

$$i(t) = (60 - t)^2 + (60 - t) \sin(\sqrt{t})$$

and the resistance is a function of the current:

$$R = 10i + 2i^{2/3}$$

Compute the average voltage over $t = 0$ to 60 using the composite Simpson's 1/3 rule.

20.14 If a capacitor initially holds no charge, the voltage across it as a function of time can be computed as

$$V(t) = \frac{1}{C} \int_0^t i(t) \, dt$$

Use MATLAB to fit these data with a fifth-order polynomial. Then, use a numerical integration function along with a value of $C = 10^{-5}$ farad to generate a plot of voltage versus time.

t, s	0	0.2	0.4	0.6
i, 10^{-3} A	0.2	0.3683	0.3819	0.2282

t, s	0.8	1	1.2
i, 10^{-3} A	0.0486	0.0082	0.1441

20.15 The work done on an object is equal to the force times the distance moved in the direction of the force. The velocity of an object in the direction of a force is given by

$$\upsilon = 4t \qquad\qquad 0 \le t \le 5$$
$$\upsilon = 20 + (5 - t)^2 \qquad 5 \le t \le 15$$

where υ is in m/s. Determine the work if a constant force of 200 N is applied for all t.

20.16 A rod subject to an axial load (Fig. P20.16a) will be deformed, as shown in the stress-strain curve in Fig. P20.16b. The area under the curve from zero stress out to the point of rupture is called the *modulus of toughness* of the material. It provides a measure of the energy per unit volume required to cause the material to rupture. As such, it is representative of the material's ability to withstand an impact load. Use numerical integration to compute the modulus of toughness for the stress-strain curve seen in Fig. P20.16b.

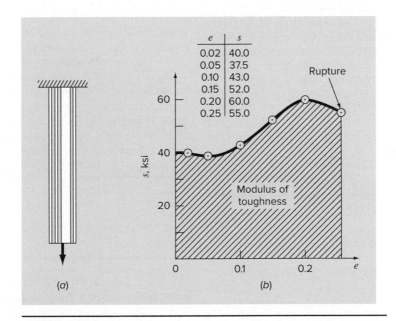

e	s
0.02	40.0
0.05	37.5
0.10	43.0
0.15	52.0
0.20	60.0
0.25	55.0

FIGURE P20.16
(*a*) A rod under axial loading and (*b*) the resulting stress-strain curve, where stress is in kips per square inch (10^3 lb/in^2), and strain is dimensionless.

FIGURE P20.17

20.17 If the velocity distribution of a fluid flowing through a pipe is known (Fig. P20.17), the flow rate Q (i.e., the volume of water passing through the pipe per unit time) can be computed by $Q = \int v \, dA$, where v is the velocity, and A is the pipe's cross-sectional area. (To grasp the meaning of this relationship physically, recall the close connection between summation and integration.) For a circular pipe, $A = \pi r^2$ and $dA = 2\pi r \, dr$. Therefore,

$$Q = \int_0^r v(2\pi r) \, dr$$

where r is the radial distance measured outward from the center of the pipe. If the velocity distribution is given by

$$v = 2\left(1 - \frac{r}{r_0}\right)^{1/6}$$

where r_0 is the total radius (in this case, 3 cm), compute Q using the composite trapezoidal rule. Discuss the results.

20.18 Using the following data, calculate the work done by stretching a spring that has a spring constant of $k = 300$ N/m to $x = 0.35$ m. To do this, first fit the data with a polynomial and then integrate the polynomial numerically to compute the work:

F, $10^3 \cdot$ N	0	0.01	0.028	0.046
x, m	0	0.05	0.10	0.15
F, $10^3 \cdot$ N	0.063	0.082	0.11	0.13
x, m	0.20	0.25	0.30	0.35

20.19 Evaluate the vertical distance traveled by a rocket if the vertical velocity is given by

$$v = 11t^2 - 5t \qquad 0 \le t \le 10$$
$$v = 1100 - 5t \qquad 10 \le t \le 20$$
$$v = 50t + 2(t - 20)^2 \qquad 20 \le t \le 30$$

20.20 The upward velocity of a rocket can be computed by the following formula:

$$v = u \ln\left(\frac{m_0}{m_0 - qt}\right) - gt$$

where v = upward velocity, u = velocity at which fuel is expelled relative to the rocket, m_0 = initial mass of the rocket at time $t = 0$, q = fuel consumption rate, and g = downward acceleration of gravity (assumed constant = 9.81 m/s²). If $u = 1850$ m/s, $m_0 = 160{,}000$ kg, and $q = 2500$ kg/s, determine how high the rocket will fly in 30 s.

20.21 The normal distribution is defined as

$$f(x) = \frac{1}{\sqrt{2\pi}} e^{-x^2/2}$$

(a) Use MATLAB to integrate this function from $x = -1$ to 1 and from -2 to 2.

(b) Use MATLAB to determine the inflection points of this function.

20.22 Use Romberg integration to evaluate

$$\int_0^2 \frac{e^x \sin x}{1 + x^2} \, dx$$

to an accuracy of $\varepsilon_s = 0.5\%$. Your results should be presented in the form of Fig. 20.1.

20.23 Recall that the velocity of the free-falling bungee jumper can be computed analytically as [Eq. (1.9)]:

$$v(t) = \sqrt{\frac{gm}{c_d}} \tanh\left(\sqrt{\frac{gc_d}{m}} \, t\right)$$

where $v(t)$ = velocity (m/s), t = time (s), g = 9.81 m/s², m = mass (kg), c_d = drag coefficient (kg/m).

(a) Use Romberg integration to compute how far the jumper travels during the first 8 seconds of free fall given $m = 80$ kg and $c_d = 0.2$ kg/m. Compute the answer to $\varepsilon_s = 1\%$.

(b) Perform the same computation with `integral`.

20.24 Prove that Eq. (20.30) is equivalent to Boole's rule.

20.25 As specified in the following table, the earth's density varies as a function of the distance from its center ($r = 0$):

r, km	0	1100	1500	2450	3400	3630	4500
ρ, g/cm³	13	12.4	12	11.2	9.7	5.7	5.2

r, km	5380	6060	6280	6380
ρ, g/cm³	4.7	3.6	3.4	3

Develop a script to fit these data with `interp1` using the `pchip` option. Generate a plot showing the resulting fit along with the data points. Then use one of MATLAB's integration functions to estimate the earth's mass (in metric tonnes) by integrating the output of the `interp1` function.

20.26 Develop an M-file function to implement Romberg integration based on Fig. 20.2. Test the function by using it to determine the integral of the polynomial from Example 20.1. Then use it to solve Prob. 20.1.

20.27 Develop an M-file function to implement adaptive quadrature based on Fig. 20.6. Test the function by using it to determine the integral of the polynomial from Example 20.1. Then use it to solve Prob. 20.20.

20.28 The average flow in a river channel, Q (m³/s), with an irregular cross-section can be computed as the integral of the product of velocity and depth

$$Q = \int_0^B U(y)H(y)\,dy$$

where $U(y)$ = water velocity (m/s) at distance y (m) from the bank, and $H(y)$ = water depth (m/s) at distance y from the bank. Use integral along with spline fits of U and H to the following data collected at different distances across the channel to estimate the flow.

y (m)	H (m)	Y (m)	U (m/s)
0	0	0	0
1.1	0.21	1.6	0.08
2.8	0.78	4.1	0.61
4.6	1.87	4.8	0.68
6	1.44	6.1	0.55
8.1	1.28	6.8	0.42
9	0.2	9	0

20.29 Use the two-point Gauss quadrature approach to estimate the average value of the following function between $a = 1$ and $b = 5$

$$f(x) = \frac{2}{1 + x^2}$$

20.30 Evaluate the following integral

$$I = \int_0^4 x^3 \, dx$$

(a) Analytically.
(b) Using the MATLAB integral function.
(c) Using Monte Carlo integration.

20.31 The MATLAB humps function defines a curve with has 2 maxima (peaks) of unequal height over the interval $0 \le x \le 2$. Develop a MATLAB script to determine the integral over the interval with (a) the MATLAB integral function and (b) Monte Carlo integration.

20.32 Evaluate the following double integral

$$I = \int_0^2 \int_{-3}^1 y^4 (x^2 + xy) \, dx\,dy$$

(a) Using a single application of Simpson's 1/3 rule across each dimension.
(b) Check your results with the integral2 function.

21

Numerical Differentiation

CHAPTER OBJECTIVES

The primary objective of this chapter is to introduce you to numerical differentiation. Specific objectives and topics covered are

- Understanding the application of high-accuracy numerical differentiation formulas for equispaced data.
- Knowing how to evaluate derivatives for unequally spaced data.
- Understanding how Richardson extrapolation is applied for numerical differentiation.
- Recognizing the sensitivity of numerical differentiation to data error.
- Knowing how to evaluate derivatives in MATLAB with the diff and gradient functions.
- Knowing how to generate contour plots and vector fields with MATLAB.

YOU'VE GOT A PROBLEM

R ecall that the velocity of a free-falling bungee jumper as a function of time can be computed as

$$v(t) = \sqrt{\frac{gm}{c_d}} \tanh\left(\sqrt{\frac{gc_d}{m}}\,t\right) \tag{21.1}$$

At the beginning of Chap. 19, we used calculus to integrate this equation to determine the vertical distance z the jumper has fallen after a time t.

$$z(t) = \frac{m}{c_d} \ln\left[\cosh\left(\sqrt{\frac{gc_d}{m}}\,t\right)\right] \tag{21.2}$$

Now suppose that you were given the reverse problem. That is, you were asked to determine velocity based on the jumper's position as a function of time. Because it is the inverse of integration, differentiation could be used to make the determination:

$$v(t) = \frac{dz(t)}{dt} \tag{21.3}$$

Substituting Eq. (21.2) into Eq. (21.3) and differentiating would bring us back to Eq. (21.1).

Beyond velocity, you might also be asked to compute the jumper's acceleration. To do this, we could either take the first derivative of velocity, or the second derivative of displacement:

$$a(t) = \frac{dv(t)}{dt} = \frac{d^2z(t)}{dt^2} \tag{21.4}$$

In either case, the result would be

$$a(t) = g \, \mathrm{sech}^2 \left(\sqrt{\frac{gc_d}{m}} t \right) \tag{21.5}$$

Although a closed-form solution can be developed for this case, there are other functions that may be difficult or impossible to differentiate analytically. Further, suppose that there was some way to measure the jumper's position at various times during the fall. These distances along with their associated times could be assembled as a table of discrete values. In this situation, it would be useful to differentiate the discrete data to determine the velocity and the acceleration. In both these instances, numerical differentiation methods are available to obtain solutions. This chapter will introduce you to some of these methods.

21.1   INTRODUCTION AND BACKGROUND

21.1.1 What Is Differentiation?

Calculus is the mathematics of change. Because engineers and scientists must continuously deal with systems and processes that change, calculus is an essential tool of our profession. Standing at the heart of calculus is the mathematical concept of differentiation.

According to the dictionary definition, to *differentiate* means "to mark off by differences; distinguish; . . . to perceive the difference in or between." Mathematically, the *derivative,* which serves as the fundamental vehicle for differentiation, represents the rate of change of a dependent variable with respect to an independent variable. As depicted in Fig. 21.1, the mathematical definition of the derivative begins with a difference approximation:

$$\frac{\Delta y}{\Delta x} = \frac{f(x_i + \Delta x) - f(x_i)}{\Delta x} \tag{21.6}$$

where y and $f(x)$ are alternative representatives for the dependent variable and x is the independent variable. If Δx is allowed to approach zero, as occurs in moving from Fig. 21.1a to c, the difference becomes a derivative:

$$\frac{dy}{dx} = \lim_{\Delta x \to 0} \frac{f(x_i + \Delta x) - f(x_i)}{\Delta x} \tag{21.7}$$

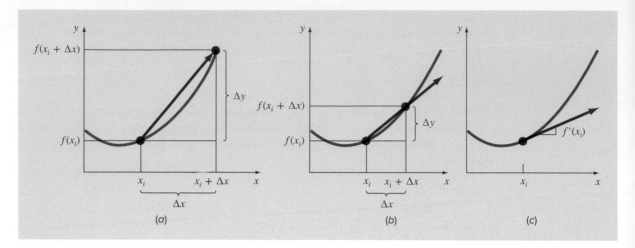

FIGURE 21.1
The graphical definition of a derivative: as Δx approaches zero in going from (a) to (c), the difference approximation becomes a derivative.

where dy/dx [which can also be designated as y' or $f'(x_i)$][1] is the first derivative of y with respect to x evaluated at x_i. As seen in the visual depiction of Fig. 21.1c, the derivative is the slope of the tangent to the curve at x_i.

The second derivative represents the derivative of the first derivative,

$$\frac{d^2 y}{dx^2} = \frac{d}{dx}\left(\frac{dy}{dx}\right)$$

(21.8)

Thus, the second derivative tells us how fast the slope is changing. It is commonly referred to as the *curvature,* because a high value for the second derivative means high curvature.

Finally, partial derivatives are used for functions that depend on more than one variable. Partial derivatives can be thought of as taking the derivative of the function at a point with all but one variable held constant. For example, given a function f that depends on both x and y, the partial derivative of f with respect to x at an arbitrary point (x, y) is defined as

$$\frac{\partial f}{\partial x} = \lim_{\Delta x \to 0} \frac{f(x + \Delta x, y) - f(x,y)}{\Delta x}$$

(21.9)

Similarly, the partial derivative of f with respect to y is defined as

$$\frac{\partial f}{\partial y} = \lim_{\Delta y \to 0} \frac{f(x, y + \Delta y) - f(x,y)}{\Delta y}$$

(21.10)

To get an intuitive grasp of partial derivatives, recognize that a function that depends on two variables is a surface rather than a curve. Suppose you are mountain climbing and have access to a function f that yields elevation as a function of longitude (the east-west oriented

[1] The form dy/dx was devised by Leibnitz, whereas y' is attributed to Lagrange. Note that Newton used the so-called dot notation: $\dot{y}$. Today, the dot notation is usually used for time derivatives.

x axis) and latitude (the north-south oriented y axis). If you stop at a particular point (x_0, y_0), the slope to the east would be $\partial f(x_0, y_0)/\partial x$, and the slope to the north would be $\partial f(x_0, y_0)/\partial y$.

21.1.2 Differentiation in Engineering and Science

The differentiation of a function has so many engineering and scientific applications that you were required to take differential calculus in your first year at college. Many specific examples of such applications could be given in all fields of engineering and science. Differentiation is commonplace in engineering and science because so much of our work involves characterizing the changes of variables in both time and space. In fact, many of the laws and other generalizations that figure so prominently in our work are based on the predictable ways in which change manifests itself in the physical world. A prime example is Newton's second law, which is not couched in terms of the position of an object but rather in its change with respect to time.

Aside from such temporal examples, numerous laws involving the spatial behavior of variables are expressed in terms of derivatives. Among the most common of these are the *constitutive laws* that define how potentials or gradients influence physical processes. For example, *Fourier's law of heat conduction* quantifies the observation that heat flows from regions of high to low temperature. For the one-dimensional case, this can be expressed mathematically as

$$q = -k \frac{dT}{dx} \qquad\qquad (21.11)$$

where $q(x)$ = heat flux (W/m^2), k = coefficient of thermal conductivity [W/(m · K)], T = temperature (K), and x = distance (m). Thus, the derivative, or *gradient,* provides a measure of the intensity of the spatial temperature change, which drives the transfer of heat (Fig. 21.2).

FIGURE 21.2
Graphical depiction of a temperature gradient. Because heat moves "downhill" from high to low temperature, the flow in (*a*) is from left to right. However, due to the orientation of Cartesian coordinates, the slope is negative for this case. Thus, a negative gradient leads to a positive flow. This is the origin of the minus sign in Fourier's law of heat conduction. The reverse case is depicted in (*b*), where the positive gradient leads to a negative heat flow from right to left.

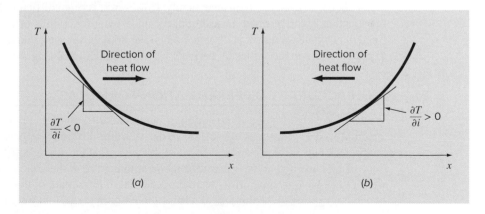

TABLE 21.1 The one-dimensional forms of some constitutive laws commonly used in engineering and science.

Law	Equation	Physical Area	Gradient	Flux	Proportionality
Fourier's law	$q = -k \dfrac{dT}{dx}$	Heat conduction	Temperature	Heat flux	Thermal Conductivity
Fick's law	$J = -D \dfrac{dc}{dx}$	Mass diffusion	Concentration	Mass flux	Diffusivity
Darcy's law	$q = -k \dfrac{dh}{dx}$	Flow through porous media	Head	Flow flux	Hydraulic Conductivity
Ohm's law	$J = -\sigma \dfrac{dV}{dx}$	Current flow	Voltage	Current flux	Electrical Conductivity
Newton's viscosity law	$\tau = \mu \dfrac{du}{dx}$	Fluids	Velocity	Shear Stress	Dynamic Viscosity
Hooke's law	$\sigma = E \dfrac{\Delta L}{L}$	Elasticity	Deformation	Stress	Young's Modulus

Similar laws provide workable models in many other areas of engineering and science, including the modeling of fluid dynamics, mass transfer, chemical reaction kinetics, electricity, and solid mechanics (Table 21.1). The ability to accurately estimate derivatives is an important facet of our capability to work effectively in these areas.

Beyond direct engineering and scientific applications, numerical differentiation is also important in a variety of general mathematical contexts including other areas of numerical methods. For example, recall that in Chap. 6 the secant method was based on a finite-difference approximation of the derivative. In addition, probably the most important application of numerical differentiation involves the solution of differential equations. We have already seen an example in the form of Euler's method in Chap. 1. In Chap. 24, we will investigate how numerical differentiation provides the basis for solving boundary-value problems of ordinary differential equations.

These are just a few of the applications of differentiation that you might face regularly in the pursuit of your profession. When the functions to be analyzed are simple, you will normally choose to evaluate them analytically. However, it is often difficult or impossible when the function is complicated. In addition, the underlying function is often unknown and defined only by measurement at discrete points. For both these cases, you must have the ability to obtain approximate values for derivatives, using numerical techniques as described next.

21.2 HIGH-ACCURACY DIFFERENTIATION FORMULAS

We have already introduced the notion of numerical differentiation in Chap. 4. Recall that we employed Taylor series expansions to derive finite-difference approximations of derivatives. In Chap. 4, we developed forward, backward, and centered difference approximations of first and higher derivatives. Remember that, at best, these estimates had errors that were $O(h^2)$—that is, their errors were proportional to the square of the step size. This level of accuracy is due to the number of terms of the Taylor series that were retained during the derivation of these formulas. We will now illustrate how high-accuracy finite-difference formulas can be generated by including additional terms from the Taylor series expansion.

For example, the forward Taylor series expansion can be written as [recall Eq. (4.13)]

$$f(x_{i+1}) = f(x_i) + f'(x_i)h + \frac{f''(x_i)}{2!}h^2 + \cdots \tag{21.12}$$

which can be solved for

$$f'(x_i) = \frac{f(x_{i+1}) - f(x_i)}{h} - \frac{f''(x_i)}{2!}h + O(h^2) \tag{21.13}$$

In Chap. 4, we truncated this result by excluding the second- and higher-derivative terms and were thus left with a forward-difference formula:

$$f'(x_i) = \frac{f(x_{i+1}) - f(x_i)}{h} + O(h) \tag{21.14}$$

In contrast to this approach, we now retain the second-derivative term by substituting the following forward-difference approximation of the second derivative [recall Eq. (4.27)]:

$$f''(x_i) = \frac{f(x_{i+2}) - 2f(x_{i+1}) + f(x_i)}{h^2} + O(h) \tag{21.15}$$

into Eq. (21.13) to yield

$$f'(x_i) = \frac{f(x_{i+1}) - f(x_i)}{h} - \frac{f(x_{i+2}) - 2f(x_{i+1}) + f(x_i)}{2h^2}h + O(h^2) \tag{21.16}$$

or, by collecting terms:

$$f'(x_i) = \frac{-f(x_{i+2}) + 4f(x_{i+1}) - 3f(x_i)}{2h} + O(h^2) \tag{21.17}$$

Notice that inclusion of the second-derivative term has improved the accuracy to $O(h^2)$. Similar improved versions can be developed for the backward and centered formulas as well as for the approximations of higher-order derivatives. The formulas are summarized in Fig. 21.3 through Fig. 21.5 along with the lower-order versions from Chap. 4. The following example illustrates the utility of these formulas for estimating derivatives.

EXAMPLE 21.1   High-Accuracy Differentiation Formulas

Problem Statement.   Recall that in Example 4.4 we estimated the derivative of

$$f(x) = -0.1x^4 - 0.15x^3 - 0.5x^2 - 0.25x + 1.2$$

at $x = 0.5$ using finite-differences and a step size of $h = 0.25$. The results are summarized in the following table. Note that the errors are based on the true value of $f'(0.5) = -0.9125$.

	Backward $O(h)$	Centered $O(h^2)$	Forward $O(h)$
Estimate	−0.714	−0.934	−1.155
ε_t	21.7%	−2.4%	−26.5%

Repeat this computation, but employ the high-accuracy formulas from Fig. 21.3 through Fig. 21.5.

First Derivative	Error
$f'(x_i) = \dfrac{f(x_{i+1}) - f(x_i)}{h}$	$O(h)$
$f'(x_i) = \dfrac{-f(x_{i+2}) + 4f(x_{i+1}) - 3f(x_i)}{2h}$	$O(h^2)$

Second Derivative	
$f''(x_i) = \dfrac{f(x_{i+2}) - 2f(x_{i+1}) + f(x_i)}{h^2}$	$O(h)$
$f''(x_i) = \dfrac{-f(x_{i+3}) + 4f(x_{i+2}) - 5f(x_{i+1}) + 2f(x_i)}{h^2}$	$O(h^2)$

Third Derivative	
$f'''(x_i) = \dfrac{f(x_{i+3}) - 3f(x_{i+2}) + 3f(x_{i+1}) - f(x_i)}{h^3}$	$O(h)$
$f'''(x_i) = \dfrac{-3f(x_{i+4}) + 14f(x_{i+3}) - 24f(x_{i+2}) + 18f(x_{i+1}) - 5f(x_i)}{2h^3}$	$O(h^2)$

Fourth Derivative	
$f''''(x_i) = \dfrac{f(x_{i+4}) - 4f(x_{i+3}) + 6f(x_{i+2}) - 4f(x_{i+1}) + f(x_i)}{h^4}$	$O(h)$
$f''''(x_i) = \dfrac{-2f(x_{i+5}) + 11f(x_{i+4}) - 24f(x_{i+3}) + 26f(x_{i+2}) - 14f(x_{i+1}) + 3f(x_i)}{h^4}$	$O(h^2)$

FIGURE 21.3
Forward finite-difference formulas: two versions are presented for each derivative. The latter version incorporates more terms of the Taylor series expansion and is, consequently, more accurate.

Solution. The data needed for this example are

$$x_{i-2} = 0 \qquad f(x_{i-2}) = 1.2$$
$$x_{i-1} = 0.25 \qquad f(x_{i-1}) = 1.1035156$$
$$x_i = 0.5 \qquad f(x_i) = 0.925$$
$$x_{i+1} = 0.75 \qquad f(x_{i+1}) = 0.6363281$$
$$x_{i+2} = 1 \qquad f(x_{i+2}) = 0.2$$

The forward difference of accuracy $O(h^2)$ is computed as (Fig. 21.3)

$$f'(0.5) = \frac{-0.2 + 4(0.6363281) - 3(0.925)}{2(0.25)} = -0.859375 \qquad \varepsilon_t = 5.82\%$$

The backward difference of accuracy $O(h^2)$ is computed as (Fig. 21.4)

$$f'(0.5) = \frac{3(0.925) - 4(1.1035156) + 1.2}{2(0.25)} = -0.878125 \qquad \varepsilon_t = 3.77\%$$

The centered difference of accuracy $O(h^4)$ is computed as (Fig. 21.5)

$$f'(0.5) = \frac{-0.2 + 8(0.6363281) - 8(1.1035156) + 1.2}{12(0.25)} = -0.9125 \qquad \varepsilon_t = 0\%$$

First Derivative | Error

$$f'(x_i) = \frac{f(x_i) - f(x_{i-1})}{h}$$ $O(h)$

$$f'(x_i) = \frac{3f(x_i) - 4f(x_{i-1}) + f(x_{i-2})}{2h}$$ $O(h^2)$

Second Derivative

$$f''(x_i) = \frac{f(x_i) - 2f(x_{i-1}) + f(x_{i-2})}{h^2}$$ $O(h)$

$$f''(x_i) = \frac{2f(x_i) - 5f(x_{i-1}) + 4f(x_{i-2}) - f(x_{i-3})}{h^2}$$ $O(h^2)$

Third Derivative

$$f'''(x_i) = \frac{f(x_i) - 3f(x_{i-1}) + 3f(x_{i-2}) - f(x_{i-3})}{h^3}$$ $O(h)$

$$f'''(x_i) = \frac{5f(x_i) - 18f(x_{i-1}) + 24f(x_{i-2}) - 14f(x_{i-3}) + 3f(x_{i-4})}{2h^3}$$ $O(h^2)$

Fourth Derivative

$$f''''(x_i) = \frac{f(x_i) - 4f(x_{i-1}) + 6f(x_{i-2}) - 4f(x_{i-3}) + f(x_{i-4})}{h^4}$$ $O(h)$

$$f''''(x_i) = \frac{3f(x_i) - 14f(x_{i-1}) + 26f(x_{i-2}) - 24f(x_{i-3}) + 11f(x_{i-4}) - 2f(x_{i-5})}{h^4}$$ $O(h^2)$

FIGURE 21.4
Backward finite-difference formulas: two versions are presented for each derivative. The latter version incorporates more terms of the Taylor series expansion and is, consequently, more accurate.

As expected, the errors for the forward and backward differences are considerably more accurate than the results from Example 4.4. However, surprisingly, the centered difference yields the exact derivative at $x = 0.5$. This is because the formula based on the Taylor series is equivalent to passing a fourth-order polynomial through the data points.

21.3 RICHARDSON EXTRAPOLATION

To this point, we have seen that there are two ways to improve derivative estimates when employing finite differences: (1) decrease the step size or (2) use a higher-order formula that employs more points. A third approach, based on Richardson extrapolation, uses two derivative estimates to compute a third, more accurate, approximation.

Recall from Sec. 20.2.1 that Richardson extrapolation provided a means to obtain an improved integral estimate by the formula [Eq. (20.4)]

$$I = I(h_2) + \frac{1}{(h_1/h_2)^2 - 1}[I(h_2) - I(h_1)] \qquad (21.18)$$

where $I(h_1)$ and $I(h_2)$ are integral estimates using two step sizes: h_1 and h_2. Because of its convenience when expressed as a computer algorithm, this formula is usually written for the case where $h_2 = h_1/2$, as in

$$I = \frac{4}{3}I(h_2) - \frac{1}{3}I(h_1) \qquad (21.19)$$

First Derivative	Error
$f'(x_i) = \dfrac{f(x_{i+1}) - f(x_{i-1})}{2h}$	$O(h^2)$
$f'(x_i) = \dfrac{-f(x_{i+2}) + 8f(x_{i+1}) - 8f(x_{i-1}) + f(x_{i-2})}{12h}$	$O(h^4)$
Second Derivative	
$f''(x_i) = \dfrac{f(x_{i+1}) - 2f(x_i) + f(x_{i-1})}{h^2}$	$O(h^2)$
$f''(x_i) = \dfrac{-f(x_{i+2}) + 16f(x_{i+1}) - 30f(x_i) + 16f(x_{i-1}) - f(x_{i-2})}{12h^2}$	$O(h^4)$
Third Derivative	
$f'''(x_i) = \dfrac{f(x_{i+2}) - 2f(x_{i+1}) + 2f(x_{i-1}) - f(x_{i-2})}{2h^3}$	$O(h^2)$
$f'''(x_i) = \dfrac{-f(x_{i+3}) + 8f(x_{i+2}) - 13f(x_{i+1}) + 13f(x_{i-1}) - 8f(x_{i-2}) + f(x_{i-3})}{8h^3}$	$O(h^4)$
Fourth Derivative	
$f''''(x_i) = \dfrac{f(x_{i+2}) - 4f(x_{i+1}) + 6f(x_i) - 4f(x_{i-1}) + f(x_{i-2})}{h^4}$	$O(h^2)$
$f''''(x_i) = \dfrac{-f(x_{i+3}) + 12f(x_{i+2}) - 39f(x_{i+1}) + 56f(x_i) - 39f(x_{i-1}) + 12f(x_{i-2}) - f(x_{i-3})}{6h^4}$	$O(h^4)$

FIGURE 21.5
Centered finite-difference formulas: two versions are presented for each derivative. The latter version incorporates more terms of the Taylor series expansion and is, consequently, more accurate.

In a similar fashion, Eq. (21.19) can be written for derivatives as

$$D = \frac{4}{3} D(h_2) - \frac{1}{3} D(h_1) \tag{21.20}$$

For centered difference approximations with $O(h^2)$, the application of this formula will yield a new derivative estimate of $O(h^4)$.

EXAMPLE 21.2   Richardson Extrapolation

Problem Statement. Using the same function as in Example 21.1, estimate the first derivative at $x = 0.5$ employing step sizes of $h_1 = 0.5$ and $h_2 = 0.25$. Then use Eq. (21.20) to compute an improved estimate with Richardson extrapolation. Recall that the true value is -0.9125.

Solution. The first-derivative estimates can be computed with centered differences as

$$D(0.5) = \frac{0.2 - 1.2}{1} = -1.0 \qquad \varepsilon_t = -9.6\%$$

and

$$D(0.25) = \frac{0.6363281 - 1.103516}{0.5} = -0.934375 \qquad \varepsilon_t = -2.4\%$$

The improved estimate can be determined by applying Eq. (21.20) to give

$$D = \frac{4}{3}(-0.934375) - \frac{1}{3}(-1) = -0.9125$$

which for the present case is exact.

The previous example yielded an exact result because the function being analyzed was a fourth-order polynomial. The exact outcome was due to the fact that Richardson extrapolation is actually equivalent to fitting a higher-order polynomial through the data and then evaluating the derivatives by centered divided differences. Thus, the present case matched the derivative of the fourth-order polynomial precisely. For most other functions, of course, this would not occur, and our derivative estimate would be improved but not exact. Consequently, as was the case for the application of Richardson extrapolation, the approach can be applied iteratively using a Romberg algorithm until the result falls below an acceptable error criterion.

21.4   DERIVATIVES OF UNEQUALLY SPACED DATA

The approaches discussed to this point are primarily designed to determine the derivative of a given function. For the finite-difference approximations of Sec. 21.2, the data had to be evenly spaced. For the Richardson extrapolation technique of Sec. 21.3, the data also had to be evenly spaced and generated for successively halved intervals. Such control of data spacing is usually available only in cases where we can use a function to generate a table of values.

In contrast, empirically derived information—that is, data from experiments or field studies—are often collected at unequal intervals. Such information cannot be analyzed with the techniques discussed to this point.

One way to handle nonequispaced data is to fit a Lagrange interpolating polynomial [recall Eq. (17.21)] to a set of adjacent points that bracket the location value at which you want to evaluate the derivative. Remember that this polynomial does not require that the points be equispaced. The polynomial can then be differentiated analytically to yield a formula that can be used to estimate the derivative.

For example, you can fit a second-order Lagrange polynomial to three adjacent points (x_0, y_0), (x_1, y_1), and (x_2, y_2). Differentiating the polynomial yields:

$$f'(x) = f(x_0)\frac{2x - x_1 - x_2}{(x_0 - x_1)(x_0 - x_2)} + f(x_1)\frac{2x - x_0 - x_2}{(x_1 - x_0)(x_1 - x_2)}$$

$$+ f(x_2)\frac{2x - x_0 - x_1}{(x_2 - x_0)(x_2 - x_1)} \tag{21.21}$$

where x is the value at which you want to estimate the derivative. Although this equation is certainly more complicated than the first-derivative approximation from Fig. 21.3 through Fig. 21.5, it has some important advantages. First, it can provide estimates anywhere within the range prescribed by the three points. Second, the points themselves do not have to be equally spaced. Third, the derivative estimate is of the same accuracy as the centered difference [Eq. (4.25)]. In fact, for equispaced points, Eq. (21.21) evaluated at $x = x_1$ reduces to Eq. (4.25).

EXAMPLE 21.3   Differentiating Unequally Spaced Data

Problem Statement. As in Fig. 21.6, a temperature gradient can be measured down into the soil. The heat flux at the soil-air interface can be computed with Fourier's law (Table 21.1):

$$q(z = 0) = -k \left.\frac{dT}{dz}\right|_{z=0}$$

where $q(z)$ = heat flux (W/m²), k = coefficient of thermal conductivity for soil [= 0.5 W/ (m · K)], T = temperature (K), and z = distance measured down from the surface into the soil (m). Note that a positive value for flux means that heat is transferred from the air to the soil. Use numerical differentiation to evaluate the gradient at the soil-air interface and employ this estimate to determine the heat flux into the ground.

Solution. Equation (21.21) can be used to calculate the derivative at the air-soil interface as

$$f'(0) = 13.5 \frac{2(0) - 0.0125 - 0.0375}{(0 - 0.0125)(0 - 0.0375)} + 12 \frac{2(0) - 0 - 0.0375}{(0.0125 - 0)(0.0125 - 0.0375)}$$

$$+ 10 \frac{2(0) - 0 - 0.0125}{(0.0375 - 0)(0.0375 - 0.0125)}$$

$$= -1440 + 1440 - 133.333 = -133.333 \text{ K/m}$$

which can be used to compute

$$q(z = 0) = -0.5 \frac{\text{W}}{\text{m K}} \left(-133.333 \frac{\text{K}}{\text{m}}\right) = 66.667 \frac{\text{W}}{\text{m}^2}$$

FIGURE 21.6
Temperature versus depth into the soil.

21.5   DERIVATIVES AND INTEGRALS FOR DATA WITH ERRORS

Aside from unequal spacing, another problem related to differentiating empirical data is that these data usually include measurement error. A shortcoming of numerical differentiation is that it tends to amplify errors in the data.

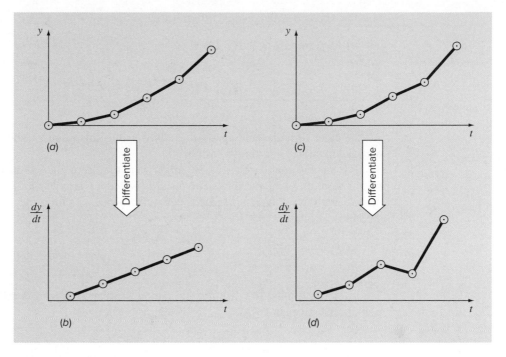

FIGURE 21.7
Illustration of how small data errors are amplified by numerical differentiation: (a) data with no error, (b) the resulting numerical differentiation of curve (a), (c) data modified slightly, and (d) the resulting differentiation of curve (c) manifesting increased variability. In contrast, the reverse operation of integration [moving from (d) to (c) by taking the area under (d)] tends to attenuate or smooth data errors.

Fig. 21.7a shows smooth, error-free data that when numerically differentiated yield a smooth result (Fig. 21.7b). In contrast, Fig. 21.7c uses the same data, but with alternating points raised and lowered slightly. This minor modification is barely apparent from Fig. 21.7c. However, the resulting effect in Fig. 21.7d is significant.

The error amplification occurs because differentiation is subtractive. Hence, random positive and negative errors tend to add. In contrast, the fact that integration is a summing process makes it very forgiving with regard to uncertain data. In essence, as points are summed to form an integral, random positive and negative errors cancel out.

As might be expected, the primary approach for determining derivatives for imprecise data is to use least-squares regression to fit a smooth, differentiable function to the data. In the absence of any other information, a lower-order polynomial regression might be a good first choice. Obviously, if the true functional relationship between the dependent and independent variable is known, this relationship should form the basis for the least-squares fit.

21.6 PARTIAL DERIVATIVES

Partial derivatives along a single dimension are computed in the same fashion as ordinary derivatives. For example, suppose that we want to determine to partial derivatives for a two-dimensional function $f(x, y)$. For equally spaced data, the partial first derivatives can be approximated with centered differences:

$$\frac{\partial f}{\partial x} = \frac{f(x + \Delta x, y) - f(x - \Delta x, y)}{2\Delta x} \tag{21.22}$$

$$\frac{\partial f}{\partial y} = \frac{f(x, y + \Delta y) - f(x, y - \Delta y)}{2\Delta y} \tag{21.23}$$

All the other formulas and approaches discussed to this point can be applied to evaluate partial derivatives in a similar fashion.

For higher-order derivatives, we might want to differentiate a function with respect to two or more different variables. The result is called a *mixed partial derivative*. For example, we might want to take the partial derivative of $f(x, y)$ with respect to both independent variables

$$\frac{\partial^2 f}{\partial x \partial y} = \frac{\partial}{\partial x}\left(\frac{\partial f}{\partial y}\right) \tag{21.24}$$

To develop a finite-difference approximation, we can first form a difference in x of the partial derivatives in y:

$$\frac{\partial^2 f}{\partial x \partial y} = \frac{\frac{\partial f}{\partial y}(x + \Delta x, y) - \frac{\partial f}{\partial y}(x - \Delta x, y)}{2\Delta x} \tag{21.25}$$

Then, we can use finite differences to evaluate each of the partials in y:

$$\frac{\partial^2 f}{\partial x \partial y} = \frac{\dfrac{f(x + \Delta x, y + \Delta y) - f(x + \Delta x, y - \Delta y)}{2\Delta y} - \dfrac{f(x - \Delta x, y + \Delta y) - f(x - \Delta x, y - \Delta y)}{2\Delta y}}{2\Delta x} \tag{21.26}$$

Collecting terms yields the final result

$$\frac{\partial^2 f}{\partial x \partial y} = \frac{f(x + \Delta x, y + \Delta y) - f(x + \Delta x, y - \Delta y) - f(x - \Delta x, y + \Delta y) + f(x - \Delta x, y - \Delta y)}{4\Delta x \Delta y} \tag{21.27}$$

21.7 NUMERICAL DIFFERENTIATION WITH MATLAB

MATLAB software has the ability to determine the derivatives of data based on two built-in functions: `diff` and `gradient`.

21.7.1 MATLAB Function: `diff`

When it is passed a one-dimensional vector of length n, the `diff` function returns a vector of length $n - 1$ containing the differences between adjacent elements. As described in the following example, these can then be employed to determine finite-difference approximations of first derivatives.

EXAMPLE 21.4   Using diff for Differentiation

Problem Statement. Explore how the MATLAB diff function can be employed to differentiate the function

$$f(x) = 0.2 + 25x - 200x^2 + 675x^3 - 900x^4 + 400x^5$$

from $x = 0$ to 0.8. Compare your results with the exact solution:

$$f'(x) = 25 - 400x^2 + 2025x^2 - 3600x^3 + 2000x^4$$

Solution. We can first express $f(x)$ as an anonymous function:

```
>> f=@(x) 0.2+25*x-200*x.^2+675*x.^3-900*x.^4+400*x.^5;
```

We can then generate a series of equally spaced values of the independent and dependent variables:

```
>> x=0:0.1:0.8;
>> y=f(x);
```

The diff function can be used to determine the differences between adjacent elements of each vector. For example,

```
>> diff(x)

ans =
  Columns 1 through 5
    0.1000    0.1000    0.1000    0.1000    0.1000
  Columns 6 through 8
    0.1000    0.1000    0.1000
```

As expected, the result represents the differences between each pair of elements of x. To compute divided-difference approximations of the derivative, we merely perform a vector division of the y differences by the x differences by entering

```
>> d=diff(y)./diff(x)

d =
  Columns 1 through 5
    10.8900   -0.0100    3.1900    8.4900    8.6900
  Columns 6 through 8
    1.3900  -11.0100  -21.3100
```

Note that because we are using equally spaced values, after generating the x values, we could have simply performed the above computation concisely as

```
>> d=diff(f(x))/0.1;
```

   The vector d now contains derivative estimates corresponding to the midpoint between adjacent elements. Therefore, in order to develop a plot of our results, we must first generate a vector holding the x values for the midpoint of each interval:

```
>> n=length(x);
>> xm=(x(1:n-1)+x(2:n))./2;
```

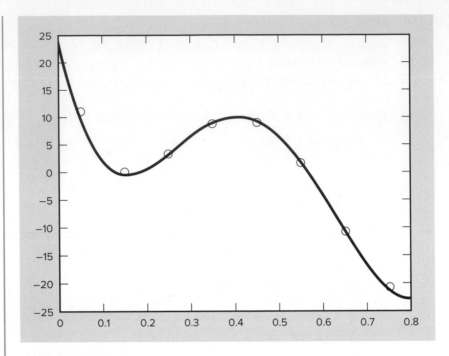

FIGURE 21.8
Comparison of the exact derivative (line) with numerical estimates (circles) computed with MATLAB's `diff` function.

As a final step, we can compute values for the analytical derivative at a finer level of resolution to include on the plot for comparison.

```
>> xa=0:.01:.8;
>> ya=25-400*xa+3*675*xa.^2-4*900*xa.^3+5*400*xa.^4;
```

A plot of the numerical and analytical estimates can be generated with

```
>> plot(xm,d,'o',xa,ya)
```

As displayed in Fig. 21.8, the results compare favorably for this case.

Note that aside from evaluating derivatives, the `diff` function comes in handy as a programming tool for testing certain characteristics of vectors. For example, the following statement displays an error message and terminates an M-file if it determines that a vector x has unequal spacing:

```
if any(diff(diff(x))~=0), error('unequal spacing'), end
```

Another common use is to detect whether a vector is in ascending or descending order. For example, the following code rejects a vector that is not in ascending order (i.e., monotonically increasing):

```
if any(diff(x)<=0), error('not in ascending order'), end
```

21.7.2 MATLAB Function: gradient

The gradient function also returns differences. However, it does so in a manner that is more compatible with evaluating derivatives at the values themselves rather than in the intervals between values. A simple representation of its syntax is

>> fx = gradient(f)

where f = a one-dimensional vector of length n, and fx is a vector of length n containing differences based on f. Just as with the diff function, the first value returned is the difference between the first and second value. However, for the intermediate values, a centered difference based on the adjacent values is returned

$$diff_i = \frac{f_{i+1} - f_{i-1}}{2} \tag{21.28}$$

The last value is then computed as the difference between the final two values. Hence, the results are akin to using centered differences for all the intermediate values, with forward and backward differences at the ends.

Note that the spacing between points is assumed to be one. If the vector represents equally spaced data, the following version divides all the results by the interval and hence returns the actual values of the derivatives,

>> fx = gradient(f, h)

where h = the spacing between points.

EXAMPLE 21.5   Using gradient for Differentiation

Problem Statement. Use the gradient function to differentiate the same function that we analyzed in Example 21.4 with the diff function.

Solution. In the same fashion as Example 21.4, we can generate a series of equally spaced values of the independent and dependent variables:

```
>> f=@(x) 0.2+25*x-200*x.^2+675*x.^3-900*x.^4+400*x.^5;
>> x=0:0.1:0.8;
>> y=f(x);
```

We can then use the gradient function to determine the derivatives as

```
>> dy=gradient(y,0.1)

dy =
  Columns 1 through 5
    10.8900     5.4400     1.5900     5.8400     8.5900
  Columns 6 through 9
     5.0400    -4.8100   -16.1600   -21.3100
```

As in Example 21.4, we can generate values for the analytical derivative and display both the numerical and analytical estimates on a plot:

```
>> xa=0:.01:.8;
>> ya=25-400*xa+3*675*xa.^2-4*900*xa.^3+5*400*xa.^4;
>> plot(x,dy,'o', xa,ya)
```

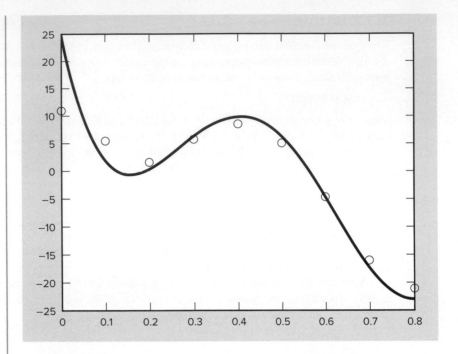

FIGURE 21.9
Comparison of the exact derivative (line) with numerical estimates (circles) computed with
MATLAB's gradient function.

As displayed in Fig. 21.9, the results are not as accurate as those obtained with the diff
function in Example 21.4. This is due to the fact that gradient employs intervals that are
two times (0.2) as wide as for those used for diff (0.1).

Beyond one-dimensional vectors, the gradient function is particularly well suited for
determining the partial derivatives of matrices. For example, for a two-dimensional matrix,
f, the function can be invoked as

    [fx, fy] = gradient(f, h)

where fx corresponds to the differences in the x (column) direction, and fy corresponds
to the differences in the y (row) direction, and h = the spacing between points. If h is
omitted, the spacing between points in both dimensions is assumed to be one. In the next
section, we will illustrate how gradient can be used to visualize vector fields.

21.8 CASE STUDY   VISUALIZING FIELDS

Background. Beyond the determination of derivatives in one dimension, the gradient function is also quite useful for determining partial derivatives in two or more dimensions. In particular, it can be used in conjunction with other MATLAB functions to produce visualizations of vector fields.

To understand how this is done, we can return to our discussion of partial derivatives at the end of Sec. 21.1.1. Recall that we used mountain elevation as an example of a two-dimensional function. We can represent such a function mathematically as

$$z = f(x, y)$$

where z = elevation, x = distance measured along the east-west axis, and y = distance measured along the north-south axis.

For this example, the partial derivatives provide the slopes in the directions of the axes. However, if you were mountain climbing, you would probably be much more interested in determining the direction of the maximum slope. If we think of the two partial derivatives as component vectors, the answer is provided very neatly by

$$\nabla f = \frac{\partial f}{\partial x} \boldsymbol{i} + \frac{\partial f}{\partial y} \boldsymbol{j}$$

where ∇f is referred to as the *gradient* of f. This vector, which represents the steepest slope, has a magnitude

$$\sqrt{\left(\frac{\partial f}{\partial x}\right)^2 + \left(\frac{\partial f}{\partial y}\right)^2}$$

and a direction

$$\theta = \tan^{-1} = \left(\frac{\partial f/\partial y}{\partial f/\partial x}\right)$$

where θ = the angle measured counterclockwise from the x axis.

Now suppose that we generate a grid of points in the x-y plane and used the foregoing equations to draw the gradient vector at each point. The result would be a field of arrows indicating the steepest route to the peak from any point. Conversely, if we plotted the negative of the gradient, it would indicate how a ball would travel as it rolled downhill from any point.

Such graphical representations are so useful that MATLAB has a special function, called quiver, to create such plots. A simple representation of its syntax is

```
quiver(x,y,u,v)
```

where x and y are matrices containing the position coordinates and u and v are matrices containing the partial derivatives. The following example demonstrates the use of quiver to visualize a field.

Employ the gradient function to determine to partial derivatives for the following two-dimensional function:

$$f(x, y) = y - x - 2x^2 - 2xy - y^2$$

21.8 CASE STUDY   continued

from $x = -2$ to 2 and $y = 1$ to 3. Then use quiver to superimpose a vector field on a contour plot of the function.

Solution. We can first express $f(x, y)$ as an anonymous function

```
>> f=@(x,y) y-x-2.*x.^2-2.*x.*y-y.^2;
```

A series of equally spaced values of the independent and dependent variables can be generated as

```
>> [x,y]=meshgrid(-2:.25:0, 1:.25:3);
>> z=f(x,y);
```

The gradient function can be employed to determine the partial derivatives:

```
>> [fx,fy]=gradient(z,0.25);
```

We can then develop a contour plot of the results:

```
>> cs=contour(x,y,z);clabel(cs);hold on
```

As a final step, the resultant of the partial derivatives can be superimposed as vectors on the contour plot:

```
>> quiver(x,y,-fx,-fy);hold off
```

FIGURE 21.10
MATLAB generated contour plot of a two-dimensional function with the resultant of the partial derivatives displayed as arrows.

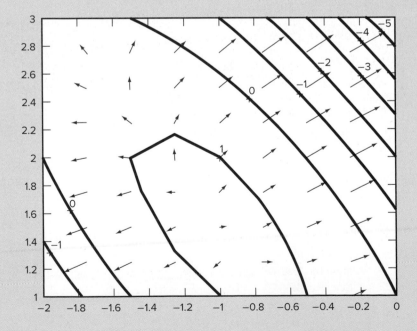

Note that we have displayed the negative of the resultants, in order that they point "downhill."

The result is shown in Fig. 21.10. The function's peak occurs at $x = -1$ and $y = 1.5$ and then drops away in all directions. As indicated by the lengthening arrows, the gradient drops off more steeply to the northeast and the southwest.

PROBLEMS

21.1 Compute forward and backward difference approximations of $O(h)$ and $O(h^2)$, and central difference approximations of $O(h^2)$ and $O(h^4)$ for the first derivative of $y = \sin x$ at $x = \pi/4$ using a value of $h = \pi/12$. Estimate the true percent relative error ε_t for each approximation.

21.2 Use centered difference approximations to estimate the first and second derivatives of $y = e^x$ at $x = 2$ for $h = 0.1$. Employ both $O(h^2)$ and $O(h^4)$ formulas for your estimates.

21.3 Use a Taylor series expansion to derive a centered finite-difference approximation to the third derivative that is second-order accurate. To do this, you will have to use four different expansions for the points x_{i-2}, x_{i-1}, x_{i+1}, and x_{i+2}. In each case, the expansion will be around the point x_i. The interval Δx will be used in each case of $i - 1$ and $i + 1$, and $2\Delta x$ will be used in each case of $i - 2$ and $i + 2$. The four equations must then be combined in a way to eliminate the first and second derivatives. Carry enough terms along in each expansion to evaluate the first term that will be truncated to determine the order of the approximation.

21.4 Use Richardson extrapolation to estimate the first derivative of $y = \cos x$ at $x = \pi/4$ using step sizes of $h_1 = \pi/3$ and $h_2 = \pi/6$. Employ centered differences of $O(h^2)$ for the initial estimates.

21.5 Repeat Prob. 21.4, but for the first derivative of $\ln x$ at $x = 5$ using $h_1 = 2$ and $h_2 = 1$.

21.6 Employ Eq. (21.21) to determine the first derivative of $y = 2x^4 - 6x^3 - 12x - 8$ at $x = 0$ based on values at $x_0 = -0.5$, $x_1 = 1$, and $x_2 = 2$. Compare this result with the true value and with an estimate obtained using a centered difference approximation based on $h = 1$.

21.7 Prove that for equispaced data points, Eq. (21.21) reduces to Eq. (4.25) at $x = x_1$.

21.8 Develop an M-file to apply a Romberg algorithm to estimate the derivative of a given function.

21.9 Develop an M-file to obtain first-derivative estimates for unequally spaced data. Test it with the following data:

x	0.6	1.5	1.6	2.5	3.5
$f(x)$	0.9036	0.3734	0.3261	0.08422	0.01596

where $f(x) = 5e^{-2x} x$. Compare your results with the true derivatives.

21.10 Develop an M-file function that computes first and second derivative estimates of order $O(h^2)$ based on the formulas in Fig. 21.3 through Fig. 21.5. The function's first line should be set up as

```
function [dydx, d2ydx2] = diffeq(x,y)
```

where x and y are input vectors of length n containing the values of the independent and dependent variables, respectively, and dydx and dy2dx2 are output vectors of length n containing the first- and second-derivative estimates at each value of the independent variable. The function should generate a plot of dydx and dy2dx2 versus x. Have your M-file return an error message if **(a)** the input vectors are not the same length or **(b)** the values for the independent variable are not equally spaced. Test your program with the data from Prob. 21.11.

21.11 The following data were collected for the distance traveled versus time for a rocket:

t, s	0	25	50	75	100	125
y, km	0	32	58	78	92	100

Use numerical differentiation to estimate the rocket's velocity and acceleration at each time.

21.12 A jet fighter's position on an aircraft carrier's runway was timed during landing:

t, s	0	0.52	1.04	1.75	2.37	3.25	3.83
x, m	153	185	208	249	261	271	273

where x is the distance from the end of the carrier. Estimate **(a)** velocity (dx/dt) and **(b)** acceleration (dv/dt) using numerical differentiation.

21.13 Use the following data to find the velocity and acceleration at $t = 10$ seconds:

Time, t, s	0	2	4	6	8	10	12	14	16
Position, x, m	0	0.7	1.8	3.4	5.1	6.3	7.3	8.0	8.4

Use second-order correct **(a)** centered finite-difference, **(b)** forward finite-difference, and **(c)** backward finite-difference methods.

21.14 A plane is being tracked by radar, and data are taken every second in polar coordinates θ and r.

t, s	200	202	204	206	208	210
θ, (rad)	0.75	0.72	0.70	0.68	0.67	0.66
r, m	5120	5370	5560	5800	6030	6240

At 206 seconds, use the centered finite-difference (second-order correct) to find the vector expressions for velocity $\vec{v}$ and acceleration $\vec{a}$. The velocity and acceleration given in polar coordinates are

$$\vec{v} = \dot{r}\vec{e}_r + r\dot{\theta}\vec{e}_\theta \quad \text{and} \quad \vec{a} = (\ddot{r} - r\dot{\theta}^2)\vec{e}_r + (r\ddot{\theta} + 2\dot{r}\dot{\theta})\vec{e}_\theta$$

21.15 Use regression to estimate the acceleration at each time for the following data with second-, third-, and fourth-order polynomials. Plot the results:

t	1	2	3.25	4.5	6	7	8	8.5	9.3	10
v	10	12	11	14	17	16	12	14	14	10

21.16 The normal distribution is defined as

$$f(x) = \frac{1}{\sqrt{2\pi}} e^{-x^2/2}$$

Use MATLAB to determine the inflection points of this function.

21.17 The following data were generated from the normal distribution:

x	-2	-1.5	-1	-0.5	0
$f(x)$	0.05399	0.12952	0.24197	0.35207	0.39894

x	0.5	1	1.5	2
$f(x)$	0.35207	0.24197	0.12952	0.05399

Use MATLAB to estimate the inflection points of these data.

21.18 Use the diff(y) command to develop a MATLAB M-file function to compute finite-difference approximations to the first and second derivative at each x value in the table below. Use finite-difference approximations that are second-order correct, $O(x^2)$:

x	0	1	2	3	4	5	6	7	8	9	10
y	1.4	2.1	3.3	4.8	6.8	6.6	8.6	7.5	8.9	10.9	10

21.19 The objective of this problem is to compare second-order accurate forward, backward, and centered finite-difference approximations of the first derivative of a function to the actual value of the derivative. This will be done for

$$f(x) = e^{-2x} - x$$

(a) Use calculus to determine the correct value of the derivative at $x = 2$.
(b) Develop an M-file function to evaluate the centered finite-difference approximations, starting with $x = 0.5$. Thus, for the first evaluation, the x values for the centered difference approximation will be $x = 2 \pm 0.5$ or $x = 1.5$ and 2.5. Then, decrease in increments of 0.1 down to a minimum value of $\Delta x = 0.01$.
(c) Repeat part **(b)** for the second-order forward and backward differences. (Note that these can be done at the same time that the centered difference is computed in the loop.)
(d) Plot the results of **(b)** and **(c)** versus x. Include the exact result on the plot for comparison.

21.20 You have to measure the flow rate of water through a small pipe. In order to do it, you place a bucket at the pipe's outlet and measure the volume in the bucket as a function of time as tabulated below. Estimate the flow rate at $t = 7$ s.

Time, s	0	1	5	8
Volume, cm³	0	1	8	16.4

21.21 The velocity v (m/s) of air flowing past a flat surface is measured at several distances y (m) away from the surface.

Use *Newton's viscosity law* to determine the shear stress τ (N/m^2) at the surface ($y = 0$),

$$\tau = \mu \frac{du}{dy}$$

Assume a value of dynamic viscosity $\mu = 1.8 \times 10^{-5}$ N·s/m^2.

y, m	0	0.002	0.006	0.012	0.018	0.024
u, m/s	0	0.287	0.899	1.915	3.048	4.299

21.22 *Fick's first diffusion law* states that

$$\text{Mass flux} = -D \frac{dc}{dx} \tag{P21.22}$$

where mass flux = the quantity of mass that passes across a unit area per unit time (g/cm^2/s), D = a diffusion coefficient (cm^2/s), c = concentration (g/cm^3), and x = distance (cm). An environmental engineer measures the following concentration of a pollutant in the pore waters of sediments underlying a lake ($x = 0$ at the sediment-water interface and increases downward):

x, cm	0	1	3
c, 10^{-6} g/cm^3	0.06	0.32	0.6

Use the best numerical differentiation technique available to estimate the derivative at $x = 0$. Employ this estimate in conjunction with Eq. (P21.22) to compute the mass flux of pollutant out of the sediments and into the overlying waters ($D = 1.52 \times 10^{-6}$ cm^2/s). For a lake with 3.6×10^6 m^2 of sediments, how much pollutant would be transported into the lake over a year's time?

21.23 The following data were collected when a large oil tanker was loading:

t, min	0	10	20	30	45	60	75
V, 10^6 barrels	0.4	0.7	0.77	0.88	1.05	1.17	1.35

Calculate the flow rate Q (i.e., dV/dt) for each time to the order of h^2.

21.24 *Fourier's law* is used routinely by architectural engineers to determine heat flow through walls. The following temperatures are measured from the surface ($x = 0$) into a stone wall:

x, m	0	0.08	0.16
T, °C	20.2	17	15

If the flux at $x = 0$ is 60 W/m^2, compute k.

21.25 The horizontal surface area A_s (m^2) of a lake at a particular depth can be computed from volume by differentiation:

$$A_s(z) = -\frac{dV}{dz}(z)$$

where V = volume (m^3) and z = depth (m) as measured from the surface down to the bottom. The average concentration of a substance that varies with depth, $\bar{c}$ (g/m^3), can be computed by integration:

$$\bar{c} = \frac{\int_0^Z c(z) A_s(z)\, dz}{\int_0^Z A_s(z)\, dz}$$

where Z = the total depth (m). Determine the average concentration based on the following data:

z, m	0	4	8	12	16
V, 10^6 m^3	9.8175	5.1051	1.9635	0.3927	0.0000
c, g/m^3	10.2	8.5	7.4	5.2	4.1

21.26 *Faraday's law* characterizes the voltage drop across an inductor as

$$V_L = L \frac{di}{dt}$$

where V_L = voltage drop (V), L = inductance (in henrys; 1 H = 1 V·s/A), i = current (A), and t = time (s). Determine the voltage drop as a function of time from the following data for an inductance of 4 H.

t	0	0.1	0.2	0.3	0.5	0.7
i	0	0.16	0.32	0.56	0.84	2.0

21.27 Based on Faraday's law (Prob. 21.26), use the following voltage data to estimate the inductance if a current of 2 A is passed through the inductor over 400 milliseconds.

t, ms	0	10	20	40	60	80	120	180	280	400
V, volts	0	18	29	44	49	46	35	26	15	7

21.28 The rate of cooling of a body (Fig. P21.28) can be expressed as

$$\frac{dT}{dt} = -k(T - T_a)$$

where T = temperature of the body (°C), T_a = temperature of the surrounding medium (°C), and k = a proportionality constant (per minute). Thus, this equation (called *Newton's law of cooling*) specifies that the rate of cooling is proportional to

FIGURE P21.28

the difference in the temperatures of the body and of the surrounding medium. If a metal ball heated to 80 °C is dropped into water that is held constant at $T_a = 20$ °C, the temperature of the ball changes, as in

Time, min	0	5	10	15	20	25
T, °C	80	44.5	30.0	24.1	21.7	20.7

Utilize numerical differentiation to determine dT/dt at each value of time. Plot dT/dt versus $T - T_a$ and employ linear regression to evaluate k.

21.29 The enthalpy of a real gas is a function of pressure as described below. The data were taken for a real fluid. Estimate the enthalpy of the fluid at 400 K and 50 atm (evaluate the integral from 0.1 to 50 atm).

$$H = \int_0^P \left(V - T \left(\frac{\partial V}{\partial T} \right)_P \right) dP$$

P, atm	$T = 350$ K	$T = 400$ K	$T = 450$ K
	V, L		
0.1	220	250	282.5
5	4.1	4.7	5.23
10	2.2	2.5	2.7
20	1.35	1.49	1.55
25	1.1	1.2	1.24
30	0.90	0.99	1.03
40	0.68	0.75	0.78
45	0.61	0.675	0.7
50	0.54	0.6	0.62

21.30 For fluid flow over a surface, the heat flux to the surface can be computed with Fourier's law: $y =$ distance normal to the surface (m). The following measurements are made for air flowing over a flat plate where $y =$ distance normal to the surface:

y, cm	0	1	3	5
T, K	900	480	270	210

If the plate's dimensions are 200 cm long and 50 cm wide, and $k = 0.028$ J/(s·m·K), **(a)** determine the flux at the surface and **(b)** the heat transfer in watts. Note that 1 J = 1 W·s.

21.31 The pressure gradient for laminar flow through a constant radius tube is given by

$$\frac{dp}{dx} = -\frac{8\mu Q}{\pi r^4}$$

where $p =$ pressure (N/m²), $x =$ distance along the tube's centerline (m), $\mu =$ dynamic viscosity (N·s/m²), $Q =$ flow (m³/s), and $r =$ radius (m).

(a) Determine the pressure drop for a 10-cm length tube for a viscous liquid ($\mu = 0.005$ N·s/m², density $= \rho = 1 \times 10^3$ kg/m³) with a flow of 10×10^{-6} m³/s and the following varying radii along its length:

x, cm	0	2	4	5	6	7	10
r, mm	2	1.35	1.34	1.6	1.58	1.42	2

(b) Compare your result with the pressure drop that would have occurred if the tube had a constant radius equal to the average radius.

(c) Determine the average Reynolds number for the tube to verify that flow is truly laminar (Re $= \rho v D/\mu < 2100$ where $v =$ velocity).

21.32 The following data for the specific heat of benzene were generated with an nth-order polynomial. Use numerical differentiation to determine n.

T, K	300	400	500	600
C_p, kJ/(kmol·K)	82.888	112.136	136.933	157.744

T, K	700	800	900	1000
C_p, kJ/(kmol·K)	175.036	189.273	200.923	210.450

21.33 The specific heat at constant pressure c_p [J/(kg·K)] of an ideal gas is related to enthalpy by

$$c_p = \frac{dh}{dT}$$

where $h =$ enthalpy (kJ/kg) and $T =$ absolute temperature (K). The following enthalpies are provided for carbon dioxide (CO_2) at several temperatures. Use these values to determine the specific heat in J/(kg · K) for each of the tabulated temperatures. Note that the atomic weights of carbon and oxygen are 12.011 and 15.9994 g/mol, respectively

T, K	750	800	900	1000
h, kJ/kmol	29,629	32,179	37,405	42,769

21.34 An nth-order rate law is often used to model chemical reactions that solely depend on the concentration of a single reactant:

$$\frac{dc}{dt} = -kc^n$$

where c = concentration (mole), t = time (min), n = reaction order (dimensionless), and k = reaction rate ($\text{min}^{-1}\ \text{mole}^{1-n}$). The *differential method* can be used to evaluate the parameters k and n. This involves applying a logarithmic transform to the rate law to yield,

$$\log\left(-\frac{dc}{dt}\right) = \log k + n \log c$$

Therefore, if the nth-order rate law holds, a plot of the $\log(-dc/dt)$ versus $\log c$ should yield a straight line with a slope of n and an intercept of $\log k$. Use the differential method and linear regression to determine k and n for the following data for the conversion of ammonium cyanate to urea:

t, min	0	5	15	30	45
c, mole	0.750	0.594	0.420	0.291	0.223

21.35 The sediment oxygen demand [SOD in units of g/(m^2 · d)] is an important parameter in determining the dissolved oxygen content of a natural water. It is measured by placing a sediment core in a cylindrical container (Fig. P21.35). After carefully introducing a layer of distilled, oxygenated water above the sediments, the container is covered to prevent gas transfer. A stirrer is used to mix the water gently, and an oxygen probe tracks how the water's oxygen concentration decreases over time. The SOD can then be computed as

$$SOD = -H\frac{do}{dt}$$

where H = the depth of water (m), o = oxygen concentration (g/m^3), and t = time (d).

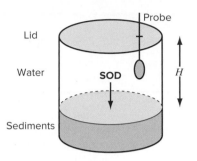

FIGURE P21.35

Based on the following data and $H = 0.1$ m, use numerical differentiation to generate plots of **(a)** SOD versus time and **(b)** SOD versus oxygen concentration:

t, d	0	0.125	0.25	0.375	0.5	0.625	0.75
o, mg/L	10	7.11	4.59	2.57	1.15	0.33	0.03

21.36 The following relationships can be used to analyze uniform beams subject to distributed loads:

$$\frac{dy}{dx} = \theta(x) \quad \frac{d\theta}{dx} = \frac{M(x)}{EI} \quad \frac{dM}{dx} = V(x) \quad \frac{dV}{dx} = -w(x)$$

where x = distance along beam (m), y = deflection (m), $\theta(x)$ = slope (m/m), E = modulus of elasticity (Pa = N/m^2), I = moment of inertia (m^4), $M(x)$ = moment (N m), $V(x)$ = shear (N), and $w(x)$ = distributed load (N/m). For the case of a linearly increasing load (recall Fig. P5.15), the slope can be computed analytically as

$$\theta(x) = \frac{w_0}{120EIL}(-5x^4 + 6L^2x^2 - L^4) \tag{P21.36}$$

Employ **(a)** numerical integration to compute the deflection (in m) and **(b)** numerical differentiation to compute the moment (in N m) and shear (in N). Base your numerical calculations on values of the slope computed with Eq. (P21.36) at equally spaced intervals of $\Delta x = 0.125$ m along a 3-m beam. Use the following parameter values in your computation: $E = 200$ GPa, $I = 0.0003\ \text{m}^4$, and $w_0 = 2.5$ kN/cm. In addition, the deflections at the ends of the beam are set at $y(0) = y(L) = 0$. Be careful of units.

21.37 You measure the following deflections along the length of a simply supported uniform beam (see Prob. 21.36)

x, m	0	0.375	0.75	1.125	1.5
y, cm	0	-0.2571	-0.9484	-1.9689	-3.2262

x, m	1.875	2.25	2.625	3
y, cm	-4.6414	-6.1503	-7.7051	-9.275

Employ numerical differentiation to compute the slope, the moment (in N m), the shear (in N), and the distributed load (in N/m). Use the following parameter values in your computation: $E = 200$ GPa and $I = 0.0003$ m^4.

21.38 Evaluate $\partial f/\partial x$, $\partial f/\partial y$ and $\partial^2 f/(\partial x \partial y)$ for the following function at $x = y = 1$ (a) analytically and (b) numerically $\Delta x = \Delta y = 0.0001$:

$$f(x, y) = 3xy + 3x - x^3 - 3y^3$$

21.39 Develop a script to generate the same computations and plots as in Sec. 21.8, but for the following functions (for $x = -3$ to 3 and $y = -3$ to 3): **(a)** $f(x, y) = e^{-(x^2+y^2)}$ and **(b)** $f(x, y) = xe^{-(x^2+y^2)}$.

21.40 Develop a script to generate the same computations and plots as in Sec. 21.8, but for the MATLAB peaks function over ranges of both x and y from -3 to 3.

21.41 The velocity (m/s) of an object at time t seconds is given by

$$v = \frac{2t}{\sqrt{1 + t^2}}$$

Using Richardson's extrapolation, find the acceleration of the particle at time $t = 5$ s using $h = 0.5$ and 0.25. Employ the exact solution to compute the true percent relative error of each estimate.

PART SIX

Ordinary Differential Equations

6.1 OVERVIEW

The fundamental laws of physics, mechanics, electricity, and thermodynamics are usually based on empirical observations that explain variations in physical properties and states of systems. Rather than describing the state of physical systems directly, the laws are usually couched in terms of spatial and temporal changes. These laws define mechanisms of change. When combined with continuity laws for energy, mass, or momentum, differential equations result. Subsequent integration of these differential equations results in mathematical functions that describe the spatial and temporal state of a system in terms of energy, mass, or velocity variations. As in Fig. PT6.1, the integration can be implemented analytically with calculus or numerically with the computer.

The free-falling bungee jumper problem introduced in Chap. 1 is an example of the derivation of a differential equation from a fundamental law. Recall that Newton's second law was used to develop an ODE describing the rate of change of velocity of a falling bungee jumper:

$$\frac{dv}{dt} = g - \frac{c_d}{m} v^2 \qquad \text{(PT6.1)}$$

where g is the gravitational constant, m is the mass, and c_d is a drag coefficient. Such equations, which are composed of an unknown function and its derivatives, are called *differential equations*. They are sometimes referred to as *rate equations* because they express the rate of change of a variable as a function of variables and parameters.

In Eq. (PT6.1), the quantity being differentiated v is called the *dependent variable*. The quantity with respect to which v is differentiated t is called the *independent variable*. When the function

573

FIGURE PT6.1
The sequence of events in the development and solution of ODEs for engineering and science. The example shown is for the velocity of the free-falling bungee jumper.

involves one independent variable, the equation is called an *ordinary differential equation* (or *ODE*). This is in contrast to a *partial differential equation* (or *PDE*) that involves two or more independent variables.

Differential equations are also classified as to their *order*. For example, Eq. (PT6.1) is called a *first-order equation* because the highest derivative is a first derivative. A *second-order equation* would include a second derivative. For example, the equation describing the position x of an unforced mass-spring system with damping is the second-order equation:

$$m\frac{d^2x}{dt^2} + c\frac{dx}{dt} + kx = 0 \tag{PT6.2}$$

where m is mass, c is a damping coefficient, and k is a spring constant. Similarly, an *n*th-order equation would include an *n*th derivative.

Higher-order differential equations can be reduced to a system of first-order equations. This is accomplished by defining the first derivative of the dependent variable as a new variable. For Eq. (PT6.2), this is done by creating a new variable v as the first derivative of displacement

$$v = \frac{dx}{dt} \tag{PT6.3}$$

where v is velocity. This equation can itself be differentiated to yield

$$\frac{dv}{dt} = \frac{d^2x}{dt^2} \tag{PT6.4}$$

Equations (PT6.3) and (PT6.4) can be substituted into Eq. (PT6.2) to convert it into a first-order equation:

$$m\frac{dv}{dt} + cv + kx = 0 \tag{PT6.5}$$

As a final step, we can express Eqs. (PT6.3) and (PT6.5) as rate equations:

$$\frac{dx}{dt} = v \tag{PT6.6}$$

$$\frac{dv}{dt} = -\frac{c}{m}v - \frac{k}{m}x \tag{PT6.7}$$

Thus, Eqs. (PT6.6) and (PT6.7) are a pair of first-order equations that are equivalent to the original second-order equation [Eq. (PT6.2)]. Because other nth-order differential equations can be similarly reduced, this part of our book focuses on the solution of first-order equations.

A solution of an ordinary differential equation is a specific function of the independent variable and parameters that satisfies the original differential equation. To illustrate this concept, let us start with a simple fourth-order polynomial,

$$y = -0.5x^4 + 4x^3 - 10x^2 + 8.5x + 1 \tag{PT6.8}$$

Now, if we differentiate Eq. (PT6.8), we obtain an ODE:

$$\frac{dy}{dx} = -2x^3 + 12x^2 - 20x + 8.5 \tag{PT6.9}$$

This equation also describes the behavior of the polynomial, but in a manner different from Eq. (PT6.8). Rather than explicitly representing the values of y for each value of x, Eq. (PT6.9) gives the rate of change of y with respect to x (i.e., the slope) at every value of x. Figure PT6.2 shows both the function and the derivative plotted versus x. Notice how the zero values of the derivatives correspond to the point at which the original function is flat—that is, where it has a zero slope. Also, the maximum absolute values of the derivatives are at the ends of the interval where the slopes of the function are greatest.

Although, as just demonstrated, we can determine a differential equation given the original function, the object here is to determine the original function given the differential equation. The original function then represents the solution.

Without computers, ODEs are usually solved analytically with calculus. For example, Eq. (PT6.9) could be multiplied by dx and integrated to yield

$$y = \int (-2x^3 + 12x^2 - 20x + 8.5) \, dx \tag{PT6.10}$$

The right-hand side of this equation is called an *indefinite integral* because the limits of integration are unspecified. This is in contrast to the *definite integrals* discussed previously in Part Five [compare Eq. (PT6.10) with Eq. (19.5)].

An analytical solution for Eq. (PT6.10) is obtained if the indefinite integral can be evaluated exactly in equation form. For this simple case, it is possible to do this with the result:

$$y = -0.5x^4 + 4x^3 - 10x^2 + 8.5x + C \tag{PT6.11}$$

which is identical to the original function with one notable exception. In the course of differentiating and then integrating, we lost the constant value of 1 in the original equation and gained the value C. This C is called a *constant of integration*. The fact that such an arbitrary constant appears indicates that the solution is not unique. In fact, it is but one of an infinite

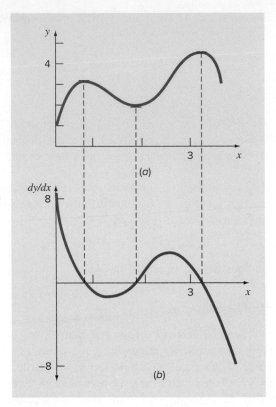

FIGURE PT6.2
Plots of (*a*) y versus x and (*b*) dy/dx versus x for the function $y = -0.5x^4 + 4x^3 - 10x^2 + 8.5x + 1$.

number of possible functions (corresponding to an infinite number of possible values of C) that satisfy the differential equation. For example, Fig. PT6.3 shows six possible functions that satisfy Eq. (PT6.11).

Therefore, to specify the solution completely, a differential equation is usually accompanied by auxiliary conditions. For first-order ODEs, a type of auxiliary condition called an initial value is required to determine the constant and obtain a unique solution. For example, the original differential equation could be accompanied by the initial condition that at $x = 0$, $y = 1$. These values could be substituted into Eq. (PT6.11) to determine $C = 1$. Therefore, the unique solution that satisfies both the differential equation and the specified initial condition is

$$y = -0.5x^4 + 4x^3 - 10x^2 + 8.5x + 1$$

Thus, we have "pinned down" Eq. (PT6.11) by forcing it to pass through the initial condition, and in so doing, we have developed a unique solution to the ODE and have come full circle to the original function [Eq. (PT6.8)].

Initial conditions usually have very tangible interpretations for differential equations derived from physical problem settings. For example, in the bungee jumper problem, the

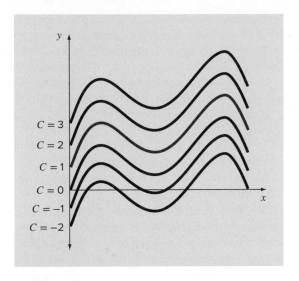

FIGURE PT6.3

Six possible solutions for the integral of $-2x^3 + 12x^2 - 20x + 8.5$. Each conforms to a different value of the constant of integration C.

initial condition was reflective of the physical fact that at time zero the vertical velocity was zero. If the bungee jumper had already been in vertical motion at time zero, the solution would have been modified to account for this initial velocity.

When dealing with an nth-order differential equation, n conditions are required to obtain a unique solution. If all conditions are specified at the same value of the independent variable (e.g., at x or $t = 0$), then the problem is called an *initial-value problem*. This is in contrast to *boundary-value problems* where specification of conditions occurs at different values of the independent variable. Chapters 22 and 23 will focus on initial-value problems. Boundary-value problems are covered in Chap. 24.

6.2 PART ORGANIZATION

Chapter 22 is devoted to one-step methods for solving initial-value ODEs. As the name suggests, *one-step methods* compute a future prediction y_{i+1}, based only on information at a single point y_i and no other previous information. This is in contrast to *multistep approaches* that use information from several previous points as the basis for extrapolating to a new value.

With all but a minor exception, the one-step methods presented in Chap. 22 belong to what are called *Runge-Kutta techniques*. Although the chapter might have been organized around this theoretical notion, we have opted for a more graphical, intuitive approach to introduce the methods. Thus, we begin the chapter with *Euler's method,* which has a very straightforward graphical interpretation. In addition, because we have already introduced Euler's method in Chap. 1, our emphasis here is on quantifying its truncation error and describing its stability.

Next, we use visually oriented arguments to develop two improved versions of Euler's method—the *Heun* and the *midpoint* techniques. After this introduction, we formally develop the concept of Runge-Kutta (or RK) approaches and demonstrate how the foregoing techniques are actually first- and second-order RK methods. This is followed by a discussion of the higher-order RK formulations that are frequently used for engineering and scientific problem solving. In addition, we cover the application of one-step methods to *systems of ODEs*. Note that all the applications in Chap. 22 are limited to cases with a fixed step size.

In *Chap. 23*, we cover more advanced approaches for solving initial-value problems. First, we describe *adaptive RK methods* that automatically adjust the step size in response to the truncation error of the computation. These methods are especially pertinent as they are employed by MATLAB to solve ODEs.

Next, we discuss *multistep methods*. As mentioned above, these algorithms retain information of previous steps to more effectively capture the trajectory of the solution. They also yield the truncation error estimates that can be used to implement step-size control. We describe a simple method—the *non-self-starting Heun* method—to introduce the essential features of the multistep approaches.

Finally, the chapter ends with a description of *stiff ODEs*. These are both individual and systems of ODEs that have both fast and slow components to their solution. As a consequence, they require special solution approaches. We introduce the idea of an *implicit solution* technique as one commonly used remedy. We also describe MATLAB's built-in functions for solving stiff ODEs.

In *Chap. 24,* we focus on two approaches for obtaining solutions to *boundary-value problems:* the *shooting* and *finite-difference methods.* Aside from demonstrating how these techniques are implemented, we illustrate how they handle *derivative boundary conditions* and *nonlinear ODEs.*

22

Initial-Value Problems

CHAPTER OBJECTIVES

The primary objective of this chapter is to introduce you to solving initial-value problems for ODEs (ordinary differential equations). Specific objectives and topics covered are

- Understanding the meaning of local and global truncation errors and their relationship to step size for one-step methods for solving ODEs.
- Knowing how to implement the following Runge-Kutta (RK) methods for a single ODE:
    Euler
    Heun
    Midpoint
    Fourth-order RK
- Knowing how to iterate the corrector of Heun's method.
- Knowing how to implement the following Runge-Kutta methods for systems of ODEs:
    Euler
    Fourth-order RK

YOU'VE GOT A PROBLEM

We started this book with the problem of simulating the velocity of a free-falling bungee jumper. This problem amounted to formulating and solving an ordinary differential equation, the topic of this chapter. Now let's return to this problem and make it more interesting by computing what happens when the jumper reaches the end of the bungee cord.

To do this, we should recognize that the jumper will experience different forces depending on whether the cord is slack or stretched. If it is slack, the situation is that of free fall where the only forces are gravity and drag. However, because the jumper can now move up as well as down, the sign of the drag force must be modified so that it always tends to retard velocity,

$$\frac{dv}{dt} = g - \text{sign}(v)\frac{c_d}{m}v^2 \tag{22.1a}$$

where v is velocity (m/s), t is time (s), g is the acceleration due to gravity (9.81 m/s^2), c_d is the drag coefficient (kg/m), and m is mass (kg). The *signum function*,[1] sign, returns a -1 or a 1 depending on whether its argument is negative or positive, respectively. Thus, when the jumper is falling downward (positive velocity, sign = 1), the drag force will be negative and hence will act to reduce velocity. In contrast, when the jumper is moving upward (negative velocity, sign = -1), the drag force will be positive so that it again reduces the velocity.

Once the cord begins to stretch, it obviously exerts an upward force on the jumper. As done previously in Chap. 8, Hooke's law can be used as a first approximation of this force. In addition, a dampening force should also be included to account for frictional effects as the cord stretches and contracts. These factors can be incorporated along with gravity and drag into a second force balance that applies when the cord is stretched. The result is the following differential equation:

$$\frac{dv}{dt} = g - \text{sign}(v)\frac{c_d}{m}v^2 - \frac{k}{m}(x - L) - \frac{\gamma}{m}v \tag{22.1b}$$

where k is the cord's spring constant (N/m), x is vertical distance measured downward from the bungee jump platform (m), L is the length of the unstretched cord (m), and γ is a dampening coefficient (N · s/m).

Because Eq. (22.1b) only holds when the cord is stretched ($x > L$), the spring force will always be negative. That is, it will always act to pull the jumper back up. The dampening force increases in magnitude as the jumper's velocity increases and always acts to slow the jumper down.

If we want to simulate the jumper's velocity, we would initially solve Eq. (22.1a) until the cord was fully extended. Then, we could switch to Eq. (22.1b) for periods that the cord is stretched. Although this is fairly straightforward, it means that knowledge of the jumper's position is required. This can be done by formulating another differential equation for distance:

$$\frac{dx}{dt} = v \tag{22.2}$$

Thus, solving for the bungee jumper's velocity amounts to solving two ordinary differential equations where one of the equations takes different forms depending on the value

---

[1] Some computer languages represent the signum function as `sgn(x)`. As represented here, MATLAB uses the nomenclature `sign(x)`.

of one of the dependent variables. Chapters 22 and 23 explore methods for solving this and similar problems involving ODEs.

22.1 OVERVIEW

This chapter is devoted to solving ordinary differential equations of the form

$$\frac{dy}{dt} = f(t, y) \tag{22.3}$$

In Chap. 1, we developed a numerical method to solve such an equation for the velocity of the free-falling bungee jumper. Recall that the method was of the general form

New value = old value + slope × step size

or, in mathematical terms,

$$y_{i+1} = y_i + \phi h \tag{22.4}$$

where the slope ϕ is called an *increment function*. According to this equation, the slope estimate of ϕ is used to extrapolate from an old value y_i to a new value y_{i+1} over a distance h. This formula can be applied step by step to trace out the trajectory of the solution into the future. Such approaches are called *one-step methods* because the value of the increment function is based on information at a single point i. They are also referred to as *Runge-Kutta methods* after the two applied mathematicians who first discussed them in the early 1900s. Another class of methods called *multistep methods* use information from several previous points as the basis for extrapolating to a new value. We will describe multistep methods briefly in Chap. 23.

All one-step methods can be expressed in the general form of Eq. (22.4), with the only difference being the manner in which the slope is estimated. The simplest approach is to use the differential equation to estimate the slope in the form of the first derivative at t_i. In other words, the slope at the beginning of the interval is taken as an approximation of the average slope over the whole interval. This approach, called Euler's method, is discussed next. This is followed by other one-step methods that employ alternative slope estimates that result in more accurate predictions.

22.2 EULER'S METHOD

The first derivative provides a direct estimate of the slope at t_i (Fig. 22.1):

$$\phi = f(t_i, y_i)$$

where $f(t_i, y_i)$ is the differential equation evaluated at t_i and y_i. This estimate can be substituted into Eq. (22.1):

$$y_{i+1} = y_i + f(t_i, y_i)h \tag{22.5}$$

This formula is referred to as *Euler's method* (or the Euler-Cauchy or point-slope method). A new value of y is predicted using the slope (equal to the first derivative at the original value of t) to extrapolate linearly over the step size h (Fig. 22.1).

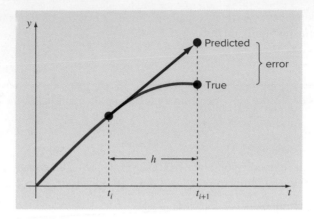

FIGURE 22.1
Euler's method.

EXAMPLE 22.1   Euler's Method

Problem Statement.   Use Euler's method to integrate $y' = 4e^{0.8t} - 0.5y$ from $t = 0$ to 4 with a step size of 1. The initial condition at $t = 0$ is $y = 2$. Note that the exact solution can be determined analytically as

$$y = \frac{4}{1.3}(e^{0.8t} - e^{-0.5t}) + 2e^{-0.5t}$$

Solution.   Equation (22.5) can be used to implement Euler's method:

$$y(1) = y(0) + f(0, 2)(1)$$

where $y(0) = 2$ and the slope estimate at $t = 0$ is

$$f(0, 2) = 4e^0 - 0.5(2) = 3$$

Therefore,

$$y(1) = 2 + 3(1) = 5$$

The true solution at $t = 1$ is

$$y = \frac{4}{1.3}(e^{0.8(1)} - e^{-0.5(1)}) + 2e^{-0.5(1)} = 6.19463$$

Thus, the percent relative error is

$$\varepsilon_t = \left| \frac{6.19463 - 5}{6.19463} \right| \times 100\% = 19.28\%$$

For the second step:

$$y(2) = y(1) + f(1, 5)(1)$$
$$= 5 + [4e^{0.8(1)} - 0.5(5)](1) = 11.40216$$

TABLE 22.1 Comparison of true and numerical values of the integral of $y' = 4e^{0.8t} - 0.5y$, with the initial condition that $y = 2$ at $t = 0$. The numerical values were computed using Euler's method with a step size of 1.

t	y_{true}	y_{Euler}	$|\varepsilon_t|$ (%)
0	2.00000	2.00000	
1	6.19463	5.00000	19.28
2	14.84392	11.40216	23.19
3	33.67717	25.51321	24.24
4	75.33896	56.84931	24.54

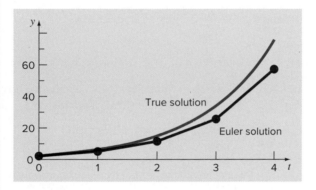

FIGURE 22.2
Comparison of the true solution with a numerical solution using Euler's method for the integral of $y' = 4e^{0.8t} - 0.5y$ from $t = 0$ to 4 with a step size of 1.0. The initial condition at $t = 0$ is $y = 2$.

The true solution at $t = 2.0$ is 14.84392 and, therefore, the true percent relative error is 23.19%. The computation is repeated, and the results compiled in Table 22.1 and Fig. 22.2. Note that although the computation captures the general trend of the true solution, the error is considerable. As discussed in the next section, this error can be reduced by using a smaller step size.

22.2.1 Error Analysis for Euler's Method

The numerical solution of ODEs involves two types of error (recall Chap. 4):

1. *Truncation,* or discretization, errors caused by the nature of the techniques employed to approximate values of y.
2. *Roundoff* errors caused by the limited numbers of significant digits that can be retained by a computer.

The truncation errors are composed of two parts. The first is a *local truncation error* that results from an application of the method in question over a single step. The second is a *propagated truncation error* that results from the approximations produced during the previous steps. The sum of the two is the total error. It is referred to as the *global trunca-tion error.*

Insight into the magnitude and properties of the truncation error can be gained by deriving Euler's method directly from the Taylor series expansion. To do this, realize that the differential equation being integrated will be of the general form of Eq. (22.3), where $dy/dt = y'$, and t and y are the independent and the dependent variables, respectively. If the solution—that is, the function describing the behavior of y—has continuous derivatives, it can be represented by a Taylor series expansion about a starting value (t_i, y_i), as in [recall Eq. (4.13)]:

$$y_{i+1} = y_i + y_i' h + \frac{y_i''}{2!} h^2 + \cdots + \frac{y_i^{(n)}}{n!} h^n + R_n \tag{22.6}$$

where $h = t_{i+1} - t_i$ and R_n = the remainder term, defined as

$$R_n = \frac{y^{(n+1)}(\xi)}{(n+1)!} h^{n+1} \tag{22.7}$$

where ξ lies somewhere in the interval from t_i to t_{i+1}. An alternative form can be developed by substituting Eq. (22.3) into Eqs. (22.6) and (22.7) to yield

$$y_{i+1} = y_i + f(t_i, y_i)h + \frac{f'(t_i, y_i)}{2!} h^2 + \cdots + \frac{f^{(n-1)}(t_i, y_i)}{n!} h^n + O(h^{n+1}) \tag{22.8}$$

where $O(h^{n+1})$ specifies that the local truncation error is proportional to the step size raised to the $(n + 1)$th power.

By comparing Eqs. (22.5) and (22.8), it can be seen that Euler's method corresponds to the Taylor series up to and including the term $f(t_i, y_i)h$. Additionally, the comparison indicates that a truncation error occurs because we approximate the true solution using a finite number of terms from the Taylor series. We thus truncate, or leave out, a part of the true solution. For example, the truncation error in Euler's method is attributable to the remaining terms in the Taylor series expansion that were not included in Eq. (22.5). Subtracting Eq. (22.5) from Eq. (22.8) yields

$$E_t = \frac{f'(t_i, y_i)}{2!} h^2 + \cdots + O(h^{n+1}) \tag{22.9}$$

where E_t = the true local truncation error. For sufficiently small h, the higher-order terms in Eq. (22.9) are usually negligible, and the result is often represented as

$$E_a = \frac{f'(t_i, y_i)}{2!} h^2 \tag{22.10}$$

or

$$E_a = O(h^2) \tag{22.11}$$

where E_a = the approximate local truncation error.

According to Eq. (22.11), we see that the local error is proportional to the square of the step size and the first derivative of the differential equation. It can also be demonstrated that the global truncation error is $O(h)$—that is, it is proportional to the step size

(Carnahan et al., 1969). These observations lead to some useful conclusions:

1. The global error can be reduced by decreasing the step size.
2. The method will provide error-free predictions if the underlying function (i.e., the solution of the differential equation) is linear, because for a straight line the second derivative would be zero.

This latter conclusion makes intuitive sense because Euler's method uses straight-line segments to approximate the solution. Hence, Euler's method is referred to as a *first-order method*.

It should also be noted that this general pattern holds for the higher-order one-step methods described in the following pages. That is, an *n*th-order method will yield perfect results if the underlying solution is an *n*th-order polynomial. Further, the local truncation error will be $O(h^{n+1})$ and the global error $O(h^n)$.

22.2.2 Stability of Euler's Method

In the preceding section, we learned that the truncation error of Euler's method depends on the step size in a predictable way based on the Taylor series. This is an accuracy issue.

The stability of a solution method is another important consideration that must be considered when solving ODEs. A numerical solution is said to be unstable if errors grow exponentially for a problem for which there is a bounded solution. The stability of a particular application can depend on three factors: the differential equation, the numerical method, and the step size.

Insight into the step size required for stability can be examined by studying a very simple ODE:

$$\frac{dy}{dt} = -ay \tag{22.12}$$

If $y(0) = y_0$, calculus can be used to determine the solution as

$$y = y_0 e^{-at}$$

Thus, the solution starts at y_0 and asymptotically approaches zero.

Now suppose that we use Euler's method to solve the same problem numerically:

$$y_{i+1} = y_i + \frac{dy_i}{dt} h$$

Substituting Eq. (22.12) gives

$$y_{i+1} = y_i - ay_i h$$

or

$$y_{i+1} = y_i (1 - ah) \tag{22.13}$$

The parenthetical quantity $1 - ah$ is called an *amplification factor*. If its absolute value is greater than unity, the solution will grow in an unbounded fashion. So clearly, the stability depends on the step size h. That is, if $h > 2/a$, $|y_i| \to \infty$ as $i \to \infty$. Based on this analysis, Euler's method is said to be *conditionally stable*.

Note that there are certain ODEs where errors always grow regardless of the method. Such ODEs are called *ill-conditioned*.

Inaccuracy and instability are often confused. This is probably because (a) both represent situations where the numerical solution breaks down and (b) both are affected by step size. However, they are distinct problems. For example, an inaccurate method can be very stable. We will return to the topic when we discuss stiff systems in Chap. 23.

22.2.3 MATLAB M-file Function: eulode

We have already developed a simple M-file to implement Euler's method for the falling bungee jumper problem in Chap. 3. Recall from Sec. 3.6, that this function used Euler's method to compute the velocity after a given time of free fall. Now, let's develop a more general, all-purpose algorithm.

Figure 22.3 shows an M-file that uses Euler's method to compute values of the dependent variable y over a range of values of the independent variable t. The name of the function holding the right-hand side of the differential equation is passed into the function as the

FIGURE 22.3
An M-file to implement Euler's method.

```
function [t,y] = eulode(dydt,tspan,y0,h,varargin)
% eulode: Euler ODE solver
%    [t,y] = eulode(dydt,tspan,y0,h,p1,p2,...):
%            uses Euler's method to integrate an ODE
% input:
%    dydt = name of the M-file that evaluates the ODE
%    tspan = [ti, tf]  where ti and tf = initial and
%            final values of independent variable
%    y0 = initial value of dependent variable
%    h = step size
%    p1,p2,... = additional parameters used by dydt
% output:
%    t = vector of independent variable
%    y = vector of solution for dependent variable

if nargin<4,error('at least 4 input arguments required'),end
ti = tspan(1);tf = tspan(2);
if ~(tf>ti),error('upper limit must be greater than lower'),end
t = (ti:h:tf)'; n = length(t);
% if necessary, add an additional value of t
% so that range goes from t = ti to tf
if t(n)<tf
  t(n+1) = tf;
  n = n+1;
end
y = y0*ones(n,1); %preallocate y to improve efficiency
for i = 1:n-1 %implement Euler's method
  y(i+1) = y(i) + dydt(t(i),y(i),varargin{:})*(t(i+1)-t(i));
end
```

variable dydt. The initial and final values of the desired range of the independent variable is passed as a vector tspan. The initial value and the desired step size are passed as y0 and h, respectively.

The function first generates a vector t over the desired range of the dependent variable using an increment of h. In the event that the step size is not evenly divisible into the range, the last value will fall short of the final value of the range. If this occurs, the final value is added to t so that the series spans the complete range. The length of the t vector is determined as n. In addition, a vector of the dependent variable y is preallocated with n values of the initial condition to improve efficiency.

At this point, Euler's method [Eq. (22.5)] is implemented by a simple loop:

```
for i = 1:n-1
  y(i+1) = y(i) + dydt(t(i),y(i),varargin{:})*(t(i+1)-t(i));
end
```

Notice how a function is used to generate a value for the derivative at the appropriate values of the independent and dependent variables. Also notice how the time step is automatically calculated based on the difference between adjacent values in the vector t.

The ODE being solved can be set up in several ways. First, the differential equation can be defined as an anonymous function object. For example, for the ODE from Example 22.1:

```
>> dydt = @(t,y) 4*exp(0.8*t) - 0.5*y;
```

The solution can then be generated as

```
>> [t,y] = eulode(dydt,[0 4],2,1);
>> disp([t,y])
```

with the result (compare with Table 22.1):

```
     0    2.0000
1.0000    5.0000
2.0000   11.4022
3.0000   25.5132
4.0000   56.8493
```

Although using an anonymous function is feasible for the present case, there will be more complex problems where the definition of the ODE requires several lines of code. In such instances, creating a separate M-file is the only option.

22.3 IMPROVEMENTS OF EULER'S METHOD

A fundamental source of error in Euler's method is that the derivative at the beginning of the interval is assumed to apply across the entire interval. Two simple modifications are available to help circumvent this shortcoming. As will be demonstrated in Sec. 22.4, both modifications (as well as Euler's method itself) actually belong to a larger class of solution techniques called Runge-Kutta methods. However, because they have very straightforward graphical interpretations, we will present them prior to their formal derivation as Runge-Kutta methods.

22.3.1 Heun's Method

One method to improve the estimate of the slope involves the determination of two derivatives for the interval—one at the beginning and another at the end. The two derivatives are then averaged to obtain an improved estimate of the slope for the entire interval. This approach, called *Heun's method,* is depicted graphically in Fig. 22.4.

Recall that in Euler's method, the slope at the beginning of an interval

$$y_i' = f(t_i, y_i) \tag{22.14}$$

is used to extrapolate linearly to y_{i+1}:

$$y_{i+1}^0 = y_i + f(t_i, y_i)h \tag{22.15}$$

For the standard Euler method we would stop at this point. However, in Heun's method the y_{i+1}^0 calculated in Eq. (22.15) is not the final answer, but an intermediate prediction. This is why we have distinguished it with a superscript 0. Equation (22.15) is called a *predictor equation.* It provides an estimate that allows the calculation of a slope at the end of the interval:

$$y_{i+1}' = f(t_{i+1}, y_{i+1}^0) \tag{22.16}$$

Thus, the two slopes [Eqs. (22.14) and (22.16)] can be combined to obtain an average slope for the interval:

$$\bar{y}' = \frac{f(t_i, y_i) + f(t_{i+1}, y_{i+1}^0)}{2}$$

This average slope is then used to extrapolate linearly from y_i to y_{i+1} using Euler's method:

$$y_{i+1} = y_i + \frac{f(t_i, y_i) + f(t_{i+1}, y_{i+1}^0)}{2} h \tag{22.17}$$

which is called a *corrector equation.*

FIGURE 22.4
Graphical depiction of Heun's method. (*a*) Predictor and (*b*) corrector.

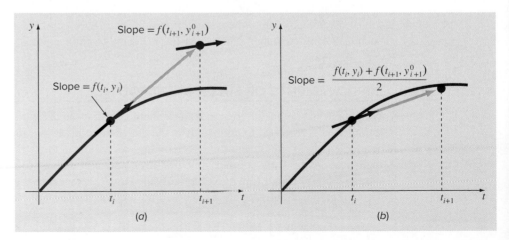

FIGURE 22.5
Graphical representation of iterating the corrector of Heun's method to obtain an improved estimate.

The Heun method is a *predictor-corrector approach*. As just derived, it can be expressed concisely as

Predictor (Fig. 22.4*a*): $y_{i+1}^0 = y_i^m + f(t_i, y_i)h$ (22.18)

Corrector (Fig. 22.4*b*): $y_{i+1}^j = y_i^m + \dfrac{f(t_i, y_i^m) + f\left(t_{i+1}, y_{i+1}^{j-1}\right)}{2} h$ (22.19)

$$(\text{for } j = 1, 2, \ldots, m)$$

Note that because Eq. (22.19) has y_{i+1} on both sides of the equal sign, it can be applied in an iterative fashion as indicated. That is, an old estimate can be used repeatedly to provide an improved estimate of y_{i+1}. The process is depicted in Fig. 22.5.

As with similar iterative methods discussed in previous sections of the book, a termination criterion for convergence of the corrector is provided by

$$|\varepsilon_a| = \left| \frac{y_{i+1}^j - y_{i+1}^{j-1}}{y_{i+1}^j} \right| \times 100\%$$

where y_{i+1}^{j-1} and y_{i+1}^j are the result from the prior and the present iteration of the corrector, respectively. It should be understood that the iterative process does not necessarily converge on the true answer but will converge on an estimate with a finite truncation error, as demonstrated in the following example.

EXAMPLE 22.2   Heun's Method

Problem Statement. Use Heun's method with iteration to integrate $y' = 4e^{0.8t} - 0.5y$ from $t = 0$ to 4 with a step size of 1. The initial condition at $t = 0$ is $y = 2$. Employ a stopping criterion of 0.00001% to terminate the corrector iterations.

Solution. First, the slope at (t_0, y_0) is calculated as

$$y_0' = 4e^0 - 0.5(2) = 3$$

Then, the predictor is used to compute a value at 1.0:

$$y_1^0 = 2 + 3(1) = 5$$

TABLE 22.2 Comparison of true and numerical values of the integral of $y' = 4e^{0.8t} - 0.5y$, with the initial condition that $y = 2$ at $t = 0$. The numerical values were computed using the Euler and Heun methods with a step size of 1. The Heun method was implemented both without and with iteration of the corrector.

				Without Iteration		With Iteration	
t	y_{true}	y_{Euler}	$\lvert\varepsilon_t\rvert$ (%)	y_{Heun}	$\lvert\varepsilon_t\rvert$ (%)	y_{Heun}	$\lvert\varepsilon_t\rvert$ (%)
0	2.00000	2.00000		2.00000		2.00000	
1	6.19463	5.00000	19.28	6.70108	8.18	6.36087	2.68
2	14.84392	11.40216	23.19	16.31978	9.94	15.30224	3.09
3	33.67717	25.51321	24.24	37.19925	10.46	34.74328	3.17
4	75.33896	56.84931	24.54	83.33777	10.62	77.73510	3.18

Note that this is the result that would be obtained by the standard Euler method. The true value in Table 22.2 shows that it corresponds to a percent relative error of 19.28%.

Now, to improve the estimate for y_{i+1} we use the value y_1^0 to predict the slope at the end of the interval

$$y_1' = f(x_1, y_1^0) = 4e^{0.8(1)} - 0.5(5) = 6.402164$$

which can be combined with the initial slope to yield an average slope over the interval from $t = 0$ to 1:

$$\bar{y}' = \frac{3 + 6.402164}{2} = 4.701082$$

This result can then be substituted into the corrector [Eq. (22.19)] to give the prediction at $t = 1$:

$$y_1^1 = 2 + 4.701082(1) = 6.701082$$

which represents a true percent relative error of -8.18%. Thus, the Heun method without iteration of the corrector reduces the absolute value of the error by a factor of about 2.4 as compared with Euler's method. At this point, we can also compute an approximate error as

$$\lvert\varepsilon_a\rvert = \left\lvert\frac{6.701082 - 5}{6.701082}\right\rvert \times 100\% = 25.39\%$$

Now the estimate of y_1 can be refined by substituting the new result back into the right-hand side of Eq. (22.19) to give

$$y_1^2 = 2 + \frac{3 + 4e^{0.8(1)} - 0.5(6.701082)}{2} 1 = 6.275811$$

which represents a true percent relative error of 1.31 percent and an approximate error of

$$\lvert\varepsilon_a\rvert = \left\lvert\frac{6.275811 - 6.701082}{6.275811}\right\rvert \times 100\% = 6.776\%$$

The next iteration gives

$$y_1^2 = 2 + \frac{3 + 4e^{0.8(1)} - 0.5(6.275811)}{2} 1 = 6.382129$$

which represents a true error of 3.03% and an approximate error of 1.666%.

The approximate error will keep dropping as the iterative process converges on a stable final result. In this example, after 12 iterations the approximate error falls below the stopping criterion. At this point, the result at $t = 1$ is 6.36087, which represents a true relative error of 2.68%. Table 22.2 shows results for the remainder of the computation along with results for Euler's method and for the Heun method without iteration of the corrector.

Insight into the local error of the Heun method can be gained by recognizing that it is related to the trapezoidal rule. In the previous example, the derivative is a function of both the dependent variable y and the independent variable t. For cases such as polynomials, where the ODE is solely a function of the independent variable, the predictor step [Eq. (22.18)] is not required and the corrector is applied only once for each iteration. For such cases, the technique is expressed concisely as

$$y_{i+1} = y_i + \frac{f(t_i) + f(t_{i+1})}{2} h \tag{22.20}$$

Notice the similarity between the second term on the right-hand side of Eq. (22.20) and the trapezoidal rule [Eq. (19.11)]. The connection between the two methods can be formally demonstrated by starting with the ordinary differential equation

$$\frac{dy}{dt} = f(t) \tag{22.21}$$

This equation can be solved for y by integration:

$$\int_{y_i}^{y_{i+1}} dy = \int_{t_i}^{t_{i+1}} f(t) \, dt \tag{22.22}$$

which yields

$$y_{i+1} - y_i = \int_{t_i}^{t_{i+1}} f(t) \, dt \tag{22.23}$$

or

$$y_{i+1} = y_i + \int_{t_i}^{t_{i+1}} f(t) \, dt \tag{22.24}$$

Now, recall that the trapezoidal rule [Eq. (19.11)] is defined as

$$\int_{t_i}^{t_{i+1}} f(t) \, dt = \frac{f(t_i) + f(t_{i+1})}{2} h \tag{22.25}$$

where $h = t_{i+1} - t_i$. Substituting Eq. (22.25) into Eq. (22.24) yields

$$y_{i+1} = y_i + \frac{f(t_i) + f(t_{i+1})}{2} h \tag{22.26}$$

which is equivalent to Eq. (22.20). For this reason, Heun's method is sometimes referred to as the trapezoidal rule.

Because Eq. (22.26) is a direct expression of the trapezoidal rule, the local truncation error is given by [recall Eq. (19.14)]

$$E_t = -\frac{f''(\xi)}{12} h^3 \tag{22.27}$$

where ξ is between t_i and t_{i+1}. Thus, the method is second order because the second derivative of the ODE is zero when the true solution is a quadratic. In addition, the local and global errors are $O(h^3)$ and $O(h^2)$, respectively. Therefore, decreasing the step size decreases the error at a faster rate than for Euler's method.

22.3.2 The Midpoint Method

Figure 22.6 illustrates another simple modification of Euler's method. Called the *midpoint method,* this technique uses Euler's method to predict a value of y at the midpoint of the interval (Fig. 22.6a):

$$y_{i+1/2} = y_i + f(t_i, y_i) \frac{h}{2} \tag{22.28}$$

Then, this predicted value is used to calculate a slope at the midpoint:

$$y'_{i+1/2} = f(t_{i+1/2}, y_{i+1/2}) \tag{22.29}$$

FIGURE 22.6
Graphical depiction of midpoint method. (a) Predictor and (b) corrector.

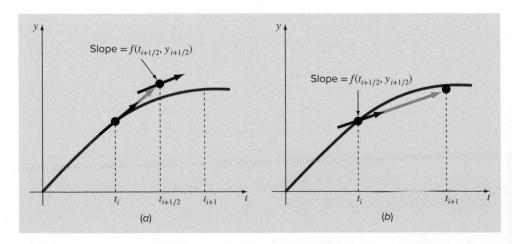

which is assumed to represent a valid approximation of the average slope for the entire interval. This slope is then used to extrapolate linearly from t_i to t_{i+1} (Fig. 22.6b):

$$y_{i+1} = y_i + f(t_{i+1/2}, y_{i+1/2})h \tag{22.30}$$

Observe that because y_{i+1} is not on both sides, the corrector [Eq. (22.30)] cannot be applied iteratively to improve the solution as was done with Heun's method.

As in our discussion of Heun's method, the midpoint method can also be linked to Newton-Cotes integration formulas. Recall from Table 19.4 that the simplest Newton-Cotes open integration formula, which is called the midpoint method, can be represented as

$$\int_a^b f(x)\, dx \cong (b - a)\, f(x_1) \tag{22.31}$$

where x_1 is the midpoint of the interval (a, b). Using the nomenclature for the present case, it can be expressed as

$$\int_{t_i}^{t_{i+1}} f(t)\, dt \cong h f(t_{i+1/2}) \tag{22.32}$$

Substitution of this formula into Eq. (22.24) yields Eq. (22.30). Thus, just as the Heun method can be called the trapezoidal rule, the midpoint method gets its name from the underlying integration formula on which it is based.

The midpoint method is superior to Euler's method because it utilizes a slope estimate at the midpoint of the prediction interval. Recall from our discussion of numerical differentiation in Sec. 4.3.4 that centered finite differences are better approximations of derivatives than either forward or backward versions. In the same sense, a centered approximation such as Eq. (22.29) has a local truncation error of $O(h^2)$ in comparison with the forward approximation of Euler's method, which has an error of $O(h)$. Consequently, the local and global errors of the midpoint method are $O(h^3)$ and $O(h^2)$, respectively.

22.4 RUNGE-KUTTA METHODS

Runge-Kutta (RK) methods achieve the accuracy of a Taylor series approach without requiring the calculation of higher derivatives. Many variations exist but all can be cast in the generalized form of Eq. (22.4):

$$y_{i+1} = y_i + \phi h \tag{22.33}$$

where ϕ is called an *increment function,* which can be interpreted as a representative slope over the interval. The increment function can be written in general form as

$$\phi = a_1 k_1 + a_2 k_2 + \cdots + a_n k_n \tag{22.34}$$

where the a's are constants and the k's are

$$k_1 = f(t_i, y_i) \tag{22.34a}$$
$$k_2 = f(t_i + p_1 h, y_i + q_{11} k_1 h) \tag{22.34b}$$
$$k_3 = f(t_i + p_2 h, y_i + q_{21} k_1 h + q_{22} k_2 h) \tag{22.34c}$$
$$\vdots$$
$$k_n = f(t_i + p_{n-1} h, y_i + q_{n-1,1} k_1 h + q_{n-1,2} k_2 h + \cdots + q_{n-1,n-1} k_{n-1} h) \tag{22.34d}$$

where the p's and q's are constants. Notice that the k's are recurrence relationships. That is, k_1 appears in the equation for k_2 which appears in the equation for k_3 and so forth. Because each k is a functional evaluation, this recurrence makes RK methods efficient for computer calculations.

Various types of RK methods can be devised by employing different numbers of terms in the increment function as specified by n. Note that the first-order RK method with $n = 1$ is, in fact, Euler's method. Once n is chosen, values for the a's, p's, and q's are evaluated by setting Eq. (22.33) equal to terms in a Taylor series expansion. Thus, at least for the lower-order versions, the number of terms n usually represents the order of the approach. For example, in Sec. 22.4.1, second-order RK methods use an increment function with two terms ($n = 2$). These second-order methods will be exact if the solution to the differential equation is quadratic. In addition, because terms with h^3 and higher are dropped during the derivation, the local truncation error is $O(h^3)$ and the global error is $O(h^2)$. In Sec. 22.4.2, the fourth-order RK method ($n = 4$) is presented for which the global truncation error is $O(h^4)$.

22.4.1 Second-Order Runge-Kutta Methods

The second-order version of Eq. (22.33) is

$$y_{i+1} = y_i + (a_1 k_1 + a_2 k_2)h \tag{22.35}$$

where

$$k_1 = f(t_i, y_i) \tag{22.35a}$$
$$k_2 = f(t_i + p_1 h, y_i + q_{11} k_1 h) \tag{22.35b}$$

The values for a_1, a_2, p_1, and q_{11} are evaluated by setting Eq. (22.35) equal to a second-order Taylor series. By doing this, three equations can be derived to evaluate the four unknown constants (see Chapra and Canale, 2010, for details). The three equations are

$$a_1 + a_2 = 1 \tag{22.36}$$
$$a_2 p_1 = 1/2 \tag{22.37}$$
$$a_2 q_{11} = 1/2 \tag{22.38}$$

Because we have three equations with four unknowns, these equations are said to be underdetermined. We, therefore, must assume a value of one of the unknowns to determine the other three. Suppose that we specify a value for a_2. Then Eqs. (22.36) through (22.38) can be solved simultaneously for

$$a_1 = 1 - a_2 \tag{22.39}$$
$$p_1 = q_{11} = \frac{1}{2a_2} \tag{22.40}$$

Because we can choose an infinite number of values for a_2, there are an infinite number of second-order RK methods. Every version would yield exactly the same results if the solution to the ODE were quadratic, linear, or a constant. However, they yield different results when (as is typically the case) the solution is more complicated. Three of the most commonly used and preferred versions are presented next.

Heun Method without Iteration ($a_2 = 1/2$). If a_2 is assumed to be $1/2$, Eqs. (22.39) and (22.40) can be solved for $a_1 = 1/2$ and $p_1 = q_{11} = 1$. These parameters, when substituted into Eq. (22.35), yield

$$y_{i+1} = y_i + \left(\frac{1}{2} k_1 + \frac{1}{2} k_2\right) h \tag{22.41}$$

where

$$k_1 = f(t_i, y_i) \tag{22.41a}$$
$$k_2 = f(t_i + h, y_i + k_1 h) \tag{22.41b}$$

Note that k_1 is the slope at the beginning of the interval and k_2 is the slope at the end of the interval. Consequently, this second-order RK method is actually Heun's technique without iteration of the corrector.

The Midpoint Method ($a_2 = 1$). If a_2 is assumed to be 1, then $a_1 = 0$, $p_1 = q_{11} = 1/2$, and Eq. (22.35) becomes

$$y_{i+1} = y_i + k_2 h \tag{22.42}$$

where

$$k_1 = f(t_i, y_i) \tag{22.42a}$$
$$k_2 = f(t_i + h/2, y_i + k_1 h/2) \tag{22.42b}$$

This is the midpoint method.

Ralston's Method ($a_2 = 3/4$). Ralston (1962) and Ralston and Rabinowitz (1978) determined that choosing $a_2 = 3/4$ provides a minimum bound on the truncation error for the second-order RK algorithms. For this version, $a_1 = 1/4$ and $p_1 = q_{11} = 2/3$, and Eq. (22.35) becomes

$$y_{i+1} = y_i + \left(\frac{1}{4} k_1 + \frac{3}{4} k_2\right) h \tag{22.43}$$

where

$$k_1 = f(t_i, y_i) \tag{22.43a}$$

$$k_2 = f\left(t_i + \frac{2}{3} h, y_i + \frac{2}{3} k_1 h\right) \tag{22.43b}$$

22.4.2 Classical Fourth-Order Runge-Kutta Method

The most popular RK methods are fourth order. As with the second-order approaches, there are an infinite number of versions. The following is the most commonly used form, and we therefore call it the *classical fourth-order RK method:*

$$y_{i+1} = y_i + \frac{1}{6} (k_1 + 2k_2 + 2k_3 + k_4)h \tag{22.44}$$

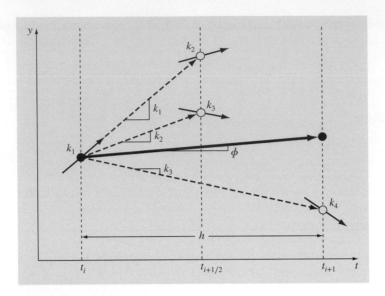

FIGURE 22.7
Graphical depiction of the slope estimates comprising the fourth-order RK method.

where

$$k_1 = f(t_i, y_i) \tag{22.44a}$$

$$k_2 = f\left(t_i + \frac{1}{2}h, y_i + \frac{1}{2}k_1h\right) \tag{22.44b}$$

$$k_3 = f\left(t_i + \frac{1}{2}h, y_i + \frac{1}{2}k_2h\right) \tag{22.44c}$$

$$k_4 = f(t_i + h, y_i + k_3h) \tag{22.44d}$$

Notice that for ODEs that are a function of t alone, the classical fourth-order RK method is similar to Simpson's 1/3 rule. In addition, the fourth-order RK method is similar to the Heun approach in that multiple estimates of the slope are developed to come up with an improved average slope for the interval. As depicted in Fig. 22.7, each of the k's represents a slope. Equation (22.44) then represents a weighted average of these to arrive at the improved slope.

EXAMPLE 22.3   Classical Fourth-Order RK Method

Problem Statement. Employ the classical fourth-order RK method to integrate $y' = 4e^{0.8t} - 0.5y$ from $t = 0$ to 1 using a step size of 1 with $y(0) = 2$.

Solution. For this case, the slope at the beginning of the interval is computed as

$$k_1 = f(0, 2) = 4e^{0.8(0)} - 0.5(2) = 3$$

This value is used to compute a value of y and a slope at the midpoint:

$$y(0.5) = 2 + 3(0.5) = 3.5$$
$$k_2 = f(0.5, 3.5) = 4e^{0.8(0.5)} - 0.5(3.5) = 4.217299$$

This slope in turn is used to compute another value of y and another slope at the midpoint:

$$y(0.5) = 2 + 4.217299(0.5) = 4.108649$$
$$k_3 = f(0.5, 4.108649) = 4e^{0.8(0.5)} - 0.5(4.108649) = 3.912974$$

Next, this slope is used to compute a value of y and a slope at the end of the interval:

$$y(1.0) = 2 + 3.912974(1.0) = 5.912974$$
$$k_4 = f(1.0, 5.912974) = 4e^{0.8(1.0)} - 0.5(5.912974) = 5.945677$$

Finally, the four slope estimates are combined to yield an average slope. This average slope is then used to make the final prediction at the end of the interval.

$$\phi = \frac{1}{6}[3 + 2(4.217299) + 2(3.912974) + 5.945677] = 4.201037$$

$$y(1.0) = 2 + 4.201037(1.0) = 6.201037$$

which compares favorably with the true solution of 6.194631 ($\varepsilon_t = 0.103\%$).

It is certainly possible to develop fifth- and higher-order RK methods. For example, Butcher's (1964) fifth-order RK method is written as

$$y_{i+1} = y_i + \frac{1}{90}(7k_1 + 32k_3 + 12k_4 + 32k_5 + 7k_6)h \tag{22.45}$$

where

$$k_1 = f(t_i, y_i) \tag{22.45a}$$

$$k_2 = f\left(t_i + \frac{1}{4}h, y_i + \frac{1}{4}k_1h\right) \tag{22.45b}$$

$$k_3 = f\left(t_i + \frac{1}{4}h, y_i + \frac{1}{8}k_1h + \frac{1}{8}k_2h\right) \tag{22.45c}$$

$$k_4 = f\left(t_i + \frac{1}{2}h, y_i - \frac{1}{2}k_2h + k_3h\right) \tag{22.45d}$$

$$k_5 = f\left(t_i + \frac{3}{4}h, y_i + \frac{3}{16}k_1h + \frac{9}{16}k_4h\right) \tag{22.45e}$$

$$k_6 = f\left(t_i + h, y_i - \frac{3}{7}k_1h + \frac{2}{7}k_2h + \frac{12}{7}k_3h - \frac{12}{7}k_4h + \frac{8}{7}k_5h\right) \tag{22.45f}$$

Note the similarity between Butcher's method and Boole's rule in Table 19.2. As expected, this method has a global truncation error of $O(h^5)$.

Although the fifth-order version provides more accuracy, notice that six function evaluations are required. Recall that up through the fourth-order versions, n function evaluations are required for an nth-order RK method. Interestingly, for orders higher than four, one or two additional function evaluations are necessary. Because the function evaluations account for the most computation time, methods of order five and higher are usually considered relatively less efficient than the fourth-order versions. This is one of the main reasons for the popularity of the fourth-order RK method.

22.5 SYSTEMS OF EQUATIONS

Many practical problems in engineering and science require the solution of a system of simultaneous ordinary differential equations rather than a single equation. Such systems may be represented generally as

$$\frac{dy_1}{dt} = f_1(t, y_1, y_2, \ldots, y_n)$$
$$\frac{dy_2}{dt} = f_2(t, y_1, y_2, \ldots, y_n)$$

$$\vdots$$

$$\frac{dy_n}{dt} = f_n(t, y_1, y_2, \ldots, y_n)$$

$$(22.46)$$

The solution of such a system requires that n initial conditions be known at the starting value of t.

An example is the calculation of the bungee jumper's velocity and position that we set up at the beginning of this chapter. For the free-fall portion of the jump, this problem amounts to solving the following system of ODEs:

$$\frac{dx}{dt} = v \tag{22.47}$$

$$\frac{dv}{dt} = g - \frac{c_d}{m}v^2 \tag{22.48}$$

If the stationary platform from which the jumper launches is defined as $x = 0$, the initial conditions would be $x(0) = v(0) = 0$.

22.5.1 Euler's Method

All the methods discussed in this chapter for single equations can be extended to systems of ODEs. Engineering applications can involve thousands of simultaneous equations. In each case, the procedure for solving a system of equations simply involves applying the one-step technique for every equation at each step before proceeding to the next step. This is best illustrated by the following example for Euler's method.

EXAMPLE 22.4   Solving Systems of ODEs with Euler's Method

Problem Statement. Solve for the velocity and position of the free-falling bungee jumper using Euler's method. Assuming that at $t = 0$, $x = v = 0$, and integrate to $t = 10$ s with a step size of 2 s. As was done previously in Examples 1.1 and 1.2, the gravitational acceleration is 9.81 m/s², and the jumper has a mass of 68.1 kg with a drag coefficient of 0.25 kg/m.

Recall that the analytical solution for velocity is [Eq. (1.9)]:

$$v(t) = \sqrt{\frac{gm}{c_d}} \tanh\left(\sqrt{\frac{gc_d}{m}}\, t\right)$$

This result can be substituted into Eq. (22.47) which can be integrated to determine an analytical solution for distance as

$$x(t) = \frac{m}{c_d} \ln\left[\cosh\left(\sqrt{\frac{gc_d}{m}}\, t\right)\right]$$

Use these analytical solutions to compute the true relative errors of the results.

Solution. The ODEs can be used to compute the slopes at $t = 0$ as

$$\frac{dx}{dt} = 0$$

$$\frac{dv}{dt} = 9.81 - \frac{0.25}{68.1}\,(0)^2 = 9.81$$

Euler's method is then used to compute the values at $t = 2$ s,

$$x = 0 + 0(2) = 0$$

$$v = 0 + 9.81(2) = 19.62$$

The analytical solutions can be computed as $x(2) = 19.16629$ and $v(2) = 18.72919$. Thus, the percent relative errors are 100% and 4.756%, respectively.

The process can be repeated to compute the results at $t = 4$ as

$$x = 0 + 19.62(2) = 39.24$$

$$v = 19.62 + \left(9.81 - \frac{0.25}{68.1}(19.62)^2\right)2 = 36.41368$$

Proceeding in a like manner gives the results displayed in Table 22.3.

TABLE 22.3 Distance and velocity of a free-falling bungee jumper as computed numerically with Euler's method.

t	x_{true}	v_{true}	x_{Euler}	v_{Euler}	$\varepsilon_t(x)$	$\varepsilon_t(v)$
0	0	0	0	0		
2	19.1663	18.7292	0	19.6200	100.00%	4.76%
4	71.9304	33.1118	39.2400	36.4137	45.45%	9.97%
6	147.9462	42.0762	112.0674	46.2983	24.25%	10.03%
8	237.5104	46.9575	204.6640	50.1802	13.83%	6.86%
10	334.1782	49.4214	305.0244	51.3123	8.72%	3.83%

Although the foregoing example illustrates how Euler's method can be implemented for systems of ODEs, the results are not very accurate because of the large step size. In addition, the results for distance are a bit unsatisfying because x does not change until the second iteration. Using a much smaller step greatly mitigates these deficiencies. As described next, using a higher-order solver provides decent results even with a relatively large step size.

22.5.2 Runge-Kutta Methods

Note that any of the higher-order RK methods in this chapter can be applied to systems of equations. However, care must be taken in determining the slopes. Figure 22.7 is helpful in visualizing the proper way to do this for the fourth-order method. That is, we first develop slopes for all variables at the initial value. These slopes (a set of k_1's) are then used to make predictions of the dependent variable at the midpoint of the interval. These midpoint values are in turn used to compute a set of slopes at the midpoint (the k_2's). These new slopes are then taken back to the starting point to make another set of midpoint predictions that lead to new slope predictions at the midpoint (the k_3's). These are then employed to make predictions at the end of the interval that are used to develop slopes at the end of the interval (the k_4's). Finally, the k's are combined into a set of increment functions [as in Eq. (22.44)] that are brought back to the beginning to make the final predictions. The following example illustrates the approach.

EXAMPLE 22.5   Solving Systems of ODEs with the Fourth-Order RK Method

Problem Statement. Use the fourth-order RK method to solve for the same problem we addressed in Example 22.4.

Solution. First, it is convenient to express the ODEs in the functional format of Eq. (22.46) as

$$\frac{dx}{dt} = f_1(t, x, v) = v$$

$$\frac{dv}{dt} = f_2(t, x, v) = g - \frac{c_d}{m}v^2$$

The first step in obtaining the solution is to solve for all the slopes at the beginning of the interval:

$$k_{1,1} = f_1(0, 0, 0) = 0$$

$$k_{1,2} = f_2(0, 0, 0) = 9.81 - \frac{0.25}{68.1}(0)^2 = 9.81$$

where $k_{i,j}$ is the ith value of k for the jth dependent variable. Next, we must calculate the first values of x and v at the midpoint of the first step:

$$x(1) = x(0) + k_{1,1}\frac{h}{2} = 0 + 0\frac{2}{2} = 0$$

$$v(1) = v(0) + k_{1,2}\frac{h}{2} = 0 + 9.81\frac{2}{2} = 9.81$$

which can be used to compute the first set of midpoint slopes:

$$k_{2,1} = f_1(1, 0, 9.81) = 9.8100$$
$$k_{2,2} = f_2(1, 0, 9.81) = 9.4567$$

These are used to determine the second set of midpoint predictions:

$$x(1) = x(0) + k_{2,1}\frac{h}{2} = 0 + 9.8100\frac{2}{2} = 9.8100$$

$$v(1) = v(0) + k_{2,2}\frac{h}{2} = 0 + 9.4567\frac{2}{2} = 9.4567$$

which can be used to compute the second set of midpoint slopes:

$$k_{3,1} = f_1(1, 9.8100, 9.4567) = 9.4567$$
$$k_{3,2} = f_2(1, 9.8100, 9.4567) = 9.4817$$

These are used to determine the predictions at the end of the interval:

$$x(2) = x(0) + k_{3,1}h = 0 + 9.4567(2) = 18.9134$$
$$v(2) = v(0) + k_{3,2}h = 0 + 9.4817(2) = 18.9634$$

which can be used to compute the endpoint slopes:

$$k_{4,1} = f_1(2, 18.9134, 18.9634) = 18.9634$$
$$k_{4,2} = f_2(2, 18.9134, 18.9634) = 8.4898$$

The values of k can then be used to compute [Eq. (22.44)]:

$$x(2) = 0 + \frac{1}{6}[0 + 2(9.8100 + 9.4567) + 18.9634] 2 = 19.1656$$

$$v(2) = 0 + \frac{1}{6}[9.8100 + 2(9.4567 + 9.4817) + 8.4898] 2 = 18.7256$$

Proceeding in a like manner for the remaining steps yields the values displayed in Table 22.4. In contrast to the results obtained with Euler's method, the fourth-order RK predictions are much closer to the true values. Further, a highly accurate, nonzero value is computed for distance on the first step.

TABLE 22.4 Distance and velocity of a free-falling bungee jumper as computed numerically with the fourth-order RK method.

t	x_{true}	v_{true}	x_{RK4}	v_{RK4}	$\varepsilon_t(x)$	$\varepsilon_t(v)$
0	0	0	0	0		
2	19.1663	18.7292	19.1656	18.7256	0.004%	0.019%
4	71.9304	33.1118	71.9311	33.0995	0.001%	0.037%
6	147.9462	42.0762	147.9521	42.0547	0.004%	0.051%
8	237.5104	46.9575	237.5104	46.9345	0.000%	0.049%
10	334.1782	49.4214	334.1626	49.4027	0.005%	0.038%

22.5.3 MATLAB M-file Function: rk4sys

Figure 22.8 shows an M-file called rk4sys that uses the fourth-order RK method to solve a system of ODEs. This code is similar in many ways to the function developed earlier (Fig. 22.3) to solve a single ODE with Euler's method. For example, it is passed the function name defining the ODEs through its argument.

FIGURE 22.8
An M-file to implement the RK4 method for a system of ODEs.

```
function [tp,yp] = rk4sys(dydt,tspan,y0,h,varargin)
% rk4sys: fourth-order Runge-Kutta for a system of ODEs
%   [t,y] = rk4sys(dydt,tspan,y0,h,p1,p2,...): integrates
%           a system of ODEs with fourth-order RK method
% input:
%   dydt = name of the M-file that evaluates the ODEs
%   tspan = [ti, tf]; initial and final times with output
%                     generated at interval of h, or
%        = [t0 t1 ... tf]; specific times where solution output
%   y0 = initial values of dependent variables
%   h = step size
%   p1,p2,... = additional parameters used by dydt
% output:
%   tp = vector of independent variable
%   yp = vector of solution for dependent variables

if nargin<4,error('at least 4 input arguments required'), end
if any(diff(tspan)<=0),error('tspan not ascending order'), end
n = length(tspan);
ti = tspan(1);tf = tspan(n);
if n == 2
  t = (ti:h:tf)'; n = length(t);
  if t(n)<tf
    t(n+1) = tf;
    n = n+1;
  end
else
  t = tspan;
end
tt = ti; y(1,:) = y0;
np = 1; tp(np) = tt; yp(np,:) = y(1,:);
i=1;
while(1)
  tend = t(np+1);
  hh = t(np+1) - t(np);
```

(Continued)

```
    if hh > h,hh = h;end
  while(1)
     if tt+hh > tend,hh = tend-tt;end
     k1 = dydt(tt,y(i,:),varargin{:})';
     ymid = y(i,:) + k1.*hh./2;
     k2 = dydt(tt+hh/2,ymid,varargin{:})';
     ymid = y(i,:) + k2*hh/2;
     k3 = dydt(tt+hh/2,ymid,varargin{:})';
     yend = y(i,:) + k3*hh;
     k4 = dydt(tt+hh,yend,varargin{:})';
     phi = (k1+2*(k2+k3)+k4)/6;
     y(i+1,:) = y(i,:) + phi*hh;
     tt = tt+hh;
     i=i+1;
     if tt> =tend,break,end
  end
  np = np+1; tp(np) = tt; yp(np,:) = y(i,:);
  if tt> =tf,break,end
end
```

FIGURE 22.8 (*Continued*)

However, it has an additional feature that allows you to generate output in two ways, depending on how the input variable tspan is specified. As was the case for Fig. 22.3, you can set tspan = [ti tf], where ti and tf are the initial and final times, respectively. If done in this way, the routine automatically generates output values between these limits at equal spaced intervals h. Alternatively, if you want to obtain results at specific times, you can define tspan = [t0,t1,...,tf]. Note that in both cases, the tspan values must be in ascending order.

We can employ rk4sys to solve the same problem as in Example 22.5. First, we can develop an M-file to hold the ODEs:

```
function dy = dydtsys(t, y)
dy = [y(2);9.81-0.25/68.1*y(2)^2];
```

where $y(1)$ = distance (x) and $y(2)$ = velocity (v). The solution can then be generated as

```
>> [t y] = rk4sys(@dydtsys,[0 10],[0 0],2);
>> disp([t' y(:,1) y(:,2)])

         0         0         0
    2.0000   19.1656   18.7256
    4.0000   71.9311   33.0995
    6.0000  147.9521   42.0547
    8.0000  237.5104   46.9345
   10.0000  334.1626   49.4027
```

We can also use tspan to generate results at specific values of the independent variable. For example,

```
>> tspan = [0 6 10];
>> [t y] = rk4sys(@dydtsys,tspan,[0 0],2);
>> disp([t' y(:,1) y(:,2)])

        0         0         0
   6.0000  147.9521   42.0547
  10.0000  334.1626   49.4027
```

22.6 CASE STUDY   PREDATOR-PREY MODELS AND CHAOS

Background. Engineers and scientists deal with a variety of problems involving systems of nonlinear ordinary differential equations. This case study focuses on two of these applications. The first relates to predator-prey models that are used to study species interactions. The second are equations derived from fluid dynamics that are used to simulate the atmosphere.

Predator-prey models were developed independently in the early part of the twentieth century by the Italian mathematician Vito Volterra and the American biologist Alfred Lotka. These equations are commonly called *Lotka-Volterra equations*. The simplest version is the following pairs of ODEs:

$$\frac{dx}{dt} = ax - bxy \tag{22.49}$$

$$\frac{dy}{dt} = -cy + dxy \tag{22.50}$$

where x and y = the number of prey and predators, respectively, a = the prey growth rate, c = the predator death rate, and b and d = the rates characterizing the effect of the predator-prey interactions on the prey death and the predator growth, respectively. The multiplicative terms (i.e., those involving xy) are what make such equations nonlinear.

An example of a simple nonlinear model based on atmospheric fluid dynamics is the *Lorenz equations* created by the American meteorologist Edward Lorenz:

$$\frac{dx}{dt} = -\sigma x + \sigma y$$

$$\frac{dy}{dt} = rx - y - xz$$

$$\frac{dz}{dt} = -bz + xy$$

Lorenz developed these equations to relate the intensity of atmospheric fluid motion x to temperature variations y and z in the horizontal and vertical directions, respectively.

22.6 CASE STUDY continued

As with the predator-prey model, the nonlinearities stem from the simple multiplicative terms: xz and xy.

Use numerical methods to obtain solutions for these equations. Plot the results to visualize how the dependent variables change temporally. In addition, graph the dependent variables versus each other to see whether any interesting patterns emerge.

Solution. The following parameter values can be used for the predator-prey simulation: $a = 1.2$, $b = 0.6$, $c = 0.8$, and $d = 0.3$. Employ initial conditions of $x = 2$ and $y = 1$ and integrate from $t = 0$ to 30, using a step size of $h = 0.0625$.

First, we can develop a function to hold the differential equations:

```
function yp = predprey(t,y,a,b,c,d)
yp = [a*y(1)-b*y(1)*y(2);-c*y(2)+d*y(1)*y(2)];
```

The following script employs this function to generate solutions with both the Euler and the fourth-order RK methods. Note that the function eulersys was based on modifying the rk4sys function (Fig. 22.8). We will leave the development of such an M-file as a homework problem. In addition to displaying the solution as a time-series plot (x and y versus t), the script also generates a plot of y versus x. Such *phase-plane* plots are often useful in elucidating features of the model's underlying structure that may not be evident from the time series.

```
h=0.0625;tspan=[0 40];y0=[2 1];
a=1.2;b=0.6;c=0.8;d=0.3;
[t y] = eulersys(@predprey,tspan,y0,h,a,b,c,d);
subplot(2,2,1);plot(t,y(:,1),t,y(:,2),'--')
legend('prey','predator');title('(a) Euler time plot')
subplot(2,2,2);plot(y(:,1),y(:,2))
title('(b) Euler phase plane plot')
[t y] = rk4sys(@predprey,tspan,y0,h,a,b,c,d);
subplot(2,2,3);plot(t,y(:,1),t,y(:,2),'--')
title('(c) RK4 time plot')
subplot(2,2,4);plot(y(:,1),y(:,2))
title('(d) RK4 phase plane plot')
```

The solution obtained with Euler's method is shown at the top of Fig. 22.9. The time series (Fig. 22.9a) indicates that the amplitudes of the oscillations are expanding. This is reinforced by the phase-plane plot (Fig. 22.9b). Hence, these results indicate that the crude Euler method would require a much smaller time step to obtain accurate results.

In contrast, because of its much smaller truncation error, the RK4 method yields good results with the same time step. As in Fig. 22.9c, a cyclical pattern emerges in time. Because the predator population is initially small, the prey grows exponentially. At a certain point, the prey become so numerous that the predator population begins to grow. Eventually, the increased predators cause the prey to decline. This decrease, in turn, leads to a decrease of the predators. Eventually, the process repeats. Notice that, as expected, the predator peak lags the prey. Also, observe that the process has a fixed period—that is, it repeats in a set time.

22.6 CASE STUDY   continued

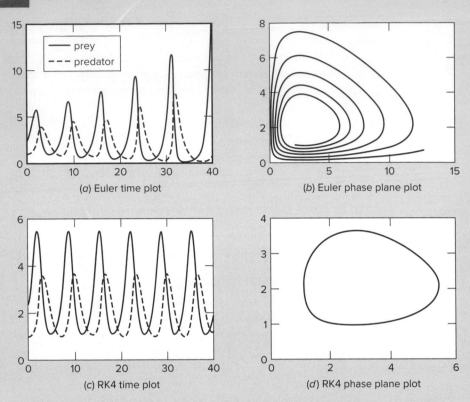

(a) Euler time plot

(b) Euler phase plane plot

(c) RK4 time plot

(d) RK4 phase plane plot

FIGURE 22.9
Solution for the Lotka-Volterra model. Euler's method (a) time-series and (b) phase-plane plots, and RK4 method (c) time-series and (d) phase-plane plots.

The phase-plane representation for the accurate RK4 solution (Fig. 22.9d) indicates that the interaction between the predator and the prey amounts to a closed counterclockwise orbit. Interestingly, there is a resting or *critical point* at the center of the orbit. The exact location of this point can be determined by setting Eqs. (22.49) and (22.50) to steady state ($dy/dt = dx/dt = 0$) and solving for $(x, y) = (0, 0)$ and $(c/d, a/b)$. The former is the trivial result that if we start with neither predators nor prey, nothing will happen. The latter is the more interesting outcome that if the initial conditions are set at $x = c/d$ and $y = a/b$, the derivatives will be zero, and the populations will remain constant.

Now, let's use the same approach to investigate the trajectories of the Lorenz equations with the following parameter values: $a = 10$, $b = 8/3$, and $r = 28$. Employ initial conditions of $x = y = z = 5$ and integrate from $t = 0$ to 20. For this case, we will use the fourth-order RK method to obtain solutions with a constant time step of $h = 0.03125$.

22.6 CASE STUDY continued

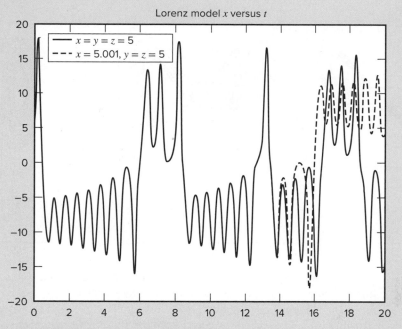

FIGURE 22.10
Time-domain representation of x versus t for the Lorenz equations. The solid time series is for the initial conditions (5, 5, 5). The dashed line is where the initial condition for x is perturbed slightly (5.001, 5, 5).

The results are quite different from the behavior of the Lotka-Volterra equations. As in Fig. 22.10, the variable x seems to be undergoing an almost random pattern of oscillations, bouncing around from negative values to positive values. The other variables exhibit similar behavior. However, even though the patterns seem random, the frequency of the oscillation and the amplitudes seem fairly consistent.

An interesting feature of such solutions can be illustrated by changing the initial condition for x slightly (from 5 to 5.001). The results are superimposed as the dashed line in Fig. 22.10. Although the solutions track on each other for a time, after about $t = 15$ they diverge significantly. Thus, we can see that the Lorenz equations are quite sensitive to their initial conditions. The term *chaotic* is used to describe such solutions. In his original study, this led Lorenz to the conclusion that long-range weather forecasts might be impossible!

The sensitivity of a dynamical system to small perturbations of its initial conditions is sometimes called the *butterfly effect*. The idea is that the flapping of a butterfly's wings might induce tiny changes in the atmosphere that ultimately leads to a large-scale weather phenomenon like a tornado.

22.6 CASE STUDY continued

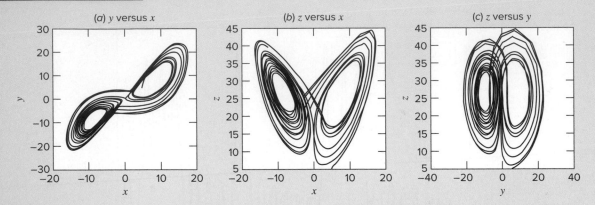

FIGURE 22.11
Phase-plane representation for the Lorenz equations. (a) xy, (b) xz, and (c) yz projections.

Although the time-series plots are chaotic, phase-plane plots reveal an underlying structure. Because we are dealing with three independent variables, we can generate projections. Figure 22.11 shows projections in the xy, xz, and the yz planes. Notice how a structure is manifest when perceived from the phase-plane perspective. The solution forms orbits around what appear to be critical points. These points are called *strange attractors* in the jargon of mathematicians who study such nonlinear systems.

Beyond the two-variable projections, MATLAB's plot3 function provides a vehicle to directly generate a three-dimensional phase-plane plot:

```
>> plot3(y(:,1),y(:,2),y(:,3))
>> xlabel('x');ylabel('y');zlabel('z');grid
```

As was the case for Fig. 22.11, the three-dimensional plot (Fig 22.12) depicts trajectories cycling in a definite pattern around a pair of critical points.

As a final note, the sensitivity of chaotic systems to initial conditions has implications for numerical computations. Beyond the initial conditions themselves, different step sizes or different algorithms (and in some cases, even different computers) can introduce small differences in the solutions. In a similar fashion to Fig. 22.10, these discrepancies will eventually lead to large deviations. Some of the problems in this chapter and in Chap. 23 are designed to demonstrate this issue.

22.6 CASE STUDY continued

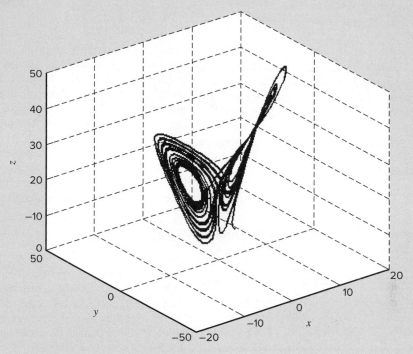

FIGURE 22.12
Three-dimensional phase-plane representation for the Lorenz equations generated with
MATLAB's **plot3** function.

PROBLEMS

22.1 Solve the following initial value problem over the
interval from $t = 0$ to 2 where $y(0) = 1$. Display all your results
on the same graph.

$$\frac{dy}{dt} = yt^2 - 1.1y$$

(a) Analytically.
(b) Using Euler's method with $h = 0.5$ and 0.25.
(c) Using the midpoint method with $h = 0.5$.
(d) Using the fourth-order RK method with $h = 0.5$.

22.2 Solve the following problem over the interval from
$x = 0$ to 1 using a step size of 0.25 where $y(0) = 1$. Display
all your results on the same graph.

$$\frac{dy}{dx} = (1 + 2x)\sqrt{y}$$

(a) Analytically.
(b) Using Euler's method.
(c) Using Heun's method without iteration.
(d) Using Ralston's method.
(e) Using the fourth-order RK method.

22.3 Solve the following problem over the interval from $t = 0$ to 3 using a step size of 0.5 where $y(0) = 1$. Display all your results on the same graph.

$$\frac{dy}{dt} = -y + t^2$$

Obtain your solutions with (a) Heun's method without iterating the corrector, (b) Heun's method with iterating the corrector until $\varepsilon_s < 0.1\%$, (c) the midpoint method, and (d) Ralston's method.

22.4 The growth of populations of organisms has many engineering and scientific applications. One of the simplest models assumes that the rate of change of the population p is proportional to the existing population at any time t:

$$\frac{dp}{dt} = k_g\, p \qquad \text{(P22.4.1)}$$

where k_g = the growth rate. The world population in millions from 1950 through 2000 was

t	1950	1955	1960	1965	1970	1975
p	2555	2780	3040	3346	3708	4087
t	1980	1985	1990	1995	2000	
p	4454	4850	5276	5686	6079	

(a) Assuming that Eq. (P22.4.1) holds, use the data from 1950 through 1970 to estimate k_g.
(b) Use the fourth-order RK method along with the results of (a) to stimulate the world population from 1950 to 2050 with a step size of 5 years. Display your simulation results along with the data on a plot.

22.5 Although the model in Prob. 22.4 works adequately when population growth is unlimited, it breaks down when factors such as food shortages, pollution, and lack of space inhibit growth. In such cases, the growth rate is not a constant, but can be formulated as

$$k_g = k_{gm}(1 - p/p_{max})$$

where k_{gm} = the maximum growth rate under unlimited conditions, p = population, and p_{max} = the maximum population. Note that p_{max} is sometimes called the *carrying capacity*. Thus, at low population density $p \ll p_{max}$, $k_g \rightarrow k_{gm}$. As p approaches p_{max}, the growth rate approaches zero. Using this growth rate formulation, the rate of change of population can be modeled as

$$\frac{dp}{dt} = k_{gm}(1 - p/p_{max})p$$

This is referred to as the *logistic model*. The analytical solution to this model is

$$p = p_0 \frac{p_{max}}{p_0 + (p_{max} - p_0)e^{-k_{gm}t}}$$

Simulate the world's population from 1950 to 2050 using (a) the analytical solution and (b) the fourth-order RK method with a step size of 5 years. Employ the following initial conditions and parameter values for your simulation: p_0 (in 1950) = 2555 million people, $k_{gm} = 0.026/\text{yr}$, and $p_{max} = 12{,}000$ million people. Display your results as a plot along with the data from Prob. 22.4.

22.6 Suppose that a projectile is launched upward from the earth's surface. Assume that the only force acting on the object is the downward force of gravity. Under these conditions, a force balance can be used to derive

$$\frac{dv}{dt} = -g(0)\frac{R^2}{(R + x)^2}$$

where v = upward velocity (m/s), t = time (s), x = altitude (m) measured upward from the earth's surface, $g(0)$ = the gravitational acceleration at the earth's surface ($\cong 9.81 \text{ m/s}^2$), and R = the earth's radius ($\cong 6.37 \times 10^6$ m). Recognizing that $dx/dt = v$, use the RK4sys M-file (Fig. 22.8) to determine the maximum height that would be obtained if $v(t = 0) = 1500$ m/s.

22.7 Solve the following pair of ODEs over the interval from $t = 0$ to 0.4 using a step size of 0.1. The initial conditions are $y(0) = 2$ and $z(0) = 4$. Obtain your solution with (a) Euler's method and (b) the fourth-order RK method. Display your results as a plot.

$$\frac{dy}{dt} = -2y + 4e^{-t}$$

$$\frac{dz}{dt} = -\frac{yz^2}{3}$$

22.8 The *van der Pol equation* is a model of an electronic circuit that arose back in the days of vacuum tubes:

$$\frac{d^2y}{dt^2} - (1 - y^2)\frac{dy}{dt} + y = 0$$

Given the initial conditions, $y(0) = y'(0) = 1$, solve this equation from $t = 0$ to 10 using Euler's method with a step size of (a) 0.2 and (b) 0.1. Plot both solutions on the same graph.

22.9 Given the initial conditions, $y(0) = 1$ and $y'(0) = 0$, solve the following initial-value problem from $t = 0$ to 4:

$$\frac{d^2y}{dt^2} + 9y = 0$$

Obtain your solutions with (a) Euler's method and (b) the fourth-order RK method. In both cases, use a step size of 0.1. Plot both solutions on the same graph along with the exact solution $y = \cos 3t$.

22.10 Develop an M-file to solve a single ODE with Heun's method with iteration. Design the M-file so that it creates a plot of the results. Test your program by using it to solve for population as described in Prob. 22.5. Employ a step size of 5 years and iterate the corrector until $\varepsilon_s < 0.1\%$.

22.11 Develop an M-file to solve a single ODE with the midpoint method. Design the M-file so that it creates a plot of the results. Test your program by using it to solve for population as described in Prob. 22.5. Employ a step size of 5 years.

22.12 Develop an M-file to solve a single ODE with the fourth-order RK method. Design the M-file so that it creates a plot of the results. Test your program by using it to solve Prob. 22.2. Employ a step size of 0.1.

22.13 Develop an M-file to solve a system of ODEs with Euler's method. Design the M-file so that it creates a plot of the results. Test your program by using it to solve Prob. 22.7 with a step size of 0.075.

22.14 Isle Royale National Park is a 210-square-mile archipelago composed of a single large island and many small islands in Lake Superior. Moose arrived around 1900, and by 1930, their population approached 3000, ravaging vegetation. In 1949, wolves crossed an ice bridge from Ontario. Since the late 1950s, the numbers of the moose and wolves have been tracked.

FIGURE P22.15

(a) Integrate the Lotka-Volterra equations (Sec. 22.6) from 1960 through 2020 using the following coefficient values: $a = 0.23$, $b = 0.0133$, $c = 0.4$, and $d = 0.0004$. Compare your simulation with the data using a time-series plot and determine the sum of the squares of the residuals between your model and the data for both the moose and the wolves.

(b) Develop a phase-plane plot of your solution.

22.15 The motion of a damped spring-mass system (Fig. P22.15) is described by the following ordinary differential equation:

$$m\frac{d^2x}{dt^2} + c\frac{dx}{dt} + kx = 0$$

where x = displacement from equilibrium position (m), t = time (s), m = 20-kg mass, and c = the damping coefficient (N · s/m). The damping coefficient c takes on three

Year	Moose	Wolves	Year	Moose	Wolves	Year	Moose	Wolves
1959	563	20	1975	1355	41	1991	1313	12
1960	610	22	1976	1282	44	1992	1590	12
1961	628	22	1977	1143	34	1993	1879	13
1962	639	23	1978	1001	40	1994	1770	17
1963	663	20	1979	1028	43	1995	2422	16
1964	707	26	1980	910	50	1996	1163	22
1965	733	28	1981	863	30	1997	500	24
1966	765	26	1982	872	14	1998	699	14
1967	912	22	1983	932	23	1999	750	25
1968	1042	22	1984	1038	24	2000	850	29
1969	1268	17	1985	1115	22	2001	900	19
1970	1295	18	1986	1192	20	2002	1100	17
1971	1439	20	1987	1268	16	2003	900	19
1972	1493	23	1988	1335	12	2004	750	29
1973	1435	24	1989	1397	12	2005	540	30
1974	1467	31	1990	1216	15	2006	450	30

FIGURE P22.16
A spherical tank.

values of 5 (underdamped), 40 (critically damped), and 200 (overdamped). The spring constant $k = 20$ N/m. The initial velocity is zero, and the initial displacement $x = 1$ m. Solve this equation using a numerical method over the time period $0 \le t \le 15$ s. Plot the displacement versus time for each of the three values of the damping coefficient on the same plot.

22.16 A spherical tank has a circular orifice in its bottom through which the liquid flows out (Fig. P22.16). The flow rate through the hole can be estimated as

$$Q_{out} = CA \sqrt{2gh}$$

where Q_{out} = outflow (m³/s), C = an empirically derived coefficient, A = the area of the orifice (m²), g = the gravitational constant (= 9.81 m/s²), and h = the depth of liquid in the tank. Use one of the numerical methods described in this chapter to determine how long it will take for the water to flow out of a 3-m diameter tank with an initial height of 2.75 m. Note that the orifice has a diameter of 3 cm and $C = 0.55$.

22.17 In the investigation of a homicide or accidental death, it is often important to estimate the time of death. From the experimental observations, it is known that the surface temperature of an object changes at a rate proportional to the difference between the temperature of the object and that of the surrounding environment or ambient temperature. This is known as Newton's law of cooling. Thus, if $T(t)$ is the temperature of the object at time t, and T_a is the constant ambient temperature:

$$\frac{dT}{dt} = -K(T - T_a)$$

where $K > 0$ is a constant of proportionality. Suppose that at time $t = 0$ a corpse is discovered and its temperature is measured to be T_o. We assume that at the time of death, the body temperature T_d was at the normal value of 37 °C. Suppose that the temperature of the corpse when it was discovered was 29.5 °C, and that two hours later, it is 23.5 °C. The ambient temperature is 20 °C.

(a) Determine K and the time of death.

(b) Solve the ODE numerically and plot the results.

22.18 The reaction $A \rightarrow B$ takes place in two reactors in series. The reactors are well mixed but are not at steady state. The unsteady-state mass balance for each stirred tank reactor is shown below:

$$\frac{dCA_1}{dt} = \frac{1}{\tau}(CA_0 - CA_1) - kCA_1$$

$$\frac{dCB_1}{dt} = -\frac{1}{\tau}CB_1 + kCA_1$$

$$\frac{dCA_2}{dt} = \frac{1}{\tau}(CA_1 - CA_2) - kCA_2$$

$$\frac{dCB_2}{dt} = \frac{1}{\tau}(CB_1 - CB_2) + kCA_2$$

where CA_0 = concentration of A at the inlet of the first reactor, CA_1 = concentration of A at the outlet of the first reactor (and inlet of the second), CA_2 = concentration of A at the outlet of the second reactor, CB_1 = concentration of B at the outlet of the first reactor (and inlet of the second), CB_2 = concentration of B in the second reactor, τ = residence time for each reactor, and k = the rate constant for reaction of A to produce B. If CA_0 is equal to 20, find the concentrations of A and B in both reactors during their first 10 minutes of operation. Use $k = 0.12$/min and $\tau = 5$ min and assume that the initial conditions of all the dependent variables are zero.

22.19 A nonisothermal batch reactor can be described by the following equations:

$$\frac{dC}{dt} = -e^{(-10/(T+273))}C$$

$$\frac{dT}{dt} = 1000e^{(-10/(T+273))}C - 10(T - 20)$$

where C is the concentration of the reactant and T is the temperature of the reactor. Initially, the reactor is at 15 °C and has a concentration of reactant C of 1.0 gmol/L. Find the concentration and temperature of the reactor as a function of time.

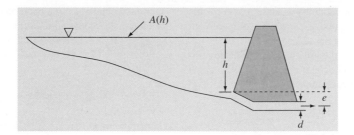

FIGURE P22.21

22.20 The following equation can be used to model the deflection of a sailboat mast subject to a wind force:

$$\frac{d^2y}{dz^2} = \frac{f(z)}{2EI}(L - z)^2$$

where $f(z)$ = wind force, E = modulus of elasticity, L = mast length, and I = moment of inertia. Note that the force varies with height according to

$$f(z) = \frac{200z}{5 + z} e^{-2z/30}$$

Calculate the deflection if $y = 0$ and $dy/dz = 0$ at $z = 0$. Use parameter values of $L = 30$, $E = 1.25 \times 10^8$, and $I = 0.05$ for your computation.

22.21 A pond drains through a pipe as shown in Fig. P22.21. Under a number of simplifying assumptions, the following differential equation describes how depth changes with time:

$$\frac{dh}{dt} = -\frac{\pi d^2}{4A(h)} \sqrt{2g(h + e)}$$

where h = depth (m), t = time (s), d = pipe diameter (m), $A(h)$ = pond surface area as a function of depth (m²), g = gravitational constant (= 9.81 m/s²), and e = depth of pipe outlet below the pond bottom (m). Based on the following area-depth table, solve this differential equation to determine how long it takes for the pond to empty, given that $h(0) = 6$ m, $d = 0.25$ m, $e = 1$ m.

h, m	6	5	4	3	2	1	0
$A(h)$, 10^4 m²	1.17	0.97	0.67	0.45	0.32	0.18	0

22.22 Engineers and scientists use mass-spring models to gain insight into the dynamics of structures under the influence of disturbances such as earthquakes. Figure P22.22 shows such a representation for a three-story building. For this case, the analysis is limited to horizontal motion of the

FIGURE P22.22

structure. Using Newton's second law, force balances can be developed for this system as

$$\frac{d^2x_1}{dt^2} = -\frac{k_1}{m_1}x_1 + \frac{k_2}{m_1}(x_2 - x_1)$$

$$\frac{d^2x_2}{dt^2} = \frac{k_2}{m_2}(x_1 - x_2) + \frac{k_3}{m_2}(x_3 - x_2)$$

$$\frac{d^2x_3}{dt^2} = \frac{k_3}{m_3}(x_2 - x_3)$$

Simulate the dynamics of this structure from $t = 0$ to 20 s, given the initial condition that the velocity of the ground floor is $dx_1/dt = 1$ m/s, and all other initial values of displacements and velocities are zero. Present your results as two time-series plots of **(a)** displacements and **(b)** velocities. In addition, develop a three-dimensional phase-plane plot of the displacements.

22.23 Repeat the same simulations as in Sec. 22.6 for the Lorenz equations but generate the solutions with the midpoint method.

22.24 Perform the same simulations as in Sec. 22.6 for the Lorenz equations but use a value of $r = 99.96$. Compare your results with those obtained in Sec. 22.6.

22.25 Figure P22.25 shows the kinetic interactions governing the concentrations of a bacteria culture and their nutrition source (substrate) in a continuously stirred flow-through bioreactor.

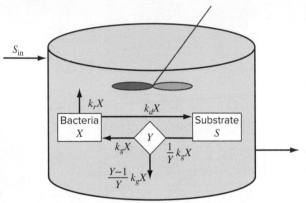

FIGURE P22.25
Continuously stirred flow-through bioreactor to grow a bacterial culture.

The mass balances for the bacteria biomass, X (gC/m^3), and the substrate concentration, S (gC/m^3), can be written as

$$\frac{dX}{dt} = \left(k_{g,max} \frac{S}{K_s + S} - k_d - k_r - \frac{1}{\tau_w} \right) X$$

$$\frac{dS}{dt} = -\frac{1}{Y} k_{g,max} \frac{S}{K_s + S} X + k_d X + \frac{1}{\tau_w} (S_{in} - S)$$

where t = time (h), $k_{g,max}$ = maximum bacterial growth rate (/d), K_s = half-saturation constant (gC/m^3), k_d = death rate (/d), k_r = respiration rate (h), Q = flow rate (m^3/h), V = reactor volume (m^3), Y = yield coefficient (gC-cell/gC-substrate), and S_{in} = inflow substrate concentration (mgC/m^3). Simulate how the substrate, bacteria, and total organic carbon ($X + S$) change over time in this reactor for three residence times: **(a)** $\tau_w = 20$ h, **(b)** $\tau_w = 10$ h, and **(c)** $\tau_w = 5$ h. Employ the following parameters for the simulation: $X(0) = 100$ gC/m^3, $S(0) = 0$, $k_{g,max} = 0.2/hr$, $K_s = 150$ gC/m^3, $k_d = k_r = 0.01/hr$, $Y = 0.5$ gC-cell/gC-substrate, $V = 0.01$ m^3, and $S_{in} = 1000$ gC/m^3, and display your results graphically.

23

Adaptive Methods and Stiff Systems

CHAPTER OBJECTIVES

The primary objective of this chapter is to introduce you to more advanced methods for solving initial-value problems for ordinary differential equations. Specific objectives and topics covered are

- Understanding how the Runge-Kutta Fehlberg methods use RK methods of different orders to provide error estimates that are used to adjust the step size.
- Familiarizing yourself with the built-in MATLAB functions for solving ODEs.
- Learning how to adjust the options for MATLAB's ODE solvers.
- Learning how to pass parameters to MATLAB's ODE solvers.
- Understanding the difference between one-step and multistep methods for solving ODEs.
- Understanding what is meant by stiffness and its implications for solving ODEs.

23.1 ADAPTIVE RUNGE-KUTTA METHODS

To this point, we have presented methods for solving ODEs that employ a constant step size. For a significant number of problems, this can represent a serious limitation. For example, suppose that we are integrating an ODE with a solution of the type depicted in Fig. 23.1. For most of the range, the solution changes gradually. Such behavior suggests that a fairly large step size could be employed to obtain adequate results. However, for a localized region from $t = 1.75$ to 2.25, the solution undergoes an abrupt change. The practical consequence of dealing with such functions is that a very small step size would be required to accurately capture the impulsive behavior. If a constant step-size algorithm were employed, the smaller step size required for the region of abrupt change would have to be applied to the entire computation. As a consequence, a much smaller step size than necessary—and, therefore, many more calculations—would be wasted on the regions of gradual change.

Algorithms that automatically adjust the step size can avoid such overkill and hence be of great advantage. Because they "adapt" to the solution's trajectory, they are said to have

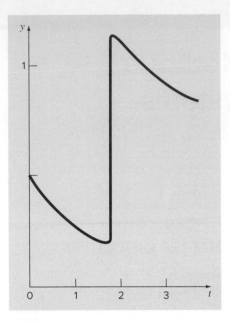

FIGURE 23.1
An example of a solution of an ODE that exhibits an abrupt change. Automatic step-size adjustment has great advantages for such cases.

---

adaptive step-size control. Implementation of such approaches requires that an estimate of the local truncation error be obtained at each step. This error estimate can then serve as a basis for either shortening or lengthening the step size.

Before proceeding, we should mention that aside from solving ODEs, the methods described in this chapter can also be used to evaluate definite integrals. The evaluation of the definite integral

$$I = \int_a^b f(x)\, dx$$

is equivalent to solving the differential equation

$$\frac{dy}{dx} = f(x)$$

for $y(b)$ given the initial condition $y(a) = 0$. Thus, the following techniques can be employed to efficiently evaluate definite integrals involving functions that are generally smooth but exhibit regions of abrupt change.

There are two primary approaches to incorporate adaptive step-size control into one-step methods. *Step halving* involves taking each step twice, once as a full step and then as two half steps. The difference in the two results represents an estimate of the local truncation error. The step size can then be adjusted based on this error estimate.

In the second approach, called *embedded RK methods*, the local truncation error is estimated as the difference between two predictions using different-order RK methods. These are currently the methods of choice because they are more efficient than step halving.

The embedded methods were first developed by Fehlberg. Hence, they are sometimes referred to as *RK-Fehlberg methods*. At face value, the idea of using two predictions of different order might seem too computationally expensive. For example, a fourth- and fifth-order prediction amounts to a total of 10 function evaluations per step [recall Eqs. (22.44) and (22.45)]. Fehlberg cleverly circumvented this problem by deriving a fifth-order RK method that employs most of the same function evaluations required for an accompanying fourth-order RK method. Thus, the approach yielded the error estimate on the basis of only six function evaluations!

23.1.1 MATLAB Functions for Nonstiff Systems

Since Fehlberg originally developed his approach, other even better approaches have been developed. Several of these are available as built-in functions in MATLAB.

ode23. The ode23 function uses the BS23 algorithm (Bogacki and Shampine, 1989; Shampine, 1994), which simultaneously uses second- and third-order RK formulas to solve the ODE and make error estimates for step-size adjustment. The formulas to advance the solution are

$$y_{i+1} = y_i + \frac{1}{9}(2k_1 + 3k_2 + 4k_3)h \tag{23.1}$$

where

$$k_1 = f(t_i, y_i) \tag{23.1a}$$

$$k_2 = f\left(t_i + \frac{1}{2}h, y_i + \frac{1}{2}k_1h\right) \tag{23.1b}$$

$$k_3 = f\left(t_i + \frac{3}{4}h, y_i + \frac{3}{4}k_2h\right) \tag{23.1c}$$

The error is estimated as

$$E_{i+1} = \frac{1}{72}(-5k_1 + 6k_2 + 8k_3 - 9k_4)h \tag{23.2}$$

where

$$k_4 = f(t_{i+1}, y_{i+1}) \tag{23.2a}$$

Note that although there appear to be four function evaluations, there are really only three because after the first step, the k_1 for the present step will be the k_4 from the previous step. Thus, the approach yields a prediction and error estimate based on three evaluations rather than the five that would ordinarily result from using second- (two evaluations) and third-order (three evaluations) RK formulas in tandem.

After each step, the error is checked to determine whether it is within a desired tolerance. If it is, the value of y_{i+1} is accepted, and k_4 becomes k_1 for the next step. If the error is too large, the step is repeated with reduced step sizes until the estimated error satisfies

$$E \leq \max(\text{RelTol} \times |y|, \text{AbsTol}) \tag{23.3}$$

where RelTol is the relative tolerance (default $= 10^{-3}$) and AbsTol is the absolute tolerance (default $= 10^{-6}$). Observe that the criteria for the relative error uses a fraction rather than a percent relative error as we have done on many occasions prior to this point.

ode45. The ode45 function uses an algorithm developed by Dormand and Prince (1980), which simultaneously uses fourth- and fifth-order RK formulas to solve the ODE and make error estimates for step-size adjustment. MATLAB recommends that ode45 is the best function to apply as a "first try" for most problems.

ode113. The ode113 function uses a variable-order Adams-Bashforth-Moulton solver. It is useful for stringent error tolerances or computationally intensive ODE functions. Note that this is a multistep method as we will describe subsequently in Sec. 23.2.

These functions can be called in a number of different ways. The simplest approach is

```
[t, y] = ode45(odefun, tspan, y0)
```

where y is the solution array where each column is one of the dependent variables and each row corresponds to a time in the column vector t, odefun is the name of the function returning a column vector of the right-hand-sides of the differential equations, tspan specifies the integration interval, and $y0 =$ a vector containing the initial values.

Note that tspan can be formulated in two ways. First, if it is entered as a vector of two numbers,

```
tspan = [ti tf];
```

the integration is performed from ti to tf. Second, to obtain solutions at specific times t0, t1, ... , tn (all increasing or all decreasing), use

```
tspan = [t0 t1 ... tn];
```

Here is an example of how ode45 can be used to solve a single ODE, $y' = 4e^{0.8t} - 0.5y$ from $t = 0$ to 4 with an initial condition of $y(0) = 2$. Recall from Example 22.1 that the analytical solution at $t = 4$ is 75.33896. Representing the ODE as an anonymous function, ode45 can be used to generate the same result numerically as

```
>> dydt=@(t,y) 4*exp(0.8*t)-0.5*y;
>> [t,y]=ode45(dydt,[0 4],2);
>> y(length(t))

ans =
   75.3390
```

As described in the following example, the ODE is typically stored in its own M-file when dealing with systems of equations.

EXAMPLE 23.1   Using MATLAB to Solve a System of ODEs

Problem Statement. Employ ode45 to solve the following set of nonlinear ODEs from $t = 0$ to 20:

$$\frac{dy_1}{dt} = 1.2y_1 - 0.6y_1\,y_2 \qquad \frac{dy_2}{dt} = -0.8y_2 + 0.3y_1\,y_2$$

where $y_1 = 2$ and $y_2 = 1$ at $t = 0$. Such equations are referred to as *predator-prey equations*.

Solution. Before obtaining a solution with MATLAB, you must create a function to compute the right-hand side of the ODEs. One way to do this is to create an M-file as in

```
function yp = predprey(t,y)
yp = [1.2*y(1)-0.6*y(1)*y(2);-0.8*y(2)+0.3*y(1)*y(2)];
```

We stored this M-file under the name: predprey.m.

Next, enter the following commands to specify the integration range and the initial conditions:

```
>> tspan = [0 20];
>> y0 = [2, 1];
```

The solver can then be invoked by

```
>> [t,y] = ode45(@predprey, tspan, y0);
```

This command will then solve the differential equations in predprey.m over the range defined by tspan using the initial conditions found in y0. The results can be displayed by simply typing

```
>> plot(t,y)
```

which yields Fig. 23.2.

In addition to a time series plot, it is also instructive to generate a *phase-plane plot*—that is, a plot of the dependent variables versus each other by

```
>> plot(y(:,1),y(:,2))
```

which yields Fig. 23.3.

FIGURE 23.2
Solution of predator-prey model with MATLAB.

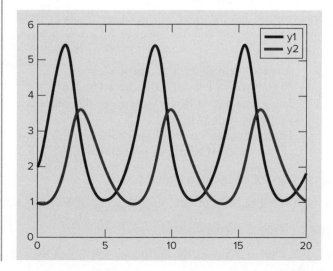

FIGURE 23.3
Phase-plane plot of predator-prey model with MATLAB.

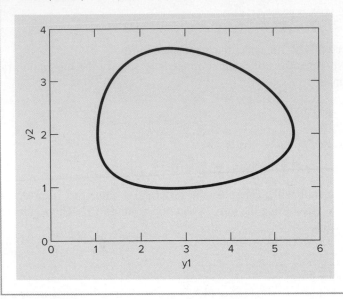

As in the previous example, the MATLAB solver uses default parameters to control various aspects of the integration. In addition, there is also no control over the differential equations' parameters. To have control over these features, additional arguments are included as in

$$[t, y] = \text{ode45}(\textit{odefun, tspan, y0, options, p1, p2,}\ldots)$$

where *options* is a data structure that is created with the *odeset* function to control features of the solution, and *p1, p2,*... are parameters that you want to pass into *odefun*.

The odeset function has the general syntax

$$\textit{options} = \text{odeset}('\textit{par}_1',\textit{val}_1,'\textit{par}_2',\textit{val}_2,\ldots)$$

where the parameter *par$_i$* has the value *val$_i$*. A complete listing of all the possible parameters can be obtained by merely entering odeset at the command prompt. Some commonly used parameters are

'RelTol'	Allows you to adjust the relative tolerance.
'AbsTol'	Allows you to adjust the absolute tolerance.
'InitialStep'	The solver automatically determines the initial step. This option allows you to set your own.
'MaxStep'	The maximum step defaults to one-tenth of the tspan interval. This option allows you to override this default.

EXAMPLE 23.2   Using odeset to Control Integration Options

Problem Statement.   Use ode23 to solve the following ODE from $t = 0$ to 4:

$$\frac{dy}{dt} = 10e^{-(t-2)^2/[2(0.075)^2]} - 0.6y$$

where $y(0) = 0.5$. Obtain solutions for the default (10^{-3}) and for a more stringent (10^{-4}) relative error tolerance.

Solution. First, we will create an M-file to compute the right-hand side of the ODE:

```
function yp = dydt(t, y)
yp = 10*exp(-(t-2)*(t-2)/(2*.075^2))-0.6*y;
```

Then, we can implement the solver without setting the options. Hence the default value for the relative error (10^{-3}) is automatically used:

```
>> ode23(@dydt, [0 4], 0.5);
```

Note that we have not set the function equal to output variables [t, y]. When we implement one of the ODE solvers in this way, MATLAB automatically creates a plot of the results displaying circles at the values it has computed. As in Fig. 23.4a, notice how ode23 takes relatively large steps in the smooth regions of the solution whereas it takes smaller steps in the region of rapid change around $t = 2$.

We can obtain a more accurate solution by using the odeset function to set the relative error tolerance to 10^{-4}:

```
>> options=odeset('RelTol',1e-4);
>> ode23(@dydt, [0, 4], 0.5, options);
```

As in Fig. 23.4b, the solver takes more small steps to attain the increased accuracy.

FIGURE 23.4
Solution of ODE with MATLAB. For (b), a smaller relative error tolerance is used and hence many more steps are taken.

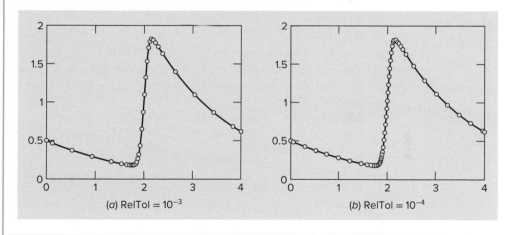

(a) RelTol = 10^{-3} (b) RelTol = 10^{-4}

23.1.2 Events

MATLAB's ODE solvers are commonly implemented for a prespecified integration interval. That is, they are often used to obtain a solution from an initial to a final value of the dependent variable. However, there are many problems where we do not know the final time.

A nice example relates to the free-falling bungee jumper that we have been using throughout this book. Suppose that the jump master inadvertently neglects to attach the cord to the jumper. The final time for this case, which corresponds to the jumper hitting the ground, is not a given. In fact, the objective of solving the ODEs would be to determine when the jumper hit the ground.

MATLAB's *events* option provides a means to solve such problems. It works by solving differential equations until one of the dependent variables reaches zero. Of course, there may be cases where we would like to terminate the computation at a value other than zero. As described in the following paragraphs, such cases can be readily accommodated.

We will use our bungee jumper problem to illustrate the approach. The system of ODEs can be formulated as

$$\frac{dx}{dt} = v$$

$$\frac{dv}{dt} = g - \frac{c_d}{m} v|v|$$

where x = distance (m), t = time (s), v = velocity (m/s) where positive velocity is in the downward direction, g = the acceleration of gravity (= 9.81 m/s^2), c_d = a second-order drag coefficient (kg/m), and m = mass (kg). Note that in this formulation, distance and velocity are both positive in the downward direction, and the ground level is defined as zero distance. For the present example, we will assume that the jumper is initially located 200 m above the ground and the initial velocity is 20 m/s in the upward direction—that is, $x(0) = -200$ and $v(0) = 20$.

The first step is to express the system of ODEs as an M-file function:

```
function dydt=freefall(t,y,cd,m)
% y(1) = x and y(2) = v
grav=9.81;
dydt=[y(2);grav-cd/m*y(2)*abs(y(2))];
```

In order to implement the event, two other M-files need to be developed. These are (1) a function that defines the event and (2) a script that generates the solution.

For our bungee jumper problem, the event function (which we have named endevent) can be written as

```
function [detect,stopint,direction]=endevent(t,y,varargin)
% Locate the time when height passes through zero
% and stop integration.
detect=y(1);    % Detect height = 0
stopint=1;      % Stop the integration
direction=0;    % Direction does not matter
```

This function is passed the values of the independent (t) and dependent variables (y) along with the model parameters (varargin). It then computes and returns three variables. The first, detect, specifies that MATLAB should detect the event when the dependent variable y(1) equals zero—that is, when the height $x = 0$. The second, stopint, is set to 1. This instructs MATLAB to stop when the event occurs. The final variable, direction, is set to 0 if all zeros are to be detected (this is the default), +1 if only the zeros where the event function increases are to be detected, and −1 if only the zeros where the event function decreases are to be detected. In our case, because the direction of the approach to zero is unimportant, we set direction to zero.[1]

---

[1] Note that, as mentioned previously, we might want to detect a nonzero event. For example, we might want to detect when the jumper reached $x = 5$. To do this, we would merely set detect = y(1) − 5.

Finally, a script can be developed to generate the solution:

```
opts=odeset('events',@endevent);
y0=[-200 -20];
[t,y,te,ye]=ode45(@freefall,[0 inf],y0,opts,0.25,68.1);
te,ye
plot(t,-y(:,1),'-',t,y(:,2),'--','LineWidth',2)
legend('Height (m)','Velocity (m/s)')
xlabel('time (s)');
ylabel('x (m) and v (m/s)')
```

In the first line, the odeset function is used to invoke the events option and specify that the event we are seeking is defined in the endevent function. Next, we set the initial conditions (y0) and the integration interval (tspan). Observe that because we do not know when the jumper will hit the ground, we set the upper limit of the integration interval to infinity. The third line then employs the ode45 function to generate the actual solution. As in all of MATLAB's ODE solvers, the function returns the answers in the vectors t and y. In addition, when the events option is invoked, ode45 can also return the time at which the event occurs (te), and the corresponding values of the dependent variables (ye). The remaining lines of the script merely display and plot the results. When the script is run, the output is displayed as

```
te =
    9.5475
ye =
    0.0000   46.2454
```

The plot is shown in Fig. 23.5. Thus, the jumper hits the ground in 9.5475 s with a velocity of 46.2454 m/s.

FIGURE 23.5
MATLAB-generated plot of the height above the ground and velocity of the free-falling bungee jumper without the cord.

23.2 MULTISTEP METHODS

The one-step methods described in the previous sections utilize information at a single point t_i to predict a value of the dependent variable y_{i+1} at a future point t_{i+1} (Fig. 23.6a). Alternative approaches, called *multistep methods* (Fig. 23.6b), are based on the insight that, once the computation has begun, valuable information from previous points is at our command. The curvature of the lines connecting these previous values provides information regarding the trajectory of the solution. Multistep methods exploit this information to solve ODEs. In this section, we will present a simple second-order method that serves to demonstrate the general characteristics of multistep approaches.

23.2.1 The Non-Self-Starting Heun Method

Recall that the Heun approach uses Euler's method as a predictor [Eq. (22.15)]:

$$y_{i+1}^0 = y_i + f(t_i, y_i)h \tag{23.4}$$

and the trapezoidal rule as a corrector [Eq. (22.17)]:

$$y_{i+1} = y_i + \frac{f(t_i, y_i) + f\left(t_{i+1}, y_{i+1}^0\right)}{2}h \tag{23.5}$$

Thus, the predictor and the corrector have local truncation errors of $O(h^2)$ and $O(h^3)$, respectively. This suggests that the predictor is the weak link in the method because it has the greatest error. This weakness is significant because the efficiency of the iterative corrector step depends on the accuracy of the initial prediction. Consequently, one way to improve Heun's method is to develop a predictor that has a local error of $O(h^3)$. This

---

FIGURE 23.6
Graphical depiction of the fundamental difference between (*a*) one-step and (*b*) multistep methods for solving ODEs.

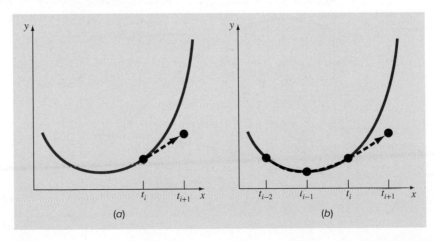

can be accomplished by using Euler's method and the slope at y_i, and extra information from a previous point y_{i-1}, as in

$$y_{i+1}^0 = y_{i-1} + f(t_i, y_i)2h \tag{23.6}$$

This formula attains $O(h^3)$ at the expense of employing a larger step size $2h$. In addition, note that the equation is not self-starting because it involves a previous value of the dependent variable y_{i-1}. Such a value would not be available in a typical initial-value problem. Because of this fact, Eqs. (23.5) and (23.6) are called the *non-self-starting Heun method*. As depicted in Fig. 23.7, the derivative estimate in Eq. (23.6) is now located at the midpoint rather than at the beginning of the interval over which the prediction is made. This centering improves the local error of the predictor to $O(h^3)$.

FIGURE 23.7
A graphical depiction of the non-self-starting Heun method. (*a*) The midpoint method that is used as a predictor. (*b*) The trapezoidal rule that is employed as a corrector.

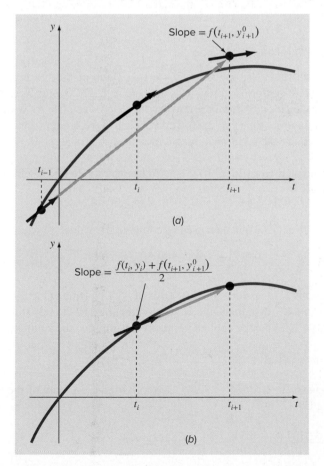

The non-self-starting Heun method can be summarized as

Predictor (Fig. 23.7a):　$y_{i+1}^0 = y_{i-1}^m + f(t_i, y_i^m)2h$　　　　　　　　　　(23.7)

Corrector (Fig. 23.7b):　$y_{i+1}^j = y_i^m + \dfrac{f(t_i, y_i^m) + f\left(t_{i+1}, y_{i+1}^{j-1}\right)}{2}h$　　　(23.8)

$$(\text{for } j = 1, 2, \ldots, m)$$

where the superscripts denote that the corrector is applied iteratively from $j = 1$ to m to obtain refined solutions. Note that y_i^m and y_{i-1}^m are the final results of the corrector iterations at the previous time steps. The iterations are terminated based on an estimate of the approximate error,

$$|\varepsilon_a| = \left| \frac{y_{i+1}^j - y_{i+1}^{j-1}}{y_{i+1}^j} \right| \times 100\% \tag{23.9}$$

When $|\varepsilon_a|$ is less than a prespecified error tolerance ε_s, the iterations are terminated. At this point, $j = m$. The use of Eqs. (23.7) through (23.9) to solve an ODE is demonstrated in the following example.

EXAMPLE 23.3　Non-Self-Starting Heun's Method

Problem Statement.　Use the non-self-starting Heun method to perform the same computations as were performed previously in Example 22.2 using Heun's method. That is, integrate $y' = 4e^{0.8t} - 0.5y$ from $t = 0$ to 4 with a step size of 1. As with Example 22.2, the initial condition at $t = 0$ is $y = 2$. However, because we are now dealing with a multistep method, we require the additional information that y is equal to -0.3929953 at $t = -1$.

Solution.　The predictor [Eq. (23.7)] is used to extrapolate linearly from $t = -1$ to 1:

$$y_1^0 = -0.3929953 + \left[4e^{0.8(0)} - 0.5(2)\right]\,2 = 5.607005$$

The corrector [Eq. (23.8)] is then used to compute the value:

$$y_1^1 = 2 + \frac{4e^{0.8(0)} - 0.5(2) + 4e^{0.8(1)} - 0.5(5.607005)}{2}\,1 = 6.549331$$

which represents a true percent relative error of -5.73% (true value $= 6.194631$). This error is somewhat smaller than the value of -8.18% incurred in the self-starting Heun.
　　Now, Eq. (23.8) can be applied iteratively to improve the solution:

$$y_1^2 = 2 + \frac{3 + 4e^{0.8(1)} - 0.5(6.549331)}{2}\,1 = 6.313749$$

which represents an error of -1.92%. An approximate estimate of the error can be determined using Eq. (23.9):

$$|\varepsilon_a| = \left| \frac{6.313749 - 6.549331}{6.313749} \right| \times 100\% = 3.7\%$$

Equation (23.8) can be applied iteratively until ε_a falls below a prespecified value of ε_s. As was the case with the Heun method (recall Example 22.2), the iterations converge on a value of 6.36087 ($\varepsilon_t = -2.68\%$). However, because the initial predictor value is more accurate, the multistep method converges at a somewhat faster rate.

For the second step, the predictor is

$$y_2^0 = 2 + \left[4e^{0.8(1)} - 0.5(6.36087)\right] 2 = 13.44346 \qquad \varepsilon_t = 9.43\%$$

which is superior to the prediction of 12.0826 ($\varepsilon_t = 18\%$) that was computed with the original Heun method. The first corrector yields 15.76693 ($\varepsilon_t = 6.8\%$), and subsequent iterations converge on the same result as was obtained with the self-starting Heun method: 15.30224 ($\varepsilon_t = -3.09\%$). As with the previous step, the rate of convergence of the corrector is somewhat improved because of the better initial prediction.

23.2.2 Error Estimates

Aside from providing increased efficiency, the non-self-starting Heun can also be used to estimate the local truncation error. As with the adaptive RK methods in Sec. 23.1, the error estimate then provides a criterion for changing the step size.

The error estimate can be derived by recognizing that the predictor is equivalent to the midpoint rule. Hence, its local truncation error is (Table 19.4)

$$E_p = \frac{1}{3} h^3 y^{(3)}(\xi_p) = \frac{1}{3} h^3 f''(\xi_p) \tag{23.10}$$

where the subscript p designates that this is the error of the predictor. This error estimate can be combined with the estimate of y_{i+1} from the predictor step to yield

$$\text{True value} = y_{i+1}^0 + \frac{1}{3} h^3 y^{(3)}(\xi_p) \tag{23.11}$$

By recognizing that the corrector is equivalent to the trapezoidal rule, a similar estimate of the local truncation error for the corrector is (Table 19.2)

$$E_c = -\frac{1}{12} h^3 y^{(3)}(\xi_c) = -\frac{1}{12} h^3 f''(\xi_c) \tag{23.12}$$

This error estimate can be combined with the corrector result y_{i+1} to give

$$\text{True value} = y_{i+1}^m - \frac{1}{12} h^3 y^{(3)}(\xi_c) \tag{23.13}$$

Equation (23.11) can be subtracted from Eq. (23.13) to yield

$$0 = y_{i+1}^m - y_{i+1}^0 - \frac{5}{12} h^3 y^{(3)}(\xi) \tag{23.14}$$

where ξ is now between t_{i-1} and t_i. Now, dividing Eq. (23.14) by 5 and rearranging the result gives

$$\frac{y_{i+1}^0 - y_{i+1}^m}{5} = -\frac{1}{12} h^3 y^{(3)}(\xi) \tag{23.15}$$

Notice that the right-hand sides of Eqs. (23.12) and (23.15) are identical, with the exception of the argument of the third derivative. If the third derivative does not vary appreciably over the interval in question, we can assume that the right-hand sides are equal, and therefore, the left-hand sides should also be equivalent, as in

$$E_c = -\frac{\overset{0}{y}_{i+1} - \overset{m}{y}_{i+1}}{5} \tag{23.16}$$

Thus, we have arrived at a relationship that can be used to estimate the per-step truncation error on the basis of two quantities that are routine by-products of the computation: the predictor (y_{i+1}^0) and the corrector (y_{i+1}^m).

EXAMPLE 23.4   Estimate of Per-Step Truncation Error

Problem Statement. Use Eq. (23.16) to estimate the per-step truncation error of Example 23.3. Note that the true values at $t = 1$ and 2 are 6.194631 and 14.84392, respectively.

Solution. At $t_{i+1} = 1$, the predictor gives 5.607005 and the corrector yields 6.360865. These values can be substituted into Eq. (23.16) to give

$$E_c = -\frac{6.360865 - 5.607005}{5} = -0.150722$$

which compares well with the exact error,

$$E_t = 6.194631 - 6.360865 = -0.1662341$$

At $t_{i+1} = 2$, the predictor gives 13.44346 and the corrector yields 15.30224, which can be used to compute

$$E_c = -\frac{15.30224 - 13.44346}{5} = -0.37176$$

which also compares favorably with the exact error, $E_t = 14.84392 - 15.30224 = -0.45831$.

The foregoing has been a brief introduction to multistep methods. Additional information can be found elsewhere (e.g., Chapra and Canale, 2010). Although they still have their place for solving certain types of problems, multistep methods are usually not the method of choice for most problems routinely confronted in engineering and science. That said, they are still used. For example, the MATLAB function ode113 is a multistep method. We have therefore included this section to introduce you to their basic principles.

23.3   STIFFNESS

Stiffness is a special problem that can arise in the solution of ordinary differential equations. A *stiff system* is one involving rapidly changing components together with slowly changing ones. In some cases, the rapidly varying components are ephemeral transients that die away quickly, after which the solution becomes dominated by the slowly varying

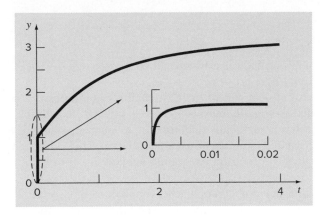

FIGURE 23.8
Plot of a stiff solution of a single ODE. Although the solution appears to start at 1, there is actually a fast transient from $y = 0$ to 1 that occurs in less than the 0.005 time unit. This transient is perceptible only when the response is viewed on the finer timescale in the inset.

components. Although the transient phenomena exist for only a short part of the integration interval, they can dictate the time step for the entire solution.

Both individual and systems of ODEs can be stiff. An example of a single stiff ODE is

$$\frac{dy}{dt} = -1000y + 3000 - 2000e^{-t} \tag{23.17}$$

If $y(0) = 0$, the analytical solution can be developed as

$$y = 3 - 0.998e^{-1000t} - 2.002e^{-t} \tag{23.18}$$

As in Fig. 23.8, the solution is initially dominated by the fast exponential term (e^{-1000t}). After a short period $(t < 0.005)$, this transient dies out and the solution becomes governed by the slow exponential (e^{-t}).

Insight into the step size required for stability of such a solution can be gained by examining the homogeneous part of Eq. (23.17):

$$\frac{dy}{dt} = -ay \tag{23.19}$$

If $y(0) = y_0$, calculus can be used to determine the solution as

$$y = y_0 e^{-at}$$

Thus, the solution starts at y_0 and asymptotically approaches zero.

Euler's method can be used to solve the same problem numerically:

$$y_{i+1} = y_i + \frac{dy_i}{dt}h$$

Substituting Eq. (23.19) gives

$$y_{i+1} = y_i - ay_i h$$

or

$$y_{i+1} = y_i (1 - ah) \tag{23.20}$$

The stability of this formula clearly depends on the step size h. That is, $|1 - ah|$ must be less than 1. Thus, if $h > 2/a$, $|y_i| \to \infty$ as $i \to \infty$.

For the fast transient part of Eq. (23.18), this criterion can be used to show that the step size to maintain stability must be $< 2/1000 = 0.002$. In addition, we should note that, whereas this criterion maintains stability (i.e., a bounded solution), an even smaller step size would be required to obtain an accurate solution. Thus, although the transient occurs for only a small fraction of the integration interval, it controls the maximum allowable step size.

Rather than using explicit approaches, implicit methods offer an alternative remedy. Such representations are called *implicit* because the unknown appears on both sides of the equation. An implicit form of Euler's method can be developed by evaluating the derivative at the future time:

$$y_{i+1} = y_i + \frac{dy_{i+1}}{dt} h$$

This is called the *backward,* or *implicit, Euler's method.* Substituting Eq. (23.19) yields

$$y_{i+1} = y_i - a y_{i+1} h$$

which can be solved for

$$y_{i+1} = \frac{y_i}{1 + ah} \tag{23.21}$$

For this case, regardless of the size of the step, $|y_i| \to 0$ as $i \to \infty$. Hence, the approach is called *unconditionally stable.*

EXAMPLE 23.5 **Explicit and Implicit Euler**

Problem Statement. Use both the explicit and implicit Euler methods to solve Eq. (23.17), where $y(0) = 0$. **(a)** Use the explicit Euler with step sizes of 0.0005 and 0.0015 to solve for y between $t = 0$ and 0.006. **(b)** Use the implicit Euler with a step size of 0.05 to solve for y between 0 and 0.4.

Solution. **(a)** For this problem, the explicit Euler's method is

$$y_{i+1} = y_i + (-1000y_i + 3000 - 2000e^{-t_i})h$$

The result for $h = 0.0005$ is displayed in Fig. 23.9a along with the analytical solution. Although it exhibits some truncation error, the result captures the general shape of the analytical solution. In contrast, when the step size is increased to a value just below the stability limit ($h = 0.0015$), the solution manifests oscillations. Using $h > 0.002$ would result in a totally unstable solution—that is, it would go infinite as the solution progressed.

(b) The implicit Euler's method is

$$y_{i+1} = y_i + (-1000y_{i+1} + 3000 - 2000e^{-t_{i+1}})h$$

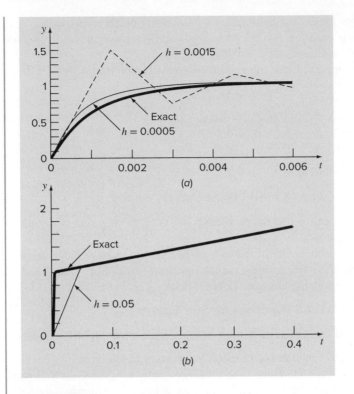

FIGURE 23.9
Solution of a stiff ODE with (*a*) the explicit and (*b*) implicit Euler methods.

Now because the ODE is linear, we can rearrange this equation so that y_{i+1} is isolated on the left-hand side:

$$y_{i+1} = \frac{y_i + 3000h - 2000he^{-t_{i+1}}}{1 + 1000h}$$

The result for $h = 0.05$ is displayed in Fig. 23.9*b* along with the analytical solution. Notice that even though we have used a much bigger step size than the one that induced instability for the explicit Euler, the numerical result tracks nicely on the analytical solution.

Systems of ODEs can also be stiff. An example is

$$\frac{dy_1}{dt} = -5y_1 + 3y_2 \tag{23.22a}$$

$$\frac{dy_2}{dt} = 100y_1 - 301y_2 \tag{23.22b}$$

For the initial conditions $y_1(0) = 52.29$ and $y_2(0) = 83.82$, the exact solution is

$$y_1 = 52.96e^{-3.9899t} - 0.67e^{-302.0101t} \tag{23.23a}$$

$$y_2 = 17.83e^{-3.9899t} + 65.99e^{-302.0101t} \tag{23.23b}$$

Note that the exponents are negative and differ by about two orders of magnitude. As with the single equation, it is the large exponents that respond rapidly and are at the heart of the system's stiffness.

An implicit Euler's method for systems can be formulated for the present example as

$$y_{1,i+1} = y_{1,i} + (-5y_{1,i+1} + 3y_{2,i+1})h \tag{23.24a}$$

$$y_{2,i+1} = y_{2,i} + (100y_{1,i+1} - 301y_{2,i+1})h \tag{23.24b}$$

Collecting terms gives

$$(1 + 5h)y_{1,i+1} - 3y_{2,i+1} = y_{1,i} \tag{23.25a}$$

$$-100y_{1,i+1} + (1 + 301h)y_{2,i+1} = y_{2,i} \tag{23.25b}$$

Thus, we can see that the problem consists of solving a set of simultaneous equations for each time step.

For nonlinear ODEs, the solution becomes even more difficult since it involves solving a system of nonlinear simultaneous equations (recall Sec. 12.2). Thus, although stability is gained through implicit approaches, a price is paid in the form of added solution complexity.

23.3.1 MATLAB Functions for Stiff Systems

MATLAB has a number of built-in functions for solving stiff systems of ODEs. These are

ode15s. This function is a variable-order solver based on numerical differentiation formulas. It is a multistep solver that optionally uses the Gear backward differentiation formulas. This is used for stiff problems of low to medium accuracy.

ode23s. This function is based on a modified Rosenbrock formula of order 2. Because it is a one-step solver, it may be more efficient than ode15s at crude tolerances. It can solve some kinds of stiff problems better than ode15s.

ode23t. This function is an implementation of the trapezoidal rule with a "free" interpolant. This is used for moderately stiff problems with low accuracy where you need a solution without numerical damping.

ode23tb. This is an implementation of an implicit Runge-Kutta formula with a first stage that is a trapezoidal rule and a second stage that is a backward differentiation formula of order 2. This solver may also be more efficient than ode15s at crude tolerances.

EXAMPLE 23.6   MATLAB for Stiff ODEs

Problem Statement.   The van der Pol equation is a model of an electronic circuit that arose back in the days of vacuum tubes,

$$\frac{d^2y_1}{dt^2} - \mu\left(1 - y_1^2\right)\frac{dy_1}{dt} + y_1 = 0 \tag{E23.6.1}$$

The solution to this equation becomes progressively stiffer as μ gets large. Given the initial conditions, $y_1(0) = dy_1/dt = 1$, use MATLAB to solve the following two cases: **(a)** for $\mu = 1$, use ode45 to solve from $t = 0$ to 20; and **(b)** for $\mu = 1000$, use ode23s to solve from $t = 0$ to 6000.

Solution. **(a)** The first step is to convert the second-order ODE into a pair of first-order ODEs by defining

$$\frac{dy_1}{dt} = y_2$$

Using this equation, Eq. (E23.6.1) can be written as

$$\frac{dy_2}{dt} = \mu \left(1 - y_1^2\right) y_2 - y_1 = 0$$

An M-file can now be created to hold this pair of differential equations:

```
function yp = vanderpol(t,y,mu)
yp = [y(2);mu*(1-y(1)^2)*y(2)-y(1)];
```

Notice how the value of μ is passed as a parameter. As in Example 23.1, ode45 can be invoked and the results plotted:

```
>> [t,y] = ode45(@vanderpol,[0 20],[1 1],[],1);
>> plot(t,y(:,1),'-',t,y(:,2),'--')
>> legend('y1','y2');
```

Observe that because we are not specifying any options, we must use open brackets [] as a place holder. The smooth nature of the plot (Fig. 23.10a) suggests that the van der Pol equation with $\mu = 1$ is not a stiff system.

(b) If a standard solver like ode45 is used for the stiff case ($\mu = 1000$), it will fail miserably (try it, if you like). However, ode23s does an efficient job:

```
>> [t,y] = ode23s(@vanderpol,[0 6000],[1 1],[],1000);
>> plot(t,y(:,1))
```

FIGURE 23.10
Solutions for van der Pol's equation. (*a*) Nonstiff form solved with **ode45** and (*b*) stiff form solved with **ode23s**.

We have only displayed the y_1 component because the result for y_2 has a much larger scale. Notice how this solution (Fig. 23.10*b*) has much sharper edges than is the case in Fig. 23.10*a*. This is a visual manifestation of the "stiffness" of the solution.

23.4   MATLAB APPLICATION: BUNGEE JUMPER WITH CORD

In this section, we will use MATLAB to solve for the vertical dynamics of a jumper connected to a stationary platform with a bungee cord. As developed at the beginning of Chap. 22, the problem consisted of solving two coupled ODEs for vertical position and velocity. The differential equation for position is

$$\frac{dx}{dt} = v \tag{23.26}$$

The differential equation for velocity is different depending on whether the jumper has fallen to a distance where the cord is fully extended and begins to stretch. Thus, if the distance fallen is less than the cord length, the jumper is only subject to gravitational and drag forces,

$$\frac{dv}{dt} = g - \text{sign}(v)\frac{c_d}{m}v^2 \tag{23.27a}$$

Once the cord begins to stretch, the spring and dampening forces of the cord must also be included:

$$\frac{dv}{dt} = g - \text{sign}(v)\frac{c_d}{m}v^2 - \frac{k}{m}(x - L)\frac{\gamma}{m}v \tag{23.27b}$$

The following example shows how MATLAB can be used to solve this problem.

EXAMPLE 23.7   Bungee Jumper with Cord

Problem Statement. Determine the position and velocity of a bungee jumper with the following parameters: $L = 30$ m, $g = 9.81$ m/s^2, $m = 68.1$ kg, $c_d = 0.25$ kg/m, $k = 40$ N/m, and $\gamma = 8$ N · s/m. Perform the computation from $t = 0$ to 50 s and assume that the initial conditions are $x(0) = v(0) = 0$.

Solution. The following M-file can be set up to compute the right-hand sides of the ODEs:

```
function dydt = bungee(t,y,L,cd,m,k,gamma)
g = 9.81;
cord = 0;
if y(1) > L %determine if the cord exerts a force
  cord = k/m*(y(1)-L)+gamma/m*y(2);
end
dydt = [y(2); g - sign(y(2))*cd/m*y(2)^2 - cord];
```

Notice that the derivatives are returned as a column vector because this is the format required by the MATLAB solvers.

Because these equations are not stiff, we can use ode45 to obtain the solutions and display them on a plot:

```
>> [t,y] = ode45(@bungee,[0 50],[0 0],[],30,0.25,68.1,40,8);
>> plot(t,-y(:,1),'-',t,y(:,2),':')
>> legend('x (m)','v (m/s)')
```

As in Fig. 23.11, we have reversed the sign of distance for the plot so that negative distance is in the downward direction. Notice how the simulation captures the jumper's bouncing motion.

FIGURE 23.11
Plot of distance and velocity of a bungee jumper.

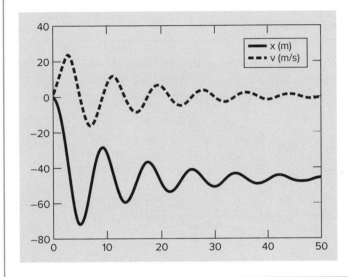

23.5 CASE STUDY   PLINY'S INTERMITTENT FOUNTAIN

Background. The Roman natural philosopher, Pliny the Elder, purportedly had an intermittent fountain in his garden. As in Fig. 23.12, water enters a cylindrical tank at a constant flow rate Q_{in} and fills until the water reaches y_{high}. At this point, water siphons out of the tank through a circular discharge pipe, producing a fountain at the pipe's exit. The fountain runs until the water level decreases to y_{low}, whereupon the siphon fills with air and the fountain stops. The cycle then repeats as the tank fills until the water reaches y_{high}, and the fountain flows again.

When the siphon is running, the outflow Q_{out} can be computed with the following formula based on *Torricelli's law:*

$$Q_{out} = C\sqrt{2gy}\,\pi r^2 \tag{23.28}$$

23.5 CASE STUDY continued

FIGURE 23.12
An intermittent fountain.

Neglecting the volume of water in the pipe, compute and plot the level of the water in the tank as a function of time over 100 seconds. Assume an initial condition of an empty tank $y(0) = 0$, and employ the following parameters for your computation:

$R_T = 0.05$ m	$r = 0.007$ m	$y_{low} = 0.025$ m
$y_{high} = 0.1$ m	$C = 0.6$	$g = 9.81$ m/s^2
$Q_{in} = 50 \times 10^{-6}$ m^3/s		

Solution. When the fountain is running, the rate of change in the tank's volume V (m^3) is determined by a simple balance of inflow minus the outflow:

$$\frac{dV}{dt} = Q_{in} - Q_{out} \tag{23.29}$$

where V = volume (m^3). Because the tank is cylindrical, $V = \pi R_t^2 y$. Substituting this relationship along with Eq. (23.28) into Eq. (23.29) gives

$$\frac{dy}{dt} = \frac{Q_{in} - C\sqrt{2gy}\,\pi r^2}{\pi R_t^2} \tag{23.30}$$

When the fountain is not running, the second term in the numerator goes to zero. We can incorporate this mechanism in the model by introducing a new dimensionless variable *siphon* that equals zero when the fountain is off and equals one when it is flowing:

$$\frac{dy}{dt} = \frac{Q_{in} - siphon \times C\sqrt{2gy}\,\pi r^2}{\pi R_t^2} \tag{23.31}$$

In the present context, *siphon* can be thought of as a switch that turns the fountain off and on. Such two-state variables are called *Boolean* or *logical variables,* where zero is equivalent to false and one is equivalent to true.

23.5 CASE STUDY continued

Next we must relate *siphon* to the dependent variable *y*. First, *siphon* is set to zero whenever the level falls below y_{low}. Conversely, *siphon* is set to one whenever the level rises above y_{high}. The following M-file function follows this logic in computing the derivative:

```
function dy = Plinyode(t,y)
global siphon
Rt = 0.05; r = 0.007; yhi = 0.1; ylo = 0.025;
C = 0.6; g = 9.81; Qin = 0.00005;
if y(1) <= ylo
  siphon = 0;
elseif y(1) >= yhi
  siphon = 1;
end
Qout = siphon * C * sqrt(2 * g * y(1)) * pi * r ^ 2;
dy = (Qin - Qout) / (pi * Rt ^ 2);
```

Notice that because its value must be maintained between function calls, siphon is declared as a global variable. Although the use of global variables is not encouraged (particularly in larger programs), it is useful in the present context.

The following script employs the built-in ode45 function to integrate Plinyode and generate a plot of the solution:

```
global siphon
siphon = 0;
tspan = [0 100]; y0 = 0;
[tp,yp]=ode45(@Plinyode,tspan,y0);
plot(tp,yp)
xlabel('time, (s)')
ylabel('water level in tank, (m)')
```

As shown in Fig. 23.13, the result is clearly incorrect. Except for the original filling period, the level seems to start emptying prior to reaching y_{high}. Similarly, when it is draining, the siphon shuts off well before the level drops to y_{low}.

At this point, suspecting that the problem demands more firepower than the trusty ode45 routine, you might be tempted to use one of the other MATLAB ODE solvers such as ode23s or ode23tb. But if you did, you would discover that although these routines yield somewhat different results, they would still generate incorrect solutions.

The difficulty arises because the ODE is discontinuous at the point that the siphon switches on or off. For example, as the tank is filling, the derivative is dependent only on the constant inflow and for the present parameters has a constant value of 6.366×10^{-3} m/s. However, as soon as the level reaches y_{high}, the outflow kicks in and the derivative abruptly drops to -1.013×10^{-2} m/s. Although the adaptive step-size routines used by MATLAB work marvelously for many problems, they often get heartburn when dealing with such discontinuities. Because they infer the behavior of the solution by comparing the results of different steps, a discontinuity represents something akin to stepping into a deep pothole on a dark street.

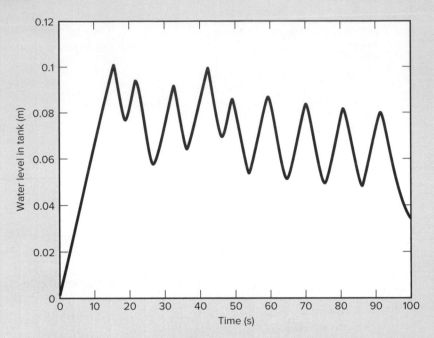

FIGURE 23.13
The level in Pliny's fountain versus time as simulated with ode45.

At this point, your first inclination might be to just give up. After all, if it's too hard for MATLAB, no reasonable person could expect you to come up with a solution. Because professional engineers and scientists rarely get away with such excuses, your only recourse is to develop a remedy based on your knowledge of numerical methods.

Because the problem results from adaptively stepping across a discontinuity, you might revert to a simpler approach and use a constant, small step size. If you think about it, that's precisely the approach you would take if you were traversing a dark, pothole-filled street. We can implement this solution strategy by merely replacing ode45 with the constant-step rk4sys function from Chap. 22 (Fig. 22.8). For the script outlined above, the fourth line would be formulated as

```
[tp,yp] = rk4sys(@Plinyode,tspan,y0,0.0625);
```

As in Fig. 23.14, the solution now evolves as expected. The tank fills to y_{high} and then empties until it reaches y_{low}, when the cycle repeats.

There are a two take-home messages that can be gleaned from this case study. First, although it's human nature to think the opposite, simpler is sometimes better. After all, to paraphrase Einstein, "Everything should be as simple as possible, but no simpler." Second, you should never blindly believe every result generated by the computer. You've probably heard the old chestnut, "garbage in, garbage out" in reference to the impact of data quality

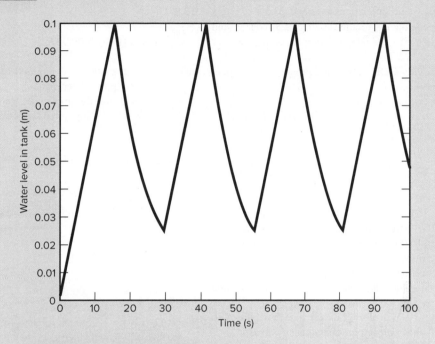

FIGURE 23.14
The level in Pliny's fountain versus time as simulated with a small, constant step size using the rk4sys function (Fig. 22.8).

on the validity of computer output. Unfortunately, some individuals think that regardless of what went in (the data) and what's going on inside (the algorithm), it's always "gospel out." Situations like the one depicted in Fig. 23.13 are particularly dangerous—that is, although the output is incorrect, it's not obviously wrong. That is, the simulation does not go unstable or yield negative levels. In fact, the solution moves up and down in the manner of an intermittent fountain, albeit incorrectly.

Hopefully, this case study illustrates that even a great piece of software such as MATLAB is not foolproof. Hence, sophisticated engineers and scientists always examine numerical output with a healthy skepticism based on their considerable experience and knowledge of the problems they are solving.

PROBLEMS

23.1 Repeat the same simulations as in Sec. 23.5 for Pliny's fountain, but generate the solutions with ode23, ode23s, and ode113. Use subplot to develop a vertical three-pane plot of the time series.

23.2 The following ODEs have been proposed as a model of an epidemic:

$$\frac{dS}{dt} = -aSI$$

$$\frac{dI}{dt} = aSI - rI$$

$$\frac{dR}{dt} = rI$$

where S = the susceptible individuals, I = the infected, R = the recovered, a = the infection rate, and r = the recovery rate. A city has 10,000 people, all of whom are susceptible.

(a) If a single infectious individual enters the city at $t = 0$, compute the progression of the epidemic until the number of infected individuals falls below 10. Use the following parameters: $a = 0.002/(\text{person} \cdot \text{week})$ and $r = 0.15/d$. Develop time-series plots of all the state variables. Also generate a phase-plane plot of S versus I versus R.

(b) Suppose that after recovery, there is a loss of immunity that causes recovered individuals to become susceptible. This reinfection mechanism can be computed as ρR, where ρ = the reinfection rate. Modify the model to include this mechanism and repeat the computations in (a) using $\rho = 0.03/d$.

23.3 Solve the following initial-value problem over the interval from $t = 2$ to 3:

$$\frac{dy}{dt} = -0.5y + e^{-t}$$

Use the non-self-starting Heun method with a step size of 0.5 and initial conditions of $y(1.5) = 5.222138$ and $y(2.0) = 4.143883$. Iterate the corrector to $\varepsilon_s = 0.1\%$. Compute the percent relative errors for your results based on the exact solutions obtained analytically: $y(2.5) = 3.273888$ and $y(3.0) = 2.577988$.

23.4 Solve the following initial-value problem over the interval from $t = 0$ to 0.5:

$$\frac{dy}{dt} = yt^2 - y$$

Use the fourth-order RK method to predict the first value at $t = 0.25$. Then use the non-self-starting Heun method to make the prediction at $t = 0.5$. Note: $y(0) = 1$.

23.5 Given

$$\frac{dy}{dt} = -100,000y + 99,999e^{-t}$$

(a) Estimate the step size required to maintain stability using the explicit Euler method.

(b) If $y(0) = 0$, use the implicit Euler to obtain a solution from $t = 0$ to 2 using a step size of 0.1.

23.6 Given

$$\frac{dy}{dt} = 30(\sin t - y) + 3 \cos t$$

If $y(0) = 0$, use the implicit Euler to obtain a solution from $t = 0$ to 4 using a step size of 0.4.

23.7 Given

$$\frac{dx_1}{dt} = 999x_1 + 1999x_2$$

$$\frac{dx_2}{dt} = -1000x_1 - 2000x_2$$

If $x_1(0) = x_2(0) = 1$, obtain a solution from $t = 0$ to 0.2 using a step size of 0.05 with the (a) explicit and (b) implicit Euler methods.

23.8 The following nonlinear, parasitic ODE was suggested by Hornbeck (1975):

$$\frac{dy}{dt} = 5(y - t^2)$$

If the initial condition is $y(0) = 0.08$, obtain a solution from $t = 0$ to 5:

(a) Analytically.

(b) Using the fourth-order RK method with a constant step size of 0.03125.

(c) Using the MATLAB function ode45.

(d) Using the MATLAB function ode23s.

(e) Using the MATLAB function ode23tb.

Present your results in graphical form.

23.9 Recall from Example 20.5 that the humps function exhibits both flat and steep regions over a relatively short x range,

$$f(x) = \frac{1}{(x - 0.3)^2 + 0.01} + \frac{1}{(x - 0.9)^2 + 0.04} - 6$$

Determine the value of the definite integral of this function between $x = 0$ and 1 using (a) the quad and (b) the ode45 functions.

23.10 The oscillations of a swinging pendulum can be simulated with the following nonlinear model:

$$\frac{d^2\theta}{dt^2} + \frac{g}{l}\sin\theta = 0$$

where θ = the angle of displacement, g = the gravitational constant, and l = the pendulum length. For small angular displacements, the $\sin\theta$ is approximately equal to θ and the model can be linearized as

$$\frac{d^2\theta}{dt^2} + \frac{g}{l}\theta = 0$$

Use ode45 to solve for θ as a function of time for both the linear and nonlinear models where $l = 0.6$ m and $g = 9.81$ m/s². First, solve for the case where the initial condition is for a small displacement ($\theta = \pi/8$ and $d\theta/dt = 0$). Then repeat the calculation for a large displacement ($\theta = \pi/2$). For each case, plot the linear and nonlinear simulations on the same plot.

23.11 Employ the events option described in Sec. 23.1.2 to determine the period of a 1-m long, linear pendulum (see description in Prob. 23.10). Compute the period for the following initial conditions: (a) $\theta = \pi/8$, (b) $\theta = \pi/4$, and (c) $\theta = \pi/2$. For all three cases, set the initial angular velocity at zero. (Hint: A good way to compute the period is to determine how long it takes for the pendulum to reach $\theta = 0$ [i.e., the bottom of its arc]). The period is equal to four times this value.

23.12 Repeat Prob. 23.11, but for the nonlinear pendulum described in Prob. 23.10.

23.13 The following system is a classic example of stiff ODEs that can occur in the solution of chemical reaction kinetics:

$$\frac{dc_1}{dt} = -0.013c_1 - 1000c_1c_3$$

$$\frac{dc_2}{dt} = -2500c_2c_3$$

$$\frac{dc_3}{dt} = -0.013c_1 - 1000c_1c_3 - 2500c_2c_3$$

Solve these equations from $t = 0$ to 50 with initial conditions $c_1(0) = c_2(0) = 1$ and $c_3(0) = 0$. If you have access to MATLAB software, use both standard (e.g., ode45) and stiff (e.g., ode23s) functions to obtain your solutions.

23.14 The following second-order ODE is considered to be stiff:

$$\frac{d^2y}{dx^2} = -1001\frac{dy}{dx} - 1000y$$

FIGURE P23.15

Solve this differential equation (a) analytically and (b) numerically for $x = 0$ to 5. For (b) use an implicit approach with $h = 0.5$. Note that the initial conditions are $y(0) = 1$ and $y'(0) = 0$. Display both results graphically.

23.15 Consider the thin rod of length l moving in the x-y plane as shown in Fig. P23.15. The rod is fixed with a pin on one end and a mass at the other. Note that $g = 9.81$ m/s² and $l = 0.5$ m. This system can be solved using

$$\ddot{\theta} - \frac{g}{l}\theta = 0$$

Let $\theta(0) = 0$ and $\dot{\theta}(0) = 0.25$ rad/s. Solve using any method studied in this chapter. Plot the angle versus time and the angular velocity versus time. (Hint: Decompose the second-order ODE.)

23.16 Given the first-order ODE:

$$\frac{dx}{dt} = -700x - 1000e^{-t}$$

$$x(t = 0) = 4$$

Solve this stiff differential equation using a numerical method over the time period $0 \le t \le 5$. Also solve analytically and plot the analytic and numerical solution for both the fast and slow transient phases of the time scale.

23.17 Solve the following differential equation from $t = 0$ to 2

$$\frac{dy}{dt} = -10y$$

with the initial condition $y(0) = 1$. Use the following techniques to obtain your solutions: (a) analytically, (b) the explicit Euler method, and (c) the implicit Euler method. For (b) and (c) use $h = 0.1$ and 0.2. Plot your results.

23.18 The Lotka-Volterra equations described in Sec. 22.6 have been refined to include additional factors that impact predator-prey dynamics. For example, over and above predation, prey population can be limited by other factors such as space. Space limitation can be incorporated into the model as a carrying capacity (recall the logistic model described in Prob. 22.5) as in

$$\frac{dx}{dt} = a\left(1 - \frac{x}{K}\right)x - bxy$$

$$\frac{dy}{dt} = -cy + dxy$$

where K = the carrying capacity. Use the same parameter values and initial conditions as in Sec. 22.6 to integrate these equations from $t = 0$ to 100 using ode45, and develop both time series and phase-plane plots of the results.
(a) Employ a very large value of $K = 10^8$ to validate that you obtain the same results as in Sec. 22.6.
(b) Compare (a) with the more realistic carrying capacity of $K = 200$. Discuss your results.

23.19 Two masses are attached to a wall by linear springs (Fig. P23.19). Force balances based on Newton's second law can be written as

$$\frac{d^2x_1}{dt^2} = -\frac{k_1}{m_1}(x_1 - L_1) + \frac{k_2}{m_1}(x_2 - x_1 - w_1 - L_2)$$

$$\frac{d^2x_2}{dt^2} = -\frac{k_2}{m_2}(x_2 - x_1 - w_1 - L_2)$$

where k = the spring constants, m = mass, L = the length of the unstretched spring, and w = the width of the mass. Compute the positions of the masses as a function of time using the following parameter values: $k_1 = k_2 = 5$, $m_1 = m_2 = 2$, $w_1 = w_2 = 5$, and $L_1 = L_2 = 2$. Set the initial conditions as $x_1 = L_1$ and $x_2 = L_1 + w_1 + L_2 + 6$. Perform the simulation from $t = 0$ to 20. Construct time-series plots of both the displacements and the velocities. In addition, produce a phase-plane plot of x_1 versus x_2.

23.20 Use ode45 to integrate the differential equations for the system described in Prob. 23.19. Generate vertically stacked subplots of displacements (top) and velocities (bottom). Employ the fft function to compute the discrete Fourier transform (DFT) of the first mass's displacement. Generate and plot a power spectrum in order to identify the system's resonant frequencies.

23.21 Perform the same computations as in Prob. 23.20 but based on the first floor of the structure in Prob. 22.22.

FIGURE P23.19

23.22 Use the approach and example outlined in Sec. 23.1.2, but determine the time, height, and velocity when the bungee jumper is the farthest above the ground, and generate a plot of the solution.

23.23 As depicted in Fig. P23.23, a double pendulum consists of a pendulum attached to another pendulum. We indicate the upper and lower pendulums by subscripts 1 and 2, respectively, and we place the origin at the pivot point of the upper pendulum with y increasing upward. We further assume that the system oscillates in a vertical plane subject to gravity, that the pendulum rods are massless and rigid, and the pendulum masses are considered to be point masses. Under these assumptions, force balances can be used to derive the following equations of motion

$$\frac{d^2\theta_1}{dt^2} = \frac{-g(2m_1 + m_2)\sin\theta_1 - m_2 g \sin(\theta_1 - 2\theta_2) - 2\sin(\theta_1 - \theta_2)m_2\left((d\theta_2/dt)^2 L_2 + (d\theta_1/dt)^2 L_1 \cos(\theta_1 - \theta_2)\right)}{L_1(2m_1 + m_2 - m_2 \cos(2\theta_1 - \theta_2))}$$

$$\frac{d^2\theta_2}{dt^2} = \frac{2\sin(\theta_1 - \theta_2)\left((d\theta_1/dt)^2 L_1(m_1 + m_2) + g(m_1 + m_2)\cos(\theta_1) + (d\theta_2/dt)^2 L_2 m_2 \cos(\theta_1 - \theta_2)\right)}{L_2(2m_1 + m_2 - m_2 \cos(2\theta_1 - \theta_2))}$$

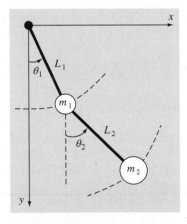

FIGURE P23.23
A double pendulum.

where the subscripts 1 and 2 designate the top and bottom pendulum, respectively, θ = angle (radians) with 0 = vertical downward and counter-clockwise positive, t = time (s), g = gravitational acceleration (= 9.81 m/s^2), m = mass (kg), and L = length (m). Note that the x and y coordinates of the masses are functions of the angles as in

$$x_1 = L_1 \sin\theta_1 \qquad y_1 = -L_1 \cos\theta_1$$

$$x_2 = x_1 + L_2 \sin\theta_2 \qquad y_2 = y_1 - L_2 \cos\theta_2$$

(a) Use ode45 to solve for the angles and angular velocities of the masses as a function of time from $t = 0$ to 40 s. Employ subplot to create a stacked plot with a time series of the angles in the top panel and a state space plot of θ_2 versus θ_1 in the bottom panel. (b) Create an animated plot depicting the motion of the pendulum. Test your code for the following:

Case 1 (small displacement): $L_1 = L_2 = 1$ m, $m_1 = m_2 = 0.25$ kg, with initial conditions: $\theta_1 = 0.5$ m and $\theta_2 = d\theta_1/dt = d\theta_2/dt = 0$.
Case 2 (large displacement): $L_1 = L_2 = 1$ m, $m_1 = 0.5$ kg, $m_2 = 0.25$ kg, with initial conditions: $\theta_1 = 1$ m and $\theta_2 = d\theta_1/dt = d\theta_2/dt = 0$.

23.24 Figure P23.24 shows the forces exerted on a hot air balloon system.

Formulate the drag force as

$$F_D = \frac{1}{2}\rho_a v^2 A\, C_d$$

where ρ_a = air density (kg/m^3), v = velocity (m/s), A = projected frontal area (m^2), and C_d = the dimensionless drag coefficient ($\cong 0.47$ for a sphere). Note also that the total mass of the balloon consists of two components:

$$m = m_G + m_P$$

where m_G = the mass of the gas inside the expanded balloon (kg), and m_P = the mass of the payload (basket, passengers, and the unexpanded balloon = 265 kg). Assume that the ideal gas law holds ($P = \rho RT$), that the balloon is a perfect sphere with a diameter of 17.3 m, and that the heated air inside the envelope is at roughly the same pressure as the outside air. Other necessary parameters are normal atmospheric pressure, $P = 101,300$ Pa; gas constant for dry air, $R = 287$ Joules/kg·K; average temperature of air inside the

FIGURE P23.24
Forces on a hot air balloon: F_B = buoyancy, F_G = weight of gas, F_P = weight of payload (including the balloon envelope), and F_D = drag. Note that the direction of the drag is downward when the balloon is rising.

balloon, $T = 100$ °C; and the normal (ambient) air density, $\rho = 1.2$ kg/m^3.
(a) Use a force balance to develop the differential equation for dv/dt as a function of the model's fundamental parameters.
(b) At steady-state, calculate the particle's terminal velocity.
(c) Use ode45 to compute the velocity and position of the balloon from $t = 0$ to 60 s given the previous parameters along with the initial condition: $v(0) = 0$. Develop a plot of your results.

23.25 Develop a MATLAB script using ode45 to compute the velocity, v, and position, z, of a hot air balloon as described in Prob. 23.24. Perform the calculation from $t = 0$ to 60 s with a step size of 1.6 s. At $z = 200$ m, assume that part of the payload (100 kg) is dropped out of the balloon. Develop a plot of your results.

23.26 You go on a two-week vacation and place your pet goldfish "Freddie" into your bathtub. Note that you dechlorinate the water first! You then place an air tight plexiglass cover over the top of the tub in order to protect Freddie from your cat, Beelzebub. You mistakenly mix one tablespoon of sugar into the tub (you thought it was fish food!).

Unfortunately, there are bacteria in the water (remember you got rid of the chlorine!), which break down the sugar consuming dissolved oxygen in the process. The oxidation reaction follows first-order kinetics with a reaction rate of $k_d = 0.15/d$. The tub initially has a sugar concentration of 20 mgO_2/L and an oxygen concentration of 8.4 mgO_2/L. Note that the mass balances for the sugar (expressed in oxygen equivalents) and dissolved oxygen can be written as

$$\frac{dL}{dt} = -k_d L$$

$$\frac{do}{dt} = -k_d L$$

where L = sugar concentration expressed as oxygen equivalents (mg/L), t = time (d), and o = dissolved oxygen concentration (mg/L). Thus, as the sugar gets oxidized, an equivalent amount of oxygen is lost from the tub. Develop a MATLAB script using ode45 to numerically compute the concentrations of sugar and oxygen as a function of time and develop plots of each versus time. Use event to automatically stop when the oxygen concentration falls below a critical oxygen level of 2 mgO_2/L.

23.27 The growth of bacteria from substrate can be represented by the following pair of differential equations

$$\frac{dX}{dt} = Y k_{max} \frac{S}{k_s + S} X$$

$$\frac{dS}{dt} = -k_{max} \frac{S}{k_s + S} X$$

where X = bacterial biomass, t = time (d), Y = a yield coefficient, k_{max} = maximum bacterial growth rate, S = substrate concentration, and k_s = half saturation constant. The parameter values are $Y = 0.75$, $k_{max} = 0.3$, and $k_s = 1 \times 10^{-4}$ and the initial conditions at $t = 0$ are $S(0) = 5$ and $X(0) = 0.05$. Note that neither X nor S can fall below zero as negative values are impossible. **(a)** Use ode23 to solve for X and S from $t = 0$ to 25. **(b)** Repeat the solution, but set the relative tolerance to 1×10^{-6}. **(c)** Keep repeating the solution with the relative tolerance set to 1×10^{-6}, but determine which of the MATLAB ode solvers (including the stiff solvers) obtains correct (i.e., positive) results. Use the tic and toc functions to determine the execution time for each option.

23.28 The oscillations of a swinging pendulum can be simulated with the following nonlinear model:

$$\frac{d^2\theta}{dt^2} = -\frac{g}{l}\sin\theta$$

where θ = the angle of displacement (radians), g = the gravitational constant (= 9.81 m/s^2), and l = the pendulum length. **(a)** Express this equation as a pair of first-order ODEs.

(b) Use ode45 to solve for θ and $d\theta/dt$ as a function of time for the case where $l = 0.65$ m and the initial conditions are $\theta = \pi/8$ and $d\theta/dt = 0$. **(c)** Generate a plot of your results, and **(d)** use the diff function to generate a plot of the angular accelerations ($d^2\theta/dt^2$) versus time based on the vector of angular velocities ($d\theta/dt$) generated in **(b)**. Use subplot to display all graphs as a single vertical three-panel plot with the top, middle, and bottom plots corresponding to θ, $d\theta/dt$, and $d^2\theta/dt^2$, respectively.

23.29 A number of individuals have made skydives from very high altitudes. Suppose that an 80-kg skydiver launches from an elevation of 36.500 km above the earth's surface. The skydiver has a projected area, $A = 0.55$ m^2; and a dimensionless drag coefficient, $C_d = 1$. Note that the gravitational acceleration, g (m/s^2), can be related to elevation by

$$g = 9.806412 - 0.000003039734z$$

where z = elevation above the earth's surface (m) and the density of air, ρ (kg/m^3), at various elevations can be tabulated as

z (km)	ρ (kg/m^3)	z (km)	ρ (kg/m^3)	z (km)	ρ (kg/m^3)
−1	1.3470	6	0.6601	25	0.04008
0	1.2250	7	0.5900	30	0.01841
1	1.1120	8	0.5258	40	0.003996
2	1.0070	9	0.4671	50	0.001027
3	0.9093	10	0.4135	60	0.0003097
4	0.8194	15	0.1948	70	8.283×10^{-5}
5	0.7364	20	0.08891	80	1.846×10^{-5}

(a) Based on a force balance between gravity and drag, derive differential equations for velocity and distance based on a force balance for the skydiver.

(b) Use a numerical method to solve for velocity and distance that terminates when the jumper reaches an elevation that is a kilometer above the earth's surface. Plot your results.

23.30 As depicted in Fig. P23.30, a parachutist jumps from an aircraft that is flying in a straight line parallel with the ground. **(a)** Using force balances derive four differential equations for the rates of change of the x and y components of distances and velocities. [Hint: Recognize that $\sin\theta = v_y/v$ and $\cos\theta = v_x/x$.] **(b)** Employ Ralston's 2nd-order method with $\Delta t = 0.25$ s to generate a solution from $t = 0$ until the parachutist hits the ground assuming that the chute never opens. The drag coefficient is 0.25 kg/m, the mass is 80 kg, and the ground is 2000 m below the initial vertical position

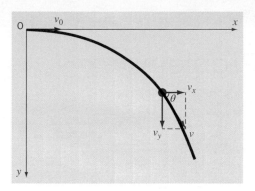

FIGURE P23.30

of the aircraft. The initial conditions are $v_x = 135$ m/s, and $v_y = x = y = 0$. **(c)** Develop a plot of position on Cartesian $(x - y)$ coordinates. **(d)** Repeat (b) and (c) but use ode45 and the events option to determine when the jumper hits the ground.

23.31 The basic differential equation of the elastic curve for a cantilever beam (Fig. P23.31) is given as

$$EI \frac{d^2 y}{dx^2} = -P(L - x)$$

where E = the modulus of elasticity and I the moment of inertia. Solve for the deflection of the beam using ode45. The following parameter values apply: $E = 2 \times 10^{11}$ Pa, $I = 0.00033$ m^4, $P = 4.5$ kN, and $L = 3$ m. Develop a plot of your results along with the analytical solution,

$$y = -\frac{PLx^2}{2EI} + \frac{Px^3}{6EI}$$

FIGURE P23.31

23.32 The following differential equations define the concentrations of three reactants in a closed system (Fig. P23.32),

$$\frac{dc_1}{dt} = -k_{12}c_1 + k_{21}c_2 + k_{31}c_3$$

$$\frac{dc_2}{dt} = k_{12}c_1 - k_{21}c_2 - k_{32}c_2$$

$$\frac{dc_3}{dt} = k_{32}c_2 - k_{31}c_3$$

An experiments with initial conditions of $c_1(0) = 100$, and $c_2(0) = c_3(0) = 0$ yields the following data:

t	1	2	3	4	5	6	8	9	10	12	15
c_1	85.3	66.6	60.6	56.1	49.1	45.3	41.9	37.8	33.7	34.4	35.1
c_2	16.9	18.7	24.1	20.9	18.9	19.9	20.6	13.9	19.1	14.5	15.4
c_3	4.7	7.9	20.1	22.8	32.5	37.7	42.4	47	50.5	52.3	51.3

Use ode45 to integrate the equations and an optimization function to estimate the values of the k's that minimize the sum of the squares of the discrepancies between the model predictions and the data. Employ initial guesses of 0.15 for all the k's.

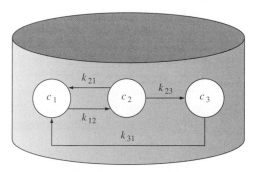

FIGURE P23.32

24

Boundary-Value Problems

CHAPTER OBJECTIVES

The primary objective of this chapter is to introduce you to solving boundary-value problems for ODEs. Specific objectives and topics covered are

- Understanding the difference between initial-value and boundary-value problems.
- Knowing how to express an nth-order ODE as a system of n first-order ODEs.
- Knowing how to implement the shooting method for linear ODEs by using linear interpolation to generate accurate "shots."
- Understanding how derivative boundary conditions are incorporated into the shooting method.
- Knowing how to solve nonlinear ODEs with the shooting method by using root location to generate accurate "shots."
- Knowing how to implement the finite-difference method.
- Understanding how derivative boundary conditions are incorporated into the finite-difference method.
- Knowing how to solve nonlinear ODEs with the finite-difference method by using root-location methods for systems of nonlinear algebraic equations.
- Familiarizing yourself with the built-in MATLAB function bvp4c for solving boundary-value ODEs.

YOU'VE GOT A PROBLEM

To this point, we have been computing the velocity of a free-falling bungee jumper by integrating a single ODE:

$$\frac{dv}{dt} = g - \frac{c_d}{m}v^2 \tag{24.1}$$

Suppose that rather than velocity, you are asked to determine the position of the jumper as a function of time. One way to do this is to recognize that velocity is the first derivative of distance:

$$\frac{dx}{dt} = v \tag{24.2}$$

Thus, by solving the system of two ODEs represented by Eqs. (24.1) and (24.2), we can simultaneously determine both the velocity and the position.

However, because we are now integrating two ODEs, we require two conditions to obtain the solution. We are already familiar with one way to do this for the case where we have values for both position and velocity at the initial time:

$$x(t = 0) = x_i$$
$$v(t = 0) = v_i$$

Given such conditions, we can easily integrate the ODEs using the numerical techniques described in Chaps. 22 and 23. This is referred to as an *initial-value problem*.

But what if we do not know values for both position and velocity at $t = 0$? Let's say that we know the initial position but rather than having the initial velocity, we want the jumper to be at a specified position at a later time. In other words:

$$x(t = 0) = x_i$$
$$x(t = t_f) = x_f$$

Because the two conditions are given at different values of the independent variable, this is called a *boundary-value problem*.

Such problems require special solution techniques. Some of these are related to the methods for initial value problems that were described in the previous two chapters. However, others employ entirely different strategies to obtain solutions. This chapter is designed to introduce you to the more common of these methods.

24.1 INTRODUCTION AND BACKGROUND

24.1.1 What Are Boundary-Value Problems?

An ordinary differential equation is accompanied by auxiliary conditions, which are used to evaluate the constants of integration that result during the solution of the equation. For an nth-order equation, n conditions are required. If all the conditions are specified at the same value of the independent variable, then we are dealing with an *initial-value problem* (Fig. 24.1a). To this point, the material in Part Six (Chaps. 22 and 23) has been devoted to this type of problem.

In contrast, there are often cases when the conditions are not known at a single point but rather are given at different values of the independent variable. Because these values are often specified at the extreme points or boundaries of a system, they are customarily referred to as *boundary-value problems* (Fig. 24.1b). A variety of significant engineering applications fall within this class. In this chapter, we discuss some of the basic approaches for solving such problems.

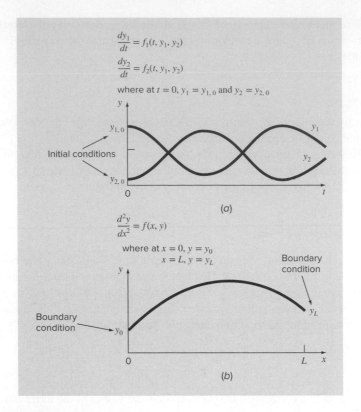

FIGURE 24.1
Initial-value versus boundary-value problems. (*a*) An initial-value problem where all the conditions are specified at the same value of the independent variable. (*b*) A boundary-value problem where the conditions are specified at different values of the independent variable.

24.1.2 Boundary-Value Problems in Engineering and Science

At the beginning of this chapter, we showed how the determination of the position and velocity of a falling object could be formulated as a boundary-value problem. For that example, a pair of ODEs was integrated in time. Although other time-variable examples can be developed, boundary-value problems arise more naturally when integrating in space. This occurs because auxiliary conditions are often specified at different positions in space.

A case in point is the simulation of the steady-state temperature distribution for a long, thin rod positioned between two constant-temperature walls (Fig. 24.2). The rod's cross-sectional dimensions are small enough so that radial temperature gradients are minimal and, consequently, temperature is a function exclusively of the axial coordinate x. Heat is transferred along the rod's longitudinal axis by conduction and between the rod and the surrounding gas by convection. For this example, radiation is assumed to be negligible.[1]

---

[1] We incorporate radiation into this problem later in this chapter in Example 24.4.

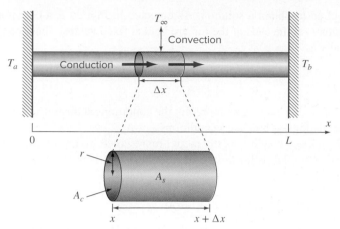

FIGURE 24.2
A heat balance for a differential element of a heated rod subject to conduction and convection.

As depicted in Fig. 24.2, a heat balance can be taken around a differential element of thickness Δx as

$$0 = q(x)A_c - q(x + \Delta x)A_c + hA_s(T_\infty - T) \tag{24.3}$$

where $q(x)$ = flux into the element due to conduction [J/(m$^2 \cdot$ s)]; $q(x + \Delta x)$ = flux out of the element due to conduction [J/(m$^2 \cdot$ s)]; A_c = cross-sectional area [m^2] $= \pi r^2$, r = the radius [m]; h = the convection heat transfer coefficient [J/(m$^2 \cdot$ K $\cdot$ s)]; A_s = the element's surface area [m^2] $= 2\pi r \Delta x$; T_∞ = the temperature of the surrounding gas [K]; and T = the rod's temperature [K].

Equation (24.3) can be divided by the element's volume ($\pi r^2 \Delta x$) to yield

$$0 = \frac{q(x) - q(x + \Delta x)}{\Delta x} + \frac{2h}{r}(T_\infty - T)$$

Taking the limit $\Delta x \rightarrow 0$ gives

$$0 = -\frac{dq}{dx} + \frac{2h}{r}(T_\infty - T) \tag{24.4}$$

The flux can be related to the temperature gradient by *Fourier's law:*

$$q = -k\frac{dT}{dx} \tag{24.5}$$

where k = the coefficient of thermal conductivity [J/(s $\cdot$ m $\cdot$ K)]. Equation (24.5) can be differentiated with respect to x, substituted into Eq. (24.4), and the result divided by k to yield,

$$0 = \frac{d^2T}{dx^2} + h'(T_\infty - T) \tag{24.6}$$

where h' = a bulk heat-transfer parameter reflecting the relative impacts of convection and conduction [m^{-2}] $= 2h/(rk)$.

Equation (24.6) represents a mathematical model that can be used to compute the temperature along the rod's axial dimension. Because it is a second-order ODE, two conditions are required to obtain a solution. As depicted in Fig. 24.2, a common case is where the temperatures at the ends of the rod are held at fixed values. These can be expressed mathematically as

$$T(0) = T_a$$
$$T(L) = T_b$$

The fact that they physically represent the conditions at the rod's "boundaries" is the origin of the terminology: boundary conditions.

Given these conditions, the model represented by Eq. (24.6) can be solved. Because this particular ODE is linear, an analytical solution is possible as illustrated in the following example.

EXAMPLE 24.1   Analytical Solution for a Heated Rod

Problem Statement. Use calculus to solve Eq. (24.6) for a 10-m rod with $h' = 0.05$ m^{-2}[$h = 1$ J/(m$^2 \cdot$ K $\cdot$ s), $r = 0.2$ m, $k = 200$ J/(s $\cdot$ m $\cdot$ K)], $T_\infty = 200$ K, and the boundary conditions:

$$T(0) = 300 \text{ K} \qquad\qquad T(10) = 400 \text{ K}$$

Solution. This ODE can be solved in a number of ways. A straightforward approach is to first express the equation as

$$\frac{d^2T}{dx^2} - h'T = -h'T_\infty$$

Because this is a linear ODE with constant coefficients, the general solution can be readily obtained by setting the right-hand side to zero and assuming a solution of the form $T = e^{\lambda x}$. Substituting this solution along with its second derivative into the homogeneous form of the ODE yields

$$\lambda^2 e^{\lambda x} - h' e^{\lambda x} = 0$$

which can be solved for $\lambda = \pm\sqrt{h'}$. Thus, the general solution is

$$T = Ae^{\lambda x} + Be^{-\lambda x}$$

where A and B are constants of integration. Using the method of undetermined coefficients we can derive the particular solution $T = T_\infty$. Therefore, the total solution is

$$T = T_\infty + Ae^{\lambda x} + Be^{-\lambda x}$$

The constants can be evaluated by applying the boundary conditions

$$T_a = T_\infty + A + B$$
$$T_b = T_\infty + Ae^{\lambda L} + Be^{-\lambda L}$$

These two equations can be solved simultaneously for

$$A = \frac{(T_a - T_\infty)e^{-\lambda L} - (T_b - T_\infty)}{e^{-\lambda L} - e^{\lambda L}}$$

$$B = \frac{(T_b - T_\infty) - (T_a - T_\infty)e^{\lambda L}}{e^{-\lambda L} - e^{\lambda L}}$$

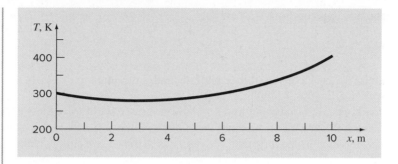

FIGURE 24.3
Analytical solution for the heated rod.

Substituting the parameter values from this problem gives $A = 20.4671$ and $B = 79.5329$. Therefore, the final solution is

$$T = 200 + 20.4671e^{\sqrt{0.05}x} + 79.5329e^{-\sqrt{0.05}x} \tag{24.7}$$

As can be seen in Fig. 24.3, the solution is a smooth curve connecting the two boundary temperatures. The temperature in the middle is depressed due to the convective heat loss to the cooler surrounding gas.

In the following sections, we will illustrate numerical approaches for solving the same problem we just solved analytically in Example 24.1. The exact analytical solution will be useful in assessing the accuracy of the solutions obtained with the approximate, numerical methods.

24.2 THE SHOOTING METHOD

The shooting method is based on converting the boundary-value problem into an equivalent initial-value problem. A trial-and-error approach is then implemented to develop a solution for the initial-value version that satisfies the given boundary conditions.

Although the method can be employed for higher-order and nonlinear equations, it is nicely illustrated for a second-order, linear ODE such as the heated rod described in the previous section:

$$0 = \frac{d^2T}{dx^2} + h'(T_\infty - T) \tag{24.8}$$

subject to the boundary conditions

$$T(0) = T_a$$
$$T(L) = T_b$$

We convert this boundary-value problem into an initial-value problem by defining the rate of change of temperature, or *gradient,* as

$$\frac{dT}{dx} = z \tag{24.9}$$

and reexpressing Eq. (24.8) as

$$\frac{dz}{dx} = -h'(T_\infty - T) \tag{24.10}$$

Thus, we have converted the single second-order equation (Eq. 24.8) into a pair of first-order ODEs [Eqs. (24.9) and (24.10)].

If we had initial conditions for both T and z, we could solve these equations as an initial-value problem with the methods described in Chaps. 22 and 23. However, because we only have an initial value for one of the variables $T(0) = T_a$ we simply make a guess for the other $z(0) = z_{a1}$ and then perform the integration.

After performing the integration, we will have generated a value of T at the end of the interval, which we will call T_{b1}. Unless we are incredibly lucky, this result will differ from the desired result T_b.

Now, let's say that the value of T_{b1} is too high ($T_{b1} > T_b$), it would make sense that a lower value of the initial slope $z(0) = z_{a2}$ might result in a better prediction. Using this new guess, we can integrate again to generate a second result at the end of the interval T_{b2}. We could then continue guessing in a trial-and-error fashion until we arrived at a guess for $z(0)$ that resulted in the correct value of $T(L) = T_b$.

At this point, the origin of the name *shooting method* should be pretty clear. Just as you would adjust the angle of a cannon in order to hit a target, we are adjusting the trajectory of our solution by guessing values of $z(0)$ until we hit our target $T(L) = T_b$.

Although we could certainly keep guessing, a more efficient strategy is possible for linear ODEs. In such cases, the trajectory of the perfect shot z_a is linearly related to the results of our two erroneous shots (z_{a1}, T_{b1}) and (z_{a2}, T_{b2}). Consequently, linear interpolation can be employed to arrive at the required trajectory:

$$z_a = z_{a1} + \frac{z_{a2} - z_{a1}}{T_{b2} - T_{b1}}(T_b - T_{b1}) \tag{24.11}$$

The approach can be illustrated by an example.

EXAMPLE 24.2   The Shooting Method for a Linear ODE

Problem Statement. Use the shooting method to solve Eq. (24.6) for the same conditions as Example 24.1: $L = 10$ m, $h' = 0.05$ m^{-2}, $T_\infty = 200$ K, $T(0) = 300$ K, and $T(10) = 400$ K.

Solution. Equation (24.6) is first expressed as a pair of first-order ODEs:

$$\frac{dT}{dx} = z$$

$$\frac{dz}{dx} = -0.05(200 - T)$$

Along with the initial value for temperature $T(0) = 300$ K, we arbitrarily guess a value of $z_{a1} = -5$ K/m for the initial value for $z(0)$. The solution is then obtained by integrating the pair of ODEs from $x = 0$ to 10. We can do this with MATLAB's ode45 function by first setting up an M-file to hold the differential equations:

```
function dy=Ex2402(x,y)
dy=[y(2);-0.05*(200-y(1))];
```

We can then generate the solution as

```
>> [t,y]=ode45(@Ex2402,[0 10],[300,-5]);
>> Tb1=y(length(y))
```

```
Tb1 =
  569.7539
```

Thus, we obtain a value at the end of the interval of $T_{b1} = 569.7539$ (Fig. 24.4a), which differs from the desired boundary condition of $T_b = 400$. Therefore, we make another guess $z_{a2} = -20$ and perform the computation again. This time, the result of $T_{b2} = 259.5131$ is obtained (Fig. 24.4b).

FIGURE 24.4
Temperature (K) versus distance (m) computed with the shooting method: (a) the first "shot," (b) the second "shot," and (c) the final exact "hit."

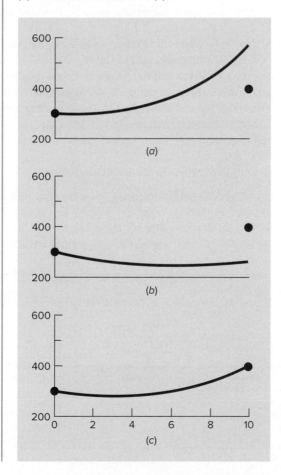

Now, because the original ODE is linear, we can use Eq. (24.11) to determine the correct trajectory to yield the perfect shot:

$$z_a = -5 + \frac{-20 - (-5)}{259.5131 - 569.7539}(400 - 569.7539) = -13.2075$$

This value can then be used in conjunction with ode45 to generate the correct solution, as depicted in Fig. 24.4c.

Although it is not obvious from the graph, the analytical solution is also plotted on Fig. 24.4c. Thus, the shooting method yields a solution that is virtually indistinguishable from the exact result.

24.2.1 Derivative Boundary Conditions

The fixed or *Dirichlet boundary condition* discussed to this point is but one of several types that are used in engineering and science. A common alternative is the case where the derivative is given. This is commonly referred to as a *Neumann boundary condition*.

Because it is already set up to compute both the dependent variable and its derivative, incorporating derivative boundary conditions into the shooting method is relatively straightforward.

Just as with the fixed-boundary condition case, we first express the second-order ODE as a pair of first-order ODEs. At this point, one of the required initial conditions, whether the dependent variable or its derivative, will be unknown. Based on guesses for the missing initial condition, we generate solutions to compute the given end condition. As with the initial condition, this end condition can either be for the dependent variable or its derivative. For linear ODEs, interpolation can then be used to determine the value of the missing initial condition required to generate the final, perfect "shot" that hits the end condition.

EXAMPLE 24.3   The Shooting Method with Derivative Boundary Conditions

Problem Statement. Use the shooting method to solve Eq. (24.6) for the rod in Example 24.1: $L = 10$ m, $h' = 0.05$ m^{-2} [$h = 1$ J/(m$^2 \cdot$ K $\cdot$ s), $r = 0.2$ m, $k = 200$ J/ (s $\cdot$ m $\cdot$ K)], $T_\infty = 200$ K, and $T(10) = 400$ K. However, for this case, rather than having a fixed temperature of 300 K, the left end is subject to convection as in Fig. 24.5.

FIGURE 24.5
A rod with a convective boundary condition at one end and a fixed temperature at the other.

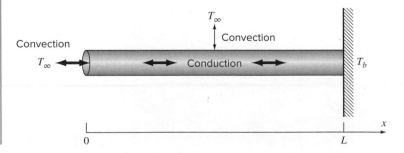

For simplicity, we will assume that the convection heat transfer coefficient for the end area is the same as for the rod's surface.

Solution. As in Example 24.2, Eq. (24.6) is first expressed as

$$\frac{dT}{dx} = z$$

$$\frac{dz}{dx} = -0.05(200 - T)$$

Although it might not be obvious, convection through the end is equivalent to specifying a gradient boundary condition. In order to see this, we must recognize that because the system is at steady state, convection must equal conduction at the rod's left boundary $(x = 0)$. Using Fourier's law [Eq. (24.5)] to represent conduction, the heat balance at the end can be formulated as

$$hA_c(T_\infty - T(0)) = -kA_c \frac{dT}{dx}(0) \tag{24.12}$$

This equation can be solved for the gradient

$$\frac{dT}{dx}(0) = \frac{h}{k}(T(0) - T_\infty) \tag{24.13}$$

If we guess a value for temperature, we can see that this equation specifies the gradient.

The shooting method is implemented by arbitrarily guessing a value for $T(0)$. If we choose a value of $T(0) = T_{a1} = 300$ K, Eq. (24.13) then yields the initial value for the gradient

$$z_{a1} = \frac{dT}{dx}(0) = \frac{1}{200}(300 - 200) = 0.5$$

The solution is obtained by integrating the pair of ODEs from $x = 0$ to 10. We can do this with MATLAB's ode45 function by first setting up an M-file to hold the differential equations in the same fashion as in Example 24.2. We can then generate the solution as

```
>> [t,y]=ode45(@Ex2402,[0 10],[300,0.5]);
>> Tb1=y(length(y))
```

```
Tb1 =
  683.5088
```

As expected, the value at the end of the interval of $T_{b1} = 683.5088$ K differs from the desired boundary condition of $T_b = 400$. Therefore, we make another guess $T_{a2} = 150$ K, which corresponds to $z_{a2} = -0.25$, and perform the computation again.

```
>> [t,y]=ode45(@Ex2402,[0 10],[150,-0.25]);
>> Tb2=y(length(y))
```

```
Tb2 =
  -41.7544
```

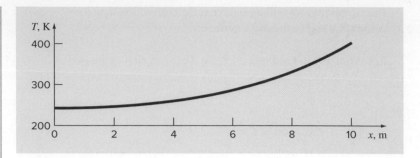

FIGURE 24.6
The solution of a second-order ODE with a convective boundary condition at one end and a fixed temperature at the other.

Linear interpolation can then be employed to compute the correct initial temperature:

$$T_a = 300 + \frac{150 - 300}{-41.7544 - 683.5088}(400 - 683.5088) = 241.3643 \text{ K}$$

which corresponds to a gradient of $z_a = 0.2068$. Using these initial conditions, ode45 can be employed to generate the correct solution, as depicted in Fig. 24.6.

Note that we can verify that our boundary condition has been satisfied by substituting the initial conditions into Eq. (24.12) to give

$$1 \frac{\text{J}}{\text{m}^2\text{K s}} \pi \times (0.2 \text{ m})^2 \times (200 \text{ K} - 241.3643 \text{ K}) = -200 \frac{\text{J}}{\text{m K s}} \pi \times (0.2 \text{ m})^2 \times 0.2068 \frac{\text{K}}{\text{m}}$$

which can be evaluated to yield -5.1980 J/s $= -5.1980$ J/s. Thus, conduction and convection are equal and transfer heat out of the left end of the rod at a rate of 5.1980 W.

24.2.2 The Shooting Method for Nonlinear ODEs

For nonlinear boundary-value problems, linear interpolation or extrapolation through two solution points will not necessarily result in an accurate estimate of the required boundary condition to attain an exact solution. An alternative is to perform three applications of the shooting method and use a quadratic interpolating polynomial to estimate the proper boundary condition. However, it is unlikely that such an approach would yield the exact answer, and additional iterations would be necessary to home in on the solution.

Another approach for a nonlinear problem involves recasting it as a roots problem. Recall that the general goal of a roots problem is to find the value of x that makes the function $f(x) = 0$. Now, let us use the heated rod problem to understand how the shooting method can be recast in this form.

First, recognize that the solution of the pair of differential equations is also a "function" in the sense that we guess a condition at the left-hand end of the rod z_a, and the integration yields a prediction of the temperature at the right-hand end T_b. Thus, we can think of the integration as

$$T_b = f(z_a)$$

That is, it represents a process whereby a guess of z_a yields a prediction of T_b. Viewed in this way, we can see that what we desire is the value of z_a that yields a specific value of T_b. If, as in the example, we desire $T_b = 400$, the problem can be posed as

$$400 = f(z_a)$$

By bringing the goal of 400 over to the right-hand side of the equation, we generate a new function $res(z_a)$ that represents the difference, or *residual,* between what we have, $f(z_a)$, and what we want, 400.

$$res(z_a) = f(z_a) - 400$$

If we drive this new function to zero, we will obtain the solution. The next example illustrates the approach.

EXAMPLE 24.4   The Shooting Method for Nonlinear ODEs

Problem Statement. Although it served our purposes for illustrating the shooting method, Eq. (24.6) was not a completely realistic model for a heated rod. For one thing, such a rod would lose heat by mechanisms such as radiation that are nonlinear.

Suppose that the following nonlinear ODE is used to simulate the temperature of the heated rod:

$$0 = \frac{d^2T}{dx^2} + h'(T_\infty - T) + \sigma''(T_\infty^4 - T^4)$$

where $\sigma' = $ a bulk heat-transfer parameter reflecting the relative impacts of radiation and conduction $= 2.7 \times 10^{-9}$ K^{-3} m^{-2}. This equation can serve to illustrate how the shooting method is used to solve a two-point nonlinear boundary-value problem. The remaining problem conditions are as specified in Example 24.2: $L = 10$ m, $h' = 0.05$ m^{-2}, $T_\infty = 200$ K, $T(0) = 300$ K, and $T(10) = 400$ K.

Solution. Just as with the linear ODE, the nonlinear second-order equation is first expressed as two first-order ODEs:

$$\frac{dT}{dx} = z$$

$$\frac{dz}{dx} = -0.05(200 - T) - 2.7 \times 10^{-9}(1.6 \times 10^9 - T^4)$$

An M-file can be developed to compute the right-hand sides of these equations:

```
function dy=dydxn(x,y)
dy=[y(2);-0.05*(200-y(1))-2.7e-9*(1.6e9-y(1)^4)];
```

Next, we can build a function to hold the residual that we will try to drive to zero as

```
function r=res(za)
[x,y]=ode45(@dydxn,[0 10],[300 za]);
r=y(length(x),1)-400;
```

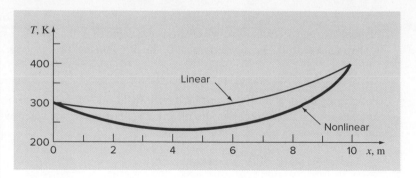

FIGURE 24.7
The result of using the shooting method to solve a nonlinear problem.

Notice how we use the ode45 function to solve the two ODEs to generate the temperature at the rod's end: y(length(x),1). We can then find the root with the fzero function:

```
>> fzero(@res,-50)

ans =
  -41.7434
```

Thus, we see that if we set the initial trajectory $z(0) = -41.7434$, the residual function will be driven to zero and the temperature boundary condition $T(10) = 400$ at the end of the rod should be satisfied. This can be verified by generating the entire solution and plotting the temperatures versus x:

```
>> [x,y]=ode45(@dydxn,[0 10],[300 fzero(@res,-50)]);
>> plot(x,y(:,1))
```

The result is shown in Fig. 24.7 along with the original linear case from Example 24.2. As expected, the nonlinear case is depressed lower than the linear model due to the additional heat lost to the surrounding gas by radiation.

24.3 FINITE-DIFFERENCE METHODS

The most common alternatives to the shooting method are finite-difference approaches. In these techniques, finite differences (Chap. 21) are substituted for the derivatives in the original equation. Thus, a linear differential equation is transformed into a set of simultaneous algebraic equations that can be solved using the methods from Part Three.

We can illustrate the approach for the heated rod model [Eq. (24.6)]:

$$0 = \frac{d^2T}{dx^2} + h'(T_\infty - T) \tag{24.14}$$

The solution domain is first divided into a series of nodes (Fig. 24.8). At each node, finite-difference approximations can be written for the derivatives in the equation. For example,

FIGURE 24.8
In order to implement the finite-difference approach, the heated rod is divided into a series of nodes.

at node i, the second derivative can be represented by (Fig. 21.5):

$$\frac{d^2T}{dx^2} = \frac{T_{i-1} - 2T_i + T_{i+1}}{\Delta x^2} \qquad (24.15)$$

This approximation can be substituted into Eq. (24.14) to give

$$\frac{T_{i-1} - 2T_i + T_{i+1}}{\Delta x^2} + h'(T_\infty - T_i) = 0$$

Thus, the differential equation has been converted into an algebraic equation. Collecting terms gives

$$-T_{i-1} + (2 + h'\Delta x^2)T_i - T_{i+1} = h'x^2 T_\infty \qquad (24.16)$$

This equation can be written for each of the $n - 1$ interior nodes of the rod. The first and last nodes T_0 and T_n, respectively, are specified by the boundary conditions. Therefore, the problem reduces to solving $n - 1$ simultaneous linear algebraic equations for the $n - 1$ unknowns.

Before providing an example, we should mention two nice features of Eq. (24.16). First, observe that since the nodes are numbered consecutively, and since each equation consists of a node (i) and its adjoining neighbors ($i - 1$ and $i + 1$), the resulting set of linear algebraic equations will be tridiagonal. As such, they can be solved with the efficient algorithms that are available for such systems (recall Sec. 9.4).

Further, inspection of the coefficients on the left-hand side of Eq. (24.16) indicates that the system of linear equations will also be diagonally dominant. Hence, convergent solutions can also be generated with iterative techniques like the Gauss-Seidel method (Sec. 12.1).

EXAMPLE 24.5 Finite-Difference Approximation of Boundary-Value Problems

Problem Statement. Use the finite-difference approach to solve the same problem as in Examples 24.1 and 24.2. Use four interior nodes with a segment length of $\Delta x = 2$ m.

Solution. Employing the parameters in Example 24.1 and $\Delta x = 2$ m, we can write Eq. (24.16) for each of the rod's interior nodes. For example, for node 1:

$$-T_0 + 2.2T_1 - T_2 = 40$$

Substituting the boundary condition $T_0 = 300$ gives

$$2.2T_1 - T_2 = 340$$

After writing Eq. (24.16) for the other interior nodes, the equations can be assembled in matrix form as

$$
\begin{bmatrix}
2.2 & -1 & 0 & 0 \\
-1 & 2.2 & -1 & 0 \\
0 & -1 & 2.2 & -1 \\
0 & 0 & -1 & 2.2
\end{bmatrix}
\begin{Bmatrix}
T_1 \\
T_2 \\
T_3 \\
T_4
\end{Bmatrix}
=
\begin{Bmatrix}
340 \\
40 \\
40 \\
440
\end{Bmatrix}
$$

Notice that the matrix is both tridiagonal and diagonally dominant.

MATLAB can be used to generate the solution:

```
>> A=[2.2 -1 0 0;
-1 2.2 -1 0;
0 -1 2.2 -1;
0 0 -1 2.2];
>> b=[340 40 40 440]';
>> T=A\b

T =
  283.2660
  283.1853
  299.7416
  336.2462
```

Table 24.1 provides a comparison between the analytical solution (Eq. 24.7) and the numerical solutions obtained with the shooting method (Example 24.2) and the finite-difference method (Example 24.5). Note that although there are some discrepancies, the numerical approaches agree reasonably well with the analytical solution. Further, the biggest discrepancy occurs for the finite-difference method due to the coarse node spacing we used in Example 24.5. Better agreement would occur if a finer nodal spacing had been used.

TABLE 24.1 Comparison of the exact analytical solution for temperature with the results obtained with the shooting and finite-difference methods.

x	Analytical Solution	Shooting Method	Finite Difference
0	300	300	300
2	282.8634	282.8889	283.2660
4	282.5775	282.6158	283.1853
6	299.0843	299.1254	299.7416
8	335.7404	335.7718	336.2462
10	400	400	400

24.3.1 Derivative Boundary Conditions

As mentioned in our discussion of the shooting method, the fixed or *Dirichlet boundary condition* is but one of several types that are used in engineering and science. A common

FIGURE 24.9
A boundary node at the left end of a heated rod. To approximate the derivative at the boundary, an imaginary node is located a distance Δx to the left of the rod's end.

alternative, called the *Neumann boundary condition*, is the case where the derivative is given.

We can use the heated rod introduced earlier in this chapter to demonstrate how a derivative boundary condition can be incorporated into the finite-difference approach:

$$0 = \frac{d^2T}{dx^2} + h'(T_\infty - T)$$

However, in contrast to our previous discussions, we will prescribe a derivative boundary condition at one end of the rod:

$$\frac{dT}{dx}(0) = T'_a$$

$$T(L) = T_b$$

Thus, we have a derivative boundary condition at one end of the solution domain and a fixed boundary condition at the other.

Just as in the previous section, the rod is divided into a series of nodes and a finite-difference version of the differential equation (Eq. 24.16) is applied to each interior node. However, because its temperature is not specified, the node at the left end must also be included. Fig. 24.9 depicts the node (0) at the left edge of a heated plate for which the derivative boundary condition applies. Writing Eq. (24.16) for this node gives

$$-T_{-1} + (2 + h'\, \Delta x^2)T_0 - T_1 = h'\, \Delta x^2 T_\infty \tag{24.17}$$

Notice that an imaginary node (-1) lying to the left of the rod's end is required for this equation. Although this exterior point might seem to represent a difficulty, it actually serves as the vehicle for incorporating the derivative boundary condition into the problem. This is done by representing the first derivative in the x dimension at (0) by the centered difference [Eq. (4.25)]:

$$\frac{dT}{dx} = \frac{T_1 - T_{-1}}{2\Delta x}$$

which can be solved for

$$T_{-1} = T_1 - 2\Delta x \frac{dT}{dx}$$

Now we have a formula for T_{-1} that actually reflects the impact of the derivative. It can be substituted into Eq. (24.17) to give

$$(2 + h'\Delta x^2)T_0 - 2T_1 = h'\Delta x^2 T_\infty - 2\Delta x \frac{dT}{dx} \tag{24.18}$$

Consequently, we have incorporated the derivative into the balance.

A common example of a derivative boundary condition is the situation where the end of the rod is insulated. In this case, the derivative is set to zero. This conclusion follows directly from Fourier's law [Eq. (24.5)], because insulating a boundary means that the heat flux (and consequently the gradient) must be zero. The following example illustrates how the solution is affected by such boundary conditions.

EXAMPLE 24.6   Incorporating Derivative Boundary Conditions

Problem Statement. Generate the finite-difference solution for a 10-m rod with $\Delta x = 2$ m, $h' = 0.05$ m^{-2}, $T_\infty = 200$ K, and the boundary conditions: $T_a' = 0$ and $T_b = 400$ K. Note that the first condition means that the slope of the solution should approach zero at the rod's left end. Aside from this case, also generate the solution for $dT/dx = -20$ at $x = 0$.

Solution. Equation (24.18) can be used to represent node 0 as

$$2.2T_0 - 2T_1 = 40$$

We can write Eq. (24.16) for the interior nodes. For example, for node 1,

$$-T_0 + 2.2T_1 - T_2 = 40$$

A similar approach can be used for the remaining interior nodes. The final system of equations can be assembled in matrix form as

$$\begin{bmatrix} 2.2 & -2 & & & \\ -1 & 2.2 & -1 & & \\ & -1 & 2.2 & -1 & \\ & & -1 & 2.2 & -1 \\ & & & -1 & 2.2 \end{bmatrix} \begin{Bmatrix} T_0 \\ T_1 \\ T_2 \\ T_3 \\ T_4 \end{Bmatrix} = \begin{Bmatrix} 40 \\ 40 \\ 40 \\ 40 \\ 440 \end{Bmatrix}$$

These equations can be solved for

$$T_0 = 243.0278$$
$$T_1 = 247.3306$$
$$T_2 = 261.0994$$
$$T_3 = 287.0882$$
$$T_4 = 330.4946$$

As displayed in Fig. 24.10, the solution is flat at $x = 0$ due to the zero derivative condition and then curves upward to the fixed condition of $T = 400$ at $x = 10$.

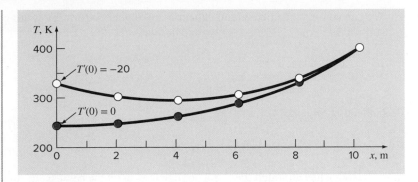

FIGURE 24.10
The solution of a second-order ODE with a derivative boundary condition at one end and a fixed boundary condition at the other. Two cases are shown reflecting different derivative values at $x = 0$.

For the case where the derivative at $x = 0$ is set to -20, the simultaneous equations are

$$
\begin{bmatrix}
2.2 & -2 & & & \\
-1 & 2.2 & -1 & & \\
& -1 & 2.2 & -1 & \\
& & -1 & 2.2 & -1 \\
& & & -1 & 2.2
\end{bmatrix}
\begin{Bmatrix}
T_0 \\
T_1 \\
T_2 \\
T_3 \\
T_4
\end{Bmatrix}
=
\begin{Bmatrix}
120 \\
40 \\
40 \\
40 \\
440
\end{Bmatrix}
$$

which can be solved for

$T_0 = 328.2710$

$T_1 = 301.0981$

$T_2 = 294.1448$

$T_3 = 306.0204$

$T_4 = 339.1002$

As in Fig. 24.10, the solution at $x = 0$ now curves downward due to the negative derivative we imposed at the boundary.

24.3.2 Finite-Difference Approaches for Nonlinear ODEs

For nonlinear ODEs, the substitution of finite differences yields a system of nonlinear simultaneous equations. Thus, the most general approach to solving such problems is to use root-location methods for systems of equations such as the Newton-Raphson method described in Sec. 12.2.2. Although this approach is certainly feasible, an adaptation of successive substitution can sometimes provide a simpler alternative.

The heated rod with convection and radiation introduced in Example 24.4 provides a nice vehicle for demonstrating this approach,

$$
0 = \frac{d^2T}{dx^2} + h'(T_\infty - T) + \sigma'(T_\infty^4 - T^4)
$$

We can convert this differential equation into algebraic form by writing it for a node i and substituting Eq. (24.15) for the second derivative:

$$0 = \frac{T_{i-1} - 2T_i + T_{i+1}}{\Delta x^2} + h'(T_\infty - T_i) + \sigma''(T_\infty^4 - T_i^4)$$

Collecting terms gives

$$-T_{i-1} + (2 + h'\Delta x^2)T_i - T_{i+1} = h'\Delta x^2 T_\infty + \sigma''\Delta x^2(T_\infty^4 - T_i^4)$$

Notice that although there is a nonlinear term on the right-hand side, the left-hand side is expressed in the form of a linear algebraic system that is diagonally dominant. If we assume that the unknown nonlinear term on the right is equal to its value from the previous iteration, the equation can be solved for

$$T_i = \frac{h'\Delta x^2 T_\infty + \sigma''\Delta x^2(T_\infty^4 - T_i^4) + T_{i-1} + T_{i+1}}{2 + h'\Delta x^2} \tag{24.19}$$

As in the Gauss-Seidel method, we can use Eq. (24.19) to successively calculate the temperature of each node and iterate until the process converges to an acceptable tolerance. Although this approach will not work for all cases, it converges for many ODEs derived from physically based systems. Hence, it can sometimes prove useful for solving problems routinely encountered in engineering and science.

EXAMPLE 24.7   The Finite-Difference Method for Nonlinear ODEs

Problem Statement. Use the finite-difference approach to simulate the temperature of a heated rod subject to both convection and radiation:

$$0 = \frac{d^2T}{dx^2} + h'(T_\infty - T) + \sigma''(T_\infty^4 - T^4)$$

where $\sigma' = 2.7 \times 10^{-9}$ K^{-3}m^{-2}, $L = 10$ m, $h' = 0.05$ m^{-2}, $T_\infty = 200$K, $T(0) = 300$ K, and T (10) = 400 K. Use four interior nodes with a segment length of $\Delta x = 2$ m. Recall that we solved the same problem with the shooting method in Example 24.4.

Solution. Using Eq. (24.19) we can successively solve for the temperatures of the rod's interior nodes. As with the standard Gauss-Seidel technique, the initial values of the interior nodes are zero with the boundary nodes set at the fixed conditions of $T_0 = 300$ and $T_5 = 400$. The results for the first iteration are

$$T_1 - \frac{0.05(2)^2\,200 + 2.7 \times 10^{-9'}(2)^2(200^4 - 0^4) + 300 + 0}{2 + 0.05(2)^2} = 159.2432$$

$$T_2 = \frac{0.05(2)^2\,200 + 2.7 \times 10^{-9'}(2)^2(200^4 - 0^4) + 159.2432 + 0}{2 + 0.05(2)^2} = 97.9674$$

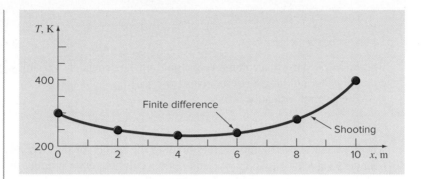

FIGURE 24.11
The filled circles are the result of using the finite-difference method to solve a nonlinear problem. The line generated with the shooting method in Example 24.4 is shown for comparison.

$$T_3 = \frac{0.05(2)^2\,200 + 2.7 \times 10^{-9'}(2)^2(200^4 - 0^4) + 97.9674 + 0}{2 + 0.05(2)^2} = 70.4461$$

$$T_4 = \frac{0.05(2)^2\,200 + 2.7 \times 10^{-9'}(2)^2(200^4 - 0^4) + 70.4461 + 400}{2 + 0.05(2)^2} = 226.8704$$

The process can be continued until we converge on the final result:

$T_0 = 300$
$T_1 = 250.4827$
$T_2 = 236.2962$
$T_3 = 245.7596$
$T_4 = 286.4921$
$T_5 = 400$

These results are displayed in Fig. 24.11 along with the result generated in Example 24.4 with the shooting method.

24.4 MATLAB FUNCTION: bvp4c

The bvp4c function solves ODE boundary-value problems by integrating a system of ordinary differential equations of the form $y' = f(x, y)$ on the interval $[a, b]$, subject to general two-point boundary conditions. A simple representation of its syntax is

```
sol = bvp4c(odefun,bcfun,solinit)
```

where sol = a structure containing the solution, odefun = the function that sets up the ODEs to be solved, bcfun = the function that computes the residuals in the boundary conditions, and solinit = a structure with fields holding an initial mesh and initial guesses for the solution.

The general format of *odefun* is

dy = odefun(x,y)

where x = a scalar, y = a column vector holding the dependent variables $[y_1; \ y_2]$, and dy = a column vector holding the derivatives $[dy_1; \ dy_2]$.

The general format of *bcfun* is

res = bcfun(ya,yb)

where *ya* and *yb* = column vectors holding the values of the dependent variables at the boundary values $x = a$ and $x = b$, and *res* = a column vector holding the residuals between the computed and the specified boundary values.

The general format of *solinit* is

solinit = bvpinit(xmesh, yinit);

where *bvpinit* = a built-in MATLAB function that creates the guess structure holding the initial mesh and solution guesses, *xmesh* = a vector holding the ordered nodes of the initial mesh, and *yinit* = a vector holding the initial guesses. Note that whereas your choices for the initial mesh and guesses will not be of great importance for linear ODEs, they can often be critical for efficiently solving nonlinear equations.

EXAMPLE 24.8 Solving a Boundary-Value Problem with bvp4c

Problem Statement. Use bvp4c to solve the following second-order ODE

$$\frac{d^2y}{dx^2} + y = 1$$

subject to the boundary conditions

$$y(0) = 1$$
$$y(\pi/2) = 0$$

Solution. First, express the second-order equation as a pair of first-order ODEs

$$\frac{dy}{dx} = z$$
$$\frac{dz}{dx} = 1 - y$$

Next, set up a function to hold the first-order ODEs

function dy = odes(x,y)
dy = [y(2); 1-y(1)];

We can now develop the function to hold the boundary conditions. This is done just like a roots problem in that we set up two functions that should be zero when the boundary conditions are satisfied. To do this, the vectors of unknowns at the left and right boundaries are defined as ya and yb. Hence, the first condition, $y(0) = 1$, can be formulated as ya(1)-1; whereas the second condition, $y(\pi/2) = 0$, corresponds to yb(1).

```
function r = bcs(ya,yb)
r = [ya(1)-1; yb(1)];
```

Finally, we can set up solinit to hold the initial mesh and solution guesses with the bvpinit function. We will arbitrarily select 10 equally spaced mesh points, and initial guesses of $y = 1$ and $z = dy/dx = -1$.

```
solinit = bvpinit(linspace(0,pi/2,10),[1,-1]);
```

The entire script to generate the solution is

```
clc
solinit = bvpinit(linspace(0,pi/2,10),[1,-1]);
sol = bvp4c(@odes,@bcs,solinit);
x = linspace(0,pi/2);
y = deval(sol,x);
plot(x,y(1,:))
```

where deval is a built-in MATLAB function which evaluates the solution of a differential equation problem with the general syntax

```
yxint = deval(sol,xint)
```

where deval evaluates the solution at all the values of the vector *xint*, and *sol* is the structure returned by the ODE problem solver (in this case, bvp4c).

When the script is run the plot below is generated. Note that the script and functions developed in this example can be applied to other boundary value problems with minor modifications. Several end-of-chapter problems are included to test your ability to do just that.

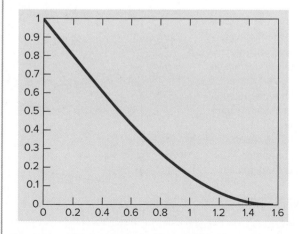

PROBLEMS

24.1 A steady-state heat balance for a rod can be represented as

$$\frac{d^2T}{dx^2} - 0.15T = 0$$

Obtain a solution for a 10-m rod with $T(0) = 240$ and $T(10) = 150$ **(a)** analytically, **(b)** with the shooting method, and **(c)** using the finite-difference approach with $\Delta x = 1$.

24.2 Repeat Prob. 24.1 but with the right end insulated and the left end temperature fixed at 240.

24.3 Use the shooting method to solve

$$7\frac{d^2y}{dx^2} - 2\frac{dy}{dx} - y + x = 0$$

with the boundary conditions $y(0) = 5$ and $y(20) = 8$.

24.4 Solve Prob. 24.3 with the finite-difference approach using $\Delta x = 2$.

24.5 The following nonlinear differential equation was solved in Examples 24.4 and 24.7.

$$0 = \frac{d^2T}{dx^2} + h'(T_\infty - T) + \sigma'' \, (T_\infty^4 - T^4) \qquad \text{(P24.5)}$$

Such equations are sometimes linearized to obtain an approximate solution. This is done by employing a first-order Taylor series expansion to linearize the quartic term in the equation as

$$\sigma'T^4 = \sigma'\bar{T}^4 + 4\sigma'\bar{T}^3(T - \bar{T})$$

where $\bar{T}$ is a base temperature about which the term is linearized. Substitute this relationship into Eq. (P24.5), and then solve the resulting linear equation with the finite-difference approach. Employ $\bar{T} = 300$, $\Delta x = 1$ m, and the parameters from Example 24.4 to obtain your solution. Plot your results along with those obtained for the nonlinear versions in Examples 24.4 and 24.7.

24.6 Develop an M-file to implement the shooting method for a linear second-order ODE with Dirichlet boundary conditions. Test the program by duplicating Prob. 24.1.

24.7 Develop an M-file to implement the finite-difference approach for solving a linear second-order ODE with Dirichlet boundary conditions. Test it by duplicating Example 24.5.

24.8 An insulated heated rod with a uniform heat source can be modeled with the *Poisson equation:*

$$\frac{d^2T}{dx^2} = -f(x)$$

Given a heat source $f(x) = 25$ °C/m^2 and the boundary conditions $T(x = 0) = 40$ °C and $T(x = 10) = 200$ °C, solve for the temperature distribution with **(a)** the shooting method and **(b)** the finite-difference method ($\Delta x = 2$).

24.9 Repeat Prob. 24.8, but for the following spatially varying heat source: $f(x) = 0.12x^3 - 2.4x^2 + 12x$.

24.10 The temperature distribution in a tapered conical cooling fin (Fig. P24.10) is described by the following differential equation, which has been nondimensionalized:

$$\frac{d^2u}{dx^2} + \left(\frac{2}{x}\right)\left(\frac{du}{dx} - pu\right) = 0$$

where u = temperature ($0 \le u \le 1$), x = axial distance ($0 \le x \le 1$), and p is a nondimensional parameter that describes the heat transfer and geometry:

$$p = \frac{hL}{k}\sqrt{1 + \frac{4}{2m^2}}$$

where h = a heat transfer coefficient, k = thermal conductivity, L = the length or height of the cone, and m = the slope of the cone wall. The equation has the boundary conditions:

$$u(x = 0) = 0 \qquad u(x = 1) = 1$$

Solve this equation for the temperature distribution using finite-difference methods. Use second-order accurate finite-difference formulas for the derivatives. Write a computer

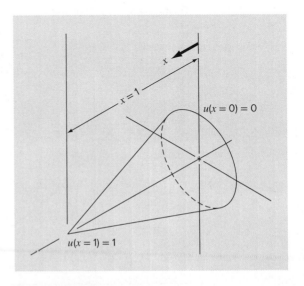

FIGURE P24.10

program to obtain the solution and plot temperature versus axial distance for various values of $p = 10, 20, 50,$ and 100.

24.11 Compound A diffuses through a 4-cm-long tube and reacts as it diffuses. The equation governing diffusion with reaction is

$$D\frac{d^2A}{dx^2} - kA = 0$$

At one end of the tube ($x = 0$), there is a large source of A that results in a fixed concentration of 0.1 M. At the other end of the tube there is a material that quickly absorbs any A, making the concentration 0 M. If $D = 1.5 \times 10^{-6}$ cm^2/s and $k = 5 \times 10^{-6}$ s^{-1}, what is the concentration of A as a function of distance in the tube?

24.12 The following differential equation describes the steady-state concentration of a substance that reacts with first-order kinetics in an axially dispersed plug-flow reactor (Fig. P24.12):

$$D\frac{d^2c}{dx^2} - U\frac{dc}{dx} - kc = 0$$

where D = the dispersion coefficient (m^2/hr), c = concentration (mol/L), x = distance (m), U = the velocity (m/hr), and k = the reaction rate (/hr). The boundary conditions can be formulated as

$$Uc_{in} = Uc(x = 0) - D\frac{dc}{dx}(x = 0)$$

$$\frac{dc}{dx}(x = L) = 0$$

where c_{in} = the concentration in the inflow (mol/L), L = the length of the reactor (m). These are called *Danckwerts boundary conditions*.

Use the finite-difference approach to solve for concentration as a function of distance given the following parameters: $D = 5000$ m^2/hr, $U = 100$ m/hr, $k = 2$/hr, $L = 100$ m, and $c_{in} = 100$ mol/L. Employ centered finite-difference approximations with $\Delta x = 10$ m to obtain your

solutions. Compare your numerical results with the analytical solution:

$$c = \frac{Uc_{in}}{(U - D\lambda_1)\lambda_2 e^{\lambda_2 L} - (U - D\lambda_2)\lambda_1 e^{\lambda_1 L}}$$
$$\times (\lambda_2 e^{\lambda_2 L}e^{\lambda_1 x} - \lambda_1 e^{\lambda_1 L}e^{\lambda_2 x})$$

where

$$\frac{\lambda_1}{\lambda_2} = \frac{U}{2D}\left(1 \pm \sqrt{1 + \frac{4kD}{U^2}}\right)$$

24.13 A series of first-order, liquid-phase reactions create a desirable product (B) and an undesirable byproduct (C):

$$A \xrightarrow{k_1} B \xrightarrow{k_2} C$$

If the reactions take place in an axially dispersed plug-flow reactor (Fig. P24.12), steady-state mass balances can be used to develop the following second-order ODEs:

$$D\frac{d^2c_a}{dx^2} - U\frac{dc_a}{dx} - k_1c_a = 0$$

$$D\frac{d^2c_b}{dx^2} - U\frac{dc_b}{dx} + k_1c_a - k_2c_b = 0$$

$$D\frac{d^2c_c}{dx^2} - U\frac{dc_c}{dx} + k_2c_b = 0$$

Use the finite-difference approach to solve for the concentration of each reactant as a function of distance given: $D = 0.1$ m^2/min, $U = 1$ m/min, $k_1 = 3$/min, $k_2 = 1$/min, $L = 0.5$ m, $c_{a,in} = 10$ mol/L. Employ centered finite-difference approximations with $\Delta x = 0.05$ m to obtain your solutions and assume Danckwerts boundary conditions as described in Prob. 24.12. Also, compute the sum of the reactants as a function of distance. Do your results make sense?

24.14 A biofilm with a thickness L_f (cm), grows on the surface of a solid (Fig. P24.14). After traversing a diffusion layer of thickness L (cm), a chemical compound A diffuses into the biofilm where it is subject to an irreversible first-order reaction that converts it to a product B.

$x = 0$ $x = L$

FIGURE P24.12
An axially dispersed plug-flow reactor.

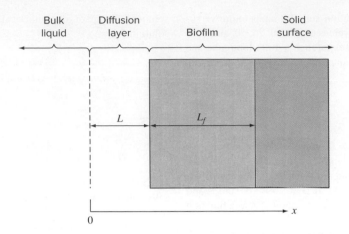

FIGURE P24.14
A biofilm growing on a solid surface.

Steady-state mass balances can be used to derive the following ordinary differential equations for compound A:

$$D\frac{d^2c_a}{dx^2} = 0 \qquad 0 \le x < L$$

$$D_f\frac{d^2c_a}{dx^2} - kc_a = 0 \qquad L \le x < L + L_f$$

where D = the diffusion coefficient in the diffusion layer = 0.08 cm²/d, D_f = the diffusion coefficient in the biofilm = 0.04 cm²/d, and k = the first-order rate for the conversion of A to B = 2000/d. The following boundary conditions hold:

$$c_a = c_{a0} \qquad \text{at } x = 0$$

$$\frac{dc_a}{dx} = 0 \qquad \text{at } x = L + L_f$$

where c_{a0} = the concentration of A in the bulk liquid = 100 mol/L. Use the finite-difference method to compute the steady-state distribution of A from $x = 0$ to $L + L_f$, where $L = 0.008$ cm and $L_f = 0.004$ cm. Employ centered finite differences with $\Delta x = 0.001$ cm.

24.15 A cable is hanging from two supports at A and B (Fig. P24.15). The cable is loaded with a distributed load whose magnitude varies with x as

$$w = w_o\left[1 + \sin\left(\frac{\pi x}{2l_A}\right)\right]$$

where w_o = 450 N/m. The slope of the cable $(dy/dx) = 0$ at $x = 0$, which is the lowest point for the cable. It is also the point where the tension in the cable is a minimum of T_o. The

FIGURE P24.15

FIGURE P24.16

differential equation which governs the cable is

$$\frac{d^2y}{dx^2} = \frac{w_o}{T_o}\left[1 + \sin\left(\frac{\pi x}{2l_A}\right)\right]$$

Solve this equation using a numerical method and plot the shape of the cable (y versus x). For the numerical solution, the value of T_o is unknown, so the solution must use an iterative technique, similar to the shooting method, to converge on a correct value of h_A for various values of T_o.

24.16 The basic differential equation of the elastic curve for a simply supported, uniformly loaded beam (Fig. P24.16) is given as

$$EI\frac{d^2y}{dx^2} = \frac{wLx}{2} - \frac{wx^2}{2}$$

where E = the modulus of elasticity and I = the moment of inertia. The boundary conditions are $y(0) = y(L) = 0$. Solve for the deflection of the beam using **(a)** the finite-differ-

ence approach ($\Delta x = 0.6$ m) and **(b)** the shooting method. The following parameter values apply: $E = 200$ GPa, $I = 30,000$ cm^4, $w = 15$ kN/m, and $L = 3$ m. Compare your numerical results to the analytical solution:

$$y = \frac{wLx^3}{12EI} - \frac{wx^4}{24EI} - \frac{wL^3x}{24EI}$$

24.17 In Prob. 24.16, the basic differential equation of the elastic curve for a uniformly loaded beam was formulated as

$$EI\frac{d^2y}{dx^2} = \frac{wLx}{2} - \frac{wx^2}{2}$$

Note that the right-hand side represents the moment as a function of x. An equivalent approach can be formulated in terms of the fourth derivative of deflection as

$$EI\frac{d^4y}{dx^4} = -w$$

For this formulation, four boundary conditions are required. For the supports shown in Fig. P24.16, the conditions are that the end displacements are zero, $y(0) = y(L) = 0$, and that the end moments are zero, $y''(0) = y''(L) = 0$. Solve for the deflection of the beam using the finite-difference approach ($\Delta x = 0.6$ m). The following parameter values apply: $E = 200$ GPa, $I = 30,000$ cm^4, $w = 15$ kN/m, and $L = 3$ m. Compare your numerical results with the analytical solution given in Prob. 24.16.

24.18 Under a number of simplifying assumptions, the steady-state height of the water table in a one-dimensional, unconfined groundwater aquifer (Fig. P24.18) can be

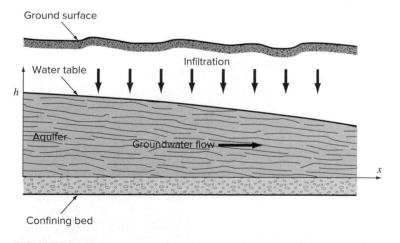

FIGURE P24.18
An unconfined or "phreatic" aquifer.

modeled with the following second-order ODE:

$$K\bar{h}\frac{d^2h}{dx^2} + N = 0$$

where x = distance (m), K = hydraulic conductivity (m/d), h = height of the water table (m), $\bar{h}$ = the average height of the water table (m), and N = infiltration rate (m/d).

Solve for the height of the water table for $x = 0$ to 1000 m where $h(0) = 10$ m and $h(1000) = 5$ m. Use the following parameters for the calculation: $K = 1$ m/d and $N = 0.0001$ m/d. Set the average height of the water table as the average of the boundary conditions. Obtain your solution with (a) the shooting method and (b) the finite-difference method ($\Delta x = 100$ m).

24.19 In Prob. 24.18, a linearized groundwater model was used to simulate the height of the water table for an unconfined aquifer. A more realistic result can be obtained by using the following nonlinear ODE:

$$\frac{d}{dx}\left(Kh\frac{dh}{dx}\right) + N = 0$$

where x = distance (m), K = hydraulic conductivity (m/d), h = height of the water table (m), and N = infiltration rate (m/d). Solve for the height of the water table for the same case as in Prob. 24.18. That is, solve from $x = 0$ to 1000 m with $h(0) = 10$ m, $h(1000) = 5$ m, $K = 1$ m/d, and $N = 0.0001$ m/d. Obtain your solution with (a) the shooting method and (b) the finite-difference method ($\Delta x = 100$ m).

24.20 Just as Fourier's law and the heat balance can be employed to characterize temperature distribution, analogous relationships are available to model field problems in other areas of engineering. For example, electrical engineers use a similar approach when modeling electrostatic fields. Under a number of simplifying assumptions, an analog of Fourier's law can be represented in one-dimensional form as

$$D = -\varepsilon\frac{dV}{dx}$$

where D is called the electric flux density vector, ε = permittivity of the material, and V = electrostatic potential. Similarly, a Poisson equation (see Prob. 24.8) for electrostatic fields can be represented in one dimension as

$$\frac{d^2V}{dx^2} = \frac{\rho_v}{\varepsilon}$$

where ρ_v = charge density. Use the finite-difference technique with $\Delta x = 2$ to determine V for a wire where $V(0) = 1000$, $V(20) = 0$, $\varepsilon = 2$, $L = 20$, and $\rho_v = 30$.

24.21 Suppose that the position of a falling object is governed by the following differential equation:

$$\frac{d^2x}{dt^2} + \frac{c}{m}\frac{dx}{dt} - g = 0$$

where c = a first-order drag coefficient = 12.5 kg/s, m = mass = 70 kg, and g = gravitational acceleration = 9.81 m/s². Use the shooting method to solve this equation for the boundary conditions:

$$x(0) = 0$$
$$x(12) = 500$$

24.22 As in Fig. P24.22, an insulated metal rod has a fixed temperature (T_0) boundary condition at its left end. On it right end, it is joined to a thin-walled tube filled with water through which heat is conducted. The tube is insulated at its right end and convects heat with the surrounding fixed-temperature air (T_∞). The convective heat flux at a location x along the tube (W/m²) is represented by

$$J_{conv} = h(T_\infty - T_2(x))$$

where h = the convection heat transfer coefficient [W/(m² · K)]. Employ the finite-difference method with $\Delta x = 0.1$ m to compute the temperature distribution for the case where both the rod and tube are cylindrical with the same radius r (m). Use the following parameters for your analysis: $L_{rod} = 0.6$ m, $L_{tube} = 0.8$ m, $T_0 = 400$ K, $T_\infty = 300$ K, $r = 3$ cm, $\rho_1 = 7870$ kg/m³, $C_{p1} = 447$ J/(kg · K), $k_1 = 80.2$ W/(m · K), $\rho_2 = 1000$ kg/m³, $C_{p2} = 4.18$ kJ/(kg · K), $k_2 = 0.615$ W/(m · K), and $h = 3000$ W/(m² · K). The subscripts designate the rod (1) and the tube (2).

24.23 Perform the same calculation as in Prob. 24.22, but for the case where the tube is also insulated (i.e., no convection) and the right-hand wall is held at a fixed boundary temperature of 200 K.

24.24 Solve the following problem with the bvp4c,

$$\frac{d^2y}{dx^2} + y = 0$$

FIGURE P24.22

subject to the following boundary conditions

$$y(0) = 1$$

$$\frac{dy}{dx}(1) = 0$$

24.25 Figure P24.25a shows a uniform beam subject to a linearly increasing distributed load. The equation for the resulting elastic curve is (see Fig. P24.25b)

$$EI\frac{d^2y}{dx^2} - \frac{w_0}{6}\left(0.6Lx - \frac{x^3}{L}\right) = 0$$

Note that the analytical solution for the resulting elastic curve is (see Fig. P24.25b)

$$y = \frac{w_0}{120\,EIL}(-x^5 + 2L^2x^3 - L^4x)$$

Use bvp4c to solve for the differential equation for the elastic curve for $L = 600$ cm, $E = 50{,}000$ kN/cm^2, $I = 30{,}000$ cm^4, and $w_0 = 2.5$ kN/cm. Then, plot both the numerical (points) and the analytical (lines) solutions on the same graph.

24.26 Use bvp4c to solve the boundary-value ordinary differential equation

$$\frac{d^2u}{dx^2} + 6\frac{du}{dx} - u = 2$$

with boundary conditions $u(0) = 10$ and $u(2) = 1$. Plot the results of u versus x.

24.27 Use bvp4c to solve the following nondimensionalized ODE that describes the temperature distribution in a circular rod with internal heat source S

$$\frac{d^2T}{dr^2} + \frac{1}{r}\frac{dT}{dr} + S = 0$$

over the range $0 \le r \le 1$, with the boundary conditions

$$T(1) = 1$$

$$\frac{dT}{dr}(0) = 0$$

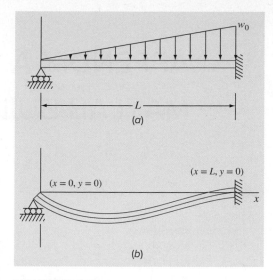

(a)

(x = L, y = 0)

(x = 0, y = 0)

(b)

FIGURE P24.25

for $S = 1$, 10, and 20 K/m^2. Plot the temperature versus radius for all three cases on the same graph.

24.28 A heated rod with a uniform heat source can be modeled with the Poisson equation,

$$\frac{d^2T}{dx^2} = -f(x)$$

Given a heat source $f(x) = 25$ and the boundary conditions, $T(0) = 40$ and $T(10) = 200$, solve for the temperature distribution with bvp4c.

24.29 Repeat Prob. 24.28, but for the following heat source: $f(x) = 0.12x^3 - 2.4x^2 + 12x$.

APPENDIX A
MATLAB BUILT-IN FUNCTIONS

APPENDIX B

MATLAB M-FILE FUNCTIONS

M-file Name	Description	Page
bisect	Root location with bisection	151
eulode	Integration of a single ordinary differential equation with Euler's method	586
fzerosimp	Brent's method for root location	180
GaussNaive	Solving linear systems with Gauss elimination without pivoting	258
GaussPivot	Solving linear systems with Gauss elimination with partial pivoting	263
GaussSeidel	Solving linear systems with the Gauss-Seidel method	310
goldmin	Minimum of one-dimensional function with golden-section search	208
incsearch	Root location with an incremental search	144
IterMeth	General algorithm for iterative calculation	105
Lagrange	Interpolation with the Lagrange polynomial	443
linregr	Fitting a straight line with linear regression	372
natspline	Cubic spline with natural end conditions	477
Newtint	Interpolation with the Newton polynomial	440
newtmult	Root location for nonlinear systems of equations	318
newtraph	Root location with the Newton-Raphson method	174
quadadapt	Adaptive quadrature	539
rk4sys	Integration of system of ODEs with 4th-order RK method	602
romberg	Integration of a function with Romberg integration	530
TableLook	Table lookup with linear interpolation	458
trap	Integration of a function with the composite trapezoidal rule	500
trapuneq	Integration of unequispaced data with the trapezoidal rule	509
Tridiag	Solving tridiagonal linear systems	266

676

APPENDIX C

INTRODUCTION TO SIMULINK

Simulink® is a graphical programming environment for modeling, simulating, and analyzing dynamic systems. In short, it allows engineers and scientists to build process models by interconnecting blocks with communication lines. Thus, it provides an easy-to-use computing framework to quickly develop dynamic process models of physical systems. Along with offering a variety of numerical integration options for solving differential equations, Simulink includes built-in features for graphical output which significantly enhance visualization of a system's behavior.

As a historical footnote, back in the Pleistocene days of analog computers (aka the 1950s), you had to design information flow diagrams that showed graphically how multiple ODEs in models were interconnected with themselves and with algebraic relationships. The diagrams also showed flaws in modeling where there was information lacking or structural defects. One of the nice features of Simulink is that it also does that. It's often beneficial to see that aspect separate from numerical methods and then merge the two in MATLAB.

As was done with Chap. 2, most of this appendix has been written as a hands-on exercise. That is, you should read it while sitting in front of your computer. The most efficient way to start learning Simulink is to actually implement it on MATLAB as you proceed through the following material.

So let's get started by setting up a simple Simulink application to solve an initial value problem for a single ODE. A nice candidate is the differential equation we developed for the velocity of the free-falling bungee jumper in Chap. 1,

$$\frac{dv}{dt} = g - \frac{c_d}{m}v^2 \tag{C.1}$$

where v = velocity (m/s), t = time (s), g = 9.81 m/s², c_d = drag coefficient (kg/m), and m = mass (kg). As in Chap. 1 use c_d = 0.25 kg/m, m = 68.1 kg, and integrate from 0 to 12 s with an initial condition of v = 0.

To generate the solution with Simulink, first launch MATLAB. You should eventually see the MATLAB window with an entry prompt, ≫, in the Command Window. After changing MATLAB's default directory, open the Simulink Library Browser using one of the following approaches:

- On the MATLAB toolbar, click the Simulink button (▦).
- At the MATLAB prompt, enter the simulink command.

The Simulink Library Browser window should appear displaying the Simulink block libraries installed on your system. Note, to keep the Library Browser above all other windows on your desktop, in the Library Browser select **View, Stay on Top.**

Click on the **New model** command button on the left of the toolbar, and an untitled Simulink model window should appear.

You will build your simulation model in the untitled window by choosing items from the Library Browser and then dragging and dropping them onto the untitled window. First, select the untitled window to activate it (clicking on the title bar is a sure way to do this) and select **Save As** from the File menu. Save the window as **Freefall** in the default directory. This file is saved automatically with an **.slx** extension. As you build your model in this window, it is a good idea to save it frequently. You can do this in three ways: the **Save** button, the **Ctrl+s** keyboard shortcut, and the menu selections, **File, Save.**

We will start by placing an integrator element (for the model's differential equation) in the **Freefall** window. To do this, you need to activate the **Library Browser** and double click on the **Commonly Used Blocks** item.

Commonly
Used Blocks

The Browser window should show something like

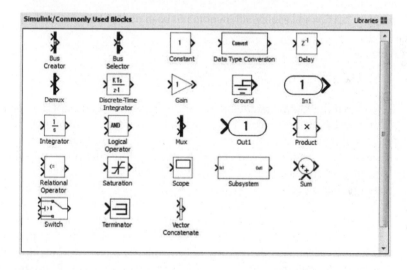

Examine the icon window until you see the Integrator icon.

Input port   Output port

Integrator

The block symbol is what will appear in your model window. Notice how the icon has both input and output ports which are used to feed values in and out of the block. The 1/s symbol represents integration in the Laplace domain. Use the mouse to drag an **Integrator** icon onto your **Freefall** window. This icon will be used to integrate the differential equations. Its input will be the differential equation [the right hand side of Eq. (C.1)] and its output will be the solution (in our example, velocity).

Next you have to build an information flow diagram that describes the differential equation and "feeds" it into the integrator. The first order of business will be to set up constant blocks to assign values to the model parameters. Drag a **Constant** icon from the **Commonly Used Blocks**[1] branch in the **Browser** window to the **Freefall** model window and place it above and to the left of the integrator.

Next, click on the **Constant Label** and change it to g. Then, double click the Constant block and the Constant Block Dialogue box will be opened. Change the default value in the Constant field to 9.81 and click OK.

Now set up Constant blocks for both the drag coefficient (cd = 0.25) and mass (m = 68.1) positioned below the g block. The result should look like

From the Math Operations Library Browser, select a Sum icon and drag it just to the left of the Integrator block. Place the mouse pointer on the output port of the sum block. Notice that the mouse pointer will change to a crosshair shape when it's on the output port. Then drag a connecting line from the output port to the Integrator block's input port. As you drag, the mouse pointer retains its crosshair shape until it is on the input port whereupon it changes to a double-lined crosshair. We have now "wired" the two blocks together with the output of the sum block feeding into the Integrator.

Notice that the sum block has two input ports into which can be fed two quantities that will be added as specified by the two positive signs inside the circular block. Recall that our differential equation consists of the difference between two quantities: $g - (c_d/m)v^2$. We therefore, have to change one of the input ports to a negative. To do this, double click on the Sum Block in order to open the Sum Block Dialogue box. Notice that the List of Signs has two plus signs (++). By changing the second to a minus sign (+−), the value entering the second input port will be subtracted from the first. After closing the Sum Block Dialogue box, the result should look like

_____

[1] All the icons in the **Commonly Used Blocks** group are available in other groups. For example, the **Constant** icon is located in the **Sources** group.

Because the first term in the differential equation is g, wire the output of the g block to the positive input port of the sum block. The system should now look like

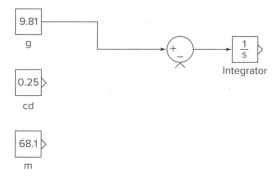

In order to construct the second term that will be subtracted from g, we must first square the velocity. This can be done by dragging a Math Functions block from the Math Operations Library Browser and positioning it to the right and below the integrator block. Double click the Math Functions icon to open the Math Functions Dialogue box and use the pull down menu to change the Math Function to square.

Before connecting the Integrator and Square blocks, it would be nice to rotate the latter so its input port is on top. To do this, select the Square block and then hit ctrl-r once. Then, we can wire the output port of the Integrator block to the input port of the square block. Because the output of the Integrator block is the solution of the differential equation, the output of the square block will be v^2. To make this clearer, double click the arrow connecting the Integrator and Square blocks and a text box will appear. Add the label, v(t), in this text box to indicate that the output of the Integrator block is the velocity. Note that you can label all the connecting wires in this way in order to better document the system diagram. At this point, it should look like

Next, drag a Divide block from the Math Operations Library Browser and position it to the right of the cd and m blocks. Notice that the Divide block has two input ports: one for the dividend (×) and one for the divisor (÷). Note that these can be switched by double clicking the Divide block to open the Divide Block Dialogue box and switching the order in the "number of inputs:" field. Wire the cd block output port to the × input port and the m block output port to the ÷ input port of the Divide block. The output port of the divide block will now carry the ratio, c_d/m.

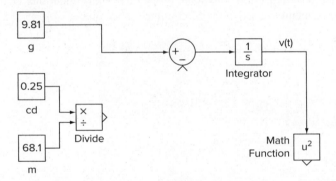

Drag a Product block from the Math Operations Library Browser and position it to the right of the divide block and just below the sum block. Use ctrl-r to rotate the product block until its output port points upward toward the sum block. Wire the divide block output port to the nearest product input port and the square Math Function block output port to the other product input port. Finally, wire the product output port to the remaining sum input port.

As displayed above, we have now successfully developed a Simulink program to generate the solution to this problem. At this point, we could run the program but we have not yet set up a way to display the output. For the present case, a simple way to do this employs a Scope Block

Scope

The Scope block displays signals with respect to simulation time. If the input signal is continuous, the Scope draws a point-to-point plot between major time step values. Drag a Scope block from the **Commonly Used Blocks** browser and position it to the right of the Integrator block. Position the mouse pointer on the Integrator's output wire (for the present case a nice position would be at the corner). Simultaneously hold down the control key and another line across to the Scope block's input port.

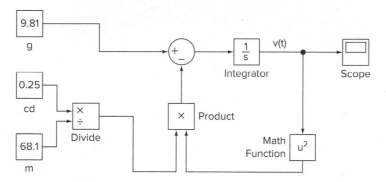

We are now ready to generate results. Before doing that, it's a good idea to save the model. Double click on the Scope block. Then click the run button, ⊙. If there are any mistakes, you will have to correct them. Once your have successfully made corrections, the program should execute and the scope display should look something like

Click on the Autoscale button, 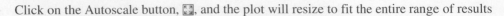, and the plot will resize to fit the entire range of results

Note that a Scope window can display multiple y-axes (graphs) with one graph per input port. All of the y-axes have a common time range on the x-axis. By selecting the parameter button on the graph window (⊚), you can use scope parameters to change graph features such as figure color and style and axis settings.

BIBLIOGRAPHY

Anscombe, F. J., "Graphs in Statistical Analysis," *Am. Stat., 27*(1):17–21, 1973.

Attaway, S., *MATLAB: A Practical Introduction to Programming and Problem Solving,* Elsevier Science, Burlington, MA, 2009.

Bogacki, P. and L. F. Shampine, "A 3(2) Pair of Runge-Kutta Formulas," *Appl. Math. Letters, 2*(1989):1–9, 1989.

Brent, R. P., *Algorithms for Minimization Without Derivatives,* Prentice Hall, Englewood Cliffs, NJ, 1973.

Butcher, J. C., "On Runge-Kutta Processes of Higher Order," *J. Austral. Math. Soc., 4*:179, 1964.

Carnahan, B., H. A. Luther, and J. O. Wilkes, *Applied Numerical Methods,* Wiley, New York, 1969.

Chapra, S. C. and R. P. Canale, *Numerical Methods for Engineers,* 6th ed., McGraw-Hill, New York, 2010.

Cooley, J. W. and J. W. Tukey, "An Algorithm for the Machine Calculation of Complex Fourier Series," *Math. Comput., 19*:297–301, 1965.

Dekker, T. J., "Finding a Zero by Means of Successive Linear Interpolation." In B. Dejon and P. Henrici (editors), *Constructive Aspects of the Fundamental Theorem of Algebra,* Wiley-Interscience, New York, 1969, pp. 37–48.

Devaney, R. L. Chaos, Fractals, and Dynamics: Computer Experiments in Mathematics, Menlo Park, CA: Addison-Wesley, 1990.

Dormand, J. R. and P. J. Prince, "A Family of Embedded Runge-Kutta Formulae," *J. Comp. Appl. Math., 6*:19–26, 1980.

Draper, N. R. and H. Smith, *Applied Regression Analysis,* 2nd ed., Wiley, New York, 1981.

Fadeev, D. K. and V. N. Fadeeva, *Computational Methods of Linear Algebra,* Freeman, San Francisco, CA, 1963.

Forsythe, G. E., M. A. Malcolm, and C. B. Moler, *Computer Methods for Mathematical Computation,* Prentice Hall, Englewood Cliffs, NJ, 1977.

Gabel, R. A. and R. A. Roberts, *Signals and Linear Systems,* Wiley, New York, 1987.

Gander, W. and W. Gautschi, *Adaptive Quadrature–Revisited, BIT Num. Math., 40*:84–101, 2000.

Gerald, C. F. and P. O. Wheatley, *Applied Numerical Analysis,* 3rd ed., Addison-Wesley, Reading, MA, 1989.

Hanselman, D. and B. Littlefield, *Mastering MATLAB 7,* Prentice Hall, Upper Saddle River, NJ, 2005.

Hayt, W. H. and J. E. Kemmerly, *Engineering Circuit Analysis,* McGraw-Hill, New York, 1986.

Heideman, M. T., D. H. Johnson, and C. S. Burrus, "Gauss and the History of the Fast Fourier Transform," *IEEE ASSP Mag., 1*(4):14 21, 1984.

Hornbeck, R. W., *Numerical Methods,* Quantum, New York, 1975.

James, M. L., G. M. Smith, and J. C. Wolford, *Applied Numerical Methods for Digital Computations with FORTRAN and CSMP,* 3rd ed., Harper & Row, New York, 1985.

Moler, C. B., *Numerical Computing with MATLAB,* SIAM, Philadelphia, PA, 2004.

Moore, H., *MATLAB for Engineers,* 2nd ed., Prentice Hall, Upper Saddle River, NJ, 2008.

Munson, B. R., D. F. Young, T. H. Okiishi, and W. D. Huebsch, *Fundamentals of Fluid Mechanics,* 6th ed., Wiley, Hoboken, NJ, 2009.

Ortega, J. M., *Numerical Analysis–A Second Course,* Academic Press, New York, 1972.

Palm, W. J. III, *A Concise Introduction to MATLAB,* McGraw-Hill, New York, 2007.

Ralston, A., "Runge-Kutta Methods with Minimum Error Bounds," *Match. Comp., 16*:431, 1962.

Ralston, A. and P. Rabinowitz, *A First Course in Numerical Analysis,* 2nd ed., McGraw-Hill, New York, 1978.

Ramirez, R. W., *The FFT, Fundamentals and Concepts,* Prentice Hall, Englewood Cliffs, NJ, 1985.

Recktenwald, G., *Numerical Methods with MATLAB,* Prentice Hall, Englewood Cliffs, NJ, 2000.

Scarborough, I. B., *Numerical Mathematical Analysis,* 6th ed., Johns Hopkins Press, Baltimore, MD, 1966.

Shampine, L. F., *Numerical Solution of Ordinary Differential Equations,* Chapman & Hall, New York, 1994.

Van Valkenburg, M. E., *Network Analysis*, Prentice Hall, Englewood Cliffs, NJ, 1974.

White, F. M., *Fluid Mechanics.* McGraw-Hill, New York, 1999.

INDEX